POLYMER SCIENCE AND TECHNOLOGY SERIES

POLYCYCLIC AROMATIC HYDROCARBONS: POLLUTION, HEALTH EFFECTS AND CHEMISTRY

POLYMER SCIENCE AND TECHNOLOGY SERIES

Advances in Polymer Latex Technology
Vikas Mittal
2009. ISBN: 978-1-60741-170-3

Oligomeric State of Substances
S.M. Mezhikovskii, A.E. Arinstein, and R.Ya. Deberdeev
2009. ISBN: 978-1-60741-344-8

Polycyclic Aromatic Hydrocarbons: Pollution, Health Effects and Chemistry
Pierre A. Haines and Milton D. Hendrickson (Editors)
2009. ISBN: 978-1-60741-462-9

POLYMER SCIENCE AND TECHNOLOGY SERIES

POLYCYCLIC AROMATIC HYDROCARBONS: POLLUTION, HEALTH EFFECTS AND CHEMISTRY

PIERRE A. HAINES
AND
MILTON D. HENDRICKSON
EDITORS

Nova Science Publishers, Inc.
New York

For permission to use material from this book please contact us:
Telephone 631-231-7269; Fax 631-231-8175
Web Site: http://www.novapublishers.com

LIBRARY OF CONGRESS CATALOGING-IN-PUBLICATION DATA

Haines, Pierre A.
Polycyclic aromatic hydrocarbons : pollution, health effects and chemistry / Pierre A. Haines and Milton D. Hendrickson.
p. cm.
Includes index.
ISBN 978-1-60741-462-9 (hardcover)
1. Polycyclic aromatic hydrocarbons--Toxicology. 2. Polycyclic aromatic hydrocarbons--Environmental aspects. I. Hendrickson, Milton D. II. Title.
RA1242.P73H35 2009
363.73'63--dc22

2009025478

Published by Nova Science Publishers, Inc. ✝ New York

CONTENTS

PREFACE

Polycyclic aromatic hydrocarbons (PAHs) are chemical compounds that consist of fused aromatic rings and do not contain heteroatoms or carry substituents. PAHs are one of the most widespread organic pollutants. Their occurrence is largely a result of anthropogenic and natural emissions such as fossil fuel-burning, motor vehicle, waste incinerator, oil refining, coke and asphalt production, and aluminum production. This book examines recent developments in the field.

Chapter 1 - Polycyclic aromatic hydrocarbons (PAHs) are one of the most important, abundant and dangerous class of persistent organic contaminants found in air, water, food and waste sites. The great concern with this class of pollutants is due to the fact that these chemicals cause various deleterious effects, which could jeopardize the genome of exposed organisms or even lead them to death. Many studies have been conducted in several laboratories distributed around the world and with different organisms to quantify the impact of those pollutants and to diagnose their action mechanisms. Thus, there is a real necessity to compile the data already described in the literature, obtained from tests in situ, in vivo and in vitro, derived from simulation experiments, occupational exposure or environmental disasters, in order to produce a report with the main effects already described for that class of pollutant. This review article intends to address the following issues: toxicity, cytotoxicity, genotoxicity, mutagenicity, teratogenicity and carcinogenicity of PAHs, besides information about estrogenicity and metabolism of these substances. In this context, it will be also presented the test systems that better responded to the assessments of the PAHs effects, in an attempt to show the most sensitive and efficient organisms to detect the contamination of these agents.

Chapter 2 - Polycyclic aromatic hydrocarbons (PAHs) are a group of about 100 different chemicals that are formed during incomplete combustion or heat-induced decomposition of organic matter. Humans can be exposed to PAHs through different routes. For non-smokers, the major route of exposure is from food with a minor contribution from inhaled air. Food can be contaminated from environmental sources, industrial food processing and from home food preparation. The European regulator has found that there is a low risk to consumers from PAHs in foods based on average dietary exposure but risk management action may be required for high dietary intake. Several PAHs have been shown to be genotoxic carcinogens. In 2002, the Scientific Committee on Food (SCF) reviewed PAH toxicity. For 15 compounds it concluded that there was clear evidence for their toxicity. In 2005, the Joint FAO/WHO Expert Committee on Food Additives (JECFA) performed a risk assessment on PAHs, and basically agreed with the SCF selection. Also the US Environmental Protection Agency has classified some PAHs as priority-pollutants based on toxicity, potential for human exposure, frequency of occurrence and the extent of information available. Seven of these PAHs (benzo(*a*)anthracene, chrysene, benzo(*a*)pyrene, benzo(*b*)fluoranthene, benzo(*k*)fluoranthene, dibenzo(*a,h*)anthracene and indeno(1,2,3-*cd*)pyrene) are considered as probable human carcinogens, but benzo(*a*)pyrene is the most studied and used as a marker for PAHs in ambient air and food. PAHs are themselves chemically inert but undergo metabolic activation in mammalian cells to diol-epoxides that bind covalently to cellular macromolecules, including DNA, causing errors in DNA replication and mutations that initiate the carcinogenic process. Other reactive intermediates can be generated by one-electron oxidation process that can result in chemically unstable alkylation of DNA, which leads to depurination, a potentially mutagenic event. In this chapter the most important effects of PAHs on animals and humans health are reported and discussed to define new possible preventive and regulatory programs for these pollutants.

Chapter 3 - Polycyclic aromatic hydrocarbons (PAHs) are a group of organic compounds composed of fused aromatic rings. Some high molecular weight PAHs are considered to be carcinogenic, mutagenic, teratogenic even at low concentrations and have received more environmental concern. PAHs are resistant to biodegradation and can bioaccumulate in the environment due to their hydrophobic nature and low solubility. PAHs are ubiquitous environmental contaminants found in air, soil and water. The main sources of PAHs in the environment and for general population exposure are natural and intentional combustion of organic matter and petroleum manipulation. Atmospheric emissions will arise from the processing of coal and petroleum products and incomplete combustion during industrial processes.

In the developing countries, present information concerning the sources and inputs of PAHs in these countries is rather limited. PAHs concentrations have been reported in a number of urban areas as well as in remote places and are consistently higher in winter than in summer, and in urban than in rural areas. The higher values are found in particular places like stack gases, tunnels of motorways, open burning sites, etc. There are no ambient air quality standards and no specific guideline value can be recommended for PAHs in air. PAHs in air are attached to particles, which may also play a role in their carcinogenicity. Although food is thought to be the major source of human exposure to PAHs, part of this contamination may arise from air pollution with PAHs.

Exposure to PAHs mainly occurs from atmospheric pollution through inhalation, ingestion and dermal contact. PAHs half-lives in humans are in fact in the range of days/hours, and metabolism is responsible of the formation of carcinogenic metabolites. Risk assessments and potency assessments of various individual and complex mixtures of PAHs have been attempted. BaP is the only PAH for which a database is available, allowing a quantitative risk assessment. PAH potencies are used to determine quantitative health risks posed by PAH exposure. The risks posed by a mixture of PAHs are based on an assumption of additively of the individual risks posed by the PAHs.

The objectives of this chapter are to: 1) report PAH concentrations in the environmental media in developing countries; 2) collate published and unpublished PAH information from environmental agencies and write a report on the state of knowledge of PAHs; 3) characterize the priority PAHs, carcinogenic risk assessment and potential health impacts.

Chapter 4 – The authors have previously shown that benzo[a]pyrene used at extremely low dose can cause weight gain in mice and that the increase in adipose tissue mass is due to inhibition of β-adrenergic stimulation of lipolysis. Moreover they have suggested that adipose tissue is a reservoir for many carcinogens including Polycyclic Aromatic Hydrocarbons (PAHs).In this chapter, the authors demonstrate that the two C4 PAHs pyrene and phenanthrene can accumulate into adipocytes, but at the difference of benzo[a]pyrene have no impact on lipolysis.Since benzo[a]pyrene has been proved to be mutagenic while C4 PAHs are tumor promoters, they consider that a mixture of the three PAHs may act as a complete carcinogen. Furthermore, because adipose tissue can be induced by benzo[a]pyrene, is a reservoir for many PAHs and other carcinogens and can stimulate proliferation of normal or cancer cells, through the secretion of adipokines by adipocytes, they therefore propose that adipose tissue is a key contributor to chemical carcinogenesis by acting not only upon initiation and promotion but also upon progression.

Chapter 5 - A review has provided the environmental sources and behavior of chlorinated 3- to 5-ring polycyclic aromatic hydrocarbons (ClPAHs). ClPAHs have been known to be ubiquitous contaminants, which can be found in urban air, vehicle exhaust gas, snow, tap water and sediments. The concentrations in the each environment are generally higher than those of chlorinated dioxins, although which are significantly lower than the concentrations of parent compounds, PAHs. Based on the environmental data and emission sources analysis for the ClPAHs, the dominant production process is by incineration of organic materials including chlorine like as the case of dioxins. On the other hand, treatment of PAHs with chlorine in water also produced ClPAHs. Indeed, such secondary reaction for ClPAHs generation could occur in the gaseous phase. For the biological effects, the toxicities such as mutagenicity and aryl hydrocarbon receptor activity of certain ClPAHs are emphasized than those of parent PAH, in which the toxicities of exposure to ClPAHs based on AhR activity were estimated to be approximately 30-50 times higher than that of dioxins. To investigate the sources and environmental behavior of ClPAHs is of great importance to assess human health risks.

Chapter 6 - Polycyclic aromatic hydrocarbons (PAHs) are a large group of organic compounds with two or more fused aromatic rings. They have received increased attention in recent years in pollution studies because some of these compounds are highly carcinogenic or mutagenic. Some, while not carcinogenic, may act as synergists. These compounds are widespread in the environment. Their occurrence is largely a result of anthropogenic and natural emissions such as fossil fuel burning, motor vehicles, waste incinerators, oil refining, coke and asphalt production, aluminum production, etc. PAHs have relatively low solubility in water but are highly lipophilic. Most of the PAHs with low vapor pressure in the air are adsorbed on particles. When dissolved in water or adsorbed on particulate matter, PAHs can undergo photodecomposition when exposed to UV light. In the atmosphere, PAHs can react with many other pollutants such as ozone, nitrogen oxides and sulfur dioxide, yielding diones, nitro- and dinitro-PAHs, and sulfonic acids, respectively. PAHs may also be degraded by some microorganisms in soils. PAHs in soils/sediments can be taken up by biota, and some bio-magnification of these toxic compounds may occur as they pass through the food chain. The US Environmental Protection Agency (EPA) has identified 16 un-substituted PAHs as priority pollutants. The European Food Security Authority (EFSA) has classified 15 PAHs by their food toxicity. Thus, potential exposure to PAHs must be assessed by testing contaminated soils, sediments, air, water and foods, and estimating the degree of intake of each of these media into the human body, as well as the overall toxicity.There are several main problems associated with the determination of PAHs in environmental and food samples: their low concentrations, the complexity of the samples, and the necessity for fast and reliable analytical methods. To achieve successful results, many extraction-separation techniques have been developed depending on the nature of the sample. An in-depth description of the most commonly-used techniques, together with recent trends in analytical methods to monitor PAHs levels in different kind of samples, will be the main focus of this chapter. Among those that will be cited are Soxhlet extraction, liquid-liquid extraction (LLE), microwave-assisted extraction (MAE), ultrasound-assisted extraction (UAE), pressurized fluid extraction (PFE), supercritical fluid extraction (SFE), solid-phase extraction (SPE), and solid-phase microextraction (SPME), mainly in combination with chromatographic techniques (HPLC and GC). Moreover, the utilization of active or passive samplers in pollution monitoring, as well as the increased utilization of semi-permeable membrane passive samplers (SPMDs), will be described.

Chapter 7 - Different configurations were studied for the screening analysis of common polycyclic aromatic hydrocarbons (PAHs) in environmental waters by liquid chromatography (LC). Analyte enrichment on a traping microcolumn connected by column switching to the analytical column, and in-valve in-tube solid-phase microextraction (IT-SPME) were tested as alternatives to conventional solid-phase extraction (SPE) with disposable cartridges. For separation conventional LC, capillary LC and LC with a monolithic column were tested and compared, whereas UV-diode array detection (DAD) or fluorescence detection were used for signal monitoring. The compounds assayed were naphthalene, anthracene, fluoranthene, benzo[*a*]anthracene, benzo[*b*]fluoranthene, benzo[*k*]fluoranthene, benzo[*a*]pyrene, dibenzo[*a,h*]anthracene and benzo[*ghi*]perylene. On the basis of the results obtained, new methods are presented for the on-line screening analysis of PAHs in several water matrices. The main advantages of the proposed procedures are: (i) the minimum off-line sample manipulation as the samples only need to be filtered; (ii) the rapidity, as the total analysis time can be reduced to less than 20 min, (iii) the improved detection limits, especially for the in-valve IT-SPME configurations, as they are compatible with the injection of very large sample volumes (up to 4.0 mL), and (iv) the total elimination of organic solvents in the sample treatment. Therefore, the proposed methods are environmentally friendly and cost-effective alternatives for the routine monitoring of PAHs in waters. The occurrence of the tested PAHs in waters of The Comunidad Valenciana (Spain) was also tested.

Chapter 8 - Polycyclic aromatic hydrocarbons (PAHs) are ubiquitous hydrophobic organic contaminants in the environment. They tend to be associated with particles and are widely transported by flooding and atmospheric pathways, resulting in elevated concentrations in sediments/soils. Coal and coal-derived particles in natural sediments/soils can act not only as strong sinks for the PAHs, but also as very important sources of PAHs in sediments/soils. The understanding of the mobility of these contaminants from the sediments/soils, especially the sequestration of PAHs by coal and coal-derived particles is very important, because they can control the transportation, bioavailability, degradation and hence the potential risk of these contaminants in the environment.

Chapter 9 - Polycyclic aromatic hydrocarbons (PAHs) are notably formed and emitted by incomplete combustion of coal, oil, petrol, wood, tobacco. In addition, they participate to soot inception. The remaining gaseous PAHs are then adsorbed onto the porous soot matrix so formed, conditioning physicochemical properties of these particles and consequently their impact on human health and environment. This has motivated various studies. However, an extensive analysis of the adsorbed PAHs onto soot particles is still challenging. The authors are developing an experimental set-up, based on the combined use of 3 techniques - laser desorption, multi-photon ionization and time-of-flight mass spectrometry, dedicated to this study with particular focus on the dependence of the adsorbed species on the stage of combustion and the nature of the fuel. The complexity of the first obtained mass spectra led us to successively investigate the laser desorption of pure PAHs then "synthetic" soot samples- obtained by adsorbing PAHs on various carbonaous substrates- to characterize the processes involved in our method, and finally combustion soot collected from various diffusion flames. They present an overview of these studies together with their first practical implications, obtained on soot collected from Diesel and Diesel surrogate flames.

Chapter 10 - Benzo[a]pyrene (BaP) is one of the most well studied PAHs. It is a product of incomplete combustion of fossil fuel. BaP can be deposited into food, soil and water. The major route of human exposure to BaP is via food consumption. BaP can interact with DNA to form bulky adducts, or biotransformed to produce free radicals. Both processes may lead to mutation and cancer. Despite the clarity of its actions, it is still difficult to establish a clear dose-response relation between BaP and cancer in human. Currently, BaP is considered class 2A carcinogen by the International Agency for Research in Cancer (IARC), a chemical with well documented evidence as an animal carcinogen, but limited evidence as human carcinogen. It is believed that only by understanding the basic difference between animal and human biochemistry will it be possible to address the toxicity of BaP in the various biological systems. In the present chapter, experiments pertaining to studying the response to BaP in human and animal cells are discussed.

Chapter 11 - Polycyclic aromatic hydrocarbons (PAHs) are well-known organic compounds derived from incomplete combustion of organic matters. PAHs have been widely studied in a broad field such as material science, industrial technologies and pharmaceutical sciences, as well known environmental pollutants, and an increasing particular concern has been paid to their adverse harmful effects to human due to the carcinogenic and mutagenic properties of many PAH species. Clearly, the concerns over adverse health effects of PAHs in human and wildlife provide the impetus for research on their sources, environmental fate and food chain transfer. High performance liquid chromatography (HPLC) is useful and conventional method to monitor PAHs. More than 60% of all HPLC separations are carried out under reversed-phase (RP) conditions and alkylated-silicas were successfully employed for the separation of mixtures of PAHs. In RP-HPLC chromatography, not only long alkyl chains but also aromatic compounds such as benzene, pyrene, and hetero aromatic selector such as vinylpyridine have been investigated for organic phase of stationary phase so far. However, development of new stationary phases has been required for separation of similar compounds such as structural isomers, positional isomers and optical isomers. In this chapter, the authors focus on the chromatographic analysis of PAHs by π-electron containing ordered and disordered polymeric organic phase-grafted silica. Polymeric phase-grafted silica stationary phases exhibits unexpected enhancement of separation for PAHs by a multiple-interaction mechanism, even if the polymer does not form any specific secondary structure. Poly(octadecylacrylate) and poly(4-vinylpyridine)-grafted mesoporous silica particles (Sil-ODA_n and Sil-VP_n, respectively) will be described as typical examples of such stationary phases. Effects of grafting method and grafting-density of polymeric phases on the separation of PAHs in HPLC will be also discussed. Extensive chromatographic studies showed that selectivity for PAHs can be enhanced by multiple π-π interactions between the π-electron containing polymeric phases and the guest PAHs molecules. The presence of aligned carbonyl groups in a crystalline state of poly(octadecylacrylate) on silica significantly enhanced the selectivity for PAHs by π-π interaction mechanism. However, Sil-VP_n recognizes molecular shape (planar and disk-like) of PAHs through π-π interactions between pyridyl groups of grafted-polymer and PAHs, along with electrostatic interactions involving a inductive effects and /or a quadrupolar effect in the normal phase HPLC. Comparison of the HPLC separations for PAHs with conventional octadecylsilylated silica (ODS) demonstrates the efficiencies and effectiveness of these π-electron containing polymeric organic phases grafted silica particles.

Chapter 12 - The use of a direct extraction device (DED) for coupling to solid phase microextraction (SPME) is described. Such a device enables the introduction of the SPME fibre in the core of solid matrixes, allowing the analysis of volatile compounds, such as polycyclic aromatic hydrocarbons (PAHs), with no or little deterioration of the solid product. The feasibility of using SPME-DED for extraction of PAHs from solid matrixes mimicking solid foodstuffs (gelatine alone, gelatine containing fat and salt, or cooked chicken patties) and from commercial smoked meat products (smoked tenderloin, smoked jowl, smoked dry cured sausage and smoked potato sausage) was tested. Moreover, the diffusion of PAHs in solid matrixes by SPME-DED was also evaluated. SPME-DED appeared as a rapid, non-destructive technique for primary screening of low molecular weight (MW) PAHs in all solid matrixes assayed, as well as for the study of the diffusion process of PAHs within solid systems, without sampling and any laboratorial manipulation.

Chapter 13 - *Cyperus laxus* and *Cyperus ligularis* are native plants growing at swamp highly contaminated with hydrocarbons. Anthracene and pyrene phytoremediation from a hydrocarbon mixture (hexadecane 50%; pristane 14%; anthracene 18% and pyrene 18%) by both species were evaluated. The aim of this chapter was to demonstrate the ability of two native plants to remove PAH from contaminated model soils under greenhouse controlled conditions. Experiments were carried out in a model system using perlite to simulate soil conditions. Initial hydrocarbon mixture concentration was 4 000 mg kg^{-1} of dry perlite for *C. laxus* and 1 000 mg kg^{-1} of dry perlite for *C. ligularis.* Plants were inoculated with hydrocarbon-degrader microorganisms. Root dry biomass, total microbial viable counts, hydrocarbon removal and hydrocarbon sorption by roots were determined during 150 days of culture. Root dry biomass from inoculated plants was 1.3 times higher than those from non inoculated plants. At the end of culture, microbial viable counts for all treatments were around 10^9 CFU g^{-1} of dry perlite for bacteria and 10^7 CFU g^{-1} of dry perlite for fungi. PAH and aliphatics were completely removed by *C. ligularis* during first 60 days of culture. *C. laxus* removed pyrene totally until 150 days of culture while anthracene only 75%. Hydrocarbons from *C. ligularis* roots were negligible. PAH sorption by roots was detected for *C. laxus* since 75 days of culture, 125 and 215 mg kg^{-1} of dry roots for pyrene and anthracene, respectively. At the end of culture, pyrene sorption was reduced (60 mg kg^{-1} of dry roots) possibly due to plant metabolism. For species, hydrocarbon removal and accumulation in roots, at 150 days of culture, was independent on the microbial inoculum added. *C. laxus* and *C. ligularis* were capable to remove efficiently anthracene and pyrene, these plants can be useful for phytoremediation of PAH contaminated soils.

Chapter 14 - Polycyclic aromatic hydrocarbons (PAH) are widespread environmental pollutants of natural and anthropogenic origins. Despite their poor water solubility, they can be taken up and bioaccumulated by plants. The effect of PAH on selected plant (*Festuca arundinacea* and *Bouteloua curtipendula*) at early grown stage was evaluated under *in vitro* conditions. Plant species were cultured on Murashige and Skoog (MS) medium with three different PAH: phenanthrene (PHE), pyrene (PYR), and benzo[a]pyrene (BaP). PAH plant accumulation and distribution were evaluated with GC and HPLC. To date, number phytoremediation mechanisms of pollutants removal are known, including: (i) degradation (phyto- and rhizodegradation), (ii) removal (phytoextraction, phytovolatilization) and (iii) immobilization (phytoestabilization). The objective of this study was to identify the phytoremediation mechanisms of selected plants use to control and/or remove PAH from MS medium. To accomplish the task, plant samples, after 90 d culture, were collected and analyzed periodically to determine the residual PAH content in MS medium, roots and plants stems. *F. arundinacea* was able to transform 40.40% of initial PHE, accumulated 6.99% in stems and almost three times in roots (20.66%). With *B. curtipendula*, total removal was 66.14% and PHE was found in roots (14.38%) and stems (6.45%). Both species were able to remove and transform PYR, higher concentrations were found in roots than in stems. BaP outlined in a different pattern: both species were able to remove BaP from MS medium, *B. curtipendula* was more efficient than *F. arundinacea* 57.41% and 38.93%, respectively; bioaccumulation of BaP was the higher while transformation the lower. The dominant mechanisms observed for PHE and PYR in both plant species were phytodegradation and phytoextraction. *B. curtipendula* employed phytoextraction for BaP; it was able to bioaccumulate 45.05% in roots, 1.8 times more than *F. arundinacea* (24.99%). Several studies investigated the effect of plant-microbe interactions on the degradation of organic contaminants. In our study, plants were able to remove and transform PAH under sterile *in vitro* conditions, *i.e.* without microorganisms.

In: Polycylic Aromatic Hydrocarbons: Pollution, Health... ISBN 978-1-60741-462-9
Editors: Pierre A. Haines and Milton D. Hendrickson

Chapter 1

A REVIEW OF THE HAZARDOUS EFFECTS OF POLYCYCLIC AROMATIC HYDROCARBONS ON LIVING ORGANISMS

Maria Aparecida Marin-Morales[1,2]*; Daniela Morais Leme[1,2,3]; Dânia Elisa Christofoletti Mazzeo[1,2]

[1]Department of Biology, Institute of Biosciences, São Paulo State University, Rio Claro, SP, Brazil

[2]Researcher of the Programa de Formação de Recursos Humanos em Geologia e Ciências Ambientais do Setor de Petróleo, Gás and Biocombustíveis – PRH05 (UNESP/MCT/FINEP/ANP)

[3]Researcher of the Fundação de Amparo à Pesquisa do Estado de São Paulo (FAPESP)

ABSTRACT

Polycyclic aromatic hydrocarbons (PAHs) are one of the most important, abundant and dangerous class of persistent organic contaminants found in air, water, food and waste sites. The great concern with this class of pollutants is due to the fact that these chemicals cause various deleterious effects, which could jeopardize the genome of exposed organisms or even lead them to death. Many studies have been conducted in several laboratories distributed around the world and with different organisms to quantify the impact of those pollutants and to diagnose their action mechanisms. Thus, there is a real necessity to compile the data already described in the literature, obtained from tests *in situ*, *in vivo* and *in vitro*, derived from simulation experiments, occupational exposure or environmental disasters, in order to produce a report with the main effects already described for that class of pollutant. This review article intends to address the following issues: toxicity, cytotoxicity, genotoxicity, mutagenicity, teratogenicity and carcinogenicity of PAHs, besides information about estrogenicity and metabolism of these substances. In this context, it will be also presented the test systems that better responded to the assessments of the PAHs effects, in an attempt to show the most sensitive and efficient organisms to detect the contamination of these agents.

* Tel: +55 19 3526 4143; Fax: +55 19 3534 0009. E-mail address: mamm@rc.unesp.br (M.A. Marin-Morales).

1. Introduction

Polycyclic aromatic hydrocarbons (PAHs) are characterized by a diversified group of organic compounds constituted of two or more condensed rings. These substances present a wide distribution in the environment, having their levels already determined in soils, drinking water, rivers, lakes, marine sediments and also in air analyses (Harvey, 1997).

PAHs are originated from several sources, being possible find them, mainly, in cigarette smoke, incomplete combustion of organic matter during industrial activities and other human activities such oil refining, fossil fuel burning resulted from vehicle traffic and coal production (Stanley, 1992; Thirman et al., 1994; Mitchell et al., 1999; Annas et al., 2000; Bihari et al., 2006). Composition and complexity of PAHs mixtures depends on the emission sources, which makes available a great variety of these substances, in different concentration levels (Pereira Netto et al., 2000).

PAHs present low solubility in water, due to its lipophilic nature and, in consequence of this characteristic, concentrations of these chemicals in sediments can be much higher than from those found in water. However, the low bioavailability of the PAHs present in sediments, reduces, significantly, the toxic potential in this environment. In the air, PAHs tend to react with the particulate matter, which can be the contamination vehicle of other sources such as soil and food, by deposition of these particles in crops, fruits and vegetables (Albers, 2002; WHO, 2000).

Due to its wide distribution in the environment, great part of the living beings is exposed to these pollutants, having some of the adverse effects already detected. Detection of PAHs hazardousness to the exposed organisms have characterized these chemicals as one of the most dangerous classes of environmental pollutants (Aina et al., 2006), being also considered as one of the most toxic among all the hydrocarbons (Catoggio, 1991).

2. Effects on the Living Beings

Toxic environmental pollutants can cause several alterations in the normal physiologic process of the living beings. To a pollutant exercise its deleterious effects in one exposed organism it must first enter in the individual to after reach its target site (Landis and Yu, 1998). Thus, assessment of the damages caused in the organisms must consider, besides the substance physico-chemical properties, the magnitude, duration and frequency of the exposition, introduction via and the susceptibility of the biological systems, intra and interspecies (Stanley, 1992, 1994; Landis and Yu, 1998; Pedrozo et al., 2002).

PAHs are characterized by their high lipophilic affinity, which allows their fast absorption by the biological membranes and, consequently, their accumulation in the tissues (Tuvikene, 1995). Biological membranes are characterized as barriers between the cell and the environment, regulating the passage of toxic agents into the cellular interior and exterior (Stanley, 1992, 1994). Independently from the introduction via used, the chemical compound must cross several cellular membranes, until reaching their target site and produce its harmful effect (Pedrozo et al., 2002).

Comprehension of the distribution and the storage places of these chemicals inside the organism is another important issue. According to Landis and Yu (1998), organs such as liver, lungs, kidney, bone and adipose tissues are considered as storage deposits. Nevertheless, the relation between the target site of the substance is not always true since the agent can be stored in one determined tissue/organ, but act in another place. In case of the PAHs, due to its lipophilic nature these chemicals are easily stored in adipose tissues (Ballantyne and Sullivan, 1992).

Living beings present certain mechanisms that help in the elimination of the toxic substances introduced in their organic system such as the biotransformation process (Guecheva and Henriques, 2003). However, for some pollutants classes such as PAHs, biotransformation can result in a more drastic effect since the reactions conjunct involved in this process lead to a formation of metabolites highly reactive, which become responsible for the toxic effect of the original substances.

2.1. Biotransformation: PAHs Bioactivation

Most living beings have the capacity to metabolize xenobiotics by the process denominated biotransformation (Tuvikene, 1995). Biotransformation is characterized as a conjunct of enzymatic reactions, responsible for the conversion of the liposoluble substances in hydrosoluble facilitating, thus, their excretion process. However, although the purpose of the xenobiotics biotransformation is detoxification not always the originated metabolite is less toxic than the own chemical. Thus, xenobiotics biotransformation can increase the toxicity of the chemical products by the formation of electrophilic metabolites, extremely reactive, which can present potential to bind, covalently, with macromolecules inside the cells with DNA, RNA and proteins, which can result in several alterations such as disturbance in the immune system, mutations and even the organism death (Stanley, 1992, 1994; Landis and Yu, 1998; Guecheva and Henriques, 2003).

PAHs constitute a group of chemical substances that characterize, perfectly, the situation exposed above. These agents per si not only confer toxic effects to the living beings, but, after the oxidative metabolization process, they become bioactivated, exerting then their deleterious effects (Tuvikene, 1995; Luczyński et al., 2005).

The main PAHs metabolization via involves the cytochrome P-450 (CYP) system, which is constituted of mixed function oxidases (MFO). These enzymes are bound to the membrane and present a wide distribution in the organism, with elevated concentration in the endoplasmic reticulum of the liver, kidney, lungs, nasal vias and intestine. Additionally, CYP enzymes are characterized by an ancient proteic family, possibly the bigger ever found in nature, with a wide distribution among bacteria, fungi, plants and animals. It is known, nowadays, that there are about 215 different families of the CYP enzymes being 72 in bacteria, 29 in lower eukaryote, 50 in plants and 67 in higher animals (Guecheva and Henriques, 2003). According to Stegeman and Hahn (1994) and Luczyński et al. (2005), isoenzymes CYP1A1 and CYP1A2 of the subfamily CYP1A are the main responsible for the biotransformation process of the PAHs.

Biotransformation process, in general occurs in two phases: phase I and phase II reactions. In the phase I, one reactive polar group such as OH, NH_2, COOH or SH, is introduced in the molecule of the chemical agent by oxidation, reduction or hydrolysis reactions, generating metabolites highly reactive (bioactivation). In phase II, occurs the detoxification process, where the metabolites generated in phase I are conjugated with endogenous substrates like glycine, cisteine, glutathione, glicuronic acid and sulphates by transferases enzymes, making them hydrophilic and passible of being excreted (Stanley, 1992, 1994; Tuvikene, 1995; Landis and Yu, 1998; Pereira Netto et al., 2000; Pedrozo et al., 2002) (Figure 1).

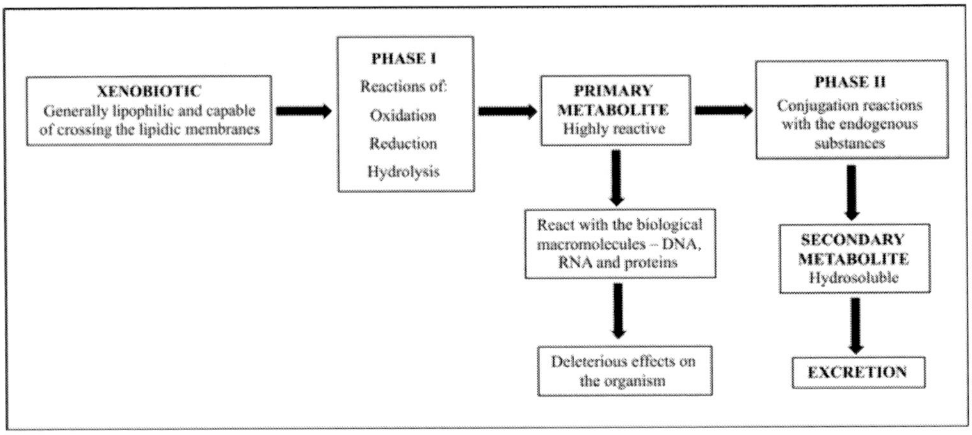

Figure 1. Representative diagram of the xenobiotic biotransformation.

Monooxigenases of the CYP1A family are the responsible for the PAHs enzymatic oxidation. Firstly, the PAHs are activated in arene oxides, in specific regions by the monooxigenases CYP action. Arene oxides can be both spontaneously isomerized in phenols or converted, enzymatically, to dihydrodiols by the epoxide hydrolases action. Dihydrodiols can suffer new oxidation forming metabolites highly reactive. Thus, this last stage plays an important role in the formation of the carcinogens, which can react together with other metabolites, with DNA, and initiate the malign transformation process of the cell (Jacob, 1996).

Among the PAHs known until now, benzo[a]pyrene is, undoubtedly, the most studied, having its metabolic via already reported. *In vitro* experiments with B[a]P have indicated that dihydrodiols is capable of binding to DNA characterizing more than75% of the metabolites formed by the CYP via (Stegman and Klopper-Sams, 1987; Jimenez and Stegeman, 1990). These experiments have still shown that an increase in the monooxygenases activity, due to exposition to environmental pollutants, can rise the quantity of reactive potentially chemicals that, consequently, can react with DNA, originating the DNA adducts.

Identification of the formed metabolites from the PAHs exposition in different groups of living beings is an important issue, since it allows a correlation between the test-system and extrapolation of the obtained results to other organisms. According to Ahokas (1979) and Ronis et al. (1992) microsomes of the trout (*Salmo trutta lacustris*), starry flounder (*Platychthys stellatus*) and coho salmon (*Oncorhynchus kisutch*) liver, metabolize B[a]P in metabolites similar to those produced by rat liver microsome. The main metabolites produced by fish microsome are 7,8 and 9,10-dihydrodiols, which are precursor of the main B[a]P carcinogenic metabolites.

Many are the metabolites originated by the PAHs biotransformation. As already exposed, some of these metabolites can be excreted by the organism or even not be reactive to the organisms not conferring, therefore, hazard to the exposed organism. However, many of them, when transformed can cause several alterations in the biological system and put in danger individuals, populations or even entire ecosystems exposed to their action.

2.2. Toxicity

Toxicity tests are reliable techniques to predict the impacts of contaminant agents over several communities of various environments. With these tests it is possible to determine in which fractions each substance can be available to exercise toxic effects on the organisms. Chemical compounds of anthropogenic origin suffer several interactions and can act in many ways on the exposed communities. Toxicity tests can be developed in laboratories (*in vivo* or *in vitro*) or in field (*in situ)* and use several methodologies since the more basic such as the dose answer curve, observation of behavior, mental, physiologic and lethal changes, even more sophisticated methods such as biochemical, cytological, genetic and teratogenic alterations. The observed effects also vary according to the characteristic of the used methodology, the biology of the specie and the type of the used compound.

The PAHs most common action mechanism that result in toxicity for the organisms is related with the interference in the cellular membranes function by the bond in the membrane hydrophobic sites, causing deformity and molecular perturbance, besides alterations in the enzymatic system (Tuvikene, 1995). As a consequence of these bonds, the membrane permeability is altered and the substances transport in and out the cells is damaged. Moreover, as many enzymatic systems are associated with structure of the membrane, other essential process for the cells can be damaged and the normal energy balance becomes debilitated (Knutzen, 1995).

Although the PAHs can result in a direct toxic effect, the major concern resides in the formation of reactive metabolites such as epoxides and dihydrodiols, which can bind to the genetic material and to the cellular proteins (Albers, 2002).

PAHs with high molecular weight do not result in high toxicity for the organisms, but the effect can be potentialized by exposition to the ultraviolet solar radiation (Arfsten et al., 1996), which leads to the formation of new photoproducts with toxicological properties different from the original molecule (Huang et al., 1993; Krylov et al., 1997). The absorption of light energy lead the PAHs molecules to an excited state, whose energy can be transferred to biological molecules, generating with this, the reactive oxygen species (ROS), intermediary reactive, free radicals or photo-metabolite that can damage these cellular constituents, resulting in toxicity (Betowski et al., 2002; Yu, 2002).

Chemicals potentially dangerous such as the PAHs and their metabolites, endanger not only human health, but also aquatic and terrestrial ecosystems by affecting, directly, the soil biota or, after lixiviation and percolation, the aquatic biota (Bispo et al., 1999).

PAHs affect the organisms in different ways, since some species are more sensitive to these contaminants than others. Thus, for a better comprehension of the PAHs toxic potentiality it must be considered their effects over several species, in different trophic levels, and in all life stages. Besides this, an important factor to be considered ís the investigation of the movement of these molecules in the environment and their action mode on the organisms. The PAHs toxic effect and action mode on plants are not very known and seem to vary according to the biologic specie and the PAH own chemical constitution (Wittig et al., 2003; Baek et al., 2004).

Smith et al. (2006) studied the toxic effects of soils contaminated with PAHs on the germination and growth of four graminaceous species and on three leguminous species. No significative effect was observed for the germination index of the studied species during the ten days of exposition, however, the subsequent growth was significantly reduced being this effect more pronounced in the leguminous.

The toxicity of soils contaminated with PAHs was evaluated by laboratory assays (*in vivo* test), using four different plant species (*Zea mays* L., *Hordeum vulgare* L., *Sinapis alba* L. and *Helianthus annus* L.). The authors considered that the root length was the parameter that best answered to the toxicity tests, being barley (*Hordeum vulgare* L.) the most sensitive specie among all the species tested, which indicates that this specie can be characterized as an efficient test-organism for PAHs toxicity assay (Smreczak and Maliszewska-Kordybach, 2003).

Studies carried out with *Brassica napus* (canola), to assess the toxic and phototoxic effects of 3 PAHs, have shown that these compounds present direct toxic action on the roots growth, not being transported to the other organs of the plant (Ren et al., 1996). The authors have also concluded that, between the tested PAHs, anthracene was the most toxic, followed by B[a]P and fluoranthene. This same toxicity order was maintained when these hydrocarbons were exposed to light.

In order to detect the toxicity of sediment samples of the Rovinj marine coast, in Croatia, contaminated with PAHs, Bihari et al. (2006) used the Microtox® test with the marine bacteria *Vibrio fisheri*. All the tested sediments were potentially toxic, nevertheless the PAHs were not considered the most toxic chemicals present.

The toxicity of lixiviate extracts of soils contaminated with PAHs extracted by water and methanol was evaluated by assays with the *Vibrio fischeri* bacteria (Microtox® test), with the alga *Pseudokirchneriella subcapitata* and with micro crustaceans *Daphnia magna* and *Thamnocephalus platyurus*. The toxicity of the lixiviated soils with water was much lower than the extracts lixiviated with methanol, mainly, for the tests carried out with bacteria and alga, which were considered by the authors as the organisms more suitable for the detection of the toxicity of soils contaminated by PAHs (Bispo et al., 1999).

Erstfeld and Snow-Aihbrook (1999) assessed the chronic effect of soil contaminated with low PAHs levels on the soil microfauna (represented by the nematode), on the mesofauna, (by the microarthropod of the Collembola and Acarina orders) and on macrofauna (by the earthworm *Eisenia fetida andrei*). The observed PAHs contamination levels (5.28 to 80.46 mg/Kg) seem to have a stimulatory effect on the growth of the invertebrates population present in the soil. According to the authors, this answer can be due to the increase in the microbial population that is capable of metabolizing the PAHs, breaking into smaller molecules that can be easily assimilated by the soil biota.

Saint-Denis et al. (1999) studied the effect of different concentrations and different exposition times to PAHs, on the earthworm *Eisenia fetida andrei*. According to the authors, this chemical was metabolized, both by the CYP via, and by the free radicals oxidation via, promoting the generation of free radicals that can stimulate the lipids peroxidation, resulting in toxicity for this organism.

PAHs tend to accumulate in the bottom sediments, which can become a constant source of toxic contaminants putting in danger, continuously, the biota health and viability, mainly, of the benthic organisms that suffer constant expositions to PAHs by the direct contact with the soil (Bihari et al., 2006; Boscolo et al., 2007).

Due to the fast accumulation and to the small metabolization capacity the bivalve mollusks have been used, frequently, as sentinel organisms for the PAHs retention (Baussant et al., 2001). According to Boscolo et al. (2007), the benthic mollusk specie *Tapes philippinarum* is a good indicator of the PAH contamination, because it concentrates and tolerates high levels of these chemicals in its tissues. PAHs sub-lethal toxic effects were registered for the mollusk *Tapes semidecussatus*, that were exposed to sediments containing high levels of these chemical substances (Byrne and O'Halloran, 2004).

Toxicity tests are very useful tools to identify the possible impacts of the PAHs on aquatic organisms. Lethality tests are considered good indicators of the toxicity of different chemicals and for different species, while the sub-lethality tests are interesting since they consider the impacts on the population by evaluating the effects on the organisms reproduction and growth (Wake, 2005).

Individual toxicity of the PAHs present in aquatic environments, for plants and animals, increase when the molecular weight of these substances are next to 202 (e.g. fluoranthene, pyrene). Values superior to this weight configure a less soluble molecule, which promotes a reduction in the PAHs concentration to lower levels than of the considered lethal (Albers, 2002). However, exposition to PAHs with high molecular weight even in small concentrations, can lead to sub-lethal effects to the organisms (Neff, 1985).

Kim et al. (2007) analyzed the PAHs toxic action on the genic expresion of the genes *recA* (related with DNA repair), *KatG* (that responds to oxidative stress), *grpE* (that is expressed in response to damages in the proteins) and *fabA* (that responds to damages in the membrane) of *Escherichia coli*, using in real time PCR (RT-PCR). The analyses showed that naphthalene and B[a]P caused damages in the *E. coli* DNA, while phenanthrene did not present any significative effect on the expression of these genes.

Reece and Burks (1985) observed that the 11 PAHs present in petroleum refinery effluents caused lethal acute effects in *Daphnia magna* and that these substances were acting in an additive or synergistic way.

Holst and Giesy (1989) studied the chronic effect of the exposition to anthracene and its photoproducts on the cladoceran *Daphnia magna* reproduction. In the experiments carried out both in the absence and in the presence of the UVR light, there was a reduction in the neonate number and an increase in the mortality number but these effects were more conspicuous when there was exposition to the light.

An important result was observed by Allred and Giesy (1985), in studies about the anthracene phototoxicity on *Daphnia pulex*. The exposition of these organisms to anthraquinone, which is the first and more stable photoproduct of the anthracene, did not result in toxicity but in the experiments where the animals were previously treated with anthracene in the presence of light and after were transferred to water, the toxicity was significative. By the results, the authors suggest that the toxicity was due to the activation, by the light energy, of the own anthracene molecule that binds to the molecules that are constituents of the *D. pulex* cells, and not of the metabolites of this substance present in the medium.

The contaminants bond to specific receptors alters the metabolic and functional capacities of the tissues, by the induction or suppression of important proteins, causing harmful alterations to the organism. Thus, the induction or suppression of proteins can be used as biomarkers of effect and/or exposition (Roos et al., 2004).

Bellas et al. (2008) studied the toxic and phototoxic potentials of 6 PAHs, individually and in mixtures, on embryos and larvae of mussels, sea urchin and ascidian. All the PAHs were toxic in the tested concentrations for sea urchin embryos and all, with the exception of fluoranthene, were toxic for mussels embryos. In relation to the ascidian embryos, only naphthalene presented significative toxic effect. However, exposition to fluorescent light increased the toxicity of phenanthrene, fluoranthene, pyrene and hydroxypyrene but not of the naphthalene and fluorene. The PAHs mixture did not present additive effect of the toxicity.

B[a]P and anthracene effects, on the activity of several enzymes of the fish *Pomatoschistus microps*, was evaluated by Vieira et al. (2008). After 96h of exposition to several concentrations, it was observed that these substances inhibited acetylcholinesterase, leading to an interruption of the cholinergic transmission and induced the activity of lactate dehydrogenase, catalase, superoxide dismutase, gluthatione peroxidase and gluthationereductase. It was observed an induction of the glutathione S-transferase activity in fishes exposed to B[a]P and an inhibition of this enzyme after exposition to anthracene, suggesting an involvement of glutathione S-transferase in the B[a]P detoxification.

Studies carried out by McCloskey and Oris (1993) showed an increase of the hematocrit, evidences of hemolysis and a decrease in the presence of hemoglobin in the blood, besides a significative enzymatic inhibition of Na, K-ATPase and Mg-ATPase in the gills of the bluegill fish, after 96 hours of exposition to anthracene and ultraviolet solar radiation. The authors concluded that this toxicity is associated with a perturbation in the cellular membrane functions, promoted by anthracene.

It was evaluated the PAHs effects on the neotropical fish *Oreochromis mossambicus*, by enzymatic bioindicators. Expositions carried out with low concentrations of PAH (<0.4-4μg/g) led to sub-lethal toxic effects, observed in the liver of this organism, identified by the increase of the activity of the serum sorbitol dehydrogenase enzyme, which is characterized to be an indicator of damages in the liver. In this way, the authors concluded that the quantification of the EROD (ethoxyresorufin-O-deethylase) and serum sorbitol dehydrogenase enzymes characterize to be an excellent tool to monitor sub-lethal effects in fishes, resulted from pollution by PAHs (Shailaja and D'Silva, 2003).

Ke et al. (2007) studied the toxic effects of the water of a lake contaminated with PAHs, using an *in vivo* test with crucian carp (*Carassius auratus*) and *in vitro* test with rat hepatoma cells H4IIE. PAHs induced activity of EROD presenting a good correlation between the *in vivo* and *in vitro* assays, indicating that there was an activation of the enzyme CYP1A1 by the aryl hydrocarbon receptor (AhR). Previous studies carried out with sediments contaminated with PAHs of the same lake (Meiliang Bay), also showed a significative induction of the EROD activities in rat cells H4IIE maintained in culture (Qiao et al., 2006).

Ingestion of contaminated soils with PAHs by rats resulted in an increase of the CYP1A1 levels in the duodenum, liver, kidney, lung and spleen, indicating that these contaminants can cross the intestinal barrier and be distributed to other regions of the body (Roos et al., 2004). B[a]P ingestion by mice resulted in acute effect in the hematopoietic system leading to a drastic alteration in the bone marrow with the almost complete destruction of the pluripotent cells (Anselstetter and Heimpel, 1986).

The target organs of the PAHs toxic action in mammals are the skin, intestine, kidney and mammary gland, beyond the hematopoietic and lymphoid tissues, immunological system and gametic tissues (Albers, 2002). PAHs phototoxic effects difficultly are observed in mammals since the skin confers a barrier against the penetration of the UV rays, blocking the absorption of the energy light by the PAHs molecules (Arfsten et al., 1996). However, according to Kornhauser et al. (1991), phototoxicity generally results in direct effects in the skin such as allergic reactions, dermatitis and erythemas.

Oral exposition to high doses of PAHs provokes several hematological effects in mammals. Aplastic anaemia, pancytopenia, drastic reduction in the leukocyte of the peripheral blood and severe diminishing in the production of red globules are effects resulted from the mammals exposition to PAHs (WHO, 2000).

Several studies have demonstrated that PAHs, principally those found in cigarettes, are strongly related with the incidence of cardiovascular diseases in humans (Korashy and El-Kadi, 2006).

In humans, acute intoxication by PAHs is related with expositions to concentrations around 100 to 200 ppm. These concentrations promote nonspecific clinic signs and symptoms such as dizziness, nausea, headaches and difficulties in the motor coordination, by the action of these substances in the nervous system. Chronis effects can lead to toxic manifestations in several organs and systems, mainly, hematopoietic, renal, hepatic and nervous (Pedrozo et al., 2002). The effects of the B[a]P inhalation by humans result in patch opacities, prominent bronchiovascular markings and pleural effusions, besides vomits with blood, problems in the respiration, pain and irritation on the chest, irritation in the throat and cough (Gupta et al. 1993).

Studies of several environmental matrices such as combustion process in general, cigarette smoke, petroleum and its derivatives, among others, have demonstrated that PAHs, present in these mixtures, are the principal responsible for the toxicity potential. As aforementioned, several studies have related adverse effects produced by the PAHs, specifically their toxic effects, but still arise the necessity to establish new techniques to this methodologic set already existent and to intensify the studies of these contaminants actions, in the attempt to solve problems resulted from their impacts on the environment and on the human health.

2.3. Cytotoxicity

Cytotoxicity is characterized by the induction of damages in the cellular structure and/or function that can lead to disturbances in the cell cycle and in its physiology resulting in, many times, in the cellular death.

The PAHs cytotoxic activity has been described in many studies carried out with several organisms including man. Many assays, principally *in vitro* tests, have been developed to evaluate the cytotoxic effects and action of the PAHs on cells. The knowledge of the chemical properties of the molecules of these hydrocarbons is very important to the comprehension of their action mechanisms on the cells.

Studies carried out with microorganisms exposed to PAHs showed that the majority of these compounds, when present in small concentrations (5 to 100 ppb) in the environment, can stimulate or inhibit the cell division of bacteria and algae. In higher concentrations (0.2 to 10 ppm) these chemicals interfere, negatively, on the cell division of these organisms being capable of even leading them to death (Albers, 2002).

Alkio et al. (2005) also proved that PAHs can promote death in plants. According to the authors, phenanthrene induced cell death in *Arabidopsis thaliana*, observed by the cytotoxicity evaluation with the trypan blue dye. The number of cells in cellular death process was dependent on the concentration and exposition time to phenanthrene. The authors suggested an inhibition in the mitotic division process, by the observation of the reduction in growth of the stem and root.

According to Leme et al. (2008), mixtures of PAHs, present in water of rivers impacted by petroleum, can induce cell death in the meristematic cells of root tips of the plant *Allium cepa.*

Babich and Borenfreund (1987), when studying the *in vitro* cytotoxicity of several PAHs in the fish bluegill BF-2 cells, observed a low cytotxicity of these substances for the tested cells. However, when the S-9 microsomal fraction was incorporated to the mixture, they identified a significant increase of the cytotoxicity that, according to the authors, was due to the formation of active metabolites. Another factor described as important to the increase of the toxicity of the PAHs was the temperature that, according to them, potentiated the cytotoxic effects of the PAHs for this culture.

The direct cytotoxic effect of sixteen PAHs was evaluated by Schirmer et al. (1998a) in a cell line of gills of the rainbow trout (RT Gill-W1) by assays with fluorescent dyes. Only the PAHs composed by 2 or 3 benzene rings (naphthalene, acenaphthylene, acenaphthene, fluorene and phenanthrene) presented cytotoxic effects on the cells without metabolic activation. In this way, the authors suggest that factors such as solubility in water and the lipophilic property are characteristics directly related with the cytotoxicity of the PAHs, which make them capable of altering the integrity of the cellular membrane when they are accumulated in it.

The capacity of a chemical to induce damages or cause cell death as a result of photochemical reactions, known as photocytotoxicity (Schirmer et al., 1997) was evaluated for PAHs in assays with RT Gill-W1 cells. Of the 16 PAHS tested by Schirmer et al. (1998b), thirteen were photocytotoxic for the cited cells. Nevertheless, tests carried out with fish by the same authors showed that only the fluoranthene and pyrene offered risks to these organisms in the environment, since the phototoxic potential for fishes were observed in concentrations below of their solubility in water. According to Schirmer et al. (1998a), naphthalene is a PAH potent in inducing direct cytotoxicity in fishes since it was also capable of causing cellular death even when it was found in concentrations below its solubility limit, being considered dangerous in environmental contaminations. The cytotoxicity of anthracene (Oris and Giesy, 1985) and fluoranthene (Weinstein et al., 1997), for fishes, was evaluated by assays developed with dorsal epidermis and gill epithelium since these systems were considered by these authors as good indicators of the photocytotoxicity induced by PAHs.

The PAHs induced cytotoxic alterations in Chinese hamster V79 cells, having their potentiality enhanced when these agents were exposed to the light (Utesch et al., 1996). Studies report that B[a]P promoted cytotoxic alterations in the cell cycle of the Chinese hamster V79 cells maintained in culture. According to Ochi et al. (1985), these cells are highly susceptible to the B[a]P cytotoxic action, when they are found in the G1/S phases or beginning of the S phase but these action decrease when the cell cycle progress. According to Pandey et al. (2006), the B[a]P cytotoxicity is dependent on the activation of the CYP. When this activation in the cell culture does not happen by the absence of monooxygenases activity no cytotoxic effect is observed (Babich et al., 1988; Lee et al., 1993).

PAHs extracted from a mixture of heavy oil were tested as regards to their potential in causing interruption of the cellular mitotic division process in lung cells of V79 Chinese hamster (Hadnagy et al., 1997). These agents showed a strong dose-dependent relation in relation to the inhibition of the mitotic index, being considered potentially cytotoxic. Besides this, these authors observed the presence of alterations in the index of the division phases, quantifying an increase of cells in the metaphase phase and a decrease in cells in the anaphase and telophase phases.

Hader et al. (1996) also reported a cytotoxic effect of the PAHs in V79 cells synchronized in a same cell division phase. After the treatment, 50% of the analyzed cells were in the G2/M phase. This high percentage in this division phase can be associated with an inhibition of the progress of the mitotic activity of these cells.

Detmar et al. (2008), when analyzing the effects of the PAHs present in the cigarette smoke on placenta cells of murine observed alterations in the cell death rates (decrease of the rates) and changes in the expression and in activity of the key proteins of the physiologic control of cell death (increase in the expression of the antiapoptotic levels and decrease of the proapoptotic levels). According to the authors, several PAHs are ligands of the aryl hydrocarbon receptors (AhR), a cellular sensor of xenbiotics responsible for the activation of the metabolic machinery that is involved in innumerous cellular process such as death, survival and cellular stress answers.

Studies showed that the induction of cytotoxicity in the bone marrow of mice caused by anthracene was dependent on the metabolic activation of the CYP1B1 (Heidel et al., 2000).

Hiura et al. (1999) suggested that organic compounds, including the PAHs, present in diesel exhaust particles can induce apoptosis in macrophages of rats and man, resulting from the formation of the reactive oxygen species.

Chou et al. (2003) observed that the PAHs present in jet fuels presented a high cytotoxicity in keratinocytes of the human epidermis leading to the death of almost all cells present in the culture after 24 hours of exposition, even when in concentrations smaller than 0.5%. The authors concluded that the increase in the cytotoxicity is related with the increase in number and size of the sequence of the aromatic rings chains.

Among the sixteen PAHs considered priority pollutants by the United States Environmental Protection Agency (USEPA), nine were considered by Wang et al. (2007) as highly photocytotoxic for keratinocytes of human skin (HaCaT). These researchers concluded that the phototoxicity of the PAHs is dependent on the molecule size and on the rings arrangement, molecules with four or more benzene rings (except chrysene) present a higher photocytotoxic potential.

Frying oil, when reused several times, can also be a source of PAHs and cause damages to the cell. Pandey et al. (2006) identified nine different types of PAHs in the reused frying oil. This study suggests that the reused oil present a significative cytotoxic effect for the human hepatoma cell line (HepG2) by the induction of the catalytic activity of the enzime CYP1A1 leading to a metabolic activation of the present PAHs.

By the neutral red assay *in vitro* in HepG2, it was possible to evaluate the effects of the PAHs over the cell viability (Babich, 1988). The cytotoxic potential was induced by B[a]P, benzo[k]fluoranthene, benzo[b]fluoranthene, chrysene, benzo[a]anthracene, pyrene, phenanthrene, fluoranthene but, however, the fluorene, anthracene, acenaphthene and acenaphthylene were not cytotoxic for the same cells.

2.4. Genotoxicity and Mutagenicity

Genotoxic agents are characterized as agents capable of inducing primary lesions on DNA, such as formation of DNA adducts, oxidation and dimerization of nitrogenous bases (Majer et al., 2005). These effects are considered transitory since they are susceptible to be repaired, in other words, they do not characterize permanent alterations in the content or structure of the genetic material of an organism (Dearfield et al., 2002). Nevertheless, if these effects are not repaired or wrongly repaired, they can progress to permanent lesions in the DNA molecule, named mutations, which characterize the mutagenic events.

Several chemical agents are considered genotoxic and mutagenic. Amongst them there are the alkylanting agents, the intercalating and the reactive oxygen species (ROS). The alkylating agents are defined, briefly, as electrophilic agents that react with DNA substituting one hydrogen atom for an alkyl group resulting in the modification of this base (Stanley, 1994). In general, in the repair process of this lesion, DNA fragmentation, induction of mutations and formation of nucleotidic bridges can occur. Adducts are defined as molecules that bind covalently to the DNA or other molecules, becoming strongly attached to them. The DNA alkylation with a posterior covalent bond can form DNA adducts, which can influence, crucially, the biological process contributing for the toxic and mutagenic effects (Friedberg et al., 1995; Sturla, 2007). The ROS, resulted from the oxidative stress, can have their production induced by several chemical agents present in the environment (Picada et al., 2003). According to Collins et al. (1997), ROS are involved in the carcinogenesis process and act on the DNA molecules.

The PAHs are substances capable of promoting several deleterious effects on organisms. The genotoxicity and mutagenicity of the PAHs can be established by different ways such as the adducts formation originated by the binding of their metabolites, resulted from the biotransformation via CYP with the DNA; by the ROS production, from the quinone reaction, product of the PAHs metabolization via CYP, with the O_2; among others.

For the detection of environmental genotoxicity and mutagenicity various short-term tests, using a great variety of organisms, have been developed. The use of this great diversity of test organisms is possible due to the high conservation of the DNA structure (Majer et al., 2005). Such characteristic is also relevant for correlation studies between the test-systems since it makes possible the extrapolation, in a cautiously way, of the results obtained in one test-system to other living beings.

Great part of the environmental samples is defined as complex mixtures since they are constituted by a diversity of chemical agents potentially dangerous. The evaluation of the pollutants present in complex mixtures is complicated due to the difficulty in recognizing the chemical group harmful to the organisms or the synergistic effect exercised by the present substances. Nonetheless, the evaluation of the complex sample is extremely important since the obtained results are more faithful to the reality of the environmental expositions.

For White (2002), the non-identification of the toxic compounds present in complex environmental samples, the paucity of knowledge of the chemical behavior of genotoxic substances present in environmental samples and the lack of scientific data, characterize a great difficulty in the risk evaluation that these substances can cause to the organisms. This author assessed the effects of PAHs present in complex mixtures estimating the genotoxicity of the PAHs individually and in 13 combinations of more than 14 PAHs by the SOS chromotest, which is based in the evaluation of the SOS functions by monitoring the levels of the enzyme ß-galactosidase in *Escherichia coli* lines with the fusion of the operon *sulA::lacZ*. The obtained results showed the highest potentiality of the mutagenic effects for PAHs mixtures, probably, equal or inferior to the sum of the hazards observed for each compound of the mixture analyzed individually.

Down to the broad distribution of the PAHs in the environment, many researchers have demonstrated a special preoccupation with the contamination by these pollutants in aquatic ecosystems since, being water an essential asset to life, any alteration in the hydric resources can expose countless organisms even those that do not have their life restricted to this environment.

Chromosome aberrations (CA) are characterized by changes in the normal structure of a chromosome or in the total number of chromosomes that can happen spontaneously or as result of exposition to physical or chemical agents (Russel, 2002). In the last 30 years, the CA test has been widely used in environmental and occupational monitoring, as biomarker of the genotoxic effects induced by carcinogens (Hagmar et al., 2004).

For the assessment of the genotoxic influence of a petroleum field in Nigeria, Odeigah et al. (1997) applied the CA test on the plant test system *A.* cepa. The authors showed that the analyzed effluent was capable of inducing CA in the meristematic cells of the *A.* cepa roots, even after carrying out several dilutions of the sample.

The micronucleus test (MN) is considered by many authors as one of the most promising techniques of evaluating the mutagenic effects induced by chemical agents (Landolt and Kocan, 1983; Heddle et al., 1983; Ribeiro, 2003; Matsumoto et al., 2006). Such event is due to the fact that MN are resulted from damages, repaired or wrongly repaired, in the parental cells (Ribeiro, 2003) being easily visualized in the daughter cells as a structure similar to the main nucleus, however, with a reduced size.

Leme and Marin-Morales (2008) evaluated the impact of petroleum hydrocarbons in a river of the North coast of São Paulo State-Brazil, after the occurrence of an oil pipeline leakage. In this work it was used the CA and MN test in two different cell types of *A. cepa* (meristematic cells and F_1) with the aim to evaluate if the genotoxic effects observed by the presence of CA in the meristematic cells progressed to mutagenic effects (presence of MN in F_1 cells). The authors carried out, together with the bioassays, chemical analyses for total petroleum hydrocarbons (TPHs) and PAHs, where they could establish a positive correlation between CA and MN with the exposition to the samples contaminated with PAHs indicating a genotoxic and mutagenic action of the PAHs. These data are also corroborated by other work carried out by Leme et al. (2008), where the authors showed that petroleum, being a complex mixture of hydrocarbons, can present both a clastogenic and an aneugenic action or even induce cellular death on meristematic cells of *A. cepa*. For Leme et al. (2008), these actions are, principally, due to the presence of PAHs detected by chemical analyses in the tested sample.

Çavaş and Ergene-Gözükara (2005) applied the MN test in the *Oreochromis niloticus* species (Pisces: Perciformes), for the assessment of a petroleum refinery effluent and a chromium processing plant. In this work, it was evaluated the presence of MN in epithelial cells of gills and erythrocytes of peripheral blood, after exposition of the fishes to different concentrations of the effluent (5, 10 and 20% v/v) for a period of 3, 6 and 9 days. The authors reported mutagenicity for both tested effluents, although the mutagenic effects observed were much higher for the petroleum refinery effluent than for the chromium processing plant, this fact can be related to the presence of PAHs in the refinery effluent samples. The authors also show a higher sensitivity of the MN test in gills, probably, due to a higher direct exposition to the environmental contaminants. Another important question reported by the authors is the values, detected by the chemical analyses, of the pollutants present in the studied effluents. Albeit the visualized levels showed to be inferior of the established by the Turkish Environmental Standards, both tested effluents were considered harmful in relation to the mutagenicity. In this way, the authors believe that the mutagenicity tests must be included in the standards for the effluents emission in water bodies for a better evaluation of the hazard caused to the living beings.

The comet assay consists, particularly, in a valuable and cheap technique since it allows the detection of different cellular answers resulting from DNA damages and repair, which can be observed by the migration, in electrophoresis, of any cellular type obtained from a cell suspension (Sasaki et al., 1997, Mitchelmore and Chipman, 1998).

Due to a great occurrence of petroleum and its derivatives spills, that impact the Brazilian rivers, and the paucity of information about the answers of neotropical fishes to these situations, Vanzella et al. (2007) developed laboratorial tests with *Prochilodus lineatus* exposing these fishes to diesel water soluble fraction (DWSF) compound that has the PAHs as one of its main constituents. *P. lineatus* specimens were exposed to DWSF for 6, 24 and 96 hours, in order to characterize the acute effect and for 15 days, for a sub-chronic answer. By the obtained results, the authors concluded that both expositions to DWSF, acute and sub-chronic, cause genotoxic and mutagenic damages in *P. lineatus* blood cells. The authors also affirm that the combination of the comet assay and MN test techniques was sensitive and adequate for the evaluation of diesel genotoxicity, due to the complementary answer generated by these two methodologies.

Hoshina et al. (2008) carried out the MN test in *O. niloticus* (Nile tilapia) erythrocytes, aiming to assess the efficiency of the treatments performed in a petroleum refinery effluent and the quality of the river that receives the effluent. The authors collected effluents samples in different phases of the industrial process, such as after the physico-chemical treatment, entrance of the stabilization pond (after the biological treatment), exit of the pond (final effluent), upstream and downstream of the effluent discharge. With the collected water samples, the authors exposed *O. niloticus* specimens for a period of 72 hours. It was observed by the realized assays that the treatments used by the refinery promoted a diminishing in the effluent mutagenic potential, however, these treatments were not totally efficient, since the final effluent still presented a potential to induct damages in the genetic material of the test organism. This can be explained, by Tatem et al. (1978) citation, which says that refinery effluents tend to present a smaller quantity of light hydrocarbons than the crude oil, but a higher quantity of PAHs.

The marine environment has been a constant target of PAHs pollution, mainly by the accidental petroleum spills and its derivatives and by the routinely contamination by fuel oils such as diesel caused by the marine traffic.

The evaluation of complex mixtures is a problematic issue, because it is difficult to identify which chemical is the responsible for the deleterious effect or recognize a synergistic effect, among the several pollutants present in the mixture. Thus, it is not easy to establish a relation between the harmful effects and the contamination by any agent, including the PAHs.

Akcha et al. (2003), to assess the genotoxic effects of the marine contamination, used the comet assay and the DNA adducts detection (nuclease P1-enhanced post-labelling technique) in hepatic tissue of female dab specimens (*Limanda limanda*), collected in different sites of the English Channel (France) east region, area known by the PAHs contamination and by polychlorinated biphenyl (PCBs). By the obtained results, the authors concluded that the comet assay is a sensitive test for environmental biomonitoring since it was observed a concordance between the percentage of the comet tail size and the degree of chemical contamination in the collection sites. In other conclusion, the authors affirm that the observed DNA adducts were not related with the exposition to PAHs, due to the knowledge of the diol epoxides migration pattern, metabolites highly electrophilics induced in dab after exposition to B[a]P, substance that was not present in any site.

Taban et al. (2004) observed, by the comet assay carried out with mussels (*Mytilus edulis* L.) and sea urchin (*Strongylocentrotus droebachiensis*), an increase in the percentage of the comet tail size, in relation to the negative control, after 4-5 days of exposition of the organisms to the crude oil. These data indicated that the comet assay is a good tool to biomonitor damages in the DNA of marine invertebrates exposed to oil contamination.

Bolognesi et al. (2006), aiming to evaluate the genotoxic risk, in a long term, caused by oil spills, used as a case study the Haven oil spill that affected all the Ligurian coast, in Italy. For this, the authors assessed the damages in the DNA and the MN frequency in several test-organisms such as mussels, oysters and benthic fishes. By the obtained results, it was observed that even after several years of the accident occurrence the genotoxic effects persisted in different levels and in different living beings indicating the existence of residual sources of mutagenic and carcinogenic agents in the local.

To investigate the effects of crude oil contamination on native marine fishes, Baršienė et al. (2006), applied the MN test in peripheral blood erythrocytes and in cephalic kidney cells of the turbot (*Scophthalmus maximus*) and of the Atlantic cod (*Gadus morua*) exposed, in *in vivo* simulations, to this pollutant. Besides the exposition to crude oil, the authors also carried out assays with the addition of alkylphenol and PAHs into the oil with the objective to estimate the potential risk caused by the produced polluted water from the petroleum platforms since the presence of these chemicals is described in these areas. The results obtained in this work indicated a mutagenicity caused by the exposition to crude oil in both cellular types and species of fishes tested. The authors reported that the MN frequency of the cephalic kidney cells was two times higher than of the peripheral blood erythrocytes, which can be due to the fact that the cephalic kidney is the main hematopoietic tissue in teleost fishes. The authors also showed that the addition of alkylphenol and PAHs elevated, significantly, the MN frequency indicating an increase in the induced mutagenicity.

The food chain is characterized as one of the contamination via by PAHs and it has been a preoccupation focus, mainly, due to the human exposition potentiality. In this way, with the purpose the evaluate the PAHs contamination by the food via, Lemiere et al. (2005), carried out a study in order to estimate the damages in mammals DNA resulted from the trophic via by the ingestion of mussels contaminated with oil (Erika ship spill, France). Contaminated mussels were offered in rats daily food supply for a period of two to four weeks. DNA damage was evaluated by the comet assay carried out with hepatic, blood and bone marrow cells of rats treated with different PAHs contamination levels (312 μg, 569 μg and 870 μg of 16 PAHs/kg dry weight). The results showed a dose-effect-time relation between the quantity of DNA damage in the rats hepatic and hematopoietic cells and the PAHs contamination levels of the mussels. The genotoxic effects of the highest tested concentrations were growing between the 15th and 30th day of the treatment. According to the authors, it seems that there is efficiency in the DNA repair capacity for lower concentrations of PAHs since there was no induction of significative damages in the hepatic and hematopoietic cells of the exposed rats to lower concentrations of PAHs for a 30 days period, opposite of the observed for the 15 days. It was concluded by this work that foods contaminated with oil can cause damages to the DNA of organisms that are in superior trophic levels by the transference via food chain.

The evaluation of sediments and soils contaminated with PAHs is very important since these compartments have been characterized as pollutants deposits, which can characterize themselves in potential sources of dangerous chemicals. Many works carried out with several test-organisms have been developed in order to estimate the PAHs effects present in soils and sediments, on exposed organisms (Bispo et al., 1999; Békaert et al., 1999; Cotelle et al., 1999; Monarca et al., 2002; Plaza et al., 2005; Bihari et al., 2006).

Bispo et al. (1999) used the Mutatox® test to evaluate the genotoxicity of industrial soils impacted by PAHs. Such test uses the mutant bacteria *Vibrio fischeri* M169 as test-organism and is based on the light emission given by the bacteria reversion to the original form thus indicating the occurrence of damages in the DNA. In this work, the bioassays were carried out with the lixiviate extract and solvent extract (specific to assess pollutants of low solubility in water) of three soil samples contaminated with PAHs. According to the authors, the different soil samples presented genotoxicity even in the absence of the exogenous metabolization system (S9 mixture), resulted from the presence of molecular structures such as the quinones and diol epoxides formed during the PAHs biodegradation process in the soil.

Békaert et al. (1999) carried out a study to evaluate the genotoxicity of aqueous lixiviate of a soil contaminated with PAHs and heavy metals, by the Ames and Mutatox® *in vitro* tests, and the MN test *in vivo* in amphibians of the *Xenopus laevis* species. The authors observed by the three tests that this contaminated soil can cause damages to the aquatic environment. The authors affirmed that biological tests are important complementary tools for the physico-chemical analyses since it allows the evaluation of the chemical substances biodisponibility and bioaccumulation, besides showing the possible interaction between the pollutants. Békaert et al. (1999) also alert for the tests that require previous filtration process of the samples before its execution. The authors affirm that this procedure can masquerade the real results by the loss of some pollutants that are important during the preparation process of the sample. The authors affirm that 0.45 μm membranes can already eliminate many micropollutants potentially genotoxic.

Monarca et al. (2002) observed by the Ames test and MN test in *Tradescantia* that soil samples of an area highly industrialized in Italy presented mutagenic activity that, according to the authors, can be related with high concentrations of PAHs and heavy metals detected in the chemical analyses of these samples.

Evaluation of contaminated soils by petroleum hydrocarbons, followed by bioremediation by the biopile aerobic process, developed by bacteria assays of short-term, SOS chromotest and *umu-test* by Plaza et al. (2005) showed that this bioremediation process can reduce in almost 81% the quantity of petroleum hydrocarbons, including the PAHs present in the samples that led to an absence of genotoxic effects for the analyzed soils, both for the tests with the presence of S9 and with absence of this metabolization system (Plaza et al., 2005).

Bihari et al. (2006) evaluated the quality of the marine sediment of the North region of the Adriatic Sea, in Croatia. For this study, superficial marine sediments were collected in eight distinct sites of the region in order to quantify and qualify chemically the PAHs and assess their toxic and genotoxic potentials. According to the authors, the chemical analyses revealed a clear difference from the natural areas to the urban-industrialized areas and harbor zones, being the latter the most contaminated with PAHs. It was not reported any genotoxic actions for the organic extracts of the sampled sediments, submitted to the *umu*-test. In this test, the genotoxicity is measured by the ß-galactosidase activity of the *Salmonella typhimurium* TA1535/psk1002 line and is expressed by this enzymatic activity induction rate, between the sediment and the negative control treatment. The DNA damage, evaluated in the mussel (*M. galloprovincialis*) haemolymph was related with the total PAHs present in four sampled sites, although, the highest verified DNA damage rate does not correspond to the sample of the site that presented the highest PAHs contamination.

It is presumed that the atmospheric pollution displays a fundamental role in the incidence of respiratory diseases and cancers. Aerotransportated particles are capable of penetrating deeply in the lungs. Amongst these particles, it is found a wide variety of chemical substances, including several mutagens and carcinogens such as the PAHs.

Du Four et al. (2004) evaluated the genotoxic and mutagenic activities of air samples collected in Belgium. The collections were carried out in two seasons of the year (summer and winter) in the vicinity of a busy street; in industrial sites and in a rural community by two sampling systems, one using quartzfilters (QF) and other with polyurethane foam (PUF). It was observed a higher mutagenicity for the air collected during winter, probably, as a consequence of the PAHs high levels registered in this season. The increase in PAHs was attributed to a greater necessity of domestic heating and the specific meteorological conditions such as solar radiation intensities and ozone concentrations. The authors related the variation in the mutagenicity rates between the summer and winter seasons with the presence of direct and indirect mutagens in the collected samples because, according to them, in winter the indirect mutagenic activities are higher than the direct activities indicating a considerable contribution of the PAHs in the observed effects. In this work it was observed that all the analyzed environments presented contamination by PAHs showing that, independently from the sampled area (rural, urban and industrial), the exposition to PAHs was always significant.

In one another evaluation the genotoxic and mutagenic activities of the Belgium air, carried out by the *Salmonella* mutagenicity standard plate-incorporation and Vitotox® assays, it was possible to observe that the highest mutagenicity values were registered out of the big cities and in the rural area, contaminated by the pollution of a chemical industry located 4 Km of distance. However, the authors reported that, based on the concentrations of 10 mutagenic PAHs and on the supposition of their specific mutagenicities additive effect, only a small percentage of the indirect mutagenic activity could be explained. The results alert for the fact that the majority of the observed mutagenic effects could be related with the action of other substances, that were not identified or measured by the chemical analyses, which shows the importance of the bioassays application since, with these tests, it is possible to monitor the pollutant and estimate the air quality (Du Four et al., 2005).

Samples of the particulate material, collected in two different region of the Mexico city – industrial and residential, were processed in two final extracts, one aqueous and one organic-soluble extract, in order to evaluate the health risks promoted by the particulate material resulted from the atmospheric pollution (Roubicek et al., 2007). By the samples chemical analyses, the authors have found that the highest PAHs concentrations were registered for the residential area, although both extracts, industrial and residential, induced a significative increase in the MN frequency, indicating the mutagenic potential for all the analyzed samples. Additionally, the authors have shown that, in the association between MN presence and chemical compounds detected, cadmium and PAHs are the most related agents, significantly, with MN induction.

Studies of the mutagenicity assessment (Ames test with the TA98 and YG1041 lines, with and without S9 addition) and the detection of DNA adducts (^{32}P-postlabelling analyses) of fractions of the particulate atmospheric material from São Paulo city- Brazil were carried out by Umbuzeiro et al. (2008). Air samples were collected in the vicinity of a highway and in a park located in the urban area of São Paulo city. The particulate material was fractioned in order to obtain the PAH, nitro-PAH and oxy-PAH fractions. The PAHs fractions, of both sites, presented insignificant mutagenicity opposite of the observed for the nitro-PAHs fraction. Both sites presented the same mutagenic activity levels, probably, due to the related effects of the nitro and oxy-PAHs action.

The use of the ^{32}P-postlabelling analyses technique for the detection of DNA adducts in HepG2 cells showed a genotoxic potential of the air samples particulate material, collected in a central area of the Prague city, Czech Republic. It was observed for several tested periods a positive correlation between B[a]P and carcinogenic PAHs present in the particulate material with the total levels of DNA adducts of the exposed cells (Sevastyanova et al., 2008). These results pointed out for the hypothesis that B[a]P and other carcinogenic PAHs are more important factors to determine the genotoxic potential. Thus, the genotoxicity estimative of the air environment and the evaluation of health risk must be based, principally, on carcinogenic PAHs concentration and in the reactivity of these substances with the biological material, while the particles mass and the particulate material type do not seem to be important factors to estimate the atmospheric genotoxicity.

To investigate the DNA adduct levels in workers exposed to traffic and in the general population of Florence city (Italy), Palli et al. (2008) carried out a study with 152 healthful adult inhabitants and 62 non-smokers workers exposed to traffic, for 1-2 weeks. In this work, it was carried out the DNA adducts evaluation, by the ^{32}P-postlabelling technique, in blood samples collected from all the sampled individuals. By the obtained results, the authors concluded that the DNA adduct levels tend to reflect the exposition levels to the particulate material being the workers exposed to the traffic the greater bearers of these alterations.

Cigarette smoke can be considered as one environmental exposition and constitute a risk factor for the development of several illnesses such as lung cancer and vascular diseases (Mitchell et al., 1999). Among the PAHs, B[a]P is a substance related with smoke that is associated with endothelial damages in blood vessels (Thirman et al., 1994). Studies to assess the B[a]P potential to induce DNA damages in human endothelial cells were carried out by Annas et al. (2000). In this work, it was evaluated the action of B[a]P different concentrations (0.1, 1.0 and 10μM) in human umbilical cord cells cultivated in medium without addition of an exogenous metabolization system, in order to verify the capacity of inducing damages in the genetic material and the biotransformation via involved in this process. The authors concluded that B[a]P induce a significative quantity of DNA damage, even in the absence of an exogenous metabolization system suggesting that the bioactivation of this agent is not totally dependent on the activity induced by CYP1A.

As explained above, several works have been developed in order to measure the genotoxic and mutagenic damages induced by the exposition to PAHs. The detection of these effects can be carried out with several techniques, using a great diversity of test-organisms, being reported, for most of the cases, positive results. In this way, we can note the real importance in the genetic risk assessment of the PAHs and the use of biological systems in the environmental monitoring of these pollutants.

2.5. Teratogenicity and Embryotoxic Effect

Teratogenic alterations are manifestations of the action of chemical agents over the fetus or over the organisms' reproduction. These actions lead to structural and/or functional malfunctions that are due to the interference in the organism development, still in the embryonary phases.

Congenital anomalies in human embryos are adverse consequences of a pregnancy that can be associated with exposition to environmental pollution (Dolk and Vrijheid, 2003). Analyses of blood samples, obtained from human umbilical cord indicated the presence of 287 chemicals, which confirmed that there is an intense exposition to agents with teratogenic potentiality during the gestational period (Houlihan et al., 2005).

Beyond their toxic, genotoxic, mutagenic and carcinogenic actions the PAHs are recognized teratogenic compounds (Wassenberg and Di Giulio, 2004). The PAHs teratogenic action has been registered for many different organisms and in the different ecosystem levels.

Studies with invertebrates, fishes, amphibians, reptiles and birds, in their majority, restrict themselves to the exposition of the eggs to PAHs mixtures or to a specific hydrocarbon. In rodents, most of the studies use intraperitoneal applications or oral ingestion in pregnant females. In humans, the studies are made, principally, by the environmental or occupational exposition to the PAHs.

Studies made with aquatic invertebrates contaminated with PAHs report a variety of alteration such as reduction in the reproductive rates, inhibition of the larvae and embryos development, besides delay in the larval emergence (Neff, 1985; Eisler, 2000). Teratogenic alterations regarding anomalies in the shell and the mantle formation were found by Wessel et al. (2007) in oysters embryos (*Crassostrea gigas*) exposed to B[a]P. According to Geffard et al. (2003), bioassays using oysters embryos and larvae contribute to a deep knowledge of the PAHs teratogenicity. Pelletier et al. (2000), studying the maternal transference of the photoreactive PAHs in benthic bivalves, observed that this transference would result in phototoxicity for the pelagic larvae, when they were exposed to ultraviolet light, which would represent a significative adverse effect over the populations of the organisms that live in sediments contaminated by this type of pollutant.

Many studies were made in order to study the PAHs influence in the embryonary development and in the reproduction of many species of fishes. Among the proved alterations, we can cite the reduction in the maturation of the gonads and eggs (Collier et al., 1993; Kirby, 1994; Ghosh and Thomas, 1995; Thomas and Budiantra, 1995); decrease of the reproductive success, with reduction of the percentage of incubated eggs and survival of the recent eclosed (Black et al., 1988; Casillas et al. 1991; Tilghman Hall and Oris, 1991; Kocan et al., 1996). Exposition to PAH, according to Collier et al. (1993), can damage the initial phases of the reproduction process of females of the English sole. Nicolas (1999) in a review article recognizes the PAHs as substances with deleterious effect in all the stages of fish vitellogenesis, which could endanger the succession of their descendants. Van Meter et al. (2006) suggest that the hydrocarbons accumulation in the embryos vitelline sac become a continuous source of teratogenicity for the larva along its development.

It was observed by Pollino and Holdway (2002), deformities in the larval stage of the fish *Melanotaenia fluviatillis* after exposition to PAHs such as edema in the pericardium, anomalies in the jaw and alteration in the flexion and curvature of the vertebrae.

Many authors consider the use of embryos of fish as an excellent test organism for the elucidation of vascular defects in vertebrates induced by chemicals potentially teratogenic, such as the PAHs (Hose et al., 1984, Antkiewicz et al., 2006; Carney et al., 2006; Goldstone and Stegeman, 2006). According to Billiard et al. (2008), one of the advantages of the use of these embryos would be the presence of a transparent chorion, which allows a direct and not invasive observation during all the embryogenesis. Bioassays developed with embryos of *Danio rerio* exposed to several solutions of organic films containing PAHs, collected in the urban area of Toronto (Canada), showed that these compounds were capable of inducing cardiovascular, hematopoiesis, neural and pigmentary abnormalities, besides behavioral alterations in these fishes (Diamond et al., 2000).

Incardona et al. (2005) affirm that embryos of fishes exposed to wheathered ANS oil and to mixtures of PAHs developed a pericardium syndrome, yolk sac edema, reduction in the craniumfacial structure and body axis defects. Many of the defects were induced by individual exposition to the tricyclic aromatic hydrocarbons that is one the most abundant class between the wheathered ANS oil, causing secondary impacts in the cardiac function of the embryo, coincident with the development of the cardiac conduction system.

Studies carried out by Incardona et al. (2006), about induction of alterations in zebrafish embryos by PAHs constituted by four aromatic rings, pointed out that chrysene did not show any embryotoxic and teratogenic effect in these embryos. The exposition to benzo[b]anthracene resulted in mortality in all the concentrations between 8 and 10 μM and defects in the development as a failure of head straightening, lack of finfold development and poor cardiac looping with pericardial edema hemorrhage, for concentrations between 2 and 6 μM. The continuous expositions to doses superior to 5 μM of pyrene resulted in systemic toxicity that occurred in the initial phases of the larval development, including dorsal curvature of the body axis, reduction in the peripheral circulation, anemia and pericardial edema.

Tuvikene (1995) reports that eggs of fishes contaminated with anthracene, after exposition to ultraviolet radiation, presented alterations in the development such as edema, malfunction of the yolk sac and deformities in the eyes. According to Tilghman Hall and Oris (1991), PAHs, specifically anthracene, can be transferred by the mother to her offspring.

Tests carried out with alevins of the rainbow trout (*Salmo gairdneri* Richardson) exposed to B[a]P showed an incidence significantly high of cranium and vertebral column deformities and microphtalmia (Hose et al., 1984). Deformities in the embrionary development such as malformation of tail and head, beyond pericardium edema, were reported by Wassenberg et al. (2002) in the marine fish *Fundulus heteroclitus*, after exposition to several concentrations of B[a]P.

Exposition of eggs of turtles (*Chelydra serpentina*) to PAHs caused significant alterations in the embryos development, varying from mild deformities (tail deformities, absence of pigmentation, malfomation of the shell), moderate (absence of the members or eyes and asynchronous development) until severe such as absence of two or more members, cerebral abnormalities and not closing of the cranium (Van Meter et al., 2006).

The effects caused by the PAHs in embryos of birds include death, abnormalities in the development and a diversity of cellular and biochemical answers (Albers, 2006). Lethal and sub-lethal effects were observed in embryos of birds in experiments carried out with eggs re-covered with PAHs fractions (Hoffman, 1990). According to Matsumoto and Kashimoto (1986), these chemicals produce embryotoxic effects in birds which vary according to the quantity and chemical structure of the PAHs. Mayura et al. (1999), when carrying out experiments with ingestion of fractionated mixtures of PAHs in egg yolks of chicken, found out severe teratogenic alterations in their embryos and according to Hoffman and Gay (1981), high rates of teratogenic deformities were observed in embryos of ducks, whose eggs were exposed to PAHs.

Feuston et al. (1997) observed deleterious effects in the development of embryos of pregnant rats, after their exposition to crude oil. According to the authors, the observed effects would be directly related with the presence of PAHs in the samples. According to Sanyal and Li (2007), introperitoneal applications of PAHs in pregnant rats lead to an efficient absorption of the mother and a more direct exposition of the fetus, resulting in toxic effects for the mother and deformities resulted from alteration in the fetus development. However, Hansen et al. (2000) reported that fetus born from females of mice submitted by gavage to carbon black oil that contains PAHs, in five different concentrations (100, 200, 300, 350, 400 mg/kg/day) did not present any structural deformation in the viscera and in the skeleton being observed only a decrease in the weight of the embryo.

Sanyal and Li (2007) observed that the prenatal exposition of rats to PAHs induced adverse effects on the blood vascular system, principally during the angiogenesis, which can lead to grave malformations and dysfunctions in vital organs. Shum et al. (1979) reported the presence of congenital defects and growth retard in fetus of mice, whose mothers were treated, orally, with doses of B[a]P between 50 and 300 mg per Kg of weight of the animal, between the seventh and tenth day of the gestation, confirming the descriptions of Nebert (1989) that classify the B[a]P as a teratogenic chemical for mice.

The presence of abnormalities in the fetus, resulted from the exposition of the mothers to PAHs, would be related to the fact that these substances are toxic to the placenta, which facilitates the transference of toxic agents for the fetus, besides causing lose of the regulatory function, with a decrease in the nutrients supplement and other placental factors essential for the correct growth and development of the fetus (Sanyal and Li, 2007).

Studies carried out by Mackenzie and Angevine (1981), exposing pregnant females of mice CD-1 to gavage of 10 mg/Kg/day of B[a]P, during 10 days, showed that the descendants presented alterations in the gonad morphology and in the development of the germ cells.

Lineages of mices Ah-responsive and Ah-non responsive, presented an increase in the abortion incidence, reabsorption of the fetus and malformations after oral administrations of B[a]P in pregnant females, being these alterations more expressive in Ah-non responsive than in embryos Ah-responsive (Legraverend et al., 1984). These authors also concluded that the teratogenicity is directly related with the maternal and/or embryo genotype controlled by the gene Ah that codifies the cytosolic Ah receptor that regulates the induction of the CYP enzymes.

Intraperitoneal injections of B[a]P in pregnant females of mice caused alterations in the follicular growth, ovulation and formation of the luteous body (Swartz and Mattison, 1985).

Prenatal exposition to B[a]P caused severe effects in the offspring of the mice such as testicular atrophy and tumors in the interstitial cells (Payne, 1958). According to Sanyal and Li (2007), the dimethylbenzo[a]anthracene is more potent in producing deleterious effects in pregnant females of rats than the B[a]P.

Exposition of pregnant rats to PAHs induced toxicity to the placenta, that resulted from abnormal development of the fetus and hemorrhages in different parts of the body, mainly intradermic and cranial, which demonstrated that the fetal vascularity is the main target of the toxic action of the PAHs (Sanyal and Li, 2007).

According to Genuis (2006), exposition to PAHs present in the environment during the human embrionary development is quite expressive. These agents penetrate in the body of the mother by different ways and can pass through the placenta and become source of deleterious alterations in the offspring (Sanyal and Li, 2007). The formation of DNA adducts, as well as the occurrence of oxidative damages in these macromolecules, are key events for the triggering of the teratogenicity molecular mechanisms (Wessel, 2007).

Perera et al. (2005), when studying pregnant women exposed to different concentrations of PAHs in 4 regions of the world, observed that the fetuses can be 10 times more susceptible to the PAHs than the mothers. It is known that some PAHs are transmitted by the mother to the fetus via placenta, causing alterations in the development, liver, lung, lymphatic and nervous system (Rice and Ward, 1982; Perera et al., 2003). Studies indicated that PAHs can induce a number of adverse effects being capable of jeopardizing even future generations (WHO, 2000). Other studies report a decrease in the head circumference of babies, whose mothers were frequently exposed to PAHs, which could affect the cognitive development and the scholar performance of these children (Hack et al., 1991; Dam et al., 1998; Perera et al., 2006).

According to Choi et al. (2006), based in studies carried out in New York (USA) and in Krakow (Poland), human exposition to low concentrations of PAHs induced adverse reproductive effects, principally in relation to the fetal growth.

2.6. Carcinogenicity

According to Pedrozo et al. (2002), carcinogenicity is the capacity of certain agents induce the formation of tumors, by the direct action in DNA, which progress to biochemical, histological, functional and clinic alterations of the target cells and tissues. Replication of the cells and tissues, that present carcinogenic adducts in their DNA, can result in gene mutations that regulate the cellular growth and differentiation occasionating a progressive genetic instability, which leads to an uncontrolled growth, lose of cellular differentiation, invasion of dedifferentiated cells in the tissues and metastasis (Stewart and Kleihues, 2003). Induction of damages in DNA, responsible for the adverse effects such as mutations and chromosome rearranges, are considered primary mechanisms of cancer induction (Jacobson-Kram et al., 1993).

PAHs carcinogenic activity is dependent on the structural characteristics of these substances, such as the number of benzene rings, fusion sites, extent of condensation, degree and local of the methylation (Ross and Nesnow, 1999). According to Jacob (1996), PAHs that contain more than 3 aromatic rings are, potentially, the most carcinogenics. According to Baird et al. (2005), PAHs are known as complete carcinogens since they can, by their action, induce tumors in the skin. However, according to Luczyński et al. (2005) and Pashin and Bakhitova (1979), PAHs do not have direct actions like carcinogens since their carcinogenic activity is promoted by metabolizations that these chemicals suffer.

There are two main PAHs metabolization process in the cells, which lead to damages in the DNA. One activation via is characterized by the dihydrodiol epoxides formation, that bind, covalently, to amino heterocyclics groups of purine bases, to form stable adducts. Another via involves the formation of cation radicals, which bind to the N7 or C8 of the purine bases to form instable adducts and generate apurine sites in the DNA, by spontaneous depurination (Melendez-Colon et al., 1999).

Pereira Netto et al. (2000) cite another two mechanisms to explain PAHs activation. The first would be related with the formation of benzilic esters, electrophilic, by a series of substitution reactions. The second would be the de-hydrogenation of dihydrodiols metabolites, which lead to the production of quinones, substances that are capable of reacting, directly, with DNA or with O_2, generating reactive oxygen species such as hidroxil radicals or superoxide anions, which attack DNA.

PAHs are metabolized by a variety of xenobiotic-metabolizing enzymes (Baird et al., 2005). CYP and epoxide hydrolase convert the PAH into PAH-diols and these products are transformed in metabolites, potentially, carcinogenic, by the CYP action, forming PAH diol-epoxides or by the Aldo-keto-reductase action, generating PAH *o*-quinones. PAHs can be also activated by CYP and by peroxidases, forming the reactive cations radicals, which bind, covalently, to DNA. Intermediary products formed are still metabolized by enzymes of phase II, resulting in metabolites more polars and soluble in water becoming ready to be excreted by the body (Shimada, 2006).

Adducts ability to bind to DNA and persist to form mutations is dependent on the adduct conformation inside the DNA sequence and of the identification of the lesion by the repair system (Wu et al., 2002). Formation of the PAH-DNA adducts can lead to a induction of mutations, which activate the protoongenes (*ras*) or inactivate the tumor suppressor genes (p53) that can trigger the tumorization initiation process (Prahalad et al., 1997; Straif et al., 2005).

PHAs induce the activation of the CYP1A1 and CYP1B1, via the aryl hydrocarbons receptor (AhR). This cytosolic receptor, after the aryl hydrocarbon nuclear translocator (ARNT) dimerization, enters in the nucleus and interact with genes that contain xenobiotic response element (XRE), resulting in the induction or hyper regulation of the proteins involved in the xenobiotics detoxification (Whitlock, 1999).

PHA-DNA adducts can be formed in several genes, however, these adducts have been found, frequently, in the *ras* genes, which when mutated are capable of transforming the cells and inducing the appearance of malign tumors (Maronpot et al., 1995). Three different types of the *ras* gene were identified and sequenced: Ha-*ras* (chromosome 11); Ki-ras (chromosome 12) and N-ras (chromosome 1). About 20% of the majority of the malign tumors in humans seems to be related with the proliferation of aberrant cells, with alteration of the cell checkpoint control and cellular differentiation (Ross and Nesnow, 1999).

There are some structural arrangements in the benzene rings that interfere in the carcinogenic properties of the PAHs and derivates. Bay region consists of 3 aromatic rings, connected in an angular way (Figure 2), creating a recoil that looks like a bay. Pseudo bay region is composed by two aromatic rings and one terminal aliphatic ring (Figure 2). A fjord region is resultant from an angular disposition of four aromatic rings (Figure 3) and, in the case of a pseudo fjord region, the terminal ring is saturated (Figure 3) (Luczyński et al., 2005).

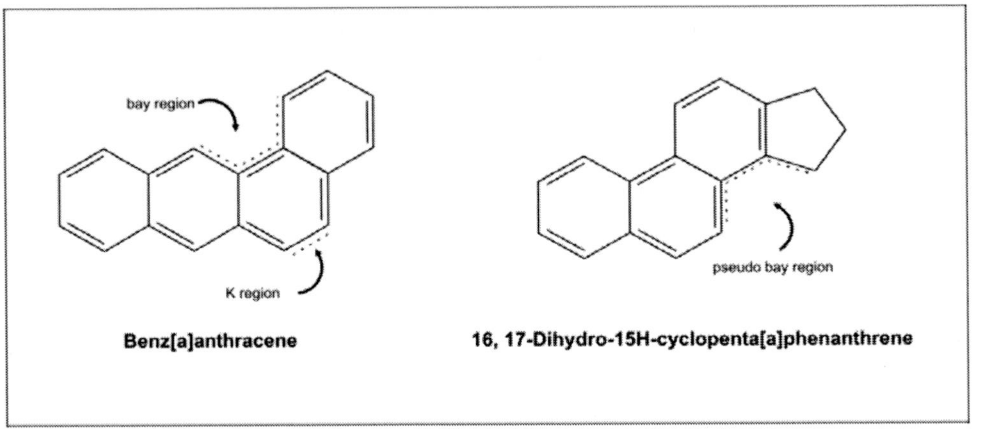

Figure 2. Scheme showing PAH with bay region, K region (e.g. Benz[a]anthracene) and showing the pseudo bay region (e.g. 16,17-Dihydro-15H-cyclopenta[a]phenanthrene).

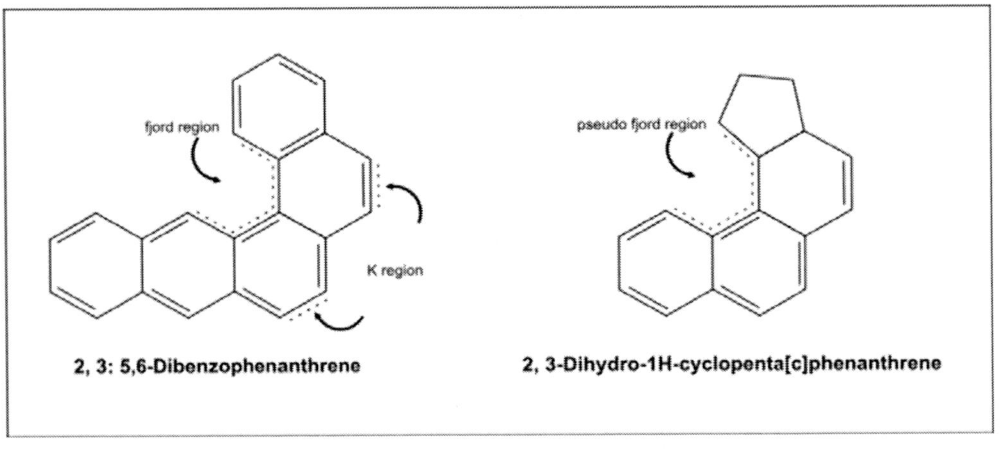

Figure 3. Scheme showing PAH with fjord region, K region (e.g. 2,3:5,6-Dibenzophenanthrene) and showing the pseudo fjord region (e.g. 2,3-Dihydro-1H-cyclopenta[c]phenanthrene).

According to Baird et al. (2005), PAHs that present a structure with bay regions are potent carcinogens but those who have fjord regions, like dibenzo[a,l]pyrene, are even more potent. Metabolites derived from PAHs with bay region present a pronunciated tendency to form carbonium ions, becoming much more reactive (Jacob, 1996). When the bay region changes to fjord region the molecule plane is also deformed. This change facilitates the access to an epoxide ring of the diolepoxide molecule, allowing the covalent bond in the DNA molecule (Luczyński et al., 2005).

Another chemical characteristic of the PAHs that contribute to their carcinogenicity is the presence of a region highly electrondense designated K region (Figure 2 and 3). This region allows the formation of the K region epoxides, which makes it a site highly reactive, for the occurrence of double definite and localized bonds. (Samanta et al., 2002; Luczyński et al., 2005), besides being the main local of metabolic reactions (Bücker et al., 1979).

Observation of the carcinogenic potentiality, individual, of several PAHs was registered by the presence of malign and benign tumors in animals, by different application vias of these compounds (oral, intraperitoneal, subcutaneous, epicutaneous, among others). The *in vitro* assays, carried out with culture animal cells and tissues, contributed, in a significative way, for the elucidation of the carcinogenic potential of these chemicals. In humans, the PAHs carcinogenic results were registered, mainly, by occupational exposition (Jacob, 1996).

The assessment and control of the human exposition to PAHs and derivates are made, generally, by the monitoring of the environmental levels of these substances, knowledge of the penetration vias in the organism, their metabolism, as well the early assessment of their biological effects (Pereira Netto et al., 2000).

Since the DNA adducts formation is a critic stage in the chemical carcinogenesis caused by the PAHs, the identification of these alterations can serve as a precocious bioindicator of cancer (Beach and Gupta, 1992). According to Pereira Netto et al. (2000), the most used assays are the ^{32}P postlabeling, imunoassays, fluorescent spectroscopy and detection of 1-hydroxypyrene. Okona-Mensah et al. (2005), suggest the use of nasal cells for the investigation of cancer induction in lungs by atmospheric pollution, as well as the use of mass spectroscopy to quantify specific adducts.

Metabolic activation vias and the types of damages induced by PAHs dibenzo[a,l]pyrene (DB[a,l]P), 7,12-dimethylbenzo[a]anthracene (DMBA) and the B[a]P in mice epidermis were evaluated by the analysis of the formation of stable adducts and AP sites in the DNA of this tissue (Melendez-Cólon, 1999). The study revealed that the topic exposition of mice to these hydrocarbons resulted in a formation dose-dependent of high levels of stable dihydrodiol epoxide-DNA adducts but lower levels of AP sites in the DNA of the epidermis cells. The registered levels for stable adducts were higher for B[a,l]P, followed by DMBA and B[a]P, for all the doses and times tested. In this way, the authors affirmed that the dihydrodiol epoxide-DNA adducts are the main responsible for the cancer initiation in skin of mice, induced by PAHs.

According to Ross and Nesnow (1999), the identification of specific mutations in the *ras* gene makes possible to diagnose which adduct induced the alterations. The authors reported that transversion mutations of the A→T type, in the codon 61, are related to the deoxyadenosime adducts action, while transversions G→T, in the codon 12, are frequently induced by the diol-epoxide, which binds to deoxyguanosine and to cyclopenta-ring oxides, inducing transversions G→C, in the codon 12.

In studies carried out about tumorigenesis of the PAHs coal tars and of the B[a]P in females of mice B6C3F1, fed with these substances, it was verified that PAHs present in the coal tars induced the formation of tumors in the small intestine, liver, lung and forestomach, besides skin sarcomas, histosarcomas and hemangiosarcomas in several organs. Evaluations made with B[a]P, showed that this substance was tumorigenic in tongue, larynx, esophagus and forestomach. The DNA adduct levels, measured in lung, liver and forestomach, were dose dependent indicating that both B[a]P and PAHs coal tars were transported to other organs, although B[a]P have not induced tumor in lung and liver of mice exposed to this PAH. B[a]P and PAHs coal tar induced mutation in the positions 1 and 2 of the guanine in the codons 12 and 13 of the *K-ras* oncogenes in forestomach tumors. Mutations in guanine were also found in lung tumors, but not in small intestine tumors. Thus, mutations in the *K-ras* gene must be involved with lung and forestomach tumors, but not in small intestine tumors (Warshawsky, 1999).

There are epidemiological evidences that suggest that occupational groups exposed frequently to complex mixtures of PAHs present a more elevated risk in the development of lung cancer (Okona-Mensah et al., 2005).

Studies carried out by Straif et al. (2005) showed that, in mice, B[a]P caused lung tumors, by the formation of diol-epoxide and skin tumors, triggered by diol-epoxide mechanisns and cations radicals. According to these same authors, few PAHs are considered carcinogens more potent than B[a]P. However, dibenzo(a,l)pyrene seems to be a carcinogenic 10 times more potent for rats than the own B[a]P (Okona-Mensah et al., 2005). Another PAH highly potent, according to Platt et al. (1990) is the dibenzo(a,h)anthracene.

Pavanello et al. (1999) used the HPLC/fluorescence method to detect the anti -BPDE–DNA adduct in mononuclear leukocytes exposed to PAHs. The authors concluded that this methodology can be used for the evaluation of the chronic exposure to PAHs and that the inhalatory via contributed, substantially, for the adducts formation in the lymphocyte plus monocyte fraction in coal factories workers and chimney cleaners.

Jacobson-Kram et al. (1993) studied the cancer risk in peripheral lymphocytes of a group of firefighters induced by occupational exposition to products potentially carcinogenic resulted from the combustion and organic matter burning, by the presence of adducts in the DNA of these lymphocytes. In relation to the PAH-DNA adducts levels and the presence of benzo[a]pyrene diol epoxide (BPDE)-DNA in the peripheral lymphocytes of these professionals, the results found were not significative, however, when other factors, such as smoke and alcohol consumption, were associated, the PAH-DNA levels were increased.

As presented above, PAHs are carcinogens in potential, both for humans and for the rest of the animals. Nevertheless, the chemical complexity of the PAHs, the diversity of the metabolic routes of the biological systems, the several enzymes involved in the metabolization process of these substances, besides the several exposition vias, to which the animals can be exposed, constitute difficulties for the assessment of the action and effects of these chemicals in the biological environment. For a better understanding of the PAHs activities it is necessary that new methodologies be applied in studies with these chemicals, besides re-evaluations that can confirm the obtained effects on some organisms.

2.7. Estrogenicity

Existence of chemical substances that are capable of inducing alterations in the hormonal homeostasis is already known for many years (Rivas et al., 1997), however, the concern with this pollutant class has been growing, due to the adverse effects related to their exposition.

According to the *U.S. Environmental Protection Agency* (USEPA), these chemicals, named endocrine disrupting chemicals (EDCs), are defined as exogenous agents that interfere in the synthesis, secretion, transport, bond, action or elimination of natural body hormones, which are responsible for the maintenance of the homeostasis, reproduction, development and/or behavior of the organism (Belgiorno et al. 2007).

It is known that chemicals suspect of causing alterations in the endocrine system such as EDCs are potentially associated with several diseases such as testicle, breast and prostate cancer, reduction in the spermatozoid rate, deformities in the reproductive organs, thyroid dysfunctions and alterations related to the neurological system (Rivas et al., 1997; Jimènez, 1997; Ghiselli and Jardim, 2007).

Many of the EDCs are synthetic organic chemicals introduced in the environment by anthropogenic discharges (Belgiorno et al., 2007). Among the EDCs acknowledged nowadays, we can cite the phthalates, pesticides, polychlorinated biphenyls (PCBs), dioxins, alkyl phenols, bisphenols, and PAHs (Jimènez, 1997; Ghiselli and Jardim, 2007; Belgiorno et al., 2007).

PAHs action like EDCs has been described in some studies, which have reported both the estrogenic and anti-estrogenic actions. According to Ghiselli and Jardim (2007), PAHs, by presenting similar structures to sexual steroids, can interfere in the natural function of the endocrine system, binding to hormonal receptors and resulting in an antagonistic effect, in other words, the occurrence of the blockage of the natural hormone action, by obstructing the interaction with its respective receptor. Additionally, PAHs action as anti-estrogens can be given by the bond to the aryl hydrocarbon receptor (AhR), triggering the induction of genes associated with the activation of this receptor (Ah-responsive genes), which results in a wide spectrum of anti-estrogenic responses, including the increase in the 17-β estradiol (E_2) metabolism and decrease of the nuclear estrogen receptor (ER) (Santodonato, 1997; Arcaro et al., 1999).

Arcaro et al. (1999) used the MCF-7 focus assay, which measures the relation estrogen-dependency in the cellular growth; recombinant human estrogen-receptor (ER) binding assay, which evaluates the binding capacity of the environmental chemicals to ER; whole-cell ER binding and 17-ß estradiol (E_2) metabolism assay, to investigate the estrogenicity and anti-estrogenitity of sediments contaminated with PAHs, originating from two localities in New York, one industrial and other non-industrial. The authors have reported the detection, by chemical analyses of 14 PAHs, being the levels observed in the industrial area much more elevated than in the non-industrial. In this work, the bioassays were carried out with fractions of the PAHs collection obtained from extracts of the sediment samples; with two reconstituted mixtures and with the 14 PAHs detected, tested separately. The authors reported anti-estrogenicity for the PAHs fractions of both tested sediments, for the reconstituted mixtures and for 8 of the 14 PAHs tested individually. The authors also investigated the processes that are involved in the anti-estrogenicity induced by the PAHs, showing a relation with, at least, two mechanisms: the first involving competition of the ER for the PAHs metabolites, showing that PAHs can exert this activity by their metabolic inactivation, and the second involving diminishing of E_2 by induction of the metabolism.

Vitellogenesis is a process hormonally controlled and that can be affected by the exposition to xenoestrogens. In most fish species, oocyte maturation process occurs seasonally, in which one spawn occurs in each annual reproductive cycle. In this way, factors that affect vitellogenesis can cause a drastic reduction in the reproductive success of the individual and, consequently, affect an entire species population. According to a review carried out by Nicolas (1999), deleterious effects in fishes vitellogenesis were reported for PAHs both for wild populations and laboratorial experiments. Amongst the effects cited, it is included reduction in the circulation of hormones and plasmatic vitelogenin (species-specific protein synthesized by hepatocytes involved in the oocyte maturation); estrogenic and anti-estrogenic effects; delays in the oocyte maturation and reduction of the reproductive success. Based in this review, Nicolas (1999) showed that PAHs action mechanisms in the vitellogenesis process in fishes are not totally clear and that the issue still requires several investigations.

Fertuck et al. (2001), to study the PAHs habilities and their monohydroxy derivates to interact with the α and β isoforms of the ER, used the *in vitro* competitive binding and the reporter gene expression techniques in breast cancer cells MCF-7. Interactions between these two different ER isoforms have presented a particular interest, due to the functions that they fulfill. ERα has showed to be critical in the fertility maintenance and in the maternal behavior. ERβ, probably, is more involved with the maintenance of the bone mineral density of the organism. From the obtained results, the authors have shown that the nonhydroxylated compounds are incapable of competing with the ER ligands, on the contrary of the monohydroxylated metabolites. For example, B[a]P is incapable of competing with the ER ligands, while its monohydroxylated derivates are capable of interacting and inducing genic expression in a much more elevated degree, and preferably with ERβ. According to the authors, taking in consideration the structural similarity of the studied chemicals with B[a]P, probably, the response obtained for the assay with MCF-7 cells is due to the metabolites produced in the PAHs bioactivation by the CYP isoenzymes, which produce monohydroxylated and other metabolites. Finally, the authors have concluded that several PAHs can interact, positively or negatively, with the *in vivo* ER signaling, probably, in a cell-specific way.

Wang et al. (2003) used the human estrogen receptor (ER) recombinant yeast bioassay to assess the estrogenic activities from different fossil fuel emissions (soot from petroleum fired car, soot from coal fire stove, soot from diesel fired agrimotor and soot from coal fired electric power station). In the present work, the authors showed a weak, but clear, estrogenic activity of the fossil fuels, including coal, petroleum and diesel. Although the estrogenicity impacts of these chemicals in the ecosystem and in human health are not fully known, the authors affirm that the estrogenic activity phenomenon, originated from pollutant sources, still requires plenty of investigation in order to evaluate the real effects of these chemical substances on the endocrine system of the exposed organisms.

Plíšková et al. (2005) assessed the estrogenic activities of some PAHs (B[a]P, benzo[a]anthracene - B[a]A, fluoranthene - Fla and dibenzo[a,l]pyrene - DB[a]P) on the cellular cycle progression and on the cellular proliferation, in the human breast carcinoma. For these analyses, it was used the E-screen assay (cellular proliferation test in MCF-7 cells); neutral red cytotoxicity assay (test to measure the cell viability using a vital dye); cell cycle and BrdU incorporation analyses; cell death detection; western blotting (that detects the p53 and β-actin proteins). According to the authors, B[a]P and B[a]A, even in low concentrations (100nM), stimulated the MCF-7 cells proliferation, in a ER-dependent way. These results were contradictory from other authors that have found an inhibition of the MCF-7 cells development focus in culture. Plíšková et al. (2005) infer that this discrepancies are related with the cell growth conditions used, which are, significantly, different from those used in post-confluent cells model and, also, different from the experimental configurations. The authors conclude that, both B[a]P and B[a]A can contribute to a tumor progression and, in final analysis, to carcinogenity of the estrogen-sensitive tissues.

Lipzig et al. (2005) investigated the metabolites formation of the B[a]P and chrysene (CHN) by the CYP microsome via in rat livers. After the metabolites akcnowledge, the authors have tested 8 mono- and tri- dihydroxylate metabolites and a mixture of 7 mono-hydroxylate metabolites, with regard to their ER affinities, using radioligand binding assay. In this regard, it was also evaluated the metabolites estrogenic activities, by the cell proliferation (modified E-screen) and gene reporter (ER-CALUX) assays, using trasnfected ER-positive

T47D stable cells (human breast adenocarcinoma cell line), which expresses the luciferase actitivity, by ER activation. From the obtained results, it was observed that B[a]P and CHN metabolites, generated by the CYP isoenzymes, presented a higher affinity by the ER than from the B[a]P and CHN. To verify the functional answer of these metabolites ligands to ER, the authors carried out a E-screen modified assay and showed that none of the B[a]P metabolites, just like this pure chemical, have the capacity of inducing cell proliferation similar to that observed to E_2 (pattern). However, by the obtained results with the ER-CALUX assay, it was observed estrogenic activity for B[a]P, CHN and some of their metabolites. For the metabolites mixture, that presented, individually, weak estrogenic activity it was observed a significative increase in the estrogenic effect. These data conffirmed the risks to animals and human health of the exposition to several environmental xenoestrogens, even if they present weak estrogenic activity when assessed individually.

Patel et al. (2006) evaluated the B[a]P effects in mRNA expression of the CYP aromatase (CYP19) and in the activity of this enzyme, by exposing embryos and adults of the *Fundulus heteroclitus* species (Pisces: Cyprinodontidae) to waters polluted by B[a]P, for a period of 10 and 15 days, respectively. CYP19 is a key steroidogenic enzyme, responsible for the conversion of androgens to estrogens, its expression in lower vertebrate is, typically, restrict to the gonads, neural tissues (brain and retina) and hypophysis. For mammals, in general, there is only one CPY19 gene with multiple promoters. On the other hand, teleosts present two distinct CYP19 genes, CYP19A1 and CYP19A2, which are regulated, differently, and codify proteins, also, with different enzymatic characteristic. CYP19A1 is expressed in the ovary and is involved with sexual differentiation and oocyte maturation, while the CYP19A2 expression occurs in elevated levels in the brain, retina and hypophysis, and is related with the neuronal differentiation process, survival, morphology, synaptogenesis and sexual behavior. From the obtained results, it was observed that AhR ligands such as B[a]P, present pontential to cause alteration in the endocrine system by modulation of the aromatase activity and this expression differs, with regards to tissue and sex. The authors concluded that B[a]P can interfere in the esteroids homeostasis and cause harmful alterations in the fish reproductive process.

Martin-Skilton et al. (2008), when studying the effects caused by the Prestige oil fuel spill, with regards to the alteration conferred to hepatic enzymatic biotransformation and to endocrine system, exposed turbot juveniles (*Scophthalmus maximus*) to a food diet with different concentration of the Prestige fuel oil, for a period of 6 weeks. To evaluate the alterations in the endocrine system, the authors measured the testosterone and E_2 levels in the blood plasma and assessed aromatase gonadal activity and glucuronodation of testosterone by liver microsomal fractions. Acute exposition to Prestige fuel oil reduced, significantly, the testosterone quantity in the plasma, which led to deleterious effects in the reproductive capacity of the exposed individuals. According to the authors, exposition to Prestige fuel oil altered the hepatic enzymatic biotransformation and the levels of the endogenous hormones in the tested organisms. The authors still affirmed the necessity of a further investigation to estimate the real effect of these chemicals over the sexual development and the reproductive process in fishes.

Kummer et al. (2008) used the *in vivo* immature rat uterotrophic assay to investigate the estrogenic and anti-estrogenic acute effects of some environmental PAHs (B[a]P, B[a]A, Fla and benzo[k]fluoranthene (B[k]F)), that were previously reported as modulators of ER-dependent genic expression and promoters of the cell proliferation in *in vitro* tests. By the analyses, B[a]P, B[a]A and Fla acted as weak estrogens in immature rat uterotrophic assay, being this estrogenicity demonstrated by the increase of the uterine weight and by a hypertrophy significative of the luminal epithelium , whose effects are, probably, mediated by ERα, which are the main ER subtype present in uterus.

Taking in consideration that PAHs can cause several deleterious effects in the functioning of the living beings endocrine system, it is important to emphasize the necessity of studies in this area, in order to proportionate a better understanding of the real consequences that these effects can provoke in organisms of the natural environments and also to the man.

3. Final Considerations

PAHs and their derivates are organic pollutants of great environmental persistence (POP-persistent organic pollutants), being that many of them can induce, in the organisms exposed, toxic, cytotoxic, genotoxic, mutagenic, teratogenic and/or carcinogenic effects, which vary according to the PAHs molecular structure (Table 1).

Table 1. Summary of the PAHs effects on several organisms.

PAHs	Organisms	Effects	References
acenaphthene	*in vitro* test with RT Gill-W1 (gill cells of rainbow trout)	cytotoxic	Schirmer et al. (1998a)
acenaphthene	*in vitro* test with HepG2 cells	non-cytotoxic	Babich (1988)
acenaphthylene	*in vitro* test with RT Gill-W1 (gill cells of rainbow trout)	cytotoxic	Schirmer et al. (1998a)
acenaphthylene	*in vitro* test with HepG2 cells	non-cytotoxic	Babich (1988)
anthracene	*in vitro* test with HepG2 cells	non-cytotoxic	Babich (1988)
anthracene	rapeseed (*Brassica napus*, canola)	photo/toxic	Ren et al. (1996)
anthracene	bluegill fish	photo/toxic	McCloskey and Oris (1993)
anthracene	fish	photocytotoxic	Oris and Giesy (1985)
anthracene	microcrustacean (*Daphnia pulex)*	phototoxic	Allred and Giesy (1985)
anthracene	fishes	teratogenic	Tuvikene (1995)
anthracene	microcrustacean *(Daphnia magna)*	toxic	Holst and Giesy (1989)
anthracene	fish (*Pomatoschistus microps)*	toxic	Vieira et al. (2008)
benzo[a]pyrene	epidermis of mice	carcinogenic	Melendez-Cólon (1999)
benzo[a]pyrene	mice	carcinogenic	Warshawsky (1999)
benzo[a]pyrene	mice	carcinogenic	Straif et al. (2005)
benzo[a]pyrene	*in vitro* test with HepG2 cells	cytotoxic	Babich (1988)
benzo[a]pyrene	*in vitro* tests with MCF-7 cells	estrogenic	Plíšková et al. (2005)
benzo[a]pyrene	*in vitro* tests with MCF-7 cells	estrogenic	Lipzig et al. (2005)
benzo[a]pyrene	*in vivo* test - immature rat uterotrophic assay	estrogenic	Kummer et al. (2008)
benzo[a]pyrene	human umbilical cord cells	genotoxic	Annas et al. (2000)
benzo[a]pyrene	*in vitro* test with Chinese hamster V79 cells	photo/cytotoxic	Utesch (1996)
benzo[a]pyrene	*in vitro* test with Chinese hamster V79 cells	photo/cytotoxic	Ochi et al. (1985)
benzo[a]pyrene	rapeseed (*Brassica napus*, canola)	photo/toxic	Ren et al. (1996)

Table 1. Summary of the PAHs effects on several organisms (Continued)

PAHs	Organisms	Effects	References
benzo[a]pyrene	alevins of the rainbow trout (*Salmo gairdneri* Richardson)	teratogenic	Hose et al. (1984)
benzo[a]pyrene	fish (*Fundulus heteroclitus*)	teratogenic	Wassenberg et al. (2002)
benzo[a]pyrene	mice	teratogenic	Shum et al. (1979)
benzo[a]pyrene	mice	teratogenic	Mackenzie and Angevine (1981)
benzo[a]pyrene	mice	teratogenic	Legraverend et al. (1984)
benzo[a]pyrene	mice	teratogenic	Payne (1958)
benzo[a]pyrene	bacteria (*Escherichia coli*)	toxic	Kim et al. (2007)
benzo[a]pyrene	fish (*Pomatoschistus microps*)	toxic	Vieira et al. (2008)
benzo[a]pyrene	mice	toxic	Anselstetter and Heimpel (1986)
benzo[a]pyrene	humans	toxic	Gupta et al. (1993)
benzo[a]anthracene	*in vitro* test with HepG2 cells	cytotoxic	Babich (1988)
benzo[a]anthracene	*in vitro* tests with MCF-7 cells	estrogenic	Plíšková et al. (2005)
benzo[a]anthracene	*in vivo* test - immature rat uterotrophic assay	estrogenic	Kummer et al. (2008)
benzo[b]anthracene	embryos of zebrafish	teratogenic	Incardona et al. (2006)
benzo[b]fluoranthene	*in vitro* test with HepG2 cells	cytotoxic	Babich (1988)
benzo[k]fluoranthene	*in vitro* test with HepG2 cells	cytotoxic	Babich (1988)
benzo[k]fluoranthene	*in vivo* test - immature rat uterotrophic assay	non-estrogenic	Kummer et al. (2008)
chrysene	*in vitro* test with HepG2 cells	cytotoxic	Babich (1988)
chrysene	*in vitro* tests with MCF-7 cells	estrogenic	Lipzig et al. (2005)
chrysene	embryos of zebrafish	non-embryotoxic/non-teratogenic	Incardona et al. (2006)
dibenz[a,h]anthracene	rat	carcinogenic	Platt et al. (1990)
dibenzo[a,l]pyrene	*in vitro* tests with MCF-7 cells	non- estrogenic	Plíšková et al. (2005)
dibenzo[a,l]pyrene	epidermis of mice	carcinogenic	Melendez-Cólon (1999)
dibenzo[a,l]pyrene	rat	carcinogenic	Okona-Mensah et al. (2005)
dimethylbenzo[a]anthracene	mice	teratogenic	Sanyal and Li (2007)
7,12-dimethylbenzo[a]anthracene	epidermis of mice	carcinogenic	Melendez-Cólon (1999)
fluoranthene	*in vitro* test with HepG2 cells	cytotoxic	Babich (1988)
fluoranthene	*in vivo* test - immature rat uterotrophic assay	estrogenic	Kummer et al. (2008)
fluoranthene	*in vitro* tests with MCF-7 cells	non-estrogenic	Plíšková et al. (2005)
fluoranthene	embryos of mussels	non-toxic	Bellas et al. (2008)
fluoranthene	rapeseed (*Brassica napus*, canola)	photo/toxic	Ren et al. (1996)
fluoranthene	embryos of sea urchin	photo/toxic	Bellas et al. (2008)
fluoranthene	*in vitro* test with RT Gill-W1 (gill cells of rainbow trout)	photocytotoxic	Schirmer et al. (1998b)
fluoranthene	fish	photocytotoxic	Weinstein et al. (1997)
fluorene	embryos of sea urchin and mussels	toxic	Bellas et al. (2008)
fluorene	*in vitro* test with RT Gill-W1 (gill cells of rainbow trout)	cytotoxic	Schirmer et al. (1998a)
fluorene	*in vitro* test with HepG2 cells	non-cytotoxic	Babich (1988)
hydroxypyrene	embryos of sea urchin and mussels	photo/toxic	Bellas et al. (2008)
naphthalene	*in vitro* test with RT Gill-W1 (gill cells of rainbow trout)	cytotoxic	Schirmer et al. (1998a)

Table 1. Summary of the PAHs effects on several organisms (Continued)

PAHs	Organisms	Effects	References
naphthalene	bacteria (*Escherichia coli*)	toxic	Kim et al. (2007)
naphthalene	embryos of sea urchin, mussels and ascidian	toxic	Bellas et al. (2008)
phenanthrene	higher plant (*Arabidopsis thaliana*)	cytotoxic	Alkio et al. (2006)
phenanthrene	*in vitro* test with RT Gill-W1 (gill cells of rainbow trout)	cytotoxic	Schirmer et al. (1998a)
phenanthrene	*in vitro* test with HepG2 cells	cytotoxic	Babich (1988)
phenanthrene	bacteria (*Escherichia coli*)	non-toxic	Kim et al. (2007)
phenanthrene	embryos of sea urchin and mussels	photo/toxic	Bellas et al. (2008)
pyrene	embryos of sea urchin and mussels	photo/toxic	Bellas et al. (2008)
pyrene	*in vitro* test with HepG2 cells	cytotoxic	Babich (1988)
pyrene	*in vitro* test with RT Gill-W1 (gill cells of rainbow trout)	photocytotoxic	Schirmer et al. (1998b)
pyrene	embryos of zebrafish	teratogenic	Incardona et al. (2006)
mixture	*in vitro* test with fish bluegill BF-2 cells	cytotoxic	Babich and Borenfreund (1987)
mixture	*in vitro* test with Chinese hamster V79 cells	cytotoxic	Hader et al. (1996)
mixture	embryos of ducks	embryotoxic /teratogenic	Hoffman and Gay (1981)
mixture	bacteria assay SOS chromotest	mutagenic	White (2002)
mixture	embryos of the fish specie *Melanotaenia fluviatillis*	teratogenic	Pollino and Holdway (2002)
mixture	turtles (*Chelydra* *serpentina*)	teratogenic	Van Meter et al. (2006)
mixture	embryos of rats	teratogenic	Sanyal and Li (2007)
mixture	humans	teratogenic	Rice and Ward (1982)
mixture	humans	teratogenic	Perera et al. (2003)
mixture	humans	teratogenic	WHO (2000)
mixture	earthworm (*Eisenia fetida* *andrei*)	toxic	Saint-Denis et al. (1999)
mixture	embryos of sea urchin, mussels and ascidian	toxic	Bellas et al. (2008)
mixture	neotropical fish (*Oreochromis mossambicus*)	toxic	Shailaja and Classy D'Silva (2003)
environmental sample	zebrafish (*Danio rerio*)	teratogenic	Diamond et al. (2000)
environmental sample - air	yeast bioassay	estrogenic	Wang et al. (2003)
environmental sample - air	HepG2 cells	genotoxic	Sevastyanova et al. (2008)
environmental sample - air	bacteria assays – *Salmonella* test and Vitotox®	mutagenic	Du Four et al. (2004)
environmental sample - air	*in vitro* tests with A549-human type II alveolar epithelial cells	mutagenic	Roubicek et al. (2007)
environmental sample - air	Ames test and DNA adducts (^{32}P-postlabelling analyses)	mutagenic	Umbuzeiro et al. (2008)
environmental sample - sediments	*in vitro* test – MCF-7 focus assay	anti-estrogenic	Arcaro et al. (1999)
environmental sample - sediments	mussel (*M. galloprovincialis*)	genotoxic	Bihari et al. (2006)
environmental sample - sediments	bacteria assay *umu*-test	non-genotoxic	Bihari et al. (2006)

Table 1. Summary of the PAHs effects on several organisms (Continued)

PAHs	Organisms	Effects	References
environmental sample - sediments	mollusk (*Tapes semidecussatus*)	toxic	Byrne and O'Halloran (2004)
environmental sample - soil	bacteria *Vibrio fischeri* M169	genotoxic	Bispo et al. (1999)
environmental sample - soil	frog (*Xenopus laevis*) and *in vitro* tests - Ames and Mutatox®	mutagenic	Békaert et al. (1999)
environmental sample - soil	higher plant (*Tradescantia*) and Ames test	mutagenic	Monarca et al. (2002)
environmental sample - soil	earthworm (*Eisenia andrei*), microarthropod and nematode	non-toxic	Erstfeld and Snow-Aihbrook (1999)
environmental sample - soil	leguminous	toxic	Smith et al. (2006)
environmental sample - soil	barley (*Hordeum vulgare* L.)	toxic	Smreczak and Maliszewska-Kordybach (2003)
environmental sample - soil	algae (*Pseudokirchneriella subcapitata*) and bacteria (*Vibrio fischeri*)	toxic	Bispo et al. (1999)
environmental sample - soil	rat	toxic	Roos et al. (2004)
environmental sample - water	higher plant (*Allium cepa*)	cytotoxic	Leme et al. (2008)
environmental sample - water	higher plant (*Allium cepa*)	genotoxic	Odeigah et al. (1997)
environmental sample - water	mussels, oysters and benthic fishes	genotoxic	Bolognesi et al. (2006)
environmental sample - water	higher plant (*Allium cepa*)	genotoxic/mutagenic	Leme and Marin-Morales (2008)
environmental sample - water	fish (*Oreochromis niloticus*)	mutagenic	Çavaş and Ergene-Gözükara (2005)
environmental sample - water	fish (*Oreochromis niloticus*)	mutagenic	Hoshina et al. (2008)
environmental sample - water	microcrustacean *(Daphnia magna)*	toxic	Reece and Burks (1985)
environmental sample - water	crucian carp (*Carassius auratus*) and *in vitro* test with rat hepatoma cells H4IIE	toxic	Ke et al. (2007)
environmental sample - water	*in vitro* test with rat hepatoma cells H4IIE	toxic	Qiao et al. (2006)
environmental sample – water/sediments	dab (*Limanda limanda*)	non-genotoxic	Akcha et al. (2003)
simulation experiment - water (crude oil)	mussel (*Mytilus edulis* L.) and sea urchin (*Strongylocentrotus droebachiensis*)	genotoxic	Taban et al. (2004)
simulation experiment – water (diesel)	neotropical fish (*Prochilodus lineatus*)	genotoxic/mutagenic	Vanzella et al. (2007)
simulation experiment - water (crude oil)	fishes turbot (*Scophthalmus maximus*) and Atlantic cod (*Gadus morua*)	mutagenic	Baršienė et al. (2006)
occupational exposure	humans	mutagenic	Palli et al. (2008)
occupacional exposure	humans	non-carcinogenic	Jacobson-Kram et al. (1993)
occupational exposure	humans	toxic	Pedrozo et al. (2002)
coal tars	mice	carcinogenic	Warshawsky (1999)
cigarette smoke	placenta cells of murine	cytotoxic	Detmar et al. (2008)
diesel exhaust particles	macrophages of rats and humans	cytotoxic	Hiura et al. (1999)
frying oil	HepG2 cells	cytotoxic	Pandey et al. (2006)
heavy oil	*in vitro* test with Chinese hamster V79 cells	cytotoxic	Hadnagy et al. (1997)
jet fuels	keratinocytes of the human epidermis	cytotoxic	Chou et al. (2003)
food chain	rat	genotoxic	Lemiere et al. (2005)
carbon black oil	mice	non-teratogenic	Hansen et al. (2000)

Table 1. Summary of the PAHs effects on several organisms (Continued)

PAHs	Organisms	Effects	References
crude oil	embryos of rats	teratogenic	Feuston et al. (1997)
wheathered ANS oil and mixtures	embryos of fishes	teratogenic	Incardona et al. (2005)

Carcinogenic and mutagenic potentialities of the different PAHs are significative for those who have more than 4 aromatic rings and even higher, for those who have 5 or 6 rings. Substitution of hydrogen for chemical groups, can also affect, drastically, the PAHs activity depending on the position where the substitution occurs and of the substituent group (Combariza et al., 1995).

Simultaneous presence of several of these compounds in the environment makes the assessment of their genotoxicities from environmental samples, very difficult, besides also difficulting the correlation studies between the exposition via and the observed carcinogenic effect (Tuvikene, 1995). These characteristics vary from one species to another but, in the human species, the respiratory via is considered the most important, particularly for individuals occupationally exposed, although, in several cases, the dermic via is so or more important (Harvey and Dune, 1978).

Another effect observed, resulted from the exposition to PAHs, is that many of them have a negative effect on the immunological system, characteristic that seems to be associated with carcinogenicity (Luster and Rosenthal, 1993; Fiala et al., 2000, 2004; Borská et al. 2006).

One of the most studied PAHs and, consequently most known in regard to their effects, is the B[a]P, which was, firstly, identified in domestic soot and, subsequently, in particulate matter. This chemical was characterized as a cancerigenous agent of large distribution (Finlayson-Pitts, 1997), however, in the 70 decade, it was acknowledged the carcinogenicity of atmospheric particles extracts, so far attributed to B[a]P, not only for this substance but, also, due to the presence of other PAHs still unknown, in regards to their action (Pitts, 1987; Finlayson-Pitts, 1997). From this epoch, much attention has been given to the assessment of PAHs in environmental and biological matrices.

The environmental importance of the PAHs and their derivates is recognized and accepted by the scientific community. Studies about this class of compounds englobe a great variety of aspects such as: reactivity of these compounds, development of analytical methodologies and field studies about concentration, synthesis, biological effects and removal process of these contaminants of the environment.

Among the PAHs, 16 (Figure 4) are considered dangerous by the USEPA, being recommended to be evaluated in environmental and biological monitoring.

Contamination of rivers, lakes, seas, forests and atmosphere can cause irreparable damages to nature and human health. PAHs malefic action on living beings can be exerted directly and, mainly, by their metabolites, many of them still unknown (Tuvikene, 1995).

The search of reliable bioindicators for exposition studies of human and all the other living beings groups to PAHs and their metabolites still constitutes an issue for many researches. As exposed in this work, some specific indicators have been proposed and used with success in assessment studies of the deleterious effects induced by PAHs in the exposed organisms.

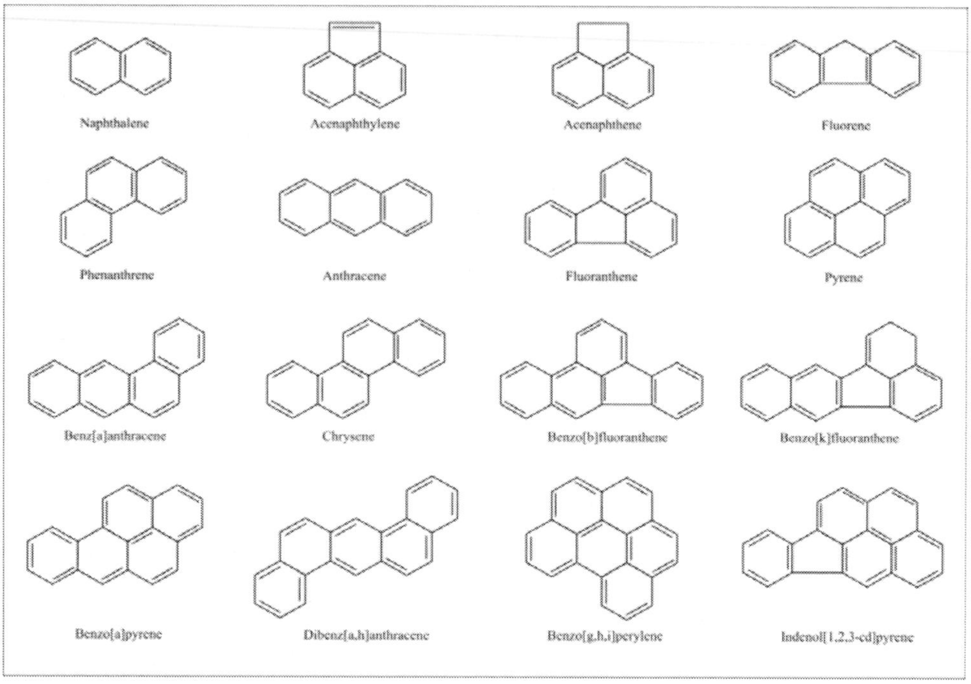

Figure 4. Chemical structure of the PAHs indicated by USEPA for environmental monitoring.

ACKNOWLEDGMENTS

The authors would like to thank the Programa de Formação de Recursos Humanos em Geologia e Ciências Ambientais do Setor de Petróleo, Gás and Biocombustíveis – PRH05 (UNESP/MCT/FINEP/ANP) for the financial support of the researches developed in the thematic of the present review.

REFERENCES

[1] Ahokas, J.T. 1979. Cytochrom P450 in fish liver microsomes and carcinogen activity. *ACS Symposium Series*, v.99, p.279-295.

[2] Aina, R.; Palin, L.; Citterio, S. 2006. Molecular evidence for benzo[a]pyrene and naphthalene genotoxicity in *Trifolium repens* L. *Chemosphere,* v.65, p.666-673.

[3] Akcha, F.; Hubert, F.V.; Pfhol-Leszkowicz, A. 2003. Potential value of the comet assay and DNA adduct measurement in dab (*Limanda limanda*) for assessment of *in situ* exposure to genotoxic compounds. *Mutation Research*, v.534, p.21-32.

[4] Albers, P.H. Petroleum and Individual Polycyclic Aromatic Hydrocarbons. In: Hoffman D.J., Rattner B.A., Jr. Burton G.A., Jr. Cairns J. (Eds.) *Handbook of Ecotoxicology* (2nd ed.). Lewis Publishers, 2002, p.341-371.

[5] Albers, P.H. 2006. Birds and polycyclic aromatic hydrocarbons. *Avian and Poultry Biology Reviews*, v.17, p.125-140.

[6] Allred, P.M.; Giesy, J.P. 1985. Solar radiation-induced toxicity of anthracene to *Daphnia pulex*. *Environmental Toxicology and Chemistry*, v.4, p.219-226.

[7] Alkio, M.; Tabuchi, T.M.; Wang, X.; Colón-Carmona, A. 2005. Stress responses to polycyclic aromatic hydrocarbons in *Arabidopsis* include growth inhibition and hypersensitive response-like symptoms. *Journal of Experimental Botany*, v.56, p.2983-2994.

[8] Annas, A.; Brittebo, E.; Hellman, B. 2000. Evaluation of benzo(a)pyrene-induced DNA damage in human endothelial cells using alkaline single cell gel electrophoresis. *Mutation Research*, v.471, p.145-155.

[9] Anselstetter, V.; Heimpel, H. 1986. Acute hematotoxicity of oral benzo[a]pyrene: the role of the Ah locus. *Acta Haematologica Basel,* v.76, p.217-223.

[10] Antkiewicz, D.S.; Peterson, R.E.; Heideman, W. 2006. Blocking expression of AHR2 and ARNT1 in zebrafish larvae protects against cardiac toxicity of 2,3,7,8-tetrachlorodibenzo-p-dioxin. *Toxicological Sciences*, v.94, p.175-182.

[11] Arcaro, K.F.; O'Keefe, P.W.; Yang, Y.; Clayton, W.; Gierthy, J.F. 1999. Antiestrogenicity of environmental polycyclic aromatic hydrocarbons in human breast cancer cells. *Toxicology*, v.133, p.115-127.

[12] Arfsten, D.P.; Schaeffer, D.J.; Mulveny, D.C. 1996. The effects of near ultraviolet radiation on the toxic effects of polycyclic aromatic hydrocarbons in animals and plants: A review. *Ecotoxicology and Environmental Safety*, v.33, p.1-24.

[13] Babich, H.; Borenfreund, E. 1987. Polycyclic aromatic hydrocarbon *in vitro* cytotoxicity to bluegill BF-2 cells: mediation by S-9 microsomal fraction and temperature. *Toxicology Letters*, v.36, p.107-116.

[14] Babich, H.; Sardana, M.K.; Borenfreund. E. 1988. Acute cytotoxicities of polynuclear aromatic hydrocarbons determined *in vitro* with the human liver tumor cell line, HepG2. *Cell Biology and Toxicology*, v.4, p.295-309.

[15] Baek K.H.; Kim, H.S.; Oh, H.M.; Yoon, B.D.; Kim, J.; Lee, I.S. 2004. Effects of crude oil, oil components, and bioremediation on plant growth. *Journal of Environmental Science and Health,* v.39, p.2473-2484.

[16] Baird, W.M.; Hooven, L.A.; Mahadevan, B. Carcinogenic Polycyclic Aromatic Hydrocarbon-DNA Adducts and Mechanism of Action. 2005. *Environmental and Molecular Mutagenesis*, v.45, p.106-114.

[17] Ballantyne, B.; Sullivan, J.B. Basic principles of toxicology. In: Sullivan J.B., Krieger G.R. (Eds.). *Hazardous materials toxicology: clinical principles of environmental health*. Baltimore: Willians & Wilkins, 1992, p.9-23.

[18] Baršienė, J.; Dedonytė, V.; Rybakovas, A.; Andreikėnaitė, L.; Andersen, O.K. 2006. Investigation of micronuclei and other nuclear abnormalities in peripheral blood and kidney of marine fish treated with crude oil. *Aquatic Toxicology*, v.78, p.99-104.

[19] Baussant, T., Sanni, S., Jonsson, G., Skadsheim, A., Børseth, J.F., 2001. Bioaccumulation of polycyclic aromatic compounds: 1. Bioconcentration in two marine species and in semipermeable membrane devices during chronic exposure to dispersed crude oil. *Environmental Toxicology and Chemistry*, v.20, p.1175-1184.

[20] Beach, A.C.; Gupta, R.C. 1992. Human biomonitoring and the 32Ppostlabeling assay. *Carcinogenesis,* v.13, p.1053-1074.

[21] Békaert, C.; Rast, C.; Ferrier, V.; Bispo, A.; Jourdain, M.J.; Vasseur P. 1999. Use of *in vitro* (Ames and Mutatox tests) and *in vivo* (Amphibian Micronucleus test) assays to

assess the genotoxicity of leachates from a contaminated soil. *Organic Geochemistry*, v.30, p.953-962.

[22] Belgiorno, V.; Rizzo, L.; Fatta, D.; Rocca, C.D.; Lofrano, G.; Nikolaou, A.; Naddeo, V.; Meric, S. 2007. Review on endocrine disrupting-emerging compounds in urban wastewater: occurrence and removal by photocatalysis and ultrasonic irradiation for wastewater reuse. *Desalination*, v.215, p.166-176.

[23] Bellas, J.; Saco-Álvarez, L.; Nieto, O.; Beiras, R. 2008. Ecotoxicological evaluation of polycyclic aromatic hydrocarbons using marine invertebrate embryo–larval bioassays. *Marine Pollution Bulletin*, v.57, p.493-502.

[24] Betowski, L.D.; Enlow, M.; Riddick, L. 2002. The phototoxicity of polycyclic aromatic hydrocarbons: a theoretical study of excited states and correlation to experiment. *Computers and chemistry*, v.26, p.371-377.

[25] Bihari, N.; Fafandel, M.; Hamer, B.; Kralj-Bilen, B. 2006. PAHs content, toxicity and genotoxicity of coastal marine sediments from the Rovinj área, Northern Adriatic, Croatia. *Science of the Total Environment*, v.366, p.602-611.

[26] Billiard, S.M.; Meyer, J.N.; Wassenberg, D.M.; Hodson, P.V.; Di Giulio, R.T. 2008. Nonadditive effects of PAHs on Early Vertebrate Development: Mechanisms and Implications for Risk Assessment. *Toxicological Sciences*, v.105, p.5-23.

[27] Bispo, A.; Jourdain, M.J.; Jauzein, M. 1999. Toxicity and genotoxicity of industrial soils polluted by polycyclic aromatic hydrocarbons (PAHs). *Organic Geochemistry*, v.30, p.947-952.

[28] Black, D.E.; Phelps, D.K.; Lapan, R.L. 1988. The effect of inherited contamination on egg and larval winter flounder Pseudopleuronectes americanus. *Marine Environmental Research,* v.25, p.45-62.

[29] Bolognesi, C.; Perrone, E.; Roggieri, P.; Sciutto, A. 2006. Bioindicators in monitoring long term genotoxic impact of oil spill: Haven case study. *Marine Environmental Research*, v.62, p.287-291.

[30] Boscolo, R.; Cacciatore, F.; Giovanardi, O. 2007. Polycyclic aromatic hydrocarbons (PAHs) in transplanted Manila clams (*Tapes philippinarum*) from the Lagoon of Venice as assessed by PAHs/shell weight index: A preliminary study. *Marine Pollution Bulletin,* v.55, p.485-493.

[31] Borská, L.; Fiala, Z.; Krejsek, J.; Hamáková, K.; Andrýs, C.; Šmejkalová, J.; Vokurková, D.; Kremláček, J. 2006. Cytogenetic and Immunological Changes after Dermal Exposure to Polycyclic Aromatic Hydrocarbons and UV Radiation. *Physiological Research*, v.55, p.317-323.

[32] Bücker, M.; Glatt, H.R.; Platt, K.L.; Avnir, D.; Ittah, Y.; Blum, J.; Oesch, F. 1979. Mutagenicity of phenanthrene and phenanthrene K-region derivatives. *Mutation Research*, v. 66, p.337-48.

[33] Byrne, P.; O'Halloran, J. 2004. The impact of ballast water effluent on the Manila Clam, *Tapes semidecussatus*. *Ecotoxicology*, v.3, p.311-322.

[34] Carney, S.A.; Prasch, A.L.; Heideman, W.; Peterson, R.E. 2006. Understanding dioxin developmental toxicity using the zebrafish model. Birth defects research. Part A, *Clinical and molecular teratology*, v.76, p.7-18.

[35] Casillas, E.; Misitano, D.; Johnson, L.L.; Rhodes, L.D.; Collier, T.K.; Stein, J.E.; Mccain, B.B.; Varanasi, U. 1991. Inducibility of spawning and reproductive success of

female English sole (*Parophrys vetulus*) from urban and nonurban areas of Puget Sound. *Marine Environmental Research*, v.31, p.99-122.

[36] Catoggio, J.A. Other organic toxic substances. *In: Matsui S. (Ed.). Guidelines of Lake Management - Toxic Substances Management in Lakes and Reservoirs*. 1991, v.4, p.113-126.

[37] Çavaş, T.; Ergene-Gözükara, S. 2005. Induction of micronuclei and nuclear abnormalities in *Oreochromis niloticus* following exposure to petroleum refinery and chromium processing plant effluents. *Aquatic Toxicology*, v.74, p.264-271.

[38] Choi, H.; Jedrychowski, W.; Spengler, J.; Camann, D.E.; Whyatt, R.M.; Rauh, V.; Tsai, W.-Y.; Perera, F.P. 2006. International Studies of Prenatal Exposure to Polycyclic Aromatic Hydrocarbons and Fetal Growth. *Environmental Health Perspectives*, v.114, p. 1744-1750.

[39] Chou, C.C.; Riviere, J.E.; Monteiro-Riviere, N.A. 2003. The cytotoxicity of jet fuel aromatic hydrocarbons and dose-related interleukin-8 release from human epidermal keratinocytes. *Archives of Toxicology*, v.77, p.384-391.

[40] Collier, T.K.; Stein, J.E.; Sanborn, H.R.; Hom, T.; Myers, M.S.; Varanasi, U. 1993. A field study of the relationship between bioindicators of maternal contaminant exposure and egg and larval viability of English sole (*Parophrys vetulus*). *Marine Environmental Research*, v.35, p.171-175.

[41] Collins, A.; Cadet, J.; Epe, B.; Gedik, C. 1997. Problems in the measurement of 8-oxoguanine in human DNA. Report of a workshop, DNA oxidation. *Carcinogenesis*, v. 18, p. 1833-1836.

[42] Combariza, J.E.; Hajos, A.K.D.; Winston, G.W. 1995. Semiempirical calculations of nitrated pyrenes and their reduced metabolites: relationship to characteristics of bioactivation. *The Journal of Physical Chemistry*, v. 99, p.14539-14543.

[43] Cotelle, S.; Masfaraud, J.-F.; Férard, J.-F. 1999. Assessment of the genotoxicity of contaminated soil with the *Allium*/*Vicia*-micronucleus and *Tradescantia*-micronucleus assays. *Mutation Research*, v.426, p.167-171.

[44] Dam, K.; Seidler, F.J.; Slotkin, T.A. 1998. Developmental neurotoxicity of chlorpyrifos: delayed targeting of DNA synthesis after repeated administration. *Developmental Brain Research* v.108, p.39-45.

[45] Dearfield, K.L.; Cimino, M.C.; Mccarroll, N.E.; Mauer, I.; Valcovic, L.R. 2002. Genotoxicity risk assessment a proposed classification strategy. *Mutation Research*, v.521, p.121-135.

[46] Detmar, J.; Rennie, M.Y.; Whiteley, K.J.; Qu, D.; Taniuchi, Y.; Shang, X.; Casper, R.F.; Adamson, S.L.; Sled, J.G.; Jurisicova, A. 2008. Fetal growth restriction triggered by polycyclic aromatic hydrocarbons is associated with altered placental vasculature and AhR-dependent changes in cell death. *American Journal of Physiology - Endocrinology and Metabolism,* v.295, p.519-530.

[47] Diamond, M.L.; Gingrich, S.E.; Fertuck, K.; McCarry, B.E.; Stern, G.A. 2000. Evidence for Organic Film on an Impervious Urban Surface: Characterization and Potential Teratogenic Effects. *Environmental Science and Technology*, v.34, p.2900-2908.

[48] Dolk, H.; Vrijheid, M. 2003. The impact of environmental pollution on congenital anomalies. *British Medical Bulletin*, v.68, p.25-45.

[49] Du Four, V.A.; Van Larebeke, N.; Janssen, C.R. 2004. Genotoxic and mutagenic activity of environmental air samples inFlanders, Belgium. *Mutation Research*, v.558, p.155-167.

[50] Du Four, V.A.; Janssen, C.R.; Brits, E.; Van Larebeke, N. 2005. Genotoxic and mutagenic activity of environmental air samples from different rural, urban and industrial sites in Flanders, Belgium. *Mutation Research*, v.588, p.106-117.

[51] Eisler, R. Polycyclic aromatic hydrocarbons. *In: Handbook of Chemical Risk Assessment: Health Hazards to Human, Plants and Animals*. Boca Raton: Lewis Publishers, 2000, p.1343-1412.

[52] Erstfeld, K. M.; Snow-Ashbrook, J. 1999. Effects of chronic low-level PAH contamination on soil invertebrate communities. *Chemosphere,* v.39, p.2117-2139.

[53] Fertuck, K.C.; Kumar, S.; Sikka, H.C.; Matthews, J.B.; Zacharewski. 2001. Interaction of PAH-related compounds with the α e β isoforms of the estrogen receptor. *Toxicology Letters,* v.121, p.167-177.

[54] Feuston; M.H.; Hamilton, C.E.; Schreiner, C.A.; Mackerer, C.R. 1997. Developmental Toxicity of Dermally Applied Crude Oils In Rats. *Journal of Toxicology and Environmental Health*, v.52, p.79-93.

[55] Fiala, Z.; Borská, L.; Vyskočil, A.; Mašín, V. 2000. Polycyclic aromatic hydrocarbons II. Toxic effects. Acta Medica (Hradec Králové), v.43, p.S37-S61.

[56] Fiala, Z.; Borská, L.; Hamáková, K.; Kraják, V. 2004. Excretion of selected urinary metabolites after dermal exposure to polycyclic aromatic hydrocarbons (PAHs). *Physiological Research*, v.53, 50P.

[57] Finlayson-Pitts, B.J.; Pitts, J.N. Jr. 1997. Tropospheric air pollution: ozone, airborne toxics, polycyclic aromatic hydrocarbons, and particulates. *Science,* v.276, p.1045–1051.

[58] Friedberg, E.C.; Walker, G.C.; Siede, W. *DNA Repair and Mutagenesis*, Washington: ASM Press, 1995.

[59] Geffard, O.; Geffard, A.; His, E.; Budzinski, H. 2003. Assessment of the bioavailability and toxicity of sediment-associated polycyclic aromatic hydrocarbons and heavy metals applied to Crassostrea gigas embryos and larvae. *Marine Pollution Bulletin*, v.46, p.481-490.

[60] Genuis, S.J. 2006. The chemical erosion of human health: adverse environmental exposure and in-utero pollution determinants of congenital disorders and chronic diseases. *Journal of Perinatal Medicine*, v.34, p.185-195.

[61] Ghiselli, G.; Jardim, W.F. 2007. Interferentes endócrinos no ambiente. Química Nova, v.30, p. 695-706.

[62] Ghosh, S.; Thomas, P. 1995. Antagonistic effects of xenobiotics on steroid-induced final maturation of Atlantic croaker oocytes *in vitro*. *Marine Environmental Research*, v.39, p.159-163.

[63] Goldstone, H.M.; Stegeman, J.J. 2006. Molecular mechanisms of 2,3,7,8-tetrachlorodibenzo-p-dioxin cardiovascular embryotoxicity. *Drug Metabolism Reviews*, v.38, p.261-289.

[64] Guecheva, T.N.; Henriques, J.A.P. Metabolismo de xenobióticos citocromo P-450. In: Silva J., Erdtmann B., Henriques J.A.P. (Eds.). Genética Toxicológica. Porto Alegre: Alcance, 2003, p.225-243.

[65] Gupta, P.; Banerjee, D.K.; Bhargava, S.K.; Kaul, R.; Shankar, V.R. 1993. Prevalence of impaired lung function in rubber manufacturing factory workers exposed to benzo[a]pyrene and respirable particulate matter. *Indoor and Built Environment*, v.2, p.26-31.

[66] Hack, M.; Breslau, N.; Weissman, B.; Aram, D.; Klein, N.; Borawski, E. 1991. Effect of very low birth weight and subnormal head size on cognitive ability at school age. *The New England Journal of Medicine,* v.325, p.231-237.

[67] Hader, C.; Hadnagy, W.; Seemayer, N.H. 1996. A rapid method for detection of nongenotoxic carcinogens of environmental pollutants using synchronized V79 cells and flow cytometry. *Toxicology Letters*, v.88, p.99-108.

[68] Hadnagy, W.; Sust, H.; Seemayer, N.H. 1997. A rapid method for detection of nongenotoxic carcinogens of environmental pollutants using synchronized V79 cells and flow cytometry. *Toxicology in Vitro*, v.11, p.703-709.

[69] Hagmar, L.; Stromberg, U.; Bonassi, S.; Hansteen, I.-L.; Knudsen, L.E.; Lindholm, C.; Norppa, H. 2004. Impact of types of lymphocyte chromosomal aberrations on human cancer risk: result from Nordic and Italian cohorts. *Cancer Research*, v.64, p.2258-2263.

[70] Hansen, J.M.; Reynolds, P.R.; Booth, G.M.; Schaalje, G.B.; Seegmiller, R.E. 2000. Developmental Toxicity of Carbon Black Oil in Mice. *Teratology,* v.62, p.227-232.

[71] Harvey, R.G. Polycyclic Aromatic Hydrocarbons. New York: Wiley-VCH, 1997.

[72] Harvey, R.G.; Dunne, F.B. 1978. Multiple regions of metabolic activation of carcinogenic hydrocarbons. *Nature*, v.273, p.566–568.

[73] Heddle, J.A.; Hite, M.; Irkhart, B.; Macgregor, J.T.E.; Salamone, M.F. 1983. The induction of micronuclei as a mensure of genotoxicity – a mensure of the US environmental protection agency gene-tox program. *Mutation Research*, v.123, p.61-118.

[74] Heidel, S.M.; MacWilliams, P.S.; Baird, W.M.; Dashwood, W.M.; Buters, J.T.M.; Gonzalez, F.J.; Larsen, M.C.; Czuprynski, C.J.; Jefcoate, C.R. 2000. Cytochrome P4501B1 Mediates Induction of Bone Marrow Cytotoxicity and Preleukemia Cells in Mice Treated with 7,12-Dimethylbenz[a]anthracene. *Cancer Research*, v.60, p.3454-3460.

[75] Hiura, T.S.; Kaszubowski, M.P.; Li, N.; Nel, A.E. 1999. Chemicals in Diesel Exhaust Particles Generate Reactive Oxygen Radicals and Induce Apoptosis in Macrophages. *The Journal of Immunology*, v.163, p.5582-5591.

[76] Hoffman, D.J. 1990. Embryotoxicity and teratogenicity of environmental contaminants to bird eggs. *Reviews of Environmental Contamination and Toxicology*, v.115, p.39-89.

[77] Hoffman, D.J.; Gay, M.L. 1981. Embryotoxic effects of benzo[a]pyrene, chrysene, and 7,12-dimethylbenz[a]anthracene in petroleum hydrocarbon mixtures in mallard ducks. *Journal of Toxicology and Environmental Health*, v.7, p. 775-787.

[78] Holst, L. L.; Giesy, J. P. 1989. Chronic effects of photoenhanced toxicity of anthracene on *Daphnia magna* reproduction. *Environmental Toxicology and Chemistry* v.8, p.933-942.

[79] Hose, J.E.; Hannaht, J.B.; Puffer, H.W.; Landolt, M.L. 1984. Histologic and Skeletal Abnormalities in Benzo(a)pyrene-treated Rainbow Trout Alevins. *Archives of environmental contamination and toxicology*, v.13, p.675-684.

[80] Hoshina, M.M.; Angelis, D.F.; Marin-Morales, M.A. 2008. Induction of micronucleus and nuclear alterations in fish (*Oreochromis niloticus*) by a petroleum refinery effluent. *Mutation Research*, v.656, p.44-48.

[81] Houlihan, J.; Kropp, T.; Wiles, R.; Gray, S.; Cambell, C. 2005. Body burden (the pollution in the newborn). A benchmark investigation of industrial chemicals, pollutants and pesticides in human umbilical cord blood. *Environmental Working Group*. Available from: http://www.ewg.org/reorts/bodyburden2.

[82] Huang, X.D.; Dixon, D.G.; Greenberg, B.M. 1993. Impact of UV radiation and photomodification on the toxicity of PAHs to the higher plant *Lemna gibba* (Duckweed). *Environmental Toxicology and Chemistry*, v.12, p.1067-1077.

[83] Incardona, J.P.; Carls, M.G.; Teraoka, H.; Sloan, C.A.; Collier, T.K.; Scholz, N.L. 2005. Aryl hydrocarbon receptor-independent toxicity of weathered crude oil during fish development. *Environmental Health Perspective*, v.113, p.1755-1762.

[84] Incardona, J.P.; Day, H.L.; Collier, T.K.; Scholz, N.L. 2006. Developmental toxicity of 4-ring polycyclic aromatic hydrocarbons in zebrafish is differentially dependent on AH receptor isoforms and hepatic cytochrome P4501A metabolism. *Toxicology and Applied Pharmacology*, v.217, p.308-321.

[85] Jacob, J. 1996. The significance of polycyclic aromatic hydrocarbons as environmental carcinogens. *Pure and Applied Chemistry*, v.68, p.301-308.

[86] Jacobson-Kram, D.; Albertini, R.J.; Branda, R.F.; Falta, M.T.; Lype, R.T.; Kolodner, K.; Liou, S.-H.; McDiarmid, M.A.; Morris, M.; Nicklas, J.A.; O'Neill, J.P.; Poirier, M.C.; Putman, D.; Strickland, P.T.; Williams, J.R.; Xiao, S. 1993. Measurement of Chromosomal Aberrations, Sister Chromatid Exchange, hprt Mutations, and DNA Adducts in Peripheral Lymphocytes of Human Populations at Increased Risk for Cancer. *Environmental Health Perspectives Supplements*, v.101, p.121-125.

[87] Jimènez, B. 1997. Environmental effects of endocrine disruptors and current methodologies for assessing wildlife health effects. *Trends in Analytical Chemistry*, v.16, p.596-606.

[88] Jimenez, B.D.; Stegeman, J.J. 1990. Detoxication enzymes as indicator of environmental stress on fish. Biological Indicators of stress in Fish. *American Fisheries Society Symposium*, v.8, p.67-79.

[89] Ke, R.; Li, J.; Qiao, M.; Xu, Y.; Wang, Z. 2007. Using Semipermeable Membrane Devices, Bioassays, and Chemical Analysis for Evaluation of Bioavailable Polycyclic Aromatic Hydrocarbons in Water. Archives of Environmental Contamination and *Toxicology,* v.53, p.313-320.

[90] Kim, Y.S.; Min, J.; Hong, H.N.; Park, J.H.; Park, K.S.; Gu, M.B. 2007. Gene expression analysis and classification of mode of toxicity of polycyclic aromatic hydrocarbons (PAHs) in *Escherichia coli. Chemosphere,* v.66, p.1243-1248.

[91] Kirby, S. 1994. Oocyte maturation in female winter flounder (*Pseudopleuronectes americanus*) exposed to hydrocarbon contamination in Sydney Harbour, Nova Scotia. Honors Bachelor of Science, Biology Department, Dalhousie University, N.S., Canada, 23 p.

[92] Knutzen, J. 1995. Effects on marine organisms from polycyclic aromatic hydrocarbons (PAH) and other constituents of waste water from aluminium smelters with examples from Norway. *The Science of the Total Environment*, v.163, p.107-122.

[93] Kocan, R.M.; Marty, G.D.; Okihiro, M.S.; Brown, E.D.; Baker, T.T. 1996. Reproductive success and histopathology of individual Prince William Sound pacific herring 3 years after the Exxon Valdez oil spill. *Canadian Journal of Fisheries and Aquatic Sciences*, v.53, p.2388-2393.

[94] Korashy, H.M.; El-Kadi, A.O. 2006. The role of aryl hydrocarbon receptor in the pathogenesis of cardiovascular diseases. *Drug Metabolism Reviews*, v.38, p.411-450.

[95] Kornhauser, A.; Wamer, W.G.; Lambert, L.L. Light-induced dermal toxicity: Effects on the cellular and the molecular level. In: Marzulli, F.N.; Maibach, H.I. (Eds.). *Dermatotoxicology* (4th ed.). New York: Hemisphere, 1991, p. 527-569.

[96] Krylov, S.N.; Huang, X.D.; Zeiler, L.F.; Dixon, D.G.; Greenberg, B.M. 1997. Mechanistic quantitative structure-activity relationship model for the photoinduced toxicity of polycyclic aromatic hydrocarbons. I. Physical model based on chemical kinetics in a two-compartment system. *Environmental Toxicology and Chemistry*, v.16, p.2283-2295.

[97] Kummer, V.; Maskova, J.; Zraly, Z.; Neca, J.; Simeckova, P.; Vondracek, J.; Machala, M. 2008. Estrogenic activity of environmental polycyclic aromatic hydrocarbons in uterus of immature Wistar rats. *Toxicology Letters*, v.180, p.212-221.

[98] Landis, W.G.; Yu, M.-H. *Environmental Toxicology.* 2nd ed. USA: Lewis Publishers, 1998.

[99] Landolt, M.L.; Kocan, R.M. Fish cell cytogenetics: a measure of the genotoxic effects of environmental pollutants. *In: Nriagu, J.R. Aquatic Toxicology*. New York: J. Wiley, 1983, p.336-353.

[100] Lee, L.E.J.; Clemons, J.H.; Bechtel, D.G.; Caldwell, S.J.; Han, K.B.; Pastischmiak-Arts, M.; Mosser, D.D.; Bols, N.C. 1993. Development and characterization of rainbow trout liver cell line expressing cytochrome P 450-dependent monooxygenases activity. *Cell Biology and Toxicology*, v.9, p.279-294.

[101] Legraverend, C.; Guenther, T.M.; Nebert, D.W. 1984. Importance of the route of administration for genetic differences in benzo[a]pyrene-induced in utero toxicity and teratogenicity. *Teratology,* v.29, p.35-47.

[102] Leme, D.M.; Angelis, D.F.; Marin-Morales, M.A. 2008. Action mechanisms of petroleum hydrocarbons present in waters impacted by an oil spill on the genetic material of *Allium cepa* root cells. *Aquatic Toxicology*, v.88, p.214-219.

[103] Leme, D.M.; Marin-Morales, M.A. 2008. Chromosome aberration and micronucleus frequencies in *Allium cepa* cells exposed to petroleum polluted water - a case study. *Mutation Research-Genetic Toxicology and Environmental Mutagenesis*, v.650, p.80-86.

[104] Lemiere, S.; Cossu-Leguille, C.; Bispo, A.; Jourdain, M.-J.; Lanhers, M.-C.; Burnel, D.; Vasseur, P. 2005. DNA damage measured by single-cell gel electrophoresis (Comet) assay in mamals fed with mussels contaminated by the 'Erika'oil-spill. *Mutation Research*, v.581, p.11-21.

[105] Lipzig, M.M.H.; Vermeulen, N.P.E.; Gusinu, R.; Legler, J.; Frank, H.; Seidel, A.; Meerman, J.H.N. 2005. Formation of estrogenic metabolites of benzo[a]pyrene and chrysene by cytochrome P450 activity and their combined and supra-maximal estrogenic activity. *Environmental Toxicology and Pharmacology*, v.19, p.41-55.

[106] Luczyński, M.K.; Góra, M.; Brzuzan, P.; Wilamowski, J.; Kozik, B. 2005. Oxidative metabolism, mutagenic and carcinogenic properties of some polycyclic aromatic hydrocarbons. *Environmental Biotechnology*, v.1, p.16-28.

[107] Luster, M.I.; Rosenthal, G.J. 1993. Chemical agents and the immune response. *Environmental Health Perspectives*, v.101, p.219-226.

[108] Mackenzie KM, Angevine DM. 1981. Infertility in mice exposed in utero to benzo[a]pyrene. *Biol Reprod*, v.24, p.183-191.

[109] Majer, B.J.; Grummt, T.; Uhl, M.; Knasmüller, S. 2005. Use of plant bioassay for the detection of genotoxins in the aquatic environment. *Acta Hydrochimica et Hydrobiologica*, v.33, p.45-55.

[110] Maronpot, R.R.; Fox, T.; Malarkey, D.E.; Goldsworthy, T.L. 1995. Mutations in the ras proto-oncogene: clues to etiology and molecular pathogenesis of mouse liver tumors. *Toxicology,* v.101, p.125-156.

[111] Martin-Skilton, R.; Saborido-Rey, F.; Porte, C. 2008. Endocrine alteration and other biochemical responses in juvenile turbot exposed to the Prestige fuel oil. *Science of the Total Environment*, v.404, p.68-76.

[112] Matsumoto, H.; Kashimoto, T. 1986. Embryotoxicity of organic extracts from airborne particulates in ambient air in the chicken embryo. *Archives of Environmental Contamination and Toxicology*, v.15, p.447-452.

[113] Matsumoto, S.T.; Mantovani, M.S.; Malaguttii, M.I.A.; Dias, A.L.; Fonseca, I.C.; Marin-Morales, M.A. 2006. Genotoxicity and mutagenicity of water contaminated with tannery effluents, as evaluated by the micronucleus test and comet assay using the fish *Oreochromis niloticus* and chromosome aberrations in onion root-tips. *Genetics and Molecular Biology*, v.29, p.148-158.

[114] Mayura, K.; Huebner, H.J.; Dwyer, M. R.; McKenzie, K.S.; Donnelly, K.C.; Kubena, L.F.; Phillips, T.D. 1999. Multi-bioassay approach for assessing the potency of complex mixtures of polycyclic aromatic hydrocarbons. *Chemosphere*, v.38, p.1721-1732.

[115] McCioskey, J.T.; Oris, J.P. 1993. Effect of anthracene and solar ultraviolet radiation exposure on gill ATPase and selected hematologic measurements in the bluegill sunfish (*Lepomis maerunrs*). *Aquatic Toxicology,* v.24, p.207-218.

[116] Melendez-Colon, V. J.; Luch, A.; Seidel, A.; Baird, W.B. 1999. Cancer initiation by polycyclic aromatic hydrocarbons results from formation of stable DNA adducts rather than apurinic sites. *Carcinogenesis*, v.20, p.1885-1891.

[117] Mitchell, B.E.; Sobel, H.L.; Alexander, M.H. 1999. The adverse health effects of tobacco and tobacco-related products. *Primary Care: Clinics in Office Practice*, v.26, p.463-498.

[118] Mitchelmore, C.L.; Chipman, J.K. 1998. DNA strand breakage in aquatic organisms and the potencial value of the comet assay in environmental monitoring. *Mutation* Research, v.399, p.135-147.

[119] Monarca, S.; Feretti, D.; Zerbini, I.; Alberti, A.; Zani, C.; Resola, S.; Gelatti, U.; Nardi, G. 2002. Soil contamination detected using bacterial and plant mutagenicity testes and chemicals analyses. *Environmental Research Section*, v.88, p.64-69.

[120] Nebert, D.W. 1989. The Ah Locus: genetic differences in toxicity, cancer, mutation, and birth defects. Critical Reviews in Toxicology, v.20, p.153-174.

[121] Neff, J.M. Polycyclic aromatic hydrocarbons. *In: Rand, G.M.; Petrocilli, S.R. (Eds.). Fundamentals of Aquatic Toxicology: Methods and Applications*. New York: Hemisphere Publishing Corporation, 1985, p.416-454.

[122] Nicolas, J.-M. 1999. Vitellogenesis in fish and the effects of polycyclic aromatic hydrocarbon contaminants. *Aquatic Toxicology*, v.45, p.77-90.

[123] Ochi, T.; Ishiguro, T.; Ohsawa, M. 1985. The effects of cell cycle position on the cytotoxicity and mutagenicity of benzo(a)pyrene in cultured Chinese hamster cells. *Toxicology Letter*, v.29, p.183-190.

[124] Odeigah P.G.C.; Nurudeen O.; Amund O.O. 1997. Genotoxicity of Oil Field Wastewater in Nigeria. *Hereditas*, v.126, p.161-167.

[125] Okona-Mensah, K.B.; Battershill, J.; Boobis, A.; Fielder, R. 2005. An approach to investigating the importance of high potency polycyclic aromatic hydrocarbons (PAHs) in the induction of lung cancer by air pollution. *Food and Chemical Toxicology*, v.43,p.1103-1116.

[126] Oris, J.T.; Giesy, J.P. 1985. The photoenhanced toxicity of anthracene to juvenile sunfish (*Lepomis* spp.). *Aquatic Toxicology*, v.6, p.133-146.

[127] Palli, D.; Saieva, C.; Munnia, A.; Peluso, M.; Grechi, D.; Zanna, I.; Caini, S.; Decarli, A.; Sera, F.; Masala, G. 2008. DNA adducts and PM10 exposure in traffic-exposed workers and urban residents from the EPIC-Florence City study. *Science of the Total Environment*, v.403, p.105-112.

[128] Pandey, M.K.; Pant, A.B.; Das, M. 2006. *In vitro* cytotoxicity of polycyclic aromatic hydrocarbon residues arising through repeated Wsh fried oil in human hepatoma Hep G2 cell line. *Toxicology in Vitro*, v.20, p.308-316.

[129] Pashin, Y.V.; Bakhitova, L.M. 1979. Mutagenic and carcinogenic properties of polycyclic aromatic hydrocarbons. *Environmental Health Perspective*, v.30, p.185-189.

[130] Patel, M.R.; Scheffler, B.E.; Wang, L.; Willet K.L. 2006. Effects of benzo(a)pyrene exposure on killifish (*Fundulus heteroclitus*) aromatase activities and mRNA. *Aquatic Toxicology*, v.77, p.267-278.

[131] Pavanello, S.; Favretto, D.; Brugnone, F.; Mastrangelo, G.; Dal Pra G.; Clonfero, E. 1999. HPLC/fluorescence determination of anti-BPDE–DNA adducts in mononuclear white blood cells from PAH-exposed humans. *Carcinogenesis*, v.20, p.431-435.

[132] Payne, S. 1958. The pathological effects of the intraperitoneal injection of 3:4-benzpyrene into rats and mice. *British journal of cancer*, v.12, p.65-72.

[133] Pedrozo, M.F.M.; Barbosa, E.M.; Corseuil, H.X.; Schneider, M.R.; Linhares, M.M. Ecotoxicologia e avaliação de risco do petróleo. Salvador: NEAMA, 2002.

[134] Pelletier, M.C.; Burgess, R. M.; Cantwell, M. G.; Serbst, J. R.; Ho, K. T.; Ryba, S. A. 2000. Importance of maternal transfer of the photoreactive polycyclic aromatic hydrocarbon fluoranthrene from benthic adult bivalves to their pelagic larvae. *Environmental Toxicology and Chemistry*, v.19, p. 2691-2698.

[135] Pereira Netto, A.D.; Moreira, J.C.; Dias, A.E.X.O.; Arbilla, G.; Ferreira, L.F.V.; Oliveira, A.S.; Barek, J. 2000. Avaliação da contaminação humana por hidrocarbonetos policíclicos aromáticos (HPAs) e seus derivados nitrados (NHPAs): uma revisão metodológica. *Química Nova*, v.23, p.765-773.

[136] Perera, F.; Rauh, V.; Tsai, W.Y. 2003. Effects of transplacental exposure to environmental pollutants on birth outcomes in a multi-ethnic population. *Environmental Health Perspectives*, v.111, p.201-205.

[137] Perera, F.; Tang, D.; Whyatt, R.; Lederman, S.A.; Jedrychowski, W. 2005. DNA Damage from Polycyclic Aromatic Hydrocarbons Measured by Benzo[a]pyrene-DNA Adducts in Mothers and Newborns from Northern Manhattan, The World Trade Center Area, Poland, and China. Cancer Epidemiology, *Biomarkers & Prevention*, v.14, p.709-714.

[138] Perera, F.P.; Rauh, V.; Whyatt, R.; Tsai, W.T.; Tang, D.; Diaz, D.; Hoepner, L.; Barr, D.; Tu, Y.H.; Camann, D.; Kinney, P. 2006. Effect of prenatal exposure to airborne polycyclic aromatic hydrocarbons on neurodevelopment in the first 3 years of life among inner-city children. *Environmental Health Perpectives*, v.114, p.1287-1292.

[139] Picada, J.N.; Kern, A.L.; Ramos, A.L.L.P.; Saffi, J. O estresse oxidativo e as defesas antioxidantes. In: Silva J., Erdtmann B., Henriques J.A.P. (Eds.). Genética Toxicológica. Porto Alegre: Alcance, 2003, p.251-264.

[140] Pitts, J.N. Jr. 1987. Nitration of gaseous polycyclic aromatic hydrocarbons in simulated and ambient urban atmospheres: a source of mutagenic nitroarenes. Atmospheric Environment, v.21, p.2531-2547.

[141] Platt, K.L.; Pfeiffer, E.; Petrovic, P.; Friesel, H.; Beermann, D.; Hecker, E.; Oesch, F. 1990. Comparative tumorigenicity of picene and dibenz[a,h]anthracene in the mouse. *Carcinogenesis;* v.11, p.1721-26.

[142] Plaza, G.; Nałęcz-Jawecki, G.; Ulfig, K.; Brigmon, R.L. 2005. Assessment of genotoxic activity of petroleum hydrocarbon-biorremediated soil. *Ecotoxicology and Environmental Safety*, v.62, p.415-420.

[143] Plíšková, M.; Vondracek, J.; Vojtesek, B.; Kozubik, A.; Machala, M. 2005. Deregulation of cell proliferation by polycyclic aromatic hydrocarbons in human breast carcinoma MCF-7 cells reflects both genotoxic and nongenotoxic events. *Toxicological Sciences,* v.83, p.246-256.

[144] Pollino, C.A.; Holdway, D.A. 2002. Toxicity testing of crude oil and related compounds using early life stages of the crimson-spotted rainbowfish (*Melanotaenia fluviatilis*). *Ecotoxicology and Environmental Safety*, v.52, p.180-189.

[145] Prahalad, A.K.; Ross, J.A.; Nelson, G.B.; Roop, B.C.; King, L.C.; Nesnow, S.; Mass, M.J. 1997. Dibenzo[a,l]pyrene-induced DNA adduction, tumorigenicity, and K-ras oncogene mutations in strain A/J mouse lung. *Carcinogenesis,* v.18, p.1955-1963.

[146] Qiao M.; Chen, Y.Y.; Zhang, Q.H.; Huang, S.B.; Ma, M.; Wang, C.X.; Wang, Z.J. 2006. Identification of Ah receptor agonists in sediment of Meiliang Bay, Taihu Lake, China. *Environmental Science and Technology*, v.40, p.1415-1419.

[147] Reece, C.H.; Burks, S.L. Isolation and chemical characterization of petroleum refinery wastewater fractions acutely lethal to *Daphnia magna*. In: Cardwell, R.D.; Purdy, R.; Bahner, R.C. Aquatic Toxicology and Hazard Assessment: Seventh Symposium. *Philadelphia: American Society for Testing and Materials*, 1985, 579p.

[148] Ren, L., Zeiler, L.F., Dixon, G., Greenberg, B.M., 1996. Photoinduced effects of polycyclic aromatic hydrocarbons on *Brassica napus* (Canola) during germination and early seedling development. *Ecotoxicology and Environmental Safety*, v.33, p.73-80.

[149] Ribeiro, L.R. Teste do micronúcleo em medula óssea de roedores *in vivo*. In: Ribeiro, L.R.; Salvadori, D.M.F.; Marques, E.K. (Eds.). Mutagênese Ambiental. Canoas: Ulbra, 2003, p.173-198.

[150] Rice, J.M.; Ward, J.M. 1982. Age dependence of susceptibility to carcinogenesis in the nervous system. *Annals of the New York Academy of Sciences*; v.381, p.274-289.

[151] Rivas, A.; Olea, N.; Olea-Serrano, F. 1997. Human exposure to endocrine-disrupting chemicals: assessing the total estrogenic xenobiotic burden. *Trends in Analytical Chemistry*, v.16, p. 613-619.

[152] Ronis, M.J.J.; Celander, M.; Förlin, L.; Badger, T.M. 1992. The use of polyclonal antibodies raised against Rat and trout cytochrome P450 CYP1A1orthologues to monitor environmental induction in the channel catfish (*Ictalurus punctatus*). *Marine Environmental Research*, v.34, p.181-188.

[153] Roos, P.H.; Tschirbsa, S.; Pfeifer, F.; Welge, P.; Hack, A.; Wilhelm, M.; Bolt, H.M. 2004. Risk potencials for humans of original and remediated PAH-contaminated soils: application of biomarkers of effect. *Toxicology*, v. 205, p.181-194.

[154] Ross, J. A.; Nesnow, S. 1999. Polycyclic aromatic hydrocarbons: correlations between DNA adducts and *ras* oncogene mutations. *Mutation Research*, v.424, p.155-166.

[155] Roubicek, D.A.; Gutiérrez-Castillo, M.E.; Sordo, M.; Cebrián-García, M.E.; Ostrosky-Wegman, P. 2007. Micronuclei induced by airborne particulate matter from Mexico City. *Mutation Research*, v.631, p.9-17.

[156] Russel, P.J. Chromosomal mutation. In: Cummings, B. (Ed.). *Genetics*. San Francisco: Pearson Education Inc, 2002, p.595-621.

[157] Saint-Denis, M.; Narbonne, J.F.; Arnaud, C.; Thybaud, E.; Ribera, D. 1999. Biochemical responses of the earthworm *Eisenia fetida andrei* exposed to contaminated artificial soil: effects of benzo(a)pyrene. *Soil Biology and Biochemistry*, v.31, p.1837-1846.

[158] Samanta, S.K.; Singh, O.V.; Jain, R.K. 2002. Polycyclic aromatic hydrocarbons: environmental pollution and bioremediation. *Trends in Biotechnology,* v.20, p. 243-248.

[159] Santodonato, J. 1997. Review of the estrogenic and antiestrogenic activity of polycyclic aromatic hydrocarbons: relationship to carcinogenicity. *Chemosphere,* v.34, p.835-848.

[160] Sanyal, M.K.; Li, Y.L. 2007. Deleterious effects of polynuclear aromatic hydrocarbon on blood vascular system of the rat fetus. Birth Defects Research Part B: *Developmental and Reprodutive Toxicology*, v.80, p.367-373.

[161] Sasaki, Y.F.; Izumiyama, F.; Nishidate, E.; Ishibashi, S.; Tsuda, S.; Matsusaka, N.; Asano, N.; Saotome, K.; Sofuni, T.; Hayashi, M. 1997. Detection of genotoxicity of polluted sea water using shellfish and alkaline single-cell gel electrophoresis (SCE) assay: a preliminary study. *Mutation Research*, v.393, p.133-139.

[162] Schirmer, K., Chan, A.G.J., Greenberg, B.M., Dixon, D.G., Bols, N.C., 1997. Methodology for demonstrating and measuring the photocytotoxicity of fluoranthene to fish cells in culture. *Toxicology in Vitro,* v.11, p.107-119.

[163] Schirmer, K.; Dixon, D.G.; Greenberg, B.M.; Bols, N.C. 1998a. Ability of 16 priority PAHs to be directly cytotoxic to a cell line from the rainbow trout gill. *Toxicology,* v.127, p.129-141.

[164] Schirmer, K.; Chan, A.G.J.; Greenberg, B.M.; Dixon, D.G.; Bols, N.C. 1998b. Ability of 16 priority PAHs to be photocytotoxic to a cell line from the rainbow trout gill. *Toxicology,* v.127, p.143-155.

[165] Sevastyanova, O.; Novakova, Z.; Hanzalova, K.; Binkova, B.; Sram, R.J.; Topinka, J. 2008. Temporal variation in the genotoxic potential of urbana ir particulate matter. *Mutation Research*, v.649, p.179-186.

[166] Shailaja, M.S.; D'Silva, C. 2003. Evaluation of impact of PAH on a tropical fish, Oreochromis mossambicus using multiple biomarkers. *Chemosphere,* v.53, p.835–841.

[167] Shimada, T. 2006. Xenobiotic-metabolizing enzymes involved in activation and detoxification of carcinogenic polycyclic aromatic hydrocarbons. *Drug Metabolism and Pharmacokinetics*, v.21, n.4, p. 257-276.

[168] Shum, S.; Jensen, N.M.; Nebert, D.W. 1979. The Murine Ah locus: In utero toxicity and teratogenesis associated with genetic differences in benzo[a]pyrene metabolism. *Teratology,* v.20, p.365-376.

[169] Smith, M.J.; Flowers, T.H.; Duncan, H.J.; Alder, J. 2006. Effects of polycyclic aromatic hydrocarbons on germination and subsequent growth of grasses and legumes in freshly contaminated soil and soil with aged PAHs residues. *Environmental Pollution*, v.141, p.519-525.

[170] Smreczak, B.; Maliszewska-Kordybach, B. 2003. Seeds germination and root growth of selected plants in PAH contaminated soil. *Fresenius Environmental Bulletin*, v.12, p. 946-949.

[171] Straif, K.; Baan, R.; Grosse, Y.; Secretan, B.; Ghissassi, F.E.; Cogliano, V. 2005. Carcinogenicity of polycyclic aromatic hydrocarbons. *Oncology,* v.6, p. 931-932.

[172] Stanley, E.M. *Toxicological Chemestry*. 2nd ed. USA: Lewis Publishers. 1992.

[173] Stanley, E.M. *Environmental Chemistry.* 6th ed. USA: Lewis Publisher. 1994.

[174] Stegeman, J.J.; Kloepper-Sams, P.J. 1987. Cytochrome P450 isoenzymes and monooxygenase activity in aquatic animals. *Environmental Health Perspectives,* v.71, p.87-95.

[175] Stegman, J.J.; Hahn, M.E. Biochemistry and molecular biology of monooxygenases: current perspectives on forms, function and regulation of cytochrome P450 in aquatic species. In: Mallins D.C., Ostrander G.K. (Eds.). *Aquatic toxicology: molecular, biochemical and cellular perspectives*. USA: CRC Press, 1994, p.87-206.

[176] Stewart, B.W.; Kleihues, P. *World Cancer Report*. Lyon: IARC Press, 2003.

[177] Sturla, S.J. 2007. DNA adduct profiles: chemical approaches to addressing the biological impact of DNA damage from small molecules. *Current Opinion in Chemical Biology*, v.11, p. 293-299.

[178] Swartz, W.J.; Mattison, D.R. 1985. Benzo(a)pyrene inhibits ovulation in C57BL/6N mice. *Anatomical Record*, v.212, p.268-276.

[179] Taban, I.C.; Bechmann, R.K.; Torgrimsen, S.; Baussant, T.; Sanni, S. 2004. Detection of DNA damage in mussels and sea urchins exposed to crude oil using comet assay. *Marine Environmental Research*, v.58, p.701-705.

[180] Tatem, H.E.; Cox, B.A.; Anderson, J.W. 1978. The toxicity of oils and petroleum hydrocarbons to estuarine crustaceans. *Estuarine, Coastal and Shelf Science*, v.6, p.365-373.

[181] Thirman, M.J.; Albrecht, J.H.; Kruger, M.A.; Ericson, R.R.; Cherwitz, D.L.; Park, S.S.; Gelboin, H.V.; Holtzman, J.L. 1994. Induction of cytochrome CYP1A1 and formation of toxic metabolites of benzo[a]pyrene by rat aorta: a possible role in atherogenesis. *Proceedings of the National Academy of Sciences of the United States of America*, v.91, p.5397-5401.

[182] Thomas, P.; Budiantara, L. 1995. Reproductive life history stages sensitive to oil and naphthalene in Atlantic croaker. *Marine Environmental Research*, v.39, p.147-150.

[183] Tilghman Hall, A.; Oris, J.T. 1991. Anthracene reduces reproductive potential and is maternally transferred during long term exposure in fathead minnows. *Aquatic Toxicology*, v.19, p.249-264.

[184] Tuvikene, A. 1995. Responses of fish to polycyclic aromatic hydrocarbons (PAHs). *Annales Zoologici Fennici*, v.32, p.295-309.

[185] Umbuzeiro, G.A.; Franco, A.; Martins, M.H.; Kummrow, F.; Carvalho, L.; Schmeiser, H.H.; Leykaulf, J.; Stibotova, M.; Claxton, L.D. 2008. Mutagenicity and DNA adduct formation of PAH, nitro-PAH, and oxi-PAH fractions of atmospheric particulate matter from São Paulo, Brazil. *Mutation Research*, v.652, p.72-80.

[186] Utesch, D.; Eray, K.; Diehl, E. 1996. Phototoxicity testing of polycyclic aromatic hydrocarbons (PAH) in mammalian cells *in vitro*. *Polycyclic Aromatic Compounds*, v.10, p.117-121.

[187] Van Meter, R.J.; Spotila, J.R.; Avery, H.W. 2006. Polycyclic aromatic hydrocarbons affect survival and development of common snapping turtle (*Chelydra serpentina*) embryos and hatchlings. *Environmental Pollution*, v.142, p. 466-475.

[188] Vanzella, T.P.; Martinez, C.B.R.; Cólus, I.M.S. 2007. Genotoxic and mutagenic effects of diesel oil water soluble fraction on a neotropical fish species. *Mutation Research*, v.631, p.36-43.

[189] Vieira, L.R.; Sousa, A.; Frasco, M.F.; Lima, I.; Morgado, F.; Guilhermino, L. 2008. Acute effects of Benzo[a]pyrene, anthracene and a fuel oil on biomarkers of the common goby *Pomatoschistus microps* (Teleostei, Gobiidae). *Science of the Total Environmental* v.395, p.87-100.

[190] Wake, H. 2005. Oil refineries: a review of their ecological impacts on the aquatic environment. *Estuarine, Coastal and Shelf Science*, v.62, p.131-140.

[191] Wang, J.; Wu, W.; Henkelmann, B.; You, L.; Kettrup, A.; Schramm, K.-W. 2003. Presence of estrogenic activity from emission of fossil fuel combustion as detected by a recombinant yeast bioassay. *Atmospheric Environment*, v.37, p.3225-3235.

[192] Wang, S.; Sheng, Y.; Feng, M.; Leszczynski, J.; Wang, L.; Tachikawa, H.; Yu, H. 2007. Light-Induced Cytotoxicity of 16 Polycyclic Aromatic Hydrocarbons on the US EPA Priority Pollutant List in Human Skin HaCaT Keratinocytes: Relationship Between Phototoxicity and Excited State Properties. *Environmental Toxicology*, v.22, p.318-327.

[193] Warshawsky, D. 1999. Polycyclic Aromatic Hydrocarbons in Carcinogenesis. *Environmental Health Perspectives*, v. 107, p. 317–319.

[194] Wassenberg, D.M.; Di Giulio, R.T. 2004. Synergistic embryotoxicity of polycyclic aromatic hydrocarbon aryl hydrocarbon receptor agonists with Cytochrome P4501A inhibitors in *Fundulus heteroclitus*. *Environmental Health Perspectives*, v.112, p.1658–1664.

[195] Wassenberg, D.M.; Swails, E.E.; Di Guilio, R.T. 2002. Effects of single and combined exposures to benzo[a]pyrene and 3,3'4,4'5-pentachlorobiphenyl on EROD activity and development in *Fundulus heteroclitus*. Marine Environmental Research, v.54, p.279-283.

[196] Weinstein, J.E.; Oris, J.T.; Taylor, D.H. 1997. An ultrastructural examination of the mode of UV-induced toxic action of fluoranthene in fathead minnows *Pimephales promelas*. *Aquatic Toxicology*, v.39, p.1-22.

[197] Wessel, N.; Rousseau, S.; Caisey, X.; Quiniou, F.; Akcha, F. 2007. Investigating the relationship between embryotoxic and genotoxic effects of benzo[a]pyrene, 17α-ethinylestradiol and endosulfan on *Crassostrea gigas* embryos. *Aquatic Toxicology*, v.85, p.133-142.

[198] White, P.A. 2002. The genotoxicity of priority polycyclic aromatic hydrocarbons in complex mixtures. *Mutation Research*, v.515, p.85-98.

[199] Whitlock, J.P. Jr. 1999. Induction of cytochrome P4501A1. *Annual Reviews in Pharmacology and Toxicology* 39:103–125.

[200] Wittig, R.; Ballach, H.J.; Kuhn, A. 2003. Exposure of the roots of *Populus nigra* L. cv. Loenen to PAHs and its effect on growth and water balance. *Environmental Science Pollution Research International*, v.10, p.235-244.

[201] [WHO] World Health Organization. Polycyclic Aromatic Hydrocarbons. *In: Air Quality Guidelines* – 2nd ed. Copenhagen, Denmark, 2000.

[202] Wu M, Yan S, Patel DJ, Geacintov NE, Broyde S. 2002. Relating repair susceptibility of carcinogen-damaged DNA with structural distortion and thermodynamic stability. *Nucleic Acid Sciences*, v.30, p.3422-3432.

[203] Yu, H. 2002. Environmental carcinogenic polycyclic aromatic hydrocarbons: photochemistry and phototoxicity. *Journal of Environmental Science and Health*, Part C: Environmental Carcinogenesis and Ecotoxicology Reviews, v.20, p.149-183.

In: Polycyclic Aromatic Hydrocarbons: Pollution, Health ... ISBN 978-1-60741-462-9
Editors: Pierre A. Haines and Milton D. Hendrickson

Chapter 2

POLYCYCLIC AROMATIC HYDROCARBONS TOXICITY IN ANIMALS AND HUMANS

Pierina Visciano and Monia Perugini
Department of Food Science, University of Teramo, Teramo, Italy

ABSTRACT

Polycyclic aromatic hydrocarbons (PAHs) are a group of about 100 different chemicals that are formed during incomplete combustion or heat-induced decomposition of organic matter. Humans can be exposed to PAHs through different routes. For non-smokers, the major route of exposure is from food with a minor contribution from inhaled air. Food can be contaminated from environmental sources, industrial food processing and from home food preparation. The European regulator has found that there is a low risk to consumers from PAHs in foods based on average dietary exposure but risk management action may be required for high dietary intake. Several PAHs have been shown to be genotoxic carcinogens. In 2002, the Scientific Committee on Food (SCF) reviewed PAH toxicity. For 15 compounds it concluded that there was clear evidence for their toxicity. In 2005, the Joint FAO/WHO Expert Committee on Food Additives (JECFA) performed a risk assessment on PAHs, and basically agreed with the SCF selection. Also the US Environmental Protection Agency has classified some PAHs as priority-pollutants based on toxicity, potential for human exposure, frequency of occurrence and the extent of information available. Seven of these PAHs (benzo(*a*)anthracene, chrysene, benzo(*a*)pyrene, benzo(*b*)fluoranthene, benzo(*k*)fluoranthene, dibenzo(*a,h*)anthracene and indeno(1,2,3-*cd*)pyrene) are considered as probable human carcinogens, but benzo(*a*)pyrene is the most studied and used as a marker for PAHs in ambient air and food. PAHs are themselves chemically inert but undergo metabolic activation in mammalian cells to diol-epoxides that bind covalently to cellular macromolecules, including DNA, causing errors in DNA replication and mutations that initiate the carcinogenic process. Other reactive intermediates can be generated by one-electron oxidation process that can result in chemically unstable alkylation of DNA, which leads to depurination, a potentially mutagenic event. In this paper the most important effects of PAHs on animals and humans health are reported and discussed to define new possible preventive and regulatory programs for these pollutants.

INTRODUCTION

Polycyclic aromatic hydrocarbons are a class of structurally different organic molecules, composed of two or more fused aromatic rings, found in a wide variety of sources. They can be present as complex mixtures in several worldwide ecosystems (air, water, sediments and soil), as a consequence of natural sources (forest fires and volcanic eruptions) and human activities (municipal waste, industrial effluents, petroleum spills, creosote oil, automobile exhaust, coal tar production). In the atmosphere PAHs tend to have low vapour pressure, so they are usually adsorbed on particulate matter. The vapour pressure of PAHs is inversely proportional to the number of rings contained and thus almost all five-ring PAH compounds are particulate bound, while three-ring PAHs are also present as vapour in the atmosphere. The solubility of PAHs in water is inversely proportional to the number of rings the PAH molecule contains. Thus, in aqueous environments, PAHs are generally found adsorbed on particulates and on humic matter, or dissolved in any oily contaminant that may be present in water, sediment and soil. PAHs are hydrophobic compounds and their persistence in the environment is mainly due to their chemical properties. Higher molecular weight PAHs exhibit greater environmental persistence than lower molecular weight PAHs due to an increase in hydrophobicity and an increase in stability with increasing mass. This relationship of environmental persistence and increasing number of benzene rings is consistent with the results of studies correlating biodegradation rates and PAH molecule size (Cerniglia, 1992).

PAHs are an important group of chemical compounds some of which are mutagenic and carcinogenic (JECFA, 2006). Furthermore many of them are also potential endocrine disruptors causing adverse effects on hormone homeostasis (Gozgit et al., 2004; Fertuck et al., 2001a). The metabolism of PAHs leads to the formation of toxic as well non-toxic metabolites. Therefore, data on the biological fate of these chemicals and on the nature of the metabolites formed by different species are needed to predict the potential hazard posed to organisms. Most of the data pertains to mammalian and fish species (Varanasi et al., 1986; Yuan et al., 1999; Viau et al., 1999; Ramesh et al., 2000; Billiard et al., 2002) and only little information is available on metabolism by other species (Honey et al., 2000).

The health effects of polycyclic aromatic hydrocarbons may be apparent under circumstances of high exposure. However, the great majority of people during their lifetime are exposed to complex mixtures of organic compounds at low concentrations. Long-term exposure may be more important in terms of overall public health. Many single PAHs are carcinogenic in rodent bioassays, and they are suspected carcinogens in humans. PAHs exhibit their biological effects through metabolic activation by cytochrome P450-related enzymes (CYP) to electrophilic species, capable of reacting with the nucleophilic sites of DNA to form adducts. The formation and persistence of carcinogen–DNA adducts are believed to be critical events for the initiation of neoplasia in target cells. The risk of mutations arising from DNA adducts does not depend only on the adduct level but also on their mutagenic potential as well as on the DNA sequence of neighbouring adducts (Binková and Šrám, 2004).

In 2008 the European Food Safety Authority (EFSA) Panel on Contaminants in the Food Chain (CONTAM Panel) reviewed the available data on occurrence and toxicity of PAHs and concluded that the risk should be based upon the PAHs (benzo(**a**)pyrene, benzo(**a**)anthracene, benzo(**b**)fluoranthene, benzo(**k**)fluoranthene, benzo(**ghi**)perylene, chrysene, dibenzo(**a,h**)anthracene and indeno(**1,2,3-cd**)pyrene) for which oral carcinogenicity data were available. The CONTAM Panel concluded that these PAHs, either individually or in a combination, are currently the only possible indicators of the carcinogenic potency of PAHs in food (EFSA, 2008).

Absorption, Metabolism and Excretion of Pahs

Humans are exposed to PAHs by various pathways. Occupational PAH exposure doses (coal tar, aluminium production, carbon electrode manufacture, coal gasification and its conversion to coke and coal tar and the processing and use of coal-tar derived products) exceed by far those encountered in the general environment. For non-smoking people with no occupational exposure to these contaminants, the main source of PAH exposure comes from ingested food and from the breathing of both indoor and outdoor air. PAHs presence in food arises from both environmental pollution (atmospheric deposition on crops, releasing in aqueous environment and adsorption on sediments) and pyrolysis of food components during cooking or food processing (grilling, roasting, frying, smoking). The two highest contributors to the dietary exposure were found to be cereals and cereal products, and seafood and seafood products (EFSA, 2008). Absorption of PAHs from the diet is determined by the size and the lipophilicity of the molecule, the presence of bile in the digestive tract, the dose ingested and the lipid content of the diet. Animals and humans have evolved inducible enzymes and transporters to facilitate the biotransformation and elimination of toxic compounds encountered in the environment. The enzymatic components of this inducible biotransformation system are well-known and include monooxygenases in the cytochrome P450 superfamily as well as conjugating enzymes such as the glutathione transferases and glucuronosyl transferases. The transporter components of this system include the ATP binding cassette (ABC) proteins, which act as efflux pumps to remove metabolites of endogenous and xenobiotic chemicals from cells (Dean et al., 2001). The component of this system consists of soluble receptors that regulate the expression of the biotransformation and transporter genes in response to environmental chemicals. These receptors include several members of the steroid/nuclear receptor superfamily (Honkakoski and Negishi, 2000) as well as the aryl hydrocarbon receptor (AhR), a member of the basic-helix-loop-helix (bHLH)–Per-ARNT-Sim (PAS) gene superfamily (Hahn, 2002). The AhR is considered an 'induction receptor' that controls the expression of aryl hydrocarbon hydroxylase activity. The mechanism by which the AhR regulates the induction of adaptive enzymes, especially CYP1A1, has been studied extensively (Withlock, 1999). Although originally thought to have a relatively narrow structural specificity, the AhR is now known to recognize an impressive range of chemical structures, including non-aromatic and non-halogenated compounds (Hahn, 2002).

It has been estimated that there are about 58 different cytochrome P450 genes within the human genome, among these enzyme families CYP 1-3 are involved in xenobiotic metabolism (Guengerich, 2001). While some of these genes are constitutively expressed, the expression of others can be induced by their substrates. Several mechanisms of induction have been characterized, and CYP1A1 has been shown to be under the control of the AhR (Safe, 1995).

For PAHs metabolic activation is required for their toxic, mutagenic and carcinogenic action (Harvey, 1991). In most cases, cytochrome P450 catalyzed epoxide formation is an initial step in the activation process, followed by formation of other, highly reactive electrophilic metabolites capable of binding to cellular macromolecules, in particular proteins and nucleic acids. In total, three principal metabolic pathways have been described for PAHs (Xue and Warschawsky, 2005). Historically, the bay region dihydrodiol epoxide pathway was the first discovered as being crucial for the mutagenicity of many PAHs (Hall and Grover, 1990; Harvey, 1991). Later, the one-electron oxidation pathway leading to reactive PAH radicals was discovered (Cavalieri and Rogan, 1992; Cavalieri and Rogan, 1995). More recently, the *ortho*-quinone pathway, also resulting in reactive eletrophilic PAH metabolites, has been described (Penning et al., 1996).

The mechanism of PAHs metabolism has been mainly investigated using benzo(*a*)pyrene (BaP) as a model compound. BaP is known to be converted to 3-hydroxybenzo(*a*)pyrene (3-OH BaP), 9-hydroxybenzo(*a*)pyrene (9-OH BaP), benzo(*a*)pyrene-4,5-dihydrodiol (BaP 4,5-diol), benzo(*a*)pyrene-7,8-dihydrodiol (BaP 7,8-diol), benzo(*a*)pyrene-9,10-dihydrodiol (BaP-9,10-diol) and benzo(*a*)pyrene quinones (BaP quinones) by mammalian and fish liver microsomes (Miranda et al., 2006). In many animal species, including humans (Shou et al., 1994), 3-OH BaP is a major metabolite of BaP, and is the major metabolite measured in the aryl hydrocarbon hydroxylase assay. In humans, CYP1A1, CYP1B1, CYP1A2, CYP2C9 and CYP2C19 are the catalysts involved in the formation of BaP 7,8-diol from BaP (Gautier et al., 1996; Kim et al., 1998; Shimada et al., 1999). BaP 7,8-diol is further oxidized to the ultimate carcinogenic metabolite, 7,8-dihydroxy-9,10-epoxy-7,8,9,10-tetrahydrobenzo(*a*)pyrene (BaPDE), mainly catalyzed by CYP1A1 (Gautier et al., 1996; Kim et al., 1998), but also by human CYPs 1B1, 1A2, 2C8, 2C9, 2C19, 2E1 and 3A4 forms (Shou et al., 1994; Gautier et al., 1996; Kim et al., 1998).

Many of the enzymes involved in the metabolism of PAHs have been shown to be polymorphic. A genetic polymorphism is a genetic difference occurring in at least 1% of the population. A genetic polymorphism in xenobiotic metabolising enzymes may change the ratio of activation/deactivation of PAHs. An increased risk of certain cancers has been linked to certain genetic polymorphisms (IARC, 1999). The exact role of genetic polymorphism on risk for PAH carcinogenesis has not been completely elucidated as compensatory mechanisms or alternative enzymatic pathways may also be involved. Currently, no clear picture of the effects of genetic polymorphisms of certain CYPs on the carcinogenic risk of humans exposed to PAHs can be derived. It seems likely that the influence of polymorphisms depends on the level of exposure being more important at low doses of exposure (SCF, 2002).

After initial metabolism, most of the detoxified portion of PAH compounds are excreted into the bile as metabolites, then subsequently eliminated in the faeces. A smaller proportion is excreted in urine. The two major metabolites of benzo(*a*)pyrene found in urine are 3-hydroxy- and 9-hydroxybenzo(*a*)pyrene, part of them being conjugated to sulphate or glucuronic acid (Bouchard and Viau, 1997; Ramesh et al., 2001). The process of excretion of PAH compounds into the intestine via the bile, reabsorption, and return to the liver by the portal circulation (enterohepatic recirculation) has been demonstrated to occur for 7,12-dimethylbenzo(*a*)anthracene, benzo(*a*)pyrene, anthracene and pyrene (Ramesh et al., 2004). Enterohepatic circulation extends the residence time of PAHs in the body and may lead to long half-lives of reactive PAH metabolites. Though enterohepatic recycling of PAHs has been studied in animal models, it has not been confirmed in human studies (de Kok and van Maanen, 2000).

Cancerogenicity and Genotoxicity

PAHs have been described as the largest group of known environmental carcinogens and are probably the most widely distributed environmental contaminants. In the past decade (WHO/IPCS, 1998; SCF, 2002; FAO/WHO, 2005) they were evaluated by the International Programme on Chemical Safety (IPCS), the Scientific Committee on Food (SCF) and by the Joint FAO/WHO Expert Committee on Food Additives (JECFA). SCF concluded that 15 PAHs, namely benzo(**a**)anthracene, benzo(**b**)fluoranthene, benzo(**j**)fluoranthene, benzo(**k**)fluoranthene, benzo(**ghi**)perylene, benzo(**a**)pyrene, chrysene, cyclopenta(**cd**)pyrene, dibenzo(**a,h**)anthracene, dibenzo(**a,e**)pyrene, dibenzo(**a,h**)pyrene, dibenzo(**a,i**)pyrene, dibenzo(**a,l**)pyrene, indeno(1,2,3-**cd**)pyrene and 5-methylchrysene show clear evidence of mutagenicity/genotoxicity in somatic cells in experimental animals **in vivo** and with the exception of benzo(**ghi**)perylene have also shown clear carcinogenic effects in various types of bioassays in experimental animals (SCF, 2002). Thus, SCF reasoned that these compounds may be regarded as potentially genotoxic and carcinogenic to humans and therefore represent a priority group in the assessment of the risk of long-term adverse health effects following dietary intake of PAHs. SCF suggested to use benzo(**a**)pyrene as a marker of occurrence and effect of the carcinogenic PAHs in food, based on examinations of PAH profiles in food and on evaluation of a carcinogenicity study of two coal tar mixtures in mice (Culp et al., 1998). Using the assessments of IPCS and SCF as starting points and taking into account newer studies, the JECFA re-evaluated PAHs in 2005. Overall, the JECFA concluded that 13 PAHs are clearly genotoxic and carcinogenic. Except benzo(**ghi**)perylene and cyclopenta(**cd**)pyrene, the compounds were the same as those stated by SCF. The JECFA also concluded that benzo(**a**)pyrene could be used as a marker of exposure to, and effect of, the 13 genotoxic and carcinogenic PAHs (JECFA, 2006).

There is a considerable amount of data on levels of PAHs in food, covering a large number of different members of the group as there are differing views on the priority to be given to the individual PAHs. The PAHs analyzed tend to be those with 4- to 5- rings as some members with these structures have been found to be carcinogens, teratogens and mutagens. However, benzo(*a*)pyrene is usually highlighted, although it comprises less than 5 percent of the total amount of PAHs present in the atmosphere, since it is thought to be the most toxic PAH. The International Agency for Research on Cancer (IARC) considers BaP a known animal carcinogen and a probable human carcinogen (Group 2A). The carcinogenicity of BaP has been attributed to the reaction of BaP metabolites, primarily the diol epoxides, with DNA. There are four enantiomers of BaP diol epoxide that can form adducts with DNA, namely: (+)-anti-BaPDE; (−)-anti-BaPDE; (+)-syn-BaPDE and (−)-syn-BaPDE (Lau and Baird, 1994). Each enantiomer can react via *cis* or *trans* addition at the C10 position of BaP with the exocyclic amino groups of guanine and adenine residues (Banasiewicz et al., 2004).

Carcinogenesis is a multi-step process, at least two stages of action of chemical compounds being distinguished in it: initiation and promotion. Certain compounds have been found to be initiators, while some others possess only promoting functions. Initiation is mediated by the interaction of an electrophilic metabolite with DNA leading to changes in

DNA structure followed by mutation. The principal promoter function consists of ''switching-off'' of the initiated cells from the regulatory control of adjacent normal cells, thus providing conditions for preferential growth of the initiated cells. The ''promotion'' stage is also referred to as epigenetic, since it is not associated with modification of the genetic apparatus and is induced by non-genotoxic compounds (Kobliakov, 1998; Rakitsky et al., 2002). In the past more attention was given to the genotoxic nature of carcinogens but little has been done to study the epigenetic effects. Recently, however, the epigenomic component of environmental carcinogens has been suggested to be very important in human carcinogenesis due to altering gene expression that leads to an imbalance of cell proliferation, differentiation and apoptosis (Thilly, 2003; Trosko and Upham, 2005). ''Complete'' carcinogens, i.e., compounds that induce tumors without the need for additional exposure, are able to stimulate both the initiation and promotion stages. Some PAHs are found to be complete carcinogens. The initiation process for PAHs is relatively well studied. The most important step is the binding of the active metabolites of PAHs to DNA, predominantly to amino groups of guanine and adenine. The major stable adduct is formed at the N2 position of desoxyguanosine. DNA adduct formation is generally regarded as one of the first steps in carcinogenicity of the mutagenic PAHs (EFSA, 2008). PAH-DNA adducts have been detected in a variety of human tissues, including target organs of PAH and tobacco-associated cancers. A large number of studies clearly show that lowering exposure to carcinogenic PAH results in decreasing PAH-DNA adduct levels (Kriek et al., 1998).

Recent developments hold at least two important potential new applications for cancer molecular epidemiology. First, the capability to sequence the genes that are part of the molecular pathological pathway leading to the initiation of tumors in individuals, provides the basis for a whole new categorization of disease outcomes. This is exemplified by the results of Denissenko and others (Denissenko et al., 1996), who studied the distribution of BaPDE adducts along exons of the p53 gene. Thus, adduct formation at specific sites appears to determine partly the p53 mutational spectrum found in lung cancer cases. The detection of mutational hotspots and genetic fingerprints caused by carcinogen-induced DNA modifications may be useful for molecular epidemiologists to disclose previous exposures and retrospectively identify causative agents. Secondly, new technology being developed will enable us to measure the cellular frequency of known types of mutations along known cancer pathways in otherwise normal tissue obtained from biopsies or samples of exfoliated normal cells from particular tissues. This approach may allow the identification of high risk subjects and early interventions through preventive measures.

Pahs as Endocrine Disruptor Compounds

The International Program for Chemical Safety defines endocrine disrupters as "exogenous substances that alter function(s) of the endocrine system and consequently cause adverse health effects in an intact organism, its progeny, or (sub)populations". Endocrine disrupters interfere with the functioning of the endocrine system in at least three possible manners: by mimicking the action of a naturally produced hormone, binding to their hormone receptors; by blocking the receptors in target cells for these hormones and therefore preventing the action of natural hormones; by altering the synthesis and function of hormone receptors and modifying the synthesis, transport, metabolism and excretion of hormones (Ropero et al., 2006).

A growing body of literature identifies PAHs as potential environmental endocrine disruptors (Santodonato, 1997). While the mutagenicity and carcinogenicity of this class of compounds has been studied extensively, more recent reports have also suggested that PAHs are suspected to exhibit both estrogenic and antiestrogenic activity. Some 4- and 5-ring PAHs can interact with the estrogen receptor (ER) and signalling pathways in a variety of test systems (Clemons et al., 1998; Arcaro et al., 1999) potentially interfering with estrogen signaling in important developmental and reproductive processes.

The antiestrogenicity of PAHs generally is based on three possible mechanisms: by binding to the Ah receptor (Clemons et al. 1998; Arcaro et al. 1999) leading to induction of Ah responsive genes that result in a broad spectrum of antiestrogenic responses (Chaloupka et al., 1992; Santodonato, 1997), by blocking activation of the estrogen receptor (ER), as reported in transfected yeast cells (Tran et al., 1996) or by binding of PAHs or metabolites to the ER resulting in an increase of estrogen catabolism. This last activity has been shown to occur in some cases (Ebright et al., 1986; Arcaro et al., 1999) but not in others (Toft and Spelsberg, 1972; Keightley and Okey, 1973). However, the most common mechanism is that PAHs bind to the AhR, and upregulate CYP1A1, which enhances the metabolism of PAHs resulting in the formation of hydroxylated metabolites that bind the ER.

The antiestrogenic effects are widely reported for BaP and benzo(*a*)anthracene (Charles et al., 2000; Fertuck et al., 2001a, 2001b; Vondracek, 2002), but some authors reported also this activity for benzo(*b*)fluoranthene, benzo(*k*)fluoranthene, chrysene, and indeno(1,2,3-*cd*)pyrene *in vitro* test systems (Tran et al., 1996; Arcaro et al., 1999). It should be highlighted that *in vitro* systems are highly artificial and the extrapolation to specific *in vivo* responses is challenging.

On the other hand, it is possible to underline an estrogenic activity of PAHs and their metabolites. One of the best described modes of action is the crosstalk between activated AhR and ER (Safe and Wormke, 2003). PAHs have been shown to be active in various ER-mediated systems that measure either the displacement of 3H-Estradiol-17 beta from the ER in a whole-cell binding assay (Arcaro et al., 1999) or the induction of ER-mediated reporter activity (Charles et al., 2000; Fertuck et al., 2001a, 2001b; Tran et al., 1996; Vondracek, 2002), thus indicating the potential role for estrogenicity.

Although certain PAHs or their metabolites may bind the ER and induce transcription in reporter gene assays, it is questionable whether they induce a full estrogen response, for example, cell proliferation and tissue restructuring, by turning on estrogen-dependent genes. For example, hydroxylated BaP metabolites that were ER agonists *in vitro* were not found to affect mouse uterine weight and uterine lactoferrin mRNA expression, two estrogen-inducible endpoints (Fertuck et al., 2001b). Nevertheless PAHs impact on ER-dependent cell proliferation in target tissues *in vivo* remains controversial.

The *in vitro* estrogenic activity of BaP and benzo(*a*)anthracene was reported in the literature (van Lipzig et al., 2005; Kummer et al., 2008), but they both also behaved as estrogen-like compounds in immature rats, as reported by Kummer and others (Kummer et al., 2008). Nevertheless, the results of these studies should be interpreted with caution as the *in vivo* assay may not reflect possible effects of chronic exposure to low doses of PAHs or impact of co-exposure to other organic pollutants being present in complex environmental mixtures (Ssempewba et al., 2004; Carpenter et al., 2002). Other authors (Kortenkamp and

Altenburger, 1999) already found that for a number of estrogenic compounds, their combined effect could be explained by 'concentration addition' based on similar action of the mixture components. Their results showed that mixtures of weakly estrogenic PAH metabolites induced significant estrogenic effects. Hence, while the risk from exposure to individual weak estrogenic compounds seems negligible, combined they may affect hormone regulation in humans and wildlife. Due to the widespread and continuous emission of the endocrine disruptors into natural, domestic and occupational environments, potential health effects as increased prevalence of cryptorchidism and hypospadias, testicular, breast and prostate cancer, endometriosis, polycysticovarian syndrome and declines in semen quality in the human population seem to be linked to environmental contaminants with hormone-like activity exposure (Foster, 2008).

Other PAHs Toxic Effects

Polycyclic aromatic hydrocarbons are potent immunosuppressive environmental contaminants. Immunotoxic effects of PAHs, which may contribute to their carcinogenic potential, have been established in various animal models and in human immune cells (van Grevenynghe et al., 2004). Inhibition of the production of immunocompetent cells (lymphocytes and monocytes) is likely one of the mechanisms contributing to the immunosuppression due to PAHs. Exposure to PAHs inhibits the differentiation of monocytes into dendritic cells and macrophages; it also induces apoptosis of both pre-T cells in the thymus and pre-B cells in the bone marrow, which may account for thymic atrophy and decreased lymphoid cell recovery from the spleen, lymph nodes, and bone marrow in PAH-exposed mice (Lutz et al., 1998; Page et al., 2002). These toxic effects towards precursors of immune cells have been linked, at least in part, to the PAHs metabolism into toxic reactive intermediates triggering apoptosis of pre-B cells (Mann et al., 1999).

Mature differentiated immune cells also constitute targets for PAHs. PAH exposure therefore likely reduces the total population of macrophagic cells by triggering their apoptotic death. Moreover, PAHs inhibit generation of macrophages from monocytes (van Grevenynghe et al., 2003). Besides benzo(*a*)pyrene, other PAHs such as 7,12-dimethylbenzo(*a*)anthracene, benzo(*a*)anthracene, dibenzo(*a,h*)anthracene, benzo(*k*)fluoranthene, benzo(*b*)fluoranthene and indeno(1,2,3-*c,d*)pyrene were found to exert toxicity towards macrophages. These PAHs are widely detected in air and food and have been reported or suspected to be carcinogenic for humans according to the International Agency for Research on Cancer; by contrast, other PAHs as anthracene, fluorene, fluoranthene and pyrene, which were found to not alter macrophagic features are considered as non-carcinogenic (IARC, 1983). Taken together, these data suggest a direct link between carcinogenicity and immunotoxicity of PAHs. In addition, these environmental contaminants reduce proliferation of lymphocytes in response to mitogenic signals (Davila et al., 1996) and diminish T cell-related cytokine production and B cell-mediated Ab secretion. They also impair some functional properties of differentiated macrophages such as cellular adhesion (van Grevenynghe et al., 2003) and cytokine production (Lyte and Bick, 1986).

Interestingly, proinflammatory cytokines, such as TNF-α or interleukin-1β have been reported to act as promoting factors in cell transformation and carcinogenesis induced by PAHs both *in vitro* and *in vivo* (Krelin et al., 2007).

In addition to the carcinogenic and immunotoxic effects, PAHs exposure has also been suggested to be involved in human cardiovascular dysfunction (Gustavsson et al., 1996; Thirman et al., 1994). It has been reported that benzo(*a*)pyrene is involved in atherogenesis (Thirman et al., 1994) and induce vascular smooth cells proliferation (Zhang and Ramos, 1997). However, the direct evidence related with the association between exposure of PAHs and cardiovascular dysfunction is still weak.

Some low molecular weight PAHs such as naphthalene, fluoranthene and fluorene induced vasorelaxation in isolated rat aorta in an endothelium dependent manner (Kang and Cheng, 1997), suggesting that the PAHs with smaller ring numbers might have direct effect on vascular endothelium. Moreover, Li and others (Li et al., 2004) showed that the above mentioned PAHs could enhance nitric oxide formation in human umbilical vein endothelial cells, as they activated and up-regulated the endothelial nitric oxide synthase expression. Another effect on vascular system is the atherosclerosis. Some reports showed that PAH-induced atherosclerosis was linked to the metabolic activation of PAHs by CYPs (Majesky et al., 1983, Paigen et al., 1986).

PAHs have been shown to cross the placenta exposing the foetus to the contaminant body burden of the mother. Consequently, a gestational exposure may result in increased adverse health outcomes, possibly affecting cognitive performance (Wormley et al., 2004). Saunders and others (Saunders et al., 2002; Saunders et al., 2003) showed that exposure to BaP and fluoranthene produced a variety of behavioral deficits specific to the nervous system, including decreased motor activity, neuromuscular, physiological and autonomic abnormalities, and decreased responsiveness to sensory stimuli. Children born in years of maximal air pollution in the Czech republic had learning disorders that were attributed to elevated levels of polyaromatic hydrocarbons in the atmosphere from the mining and combustion of coal (Otto et al., 1997). Similarly, in Poland children exposed in utero to PAHs released into the environment as a result of industrial emissions and combustion of coal had impaired neurocognitive development (Jedrychowski et al., 2003). Maternal exposure to airborne PAHs during pregnancy in the inner city African-American population of New York was associated with a reduction in head circumference in children that may correlate with poorer cognitive functioning (Perera et al., 2003).

Paternal occupational exposure to polycyclic aromatic hydrocarbons was associated with an increased risk of neuroectodermal tumors in children from Italy, France, and Spain (Cordier et al., 1997). Similarly, an association between paternal exposure to creosote and diagnosed cases of neuroblastoma in children was seen (Kerr et al., 2000).

Toxic effects of PAHs in tissues such as placenta have been demonstrated to be due to their metabolites which interact with DNA.

Neurotoxicity of PAHs was also studied in adults. Plant workers in Poland employed to process coke developed neurotoxic symptoms with vegetative dysregulation and loss of short-term memory; their prevalence depended on the level of exposure to BaP (Majchrzak et al., 1990). Neurological symptoms were also reported from a community chronically exposed to BaP, benzo(*a*)anthracene, chrysene, naphthalene, fluorene and pyrene dumped at a site in Texas from the 1960s until the 1970s (Dayal et al., 1995). Similarly, residents near a combustion site in Louisiana displayed neurophysiological and neuropsychological impairments (Kilburn and Warshaw, 1995). Zhang and others showed that benzo(*a*)pyrene could affect the autonomic nervous function of coke oven workers mainly by down-regulating the parasympathetic nervous function (Zhang et al., 2008).

Conclusion

The number of chemicals to which humans are exposed has dramatically increased in the past 100 years. Although the health effects of single contaminants may be apparent under circumstances of high exposure, most people are exposed to chemical mixtures of organics and inorganics at lower concentrations. PAHs are generally found as mixtures in the environment (e.g., products of partial combustion, coal tars, environmental cigarette smoke), and the interactions among PAHs may be quite complex. The toxicity and carcinogenicity of PAH mixtures may depend on the net activities of CYP inducers (AhR agonists or antagonists), the prevalence of enzyme inhibitors versus substrates for bioactivation, and the activation or inhibition of metabolic pathways leading to nontoxic/noncarcinogenic metabolites (Carpenter et al., 2002).

The interpretation of the studies on toxicological risk from PAH exposure is complicated by several factors. The toxicity of PAH mixtures can not be accurately predicted based on chemical analysis, due to chemical interactions or to the presence of unidentified chemicals. In addition, it is practically impossible for epidemiology to disentangle the risk posed by each individual PAH. Indeed PAHs occur invariably as mixtures, whose composition depends mainly on the raw material and the combustion circumstances, and they are often adsorbed to particles in the environment. As a consequence epidemiologic data are inadequate to evaluate the possible interaction between PAHs and particles in the induction of cancer in humans. Finally, exposure to PAHs is ubiquitous and the risk associated with complex mixtures remains a challenge to predict.

In recent years, evidence has emerged for a contribution of environmental agents to a number of diseases that have not previously been considered to have environmental causes. This is especially true for some general categories of diseases linked to endocrine disruption, as well as a number of chronic diseases. In many of these situations we have some evidence for a variety of chemicals having a relationship to the disease, but currently little or no evidence of the nature and effects of interactions among chemicals related to the disease state has been demonstrated.

References

Arcaro, K. F., O'Keefe, P. W., Yang, Y., Clayton, W. & Gierthy, J. F. (1999). Antiestrogenicity of environmental polycyclic aromatic hydrocarbons in human breast cancer cells. *Toxicology, 133,* 115–127.

Banasiewicz, M., Nelson, G., Swank, A., Grubor, N., Ross, J., Nesnow, S., Kofeler, H., Small, G. J. & Jankowiak, R. (2004). Identification and quantization of benzo(*a*)pyrene-derived DNA adducts formed at low adduction level in mice lung tissue. *Analitical Biochemistry, 334,* 390–400.

Billiard, S. M., Hahn M. E., Franks, D. G., Peterson, R. E., Bols, N. C. & Hodson, P. V. (2002). Binding of polycyclic aromatic hydrocarbons (PAHs) to teleost aryl hydrocarbon receptors. *Comparative Biochemistry and Physiology Part B, 133*, 55-68.

Binková, B. & Šrám, R. J. (2004). The genotoxic effect of carcinogenic PAHs, their artificial and environmental mixtures (EOM) on human diploid fibroblasts. *Mutation Research, 547*, 109-121.

Bouchard, M. & Viau, C. (1997). Urinary excretion of benzo(*a*)pyrene metabolites following intravenous, oral, and cutaneous benzo(*a*)pyrene administration. *Canadian Journal of Physiology and Pharmacology, 75*, 185–192.

Carpenter, D. O., Arcaro, K. & Spink, D. C. (2002). Understanding the human health effects of chemical mixtures. *Environmental Health Perspectives, 110 (Suppl. 1),* 25–42.

Cavalieri, E. L. & Rogan, E. G. (1992). The approach to understanding aromatic hydrocarbon carcinogenesis. The central role of radical cations in metabolic activation. *Pharmacology & Therapeutics, 55,* 183-194.

Cavalieri, E. L. & Rogan, E. G. (1995). Central role of radical cations in metabolic activation of polycyclic aromatic hydrocarbons. *Xenobiotica, 25*, 677-688.

Cerniglia, C. (1992). Biodegradation of polycyclic aromatic hydrocarbons. *Biodegradation, 3*, 351-368.

Chaloupka, K., Krishnan, V. & Safe, S. (1992). Polynuclear aromatic hydrocarbon carcinogens as antiestrogens in MCF-7 human breast cancer cells: role of the Ah receptor. *Carcinogenesis, 13*, 2233–2239.

Charles, G. D., Bartels, M. J., Zacharewski, T. R., Gollapudi, B. B., Freshour, N. L. & Carney, E. W. (2000). Activity of benzo(*a*)pyrene and its hydroxylated metabolites in an estrogen receptor-a reporter gene assay. *Toxicological Sciences, 55*, 320–326.

Clemons, J. H., Allan, L. M., Marvin, C. H., Wu, Z., McCarry, B. E. B. D. W. & Zacharewski, T. R. (1998). Evidence of estrogen-and TCDD-like activities in crude and fractionated extracts of PM10 air particulate material using *in vitro* gene expression assays. *Environmental Science and Technology, 32*, 1853–1860.

Cordier, S., Lefeuvre, B., Filippini, G., Peris-Bonet, R., Farinotti, M., Lovicu, G. & Mandereau, L. (1997). Parental occupation, occupational exposure to solvents and polycyclic aromatic hydrocarbons and risk of childhood brain tumors (Italy, France, Spain). *Cancer Causes Control, 8*, 688–697.

Culp, S. J., Gaylor, D. W., Sheldon, W. G., Goldstein, L. S. & Beland, F. A. (1998). A comparison of the tumours induced by coal tar and benzo(*a*)pyrene in a 2-year bioassay. *Carcinogenesis, 19,* 117-124.

Davila, D. R., Romero, D. L & Burchiel, S. W. (1996). Human T cells are highly sensitive to suppression of mitogenesis by polycyclic aromatic hydrocarbons and this effect is differentially reversed by alpha-naphthoflavone. *Toxicology and Applied Pharmacology, 139,* 333–341.

Dayal, H., Gupta, S., Trieff, N., Maierson, D. & Reich, D. (1995). Symptom clusters in a community with chronic exposure to chemicals in two superfund sites. *Archives of Environmental Health, 50*, 108– 111.

de Kok, T. M. C. M. & van Maanen, J. M. S. (2000). Evaluation of fecal mutagenicity and colorectal cancer risk. *Mutation Research, 463*, 53-101.

Dean, M., Hamon, Y & Chimini, G. (2001). The human ATP-binding cassette (ABC) transporter superfamily. *Journal of Lipid Research, 42*, 1007–1017.

Denissenko, M. F., Pao, A., Tang, M. & Pfeifer, G. P. (1996). Preferential formation of benzo(*a*)pyrene adducts at lung cancer mutational hotspots in p35. *Science, 274*, 430–432.

Ebright, R. H., Wong, J. R. & Chen, L. B. (1986). Binding of 2-hydroxybenzo(*a*)pyrene to estrogen receptors in rat cytosol. *Cancer Research, 46*, 2349–2351.

EFSA (2008). Polycyclic Aromatic Hydrocarbons in Food: Scientific Opinion of the Panel on Contaminants in the Food Chain. *The EFSA Journal, 724*, 1-114.

FAO/WHO (2005). Food and Agriculture Organization of the United Nations/World Health Organization. Joint FAO/WHO Expert Committee on Food Additives (JECFA). Sixtyfourth meeting, Rome, 8-17 February 2005. Summary and Conclusions.

Fertuck, K. C., Kumar, S., Sikka, H. C., Matthews, J. B. & Zacharewski, T. R. (2001a). Interaction of PAH-related compounds with the α and β isoforms of the estrogen receptor. *Toxicology Letters, 121,* 167–177.

Fertuck, K. C., Matthews, J. B. & Zacharewski, T. R. (2001b). Hydroxylated benzo(*a*)pyrene metabolites are responsible for *in vitro* estrogen receptor-mediated gene expression induced by benzo(*a*)pyrene, but do not elicit uterotrophic effects *in vivo*. *Toxicological Sciences, 59*, 231–240.

Foster W. G. (2008). Environmental Estrogens and Endocrine Disruption: Importance of Comparative Endocrinology. *Endocrinology, 149,* 4267–4268.

Gautier, J. C., Lecoeur, S., Cosme, J., Perret, A., Urban, P., Beaune, P. & Pompon, D. (1996). Contribution of human cytochrome P450 to benzo(*a*)pyrene and benzo(*a*)pyrene-7,8-dihydrodiol metabolism, as predicted from heterologous expression in yeast. *Pharmacogenetics, 6*, 489–949.

Gozgit, J. M., Nestor, K. M., Fasco, M. J., Pentecost, B. T. & Arcaro, K. F. (2004). Differential action of polycyclic aromatic hydrocarbons on endogenous estrogen-responsive genes and on a transfected estrogen-responsive reporter in MCF-7 cells. *Toxicology and Applied Pharmacology, 196*, 58-67.

Guengerich, F. P. (2001). Common and uncommon cytochrome P450 reactions related to metabolism and chemical toxicity. *Chemical Research in Toxicology, 14*, 611–650.

Gustavsson, P., Alfredsson, L., Brunnberg, H., Hammar, N., Jakobsson, R., Reuterwall, C. & Ostlin, P. (1996). Myocardial infarction among male bus, taxi, and lorry drivers in middle Sweden. *Occupational and Environmental Medicine, 53*, 235–240.

Hahn, M. E. (2002). Aryl hydrocarbon receptors: diversity and evolution. *Chemico-Biological Interactions, 141*, 131-160.

Hall, M. & Grover, P. L. (1990). Polycyclic aromatic hydrocarbons: metabolism, activation and tumor initiation. In: Chemical Carcinogenesis and Mutagenesis Vol. 1. (Cooper, C.S. and Grover, P.L., ed.). Springer- Verlag, Heidelberg, 327-372.

Harvey, R. G. (1991). Polycyclic aromatic hydrocarbons, Chemistry and Carcinogenicity. Cambridge University Press, Cambridge, UK, 11-87.

Honey, S., O'Keefe, P., Drahushuk, A. T., Olson, J. R., Kumar, S. & Sikka, H. C. (2000). Metabolism of benzo(*a*)pyrene by duck liver microsomes. *Comparative Biochemistry and Physiology Part C, 126*, 285-292.

Honkakoski, P. & Negishi, M. (2000). Regulation of cytochrome P450 (CYP) genes by nuclear receptors. *Biochemical Journal, 347*, 321–337.

IARC (1983). Polynuclear aromatic coumpounds, part 1: chemical, environmental and experimental data, in: IARC, Monographs on the Evolution of Carcinogenic Risks to Humans, IARC Press, Lyon, 1983, 32.

IARC (1999). Metabolic polymorphisms and susceptibility to cancer. IARC Scientific Publications no 148, International Agency for Research on Cancer, World Health Organization, Lyon, 505.

JECFA (2006). Polycyclic aromatic hydrocarbons. In: Evaluation of certain food contaminants. Sixty-fourth report of the Joint FAO/WHO Expert Committee on Food Additives. WHO Technical Report Series 930. Geneva, 61-81.

Jedrychowski, W., Whyatt, R. M., Camann, D. E., Bawle, U. V., Peki, K., Spengler, J. D., Dumyahn, T. S., Penar, A. & Perera, F. F. (2003). Effect of prenatal PAH exposure on birth outcomes and neurocognitive development in a cohort of newborns in Poland. Study design and preliminary ambient data. *International Journal of Occupational Medicine and Environmental Health, 16,* 21–29.

Kang, J. J. & Cheng, Y. W. (1997). Polycyclic aromatic hydrocarbons induced vasorelaxation through induction of nitric oxide formation in endothelium of rat aorta. *Toxicology Letters, 93*, 39–45.

Keightley, D. D. & Okey, A. B. (1973). Effects of dimethylbenz(*a*)anthracene and dihydrotestosterone on estradiol-17β binding in rat mammary cytosol fraction. *Cancer Research, 33,* 2637-2642.

Kerr, M. M., Nasca, P. C., Mundt, K., Michalec, A. M., Baptiste, M. S. & Mahoney, M. C. (2000). Parenteral occupational exposures and risk of neuroblastomas: a case control study (United States). *Cancer Causes Control, 11*, 635– 664.

Kilburn, K. H. & Warshaw, R.H. (1995). Neurotoxic effects from residential exposure to chemicals from an oil reprocessing facility and superfund site. *Neurotoxicology and Teratology, 17*, 89–102.

Kim, J. H., Stansbury, K. H., Walker, N. J., Trush, M. A., Strickland, P. T. & Sutter, T. R. (1998). Metabolism of benzo(*a*)pyrene and benzo(*a*)pyrene-7,8-diol by human cytochrome P450 1B1. *Carcinogenesis, 19,* 1847–1853.

Kobliakov, V., (1998). Inducers of the cytochrome P-450 superfamily as promoters of carcinogenesis. *Biochemistry (Moscow), 63*, 1043–1058.

Kortenkamp, A. & Altenburger, R. (1999). Approaches to assessing combination effects of oestrogenic environmental pollutants. *Science of the Total Environment, 233*, 131–140.

Krelin, Y., Voronov, E., Dotan, S., Elkabets, M., Reich, E., Fogel, M., Huszar, M., Iwakura, Y., Segal, S., Dinarello, C. A. & Apte, R. N. (2007). Interleukin- 1beta-driven inflammation promotes the development and invasiveness of chemical carcinogen-induced tumors. *Cancer Research, 67*, 1062–1071.

Kriek, E., Rojas, M., Alexandrov, K. & Bartsch, H. (1998). Polycyclic aromatic hydrocarbon-DNA adducts in humans: relevance as biomarkers for exposure and cancer risk. *Mutation Research, 400*, 215–231.

Kummer, V., Masková, J., Zraly, Z., Neca, J., Simecková, P., Vondráceka, J. & Machala, M. (2008). Estrogenic activity of environmental polycyclic aromatic hydrocarbons in uterus of immature Wistar rats. *Toxicology Letters, 180,* 212–221.

Lau, H. H. & Baird, W. M. (1994). Separation and characterization of post-labeled DNA adducts of stereoisomers of benzo(*a*)pyrene-7,8-diol-9,10-epoxide by immobilized boronate chromatography and HPLC analysis. *Carcinogenesis, 15*, 907–915.

Li, C-H, Lee, C-C, Cheng, Y-W, Juang, H-A & Kanga, J-J (2004). Activation and up-regulation of nitric oxide synthase in human umbilical vein endothelial cells by polycyclic aromatic hydrocarbons. *Toxicology Letters, 151,* 367–374.

Lutz, C. T., Browne, G. & Petzold, C. R. (1998). Methylcholanthrene causes increased thymocyte apoptosis. *Toxicology, 128*, 151–167.

Lyte, M. & Bick, P. H. (1986). Modulation of interleukin-1 production by macrophages following benzo(*a*)pyrene exposure. *International Journal of Immunopharmacology, 8*, 377–381

Majchrzak, R., Sroczynski, J. & Chelmecka, E. (1990). Evaluation of the nervous system in workers in the furnace and coal divisions of the coke-producing plants. *Medycyna Pracy, 41*, 108– 113.

Majesky, M. W., Yang, H. Y., Benditt, E. P. & Juchau, M. R. (1983). Carcinogenesis and atherogenesis: differences in monooxygenase inducibility and bioactivation of benzo(*a*)pyrene in aortic and hepatic tissues of atherosclerosis-susceptible versus resistant pigeons. *Carcinogenesis, 4*, 647–652.

Mann, K. K., Matulka, R. A., Hahn, M. E., Trombino, A. F., Lawrence, B. P., Kerkvliet, N. I. & Sherr, D. H. (1999). The role of polycyclic aromatic hydrocarbon metabolism in dimethylbenz[a]anthracene induced pre-B lymphocyte apoptosis. *Toxicology and Applied Pharmacology, 161,* 10–22.

Miranda, C. L., Chung, W. G., Wang-Buhler, J.-L., Musafia-Jeknic T., Baird, W. M. & Buhler D. R. (2006). Comparative *in vitro* metabolism of benzo[a]pyrene by recombinant zebrafish CYP1A and liver microsomes from β-naphthoflavone-treated rainbow trout. *Aquatic Toxicology, 80*, 101-108.

Otto, D., Skalik, I., Bahboli, R., Hudnell, K. & Sram, R. (1997). Neurobehavioral performance of Czech school children born in years of maximal air pollution. *Neurotoxicology, 18,* 903.

Page, T. J., O'Brien, S., Jefcoate, C. R. & Czuprynski, C. J. (2002). 7,12-Dimethylbenz(*a*)anthracene induces apoptosis in murine pre-B cells through a caspase-8-dependent pathway. *Molecular Pharmacology*, *62*, 313–319.

Paigen, B., Holmes, P. A., Morrow, A. & Mitchell, D. (1986). Effect of 3-methylcholanthrene on atherosclerosis in two congenic strains of mice with different susceptibilities to methylcholanthrene-induced tumors, *Cancer Research, 46*, 3321–3324.

Penning, T. M., Ohnishi, S. T., Ohnishi, T. & Harvey, R. G. (1996). Generation of reactive oxygen species during the enzymatic oxidation of polycyclic aromatic hydrocarbon dihydrodiols catalyzed by dihydrodiol dehydrogenase. *Chemical Research in Toxicology, 9,* 84-92.

Perera, F. P., Rauh, V., Tsai, W.-Y., Kinney, P., Camann, D., Barr, D., Bernert, T., Garfinkel, R., Tu, Y.-H., Diaz, D., Dietrich, J. & Whyatt, R. (2003). Effects of transplacental exposure to environmental pollutants on birth outcomes in a multiethnic population. *Environmental Health Perspective, 111,* 201– 205.

Rakitsky, V., Kobliakov, V., Turusov, V. (2002). Nongenotoxic (epigenetic) carcinogens: pesticides as an example. A critical review. *Teratogenesis, Carcinogenesis and Mutagenesis, 20,* 229–240.

Ramesh, A., Inyang, F., Hood, D. B. & Knuckles, M. E. (2000). Aryl hydrocarbon hydroxylase activity in F-344 rats subchronically exposed to benzo(*a*)pyrene and fluoranthene through diet. *Journal of Biochemistry and Molecular Toxicology, 14*, 155-161.

Ramesh, A., Inyang, F., Hood, D.B., Archibong, A. E., Knuckles, M. E. & Nyanda, A. M. (2001). Metabolism, bioavailability, and toxicokinetics of benzo[*a*]pyrene [B(a)P] in F-344 rats following oral administration. *Experimental and Toxicologic Pathology, 53*, 253–270.

Ramesh, A., Walker, S. A., Hood, D. B., Guillén, M. D., Schneider, K. & Weyand, E. H. (2004). Bioavailability and risk assessment of orally ingested polycyclic aromatic hydrocarbons. *International Journal of Toxicology, 23*, 301-333.

Ropero, A. B., Alonso-Magdalena, P., Ripoll, C., Fuentes, E. & Nadal, A. (2006). Rapid endocrine disruption: Environmental estrogen actions triggered outside the nucleus. *Journal of Steroid Biochemistry & Molecular Biology*, *102*, 163–169.

Safe, S. & Wormke, M. (2003). Inhibitory aryl hydrocarbon receptor-estrogen receptor a cross-talk and mechanisms of action. *Chemical Research in Toxicology, 16*, 807– 816.

Safe, S. (1995). Modulation of gene expression and endocrine response pathways by 3,4,7,8-tetrachloro-*p*-dibenzodioxin and related compounds. *Pharmacology & Therapeutics, 67*, 247–281.

Santodonato, J. (1997). Review of the estrogenic and antiestrogenic activity of polycyclic aromatic hydrocarbons: relationship to carcinogenicity. *Chemosphere, 34*, 835-848.

Saunders, C. R., Ramesh, A. & Shockley, D. C. (2002). Modulation of neurotoxic behavior in F-344 rats by temporal disposition of benzo(*a*)pyrene. *Toxicology Letters, 129*, 33– 45.

Saunders, C. R., Shockley, D. C. & Knuckles, M. E. (2003). Fluoranthene-induced neurobehavioral toxicity in F-344 rats. *International Journal of Toxicology, 22*, 263–276.

SCF (2002). Opinion of the Scientific Committee on Food on the risks to human health of polycyclic aromatic hydrocarbons in food. SCF/CS/CNTM/PAH/29 Final. Brussels, Belgium, 1-84.

Shimada, T., Gillam, E. M., Oda, Y., Tsumura, F., Sutter, T. R., Guengerich, F. P. & Inoue, K. (1999). Metabolism of benzo(*a*)pyrene to *trans*-7,8-dihydroxy-7,8-dihydrobenzo(*a*)pyrene by recombinant human cytochrome P450 1B1 and purified liver epoxide hydrolase. *Chemical Research in Toxicology, 12*, 623–629.

Shou, M., Korzekwa, K. R., Crespi, C. L., Gonzalez, F. J. & Gelboin, H. V. (1994). The role of 12 cDNA-expressed human, rodent, and rabbit cytochromes P450 in the metabolism of benzo[*a*]pyrene and benzo[*a*]pyrene *trans*-7,8-dihydrodiol. *Molecular Carcinogenesis, 10*, 159–168.

Ssempewba, J., Carpenter, D., Yilmaz, B., DeCaprio, A., O'Hehir, D. & Arcaro, K. (2004). Waste crankcase oil: an environmental contaminant with potential to modulate estrogenic responses. *Journal of Toxicology and Environmental Health, Part A, 67,* 1081–1094.

Thilly, W.G. (2003). Have environmental mutagens caused oncomutations in people? *Nature Genetics,* 255–259.

Thirman, M. J., Albrecht, J. H., Krueger, M. A., Erickson, R. R., Cherwitz, D. L., Park, S. S., Gelboin, H. V. & Holtzman, J. L. (1994). Induction of cytochrome CYP1A1 and formation of toxic metabolites of benzo[a]pyrene by rat aorta: a possible role in atherogenesis. *Proceedings of the National Academy of Science, USA, 91*, 5397–5401.

Toft, D. O. & Spelsberg, T. C. (1972). A distinction between 3-methylcholanthrene and estrogen binding in the uterus. *Cancer Research, 32,* 2743-2746.

Tran, D. Q., Ide, C. F., McLachlan, J. A. & Arnold, S. F. (1996). The anti-estrogenic activity of selected polynuclear aromatic hydrocarbons in yeast expressing human estrogen receptor. *Biochemical and Biophysical Research Communications, 229*, 102–108.

Trosko, J. E. & Upham, B. (2005). The emperor wears no clothes in the field of carcinogen risk assessment: ignored concepts in cancer risk assessment. *Mutagenesis 20,* 81–92.

van Grevenynghe, J., Rion, S., Le Ferrec, E., Le Vee, M., Amiot, L., Fauchet, R. & Fardel, O. (2003). Polycyclic aromatic hydrocarbons inhibit differentiation of human monocytes into macrophages, *The Journal of Immunology, 170*, 2374–2381.

van Grevenynghe, J., Sparfel, L., Le Vee, M., Gilot, D., Drenou, B., Fauchet, R. & Fardel, O. (2004). Cytochrome P450-dependent toxicity of environmental polycyclic aromatic hydrocarbons towards human macrophages. *Biochemical and Biophysical Research Communications, 317,* 708-716.

van Lipzig, M. M. H., Vermeulen, N. P. E., Gusinu, R., Legler, J., Frank, H., Seidel, A. & Meerman, J. H. N. (2005). Formation of estrogenic metabolites of benzo(*a*)pyrene and chrysene by cytochrome P450 activity and their combined and supra-maximal estrogenic activity. *Environmental Toxicology and Pharmacology, 19,* 41–55.

Varanasi, H., Nishimoto, M., Reichert, W. L. & Eberhalt, B. (1986). Comparative metabolism of benzo(*a*)pyrene and covalent binding to hepatic DNA in English sole, Starry Flounder, and rat. *Cancer Research, 46*, 3817-3824.

Viau, C., Bouchard, M., Carrier, G., Brunet, R. & Krishnan, K. (1999). The toxicokinetics of pyrene and its metabolites in rats. *Toxicology Letters, 108*, 201-207.

Vondracek, J. (2002). Modulation of estrogen receptor-dependent reporter construct activation and G0/G1-S-phase transition by polycyclic aromatic hydrocarbons in human breast carcinoma MCF-7 cells. *Toxicological Sciences, 70,* 193– 201.

Whitlock, J. P. (1999). Induction of cytochrome P4501A1, *Annual Review of Pharmacology and Toxicology, 39*, 103-125.

WHO/IPCS (1998). World Health Organization/International Programme on Chemical Safety. Selected Non-heterocyclic Polycyclic Aromatic Hydrocarbons. Environmental Health Criteria 202. International Programme on Chemical Safety, World Health Organization, Geneva.

Wormley, D. D., Ramesh, A & Hood, D. B. (2004). Environmental contaminant–mixture effects on CNS development, plasticity, and behaviour. *Toxicology and Applied Pharmacology, 197,* 49-65.

Xue, W. & Warschawsky, D. (2005). Metabolic activation of polycyclic and heterocyclic aromatic hydrocarbons and DNA damage: A review. *Toxicology and Applied Pharmacology, 206*, 73-93.

Yuan, Z.-X., Honey, S. A., Kumar, S. & Sikka, H. C. (1999). Comparative metabolism of dibenzo(*a,l*)pyrene by liver microsomes from rainbow trout and rat. *Aquatic Toxicology, 45*, 1-8.

Zhang, H.-M., Nie, J.-S., Wang, F., Shi, I.-T., Zhang, L., Antonucci, A., Liu, H.-J., Wang, J., Zhao, J., Zhang, Q.-L., Wang, L.-P., Song, J., Xue, C.-E, Di Gioacchino, M. & Niu, Q. (2008). Effects of benzo(*a*)pyrene on autonomic nervous system of coke oven workers. *Journal of Occupational Health, 50*, 308-316.

Zhang, Y. & Ramos, K. S. (1997). The induction of proliferation vascular smooth muscle cell phenotypes by benzo(*a*)pyrene does not involved mutational activation of ras genes. *Mutation Research, 373*, 285–292.

In: Polycyclic Aromatic Hydrocarbons: Pollution, Health … ISBN 978-1-60741-462-9
Editors: Pierre A. Haines and Milton D. Hendrickson

Chapter 3

POLYCYCLIC AROMATIC HYDROCARBONS AND CARCINOGENIC RISK ASSESSMENT IN DEVELOPING COUNTRIES

Mahmoud A. Hassanien
Air Pollution Department, National Research Center, Dokki, Cairo, Egypt

INTRODUCTION

Polycyclic aromatic hydrocarbons (PAHs) are a group of organic compounds composed of fused aromatic rings. Some high molecular weight PAHs are considered to be carcinogenic, mutagenic, teratogenic even at low concentrations and have received more environmental concern. PAHs are resistant to biodegradation and can bioaccumulate in the environment due to their hydrophobic nature and low solubility. PAHs are ubiquitous environmental contaminants found in air, soil and water. The main sources of PAHs in the environment and for general population exposure are natural and intentional combustion of organic matter and petroleum manipulation. Atmospheric emissions will arise from the processing of coal and petroleum products and incomplete combustion during industrial processes.

In the developing countries, present information concerning the sources and inputs of PAHs in these countries is rather limited. PAHs concentrations have been reported in a number of urban areas as well as in remote places and are consistently higher in winter than in summer, and in urban than in rural areas. The higher values are found in particular places like stack gases, tunnels of motorways, open burning sites, etc. There are no ambient air quality standards and no specific guideline value can be recommended for PAHs in air. PAHs in air are attached to particles, which may also play a role in their carcinogenicity. Although food is thought to be the major source of human exposure to PAHs, part of this contamination may arise from air pollution with PAHs.

Exposure to PAHs mainly occurs from atmospheric pollution through inhalation, ingestion and dermal contact. PAHs half-lives in humans are in fact in the range of days/hours, and metabolism is responsible of the formation of carcinogenic metabolites. Risk assessments and potency assessments of various individual and complex mixtures of PAHs have been attempted. BaP is the only PAH for which a database is available, allowing a quantitative risk assessment. PAH potencies are used to determine quantitative health risks posed by PAH exposure. The risks posed by a mixture of PAHs are based on an assumption of additively of the individual risks posed by the PAHs.

The objectives of this chapter are to: 1) report PAH concentrations in the environmental media in developing countries; 2) collate published and unpublished PAH information from environmental agencies and write a report on the state of knowledge of PAHs; 3) characterize the priority PAHs, carcinogenic risk assessment and potential health impacts.

Definition and Sources

Polycyclic Aromatic Hydrocarbons (PAHs) are a ubiquitous environmental contaminants found in air, soil, and water. Some of which are harmful to human and animals and considered environmental contaminants. PAHs have been shown to be persistent in various media in nature at significant levels or at levels to which exposure to human and animal life over a continued period is toxic. A lot of PAHs are recalcitrant and thus their continued presence in the nature is a constant source of risk to human, animal and plant health, where they could be carcinogenic, mutagenic or induce gentotoxic effects in cells of plants and animals. PAHs are a group of organic compounds composed of fused aromatic rings. Some high molecular weight (greater than four fused rings, HMW) PAHs are considered to be carcinogenic, mutagenic and teratogenic even at low concentrations (IARC, 1983, Ciganek et al., 2002). Because of their hydrophobic nature and low solubility, PAHs are resistant to biodegradation and can bioaccumulate in the environment through the food chain. Therefore, PAHs in the environment represent a long-term threat to human health and have received more and more environmental concern. PAHs are complex mixtures of hundreds of chemicals, including derivatives of PAHs, such as nitro-PAHs and oxygenated products, and also heterocyclic PAHs. The biological properties of the majority of these compounds are as yet unknown

There are natural and anthropogenic sources of PAHs. Natural PAHs are formed as side-products of natural and anthropogenic carbon combustion. During combustion there is the thermal alteration of organic compounds in the organic material being combusted, this alteration produces PAHs and the type of PAH produced depends on the pyrolysis temperature. Natural sources of PAHs include volcanoes, the earths' geologic permutations, and crude oil (Zander, 1980, Baek *et al.,* 1991, Juhasz et al., 2000). Other natural sources of PAHs include PAH formation during thermal and geologic production of materials, emissions from trees, burning of vegetation in forest or bush fires and biogenic precursions during early digenesis (Cerniglia 1980). Anthropogenic sources of PAHs include incomplete combustion of fuels such as wood, petroleum, oil and gas. Approximately 90% of PAHs emissions are estimated to be anthropogenic sources typically including the combustion of fossil fuel, industrial processes and domestic heating systems. Vehicular traffic is considered to be the major contributor to the urban atmosphere (Fang et al., 2004; Yang and Chen, 2004). Anthropogenic sources of PAHs account for significant amount of PAHs present in nature today, "concentration of PAH found in soils around urban areas are sometimes up to two orders of magnitude higher that those in less-developed areas (derived from forest fires and air pollution)" (Dabestani and Ivanov, 1999). Vehicle emissions alone are believed to be responsible for 35% of the total PAH emission in highly populated and industrialized urban areas of the United States (Bjoerseth et al. 1985). Anthropogenic sources are rapidly increasing the amount of PAHs in the environment as PAH accumulation rate far exceeds the degradation rate of PAHs. Industrialization of developing countries and continued industrial expansion of developed countries is projected to increase the PAH production as projections on PAH production are forecasted to increase in the coming decades (Han and Naeher, 2006). Other sources include grilling meat, incineration of wastes, and use of industrial power generators. Industrial processes can produce PAHs as can the use of chemical solvents and the various incineration, or high thermal processes. Chemical processes such as the production of coal tar, petroleum, fractional distillation of crude oil, creosote production (Harrad et al., 2006; Lin et al., 2008) and various industrial processes are significant point sources for PAHs. Most of PAHs are released into the environment from anthropogenic sources such as combustion of fossil fuels in power plants, domestic heating, waste incineration, industrial processes and, most importantly, motor vehicle exhaust (Lim et al., 1999; Marr et al., 1999; Lehndorff and Schwark, 2004; Chang et al., 2006). Additional contributions to ambient air levels arise from tobacco smoking, while the use of invented PAHs can be found in various compartments of the environment: air, surface water, sediment, soil, food and in lipid tissues of both aquatic and terrestrial organisms' sources can increase. Because of such widespread sources, PAHs are present almost everywhere (Zhang et al., 2008; Lin et al., 2008).

In 1992, the National Academy of Sciences identified PAHs as among the highest priorities for exposure research, in part because these compounds are frequently constituents of fine aerosol and some of them are mutagens and probable human carcinogens. Determining exposure to PAH is still a new area of research. It is still largely unknown in what ways they are distributed among the vapor and particulate phases in air or the aqueous and nonaqueous phases in water. Likewise, their distributions and levels in other media, such as food or soil, are largely unknown (EPA, 1997). PAH carcinogenicity and toxicity is induced during the metabolism of PAHs by living cells (Cerniglia 1984), for this reason knowledge of PAH metabolism or degradation is very important to help us assess correctly the risk involved in exposure to PAHs. Exposure to PAHs is nature is always to multiple PAHs and rarely to an individual PAH. A better assessment of the risk involved in PAH metabolism would involve understanding the degradation of PAHs in mixtures as it occurs in nature, the complex substrate interactions and its effects on the magnitude of risk assessed.

DESCRIPTION OF PAHS

PAHs are organic chemicals containing two or more fused aromatic (benzene) rings (Kanaly et al. 2000; Juhasz et al. 2000). They range from the simple two (2) fused benzene-ring, naphthalene, to complex hydrocarbons such as the ten (7) fused ring, coronene. Parent compound PAHs are fused benzene rings without any chemical moiety or group added to them. They are the basic skeletal structure of PAHs and serve as the template for modified PAHs. PAHs usually exist in the modified form different from the parent compound through the addition of different chemical groups to the parent compound such as the alkyl, nitro and sulfur groups. These additions can significantly change the character of the parent compound they can confer previously unknown properties to the parent compound such as increased or decreased solubility, generally affecting properties already present in the parent compound such as hydrophobicity. Addition of these groups considerably modify action and effects due changes in the physical, stearic and chemical properties of the PAH, in effect changing the metabolism and actions of the parent compound.

Table 1. Classification of polycyclic aromatic hydrocarbons.

Low Molecular Weight; LMW (PAHs). *Two or three rings*	Middle Molecular Weight MMW (PAHs). *Four or five rings*	High Molecular Weight, HMW (PAHs). *Six or seven rings*
Naphthlene , Acenaphtylene, Acenaphthene, Fluorene, Phenanthrene, Anthracene.	Fluoranthene , Pyrene, Benzo(a)anthracene [a], Chrysene [a],	Benzo(b)fluoranthene[a], Benzo(k)fluoranthene , Benzo(e)pyrene, Benzo(a)pyrene[a], Perylene, Indeno(1,2,3-cd)pyrene[a], Dibenz(a,h)anthrathene[a], Benzo(b)chrycene, Benzo(ghi)perylene[a], Coronene.

[a]Carcinogenic polycyclic aromatic hydrocarbon compounds
(IARC, 1983 and Kipopoulou et al., 1999, Hassanien, 2006)

CHEMICAL AND PHYSICAL CHARACTERISTICS

The fused benzene rings of PAHs could be in linear, angular, or clustered arrangements (Kanaly et al. 2000). PAHs with less than three benzene rings are termed low molecular weight PAHs (LMW PAHs) while those with more than three benzene rings are called high molecular weight PAHs (HMW PAHs) as shown in Table 1. The chemical properties, and hence environmental fate of an individual PAH, is dependent in part upon its molecular size i.e. number of aromatic rings, this determines its solubility, octanol water partitioning co-efficient (Kow), configuration and electrochemical stability (Dabestani and Ivanov, 1999, Kanaly et al. 2000, Juhasz et al. 2000).

This general grouping of PAHs as HMW and LMW PAHs becomes useful because each group member share similar characteristics, though with slight variations. The physico-chemical and environmental properties interactions of parent compounds of PAHs in both groups are characteristic of the groups, in the low molecular PAH group behavior within similar range includes, low boiling and melting points, high water solubility which translates into them being easily degradable, similar octanol-water partition co-efficient, similar fugacity values, vapor pressure and other properties. In the high molecular weight PAH group physico-chemical and environmental properties apart from their molecular weights are also shared such as high boiling and melting points, low solubility in water, low octanol-water partition co-efficient, not being easily degradable, higher toxicity to plants and animals, similar fugacity, vapor pressure and other physical properties. It is noticed that this grouping might not hold when parent compounds are substituted as these functional groups greatly alter the properties of the parent compound.

Generally, PAHs are hydrophobic compounds measured directly by their aqueous solubility in water (Bjoerseth 1985). PAH hydrophobicity increases with an increase in the number of fused benzene rings. PAH hydrophobicity is a major factor contributing to the persistence of PAHs in the different polar media in the environment and in lipids (Cerniglia 1984; Dabestani and Ivanov 1999; Kanaly et al. 2000; Juhasz et al 2000). The increase in size of PAHs with increasing benzene rings along with increasing angularity confers greater electrochemical stability to PAHs making them be more stable in nature (Kanaly et al. 2000).

The Octanol-water partition constant (K_{ow}) is the ratio of the concentration of a substance in octanol to that in water at equilibrium at a specified temperature (Asante-Duah 2002), (K_{ow}) is used to predict the fate of PAHs either in the environment or in biological tissues. (K_{ow}) increases with increasing benzene rings, from LMW PAHs to HMW PAHs. The (K_{ow}) is used to determine the phase which the PAH shall partition to, this ability is very useful in human exposure analysis to PAHs (Asante-Duah 2002). Measurements of (K_{ow}) of individual PAHs correlate perfectly with measurements of solubility and hydrophobicity, with increasing hydrophobicity is low solubility and an increasing Octanol-Water partition coefficient.

Sources of PAHs in Developing Countries

The main sources of PAHs in the environment are natural and intentional combustion of organic matter (e.g. forest fires and fuel combustion) and petroleum manipulation (extraction, transport and refining). The contributions, therefore, range from diffused chronic inputs (e.g. atmospheric deposition and terrestrial run-off) to large point releases (i.e. tanker accidents or accidents at coastal installations). Present information concerning the sources and inputs of PAHs in these countries is rather limited, but the Mediterranean Sea and its region has been considered to be relatively more polluted by oil than any other sea from which data are available (UNEP, 2002). Sea transportation appears to be one of the main sources of PAHs pollution into the Mediterranean. The land-based discharges are both industrial and urban (Viana et al., 2005). The number of major refineries located in the Mediterranean region is more than 40 with a combined capacity, in 2000, of about 438 million tones. An additional source of PAHs is the result of the sewage sludge management strategies. Moreover, accediental sources in the Mediterranean region showed that over the last decades, about 55000 tonnes of oil have been spilled at the Mediterranean Sea with 3 major accidents amounting to 75% of the quantity spilled. 82 accidents involving oil spills were recorded during the period Jan. 1990 to Dec. 1999 while the quantity of spilt oil was 22150 tonnes (UNEP, 2002).

Table 2. Estimated PAH inputs from diverse sources, assuming two Different PAH contents in the case of oil discharges (UNEP, 2002)

Type	Annual oil input rate (tones)	PAH content (g/tonne)	PAH input (tonne)
Shipping discharges	25000 (low estimation)	1.28	0.32
		2000	500
	500000 (high estimation)	1.28	0.64
		2000	1000
Accidents	3667	1.28	0.005
		2000	7.3
Refinery effluents	900	1.28	0.001
		2000	1.8
Urban and river discharges			30-40
Sludge sea dumping			0.2
Sludge applications in soils			1.2
Sludge landfilling			1.2
Atmospheric			48
Sludge incineration			0.2

The estimates of the relative importance of the various sources of PAHs in the region are summarised in Table 2. Atmospheric emissions appear to have a relatively high importance while the actual PAHs input from shipping activities remains uncertain. Another source, difficult to estimate, is the accidental fires which are rather current in the region.

Distribution of PAHs

Effective transport facilitates PAHs through different media. For example winds can aid in the transference of airborne PAHs from the foci point to regions thousand of miles away from the focal point depending on the wind current and strength, this is a common occurrence as seen in areas with high smoke stacks (Smith and Frankeberg 1975). Also flowing polluted streams might be vectors of pollution as they can transport the pollutants from its source to other water bodies (within medium) or could cause deposition in soil and sediments far off (intermedia) (Chiou et al. 1998), also atmospheric deposition of PAHs is a major source of soil pollution (Chiou et al. 1998; Grimmer 1983), as volatilization of LMW PAHs from the soil could be a source of atmospheric PAH. There are a lot of computational and mathematical models that effectively model transport of PAHs in a medium and through media.

Phases of PAHs and their Fate in the Environment

Historically, small quantities of PAHs have been found in samples of atmospheric smoke collected in both urban and rural parts of Wales, Cheshire and Lancashire (UK). The results showed that the concentration of hydrocarbons were much lower in rural than in urban areas (Commins, 1958). In the atmosphere, PAHs display a wide range of gas-particle partitioning characteristics. Vapor-particle partitioning in the atmosphere can be approximated from a compound's subcooled liquid vapor pressure (PL) and/or its octanole air partition coefficient (KOA). Compounds with PL greater than ca. 0.01 Pa (log KOA < 10) are vapor phase constituents; compounds with PL less than ca. 0.000001 Pa (log KOA > 12) are almost entirely particle associated, and compounds with PL (log KOA) between these values can be considered as mixed phase atmospheric constituents (Lohmann et al., 2001). PAHs with two or three rings, which have relatively low log- KOA values, are mainly associated with the vapor phase, 4-ring PAHs occur in both the gas and particle phase, whereas PAHs with five rings or more, which have high log KOA values, are mostly bound to particles. However, a higher temperature probably increases the gas partitioning of the PAHs with four rings or more (Soderstrom et al., 2005). Some high molecular weight (more than four fused rings) PAHs are considered to be carcinogenic, mutagenic and teratogenic even at low concentrations (IARC, 1983; Bostrom et al., 2002; Yu et al., 2006). However, lighter PAHs are most abundant in the urban atmosphere and react with other pollutants to form more toxic derivatives (Chang et al., 2006). Because of their hydrophobic nature and low solubility, PAHs are more resistant to biodegradation and can bioaccumulate in the environment through the food web (Ohura et al., 2004a; Conde et al., 2005; Motelay-Massei et al., 2005; Bartkow et al., 2006; Cupr et al., 2006; Huckins et al., 2006; de Souza Pereira et al., 2007; Zhu et al., 2008). Therefore, PAHs in the environment represent a long-term threat to human health and have received more and more environmental concern. Regulation of PAH emission and reliable monitoring of PAH concentration in ambient air is thus of paramount importance for public health (Lehndorff and Schwark, 2004).

Levels of PAHs in the Environmental Media

Atmospheric Air

No specific guideline value can be recommended for PAHs as such in air. These compounds are typically constituents of complex mixtures. Some PAHs are also potent carcinogens, which may interact with a number of other compounds. In addition, PAHs in air are attached to particles, which may also play a role in their carcinogenicity. Although food is thought to be the major source of human exposure to PAHs, part of this contamination may arise from air pollution with PAHs. The levels of PAHs in air should therefore be kept as low as possible. In view of the difficulties in dealing with guidelines for PAH mixtures, the advantages and disadvantages of using a single indicator carcinogen to represent the carcinogenic potential of a fraction of PAH in air were considered. Evaluation of, for example, BaP alone will probably underestimate the carcinogenic potential of airborne PAH mixtures, since co-occurring substances are also carcinogenic. Nevertheless, the well-studied common constituent of PAH mixtures, BaP, was chosen as an indicator, although the limitations and uncertainties in such an approach were recognized

The determination of atmospheric PAHs is hindered by many problems, the most common derived from sampling. Therefore, it is difficult to integrate the existing information into a coherent data set. PAHs concentrations have been reported in dry deposition in a number of urban areas as well as in remote places, including the open Mediterranean Sea and developing countries Table 3. Despite the variety of components analysed, some seasonal and spatial trends are apparent. Concentrations are consistently higher in winter than in summer, and in urban than in rural areas. The higher values are found in particular places like stack gases, tunnels of motorways, open burning sites, etc.

To compare between PAHs concentration in developed countries with other developing ones Table 4 showing data of estimated indoor and outdoor air concentrations for PAHs (ng/m3) at each concerned area using sampling rates of 2.5 m^3/day (indoor) and 3.5 m^3/day (outdoor) (Bohlin et al., 2008). Using sampling rates suggested in the literature, the sums of 13 PAHs in the different sites were estimated to be 6.1–180 ng/m^3, with phenanthrene as the predominant compound. Indoor PAH levels tended to be higher in Gothenburg and outdoor levels higher in Mexico City.

Table 3. Concentration of PAHs in aerosol samples from different cities*.

Location	Sampling period	PAHs	Conc. (ng/m^3)
Algeria			
Algiers	*Wint./sum. 1999*	*∑18 PAHs*	*59.9/13.7*
		BaP	*0.4-2.3*
Oued Smar landfill	*Wint./sum. 1999*	*∑18 PAHs*	*136.5/24.9*
		BaP	*0.3-3.9*
Egypt			
Cairo(urban area)	*Wint/sum*	*∑18 PAHs*	*14.8/7.5*
		BaP	*0.1-0.2*
Alexandria	*Summer 97-98*	*∑18 PAHs*	*32*
		BaP	*0.95*
Greece			
Thessaloniki	*1987-1988*	*∑9 PAHs*	*22-55*
		BaP	*2-6*
North Greece town	*wint/sum 96-97*	*∑13 PAHs*	*5-60/0.4-8.1*
Crete	*2000-2001*	*∑10 PAHs*	*11.3(V), 0.8(P)*
inokalia		*BaP*	*0.003(V),0.025(P)*
Italy			
Rome	*1993-1998*	*∑7 PAHs*	*5.96-7.33*
		BaP	*1.2*
Bolgan	*1991-1993*	*BaP*	*0-174/(av.=22.7)*
Naples	*1996-1997*	*PAHs*	*2-130*
		BaP	*0.31-3.8*
Genoa	*1997-1999*	*∑4 PAHs*	*0.8-19.1*
(traffic site)	*1994-1999*	*BaP*	*1.1-2.0*
Syria			
Damascus	*1999-2000*	*∑18 PAHs*	*14.8-45.7*
		BaP	*0.9-2.7*
FR Yugoslavia			
Belgrade	*2000*	*BaP*	*0.2-0.6*

*(modified from UNEP, 2002)

Phenanthrene was the predominant PAH compound and the 3–4 ring PAHs which are mainly in the gas phase contributed to >96% of ∑13PAH. The indoor levels in Mexico City ranged from 6.1 to 92 ng/m^3, with median levels being 28 ng/m^3 at the urban area, and 25 ng/m^3 at the semi-rural area. The indoor levels of PAHs tended to be higher in Gothenburg (median 37 ng/m^3) than in Mexico City and Lancaster (median 28 ng/m^3).

The estimated levels of indoor air in Kuwait, both in terms of the composition profile and ∑PAHs (1.3–16 ng/m^3). The levels herein are also in agreement with estimated indoor air levels of PAHs from an area in Sweden where domestic wood burning is widespread (20–250 ng/m^3) (Strandberg et al., 2006). The outdoor levels in Mexico City ranged from 32 to 92 ng/m^3. Highest median levels (82 ng/m^3) were found in the semi-rural area where both point sources of PAHs (i.e. main roads, backyard burning and forest fires), and wind drift from the urban plume occur. As expected, outdoor levels tended to be lower in Gothenburg and Lancaster compared to Mexico City (Coleman et al., 1997; Motelay-Massei et al., 2005).

Table 4. Indoor and outdoor concentrations of PAHs in different areas

			$\sum_{13}$ PAHs
	Range	Mean	Median
Mexico City urban			
In	6.1-92	34	28
Out	32-64	47	44
Mexico City semi-rural			
In	12-37	25	25
Out	72-92	82	82
Gottenburg			
In	14-180	69	37
Out	7.7-68	37	33
Lancaster			
In	8.5-60	30	28
Out	6.8	6.8	6.8

The levels of PAHs released into the environment from various human activities especially from industrial sources in the Niger Delta area of Nigeria shows that the most polluted medium is the air followed by the soil and water media. Comparing the study locations there is an indication that Eleme (the most industrialized) as compared to Ahoada East (one of the least industrialized) recorded PAH levels 7- million fold higher in the air. Levels of PAHs in Eleme were 3-fold higher in surface water and 7 -fold higher in soil than the levels found in Ahoada East as clearly shown in Fig. 1.

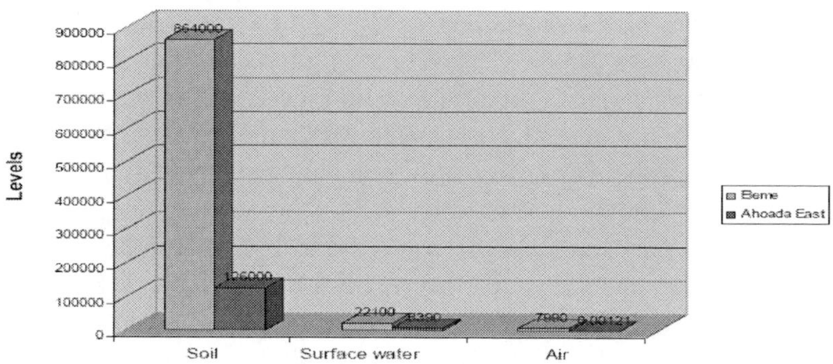

Figure 1. Environmental levels of PAH in the Niger Delta Area.

PAHs occurrence and situation in atmosphere of Egypt is well shown in the following section which represent a case study of developing country. Since the 1960s, Egypt has witnessed rapid development and industrialization, a by-product of which has been the generation of large amounts of air pollutants. As a result, air quality has decreased drastically.

In Egypt as an example of developing countries, most air toxics originate from human-made sources, including; mobile sources (e.g., cars, trucks, buses) and stationary sources (e.g., factories, refineries, power plants); as well as indoor sources (e.g., building materials and activities such as cleaning). Toxic air pollutants are emitted from many industrial and non-industrial sources; fuel combustion, open burning of organic material, solvent uses, storage and transport of petroleum products, volatilization from the use of consumer products, and various manufacturing and commercial processes. These sources are categorized as point, area or non-point, non-road mobile, and on-road mobile sources. Several new industrial areas were established such as 6th of October, 10th of Ramadan and Sadat City. These new industrial cities were founded to divert industrial development from the Nile Valley and Delta towards the adjacent desert areas.

PAHs are at the present time one of the most documented class of organic pollutants in the environment due to their toxicity carcinogenicity, and mutagenicity, PAHs continue to attract the curiosity of researchers. PAH concentrations found in the atmosphere are highly dependent of the size of particles and the highest concentrations are found in breathable particle. Certain PAHs are suspected of being carcinogens and some are associated with acute and chronic health problems. The main sources of PAHs in Egypt are diesel exhausts, the petroleum and printing industries, and road transport. However, there are few available studies on PAHs in ambient air in Cairo city.

PAH concentrations in Sadat city as an example of new industrial area were measured, beside old industrial areas in Egypt. The seasonal and annual mean concentrations of $\sum$9PAHs in Sadat City during the year 2003 were 403.79 and 143.82 ng/m^3, respectively at two measuring sites. The $\sum$9PAH reached the maximum during winter season (786.69 ng/m^3), while the minimum value was recorded during autumn season (33.09 ng/m^3). The annual average concentration of benzo(a)pyrene was 18.464 ng/m^3 at the industrial area in Sadat city during the period under study. The mean concentrations of benzo(a)pyrene in the present study were 6 times higher than the annual limit set by Italy (2.5 ng/m^3), and the current guideline limit of the Federal Environmental Agency of Germany (1 ng/m^3) (Ezzo, 2006).

In the same way, the concentration of PAHs were measured in the industrial area in Greater Cairo, the obtained values were much higher than recorded for the industrial areas for many countries such as New Jersey, Canada and Belgium. This may be due to the higher local emission of PAHs from the coke factory, iron and steel factory and other industries in and around this area. PAHs are normally associated with small particles that generally have long residence times in atmosphere and, for that reason have the potential to be transported to quite long distances (Hassan, 2006).

The annual mean concentrations of atmospheric PAH over the 4 investigated sites in Cairo from autumn 2001 to autumn 2003 were illustrated in Table 5 (Hassan, 2006). It can be noted that the highest levels were recorded in Shoubra Elkhiem (industrial area) followed by El-Tebeen (industrial area), then City center, and finally Faysal (residential area).

Worldwide, regulations and standards regarding permissible levels of PAHs in ambient air vary somewhat. The maximum permissible concentration in ambient air is 1 ng/m^3 in Soviet Union former and Hungary (Hassanien et al, 1996). In the US, no national ambient air quality standard exists. However, several states have established ambient standards. For B(a)P, these range from 0 to 100 ng/m^3 (8-hour average). In the Netherlands, the

Environmental Programme 1988-1991 recommended a draft tolerable B(a)P concentration of 5 ng/m^3 (annual average) (RIVM, 1989) see Table 6.

Table 5. Seasonal and annual mean concentrations of atmospheric PAH in different sites in Egypt

Location	PAHs	Sampling year	Recorded value (ng/m^3)	References
Cairo	∑12 PAHs	1997 Wint/Sum	22.32 (annual) 14.79/7.53	Hassanien et al, 2001
Sadat City New Indus. city	∑9 PAHs	2003 Win./sum	273.8 (annual) 786.7/145.2	Ezzo, 2006
Greater Cairo				
Soubra ElKheima (old indus. Area)		2001-2002 Wint/Sum	4460 (annual) 7040/1771	
City Center	∑16 PAHs	2001-2002 Wint/Sum	3552 (annual) 5416/1378	Hassan, 2006
Elattaba		2001-2002 Wint/Sum	3477 (annual) 248/134	
Faysal (resid..area)		2002-2003 Wint/Sum	1965 2274/1789	
Cairo (traffic	∑12 PAHs ∑carcino.PAHs	2005	37.3 (outdoor) 29.7 (indoor) 20.6 (outdoor 8.8 (indoor)	Hassanien, 2006

An average of 32 ng/m^3 of total PAHs in the ambient air found in a study in heavy traffic areas in Alexandria City (Barakat, 2002). A summary of the data obtained from analysis of polycyclic aromatic hydrocarbons in suspended dust collected during 1997 in Cairo Egypt (Hassanien et al., 2001 and recently Hassanien, 2006) is given in Table 5. It can be seen from the analysis of the 12 components that the sum of PAH concentrations at the sampling site was higher during the winter season 1997 (14.79 ng/m^3) than in summer during 1997 (7.53 ng/m^3). It was found that the carcinogenic materials constitute about 20%; 33% and 25% of the total PAH for the winter, summer and annual periods, respectively. A highly significant difference between the estimated concentrations of PAHs of winter and summer was observed. It was found that PAHs concentrations in outdoor air generally exceeded that in the corresponding indoor air. The sum of PAHs concentrations in outdoor air was 37.3 ng/m^3; the corresponding concentrations in indoor air was 29.7 ng/m^3. Carcinogenic compounds measured in both outdoor and indoor represent 20.6 and 8.8 ng/m^3, respectively with a contribution 55.2% and 29.6% from the total of PAHs. In addition, from the eighteen PAHs, fluoranthene, pyrene, chrysen, benzo(b)fluroanthene, dibenzo(a,h)anthracene, benzo(ghi)perylene, and benzo(a)pyrene were the most abundant PAHs found in indoor and outdoor air. They contributed more than 55-65%, respectively to the sum of PAHs (Hassanien, 2006).

Table 6. Ambient air quality standard for the PAHs*.

	Limit	Guide value	Measuring period
Netherlands PAH	5 ng/m^3	5 ng/m^3	Years as ng/BaP/m^3
WHO-AQG PAH		1.0 ng/m^3	Years as ng/BaP/m^3

*(Hassanien, 2006)

INDOOR VERSUS OUTDOOR PAHS LEVELS IN SOME CITIES

In recent decades, there has been a tendency toward atmospheric pollution in the developing countries due to fast urbanization and industrialization. In past years, many studies have stressed the characterization of inorganic components of aerosols in developing cities in Asia and Africa because the desert land profile for these countries which considered the largest source of particulate matter in the world (Baek et al., 1991; Maliszewska-Kordybach, 1999; Mastral and Callen, 2000; Ohura et al., 2004$_b$). Indoor emission sources of PAHs include smoking, cooking, heating, and furniture (Liu et al., 2001; Ohura et al., 2002).

Ratios between indoor and outdoor air concentrations (I/O ratios) may be used as a tool to indicate whether there are indoor sources (I/O> 1) or outdoor sources (I/O<1). Spatially and temporally consistent indoor and outdoor samples were taken at some sites in different cities. The I/O ratios were considerably >1 in Gothenburg, and generally close to 1 in Mexico City. The high I/O ratios for ∑PAH in Gothenburg together with somewhat elevated indoor levels may indicate that PAH emissions in these homes are mainly from indoor sources while the opposite pattern was observed in Mexico City (Motelay-Massei et al., 2005, Strandberg et al., 2006). An interesting observation was that low molecular weight PAHs in Gothenburg in general show an I/O ratio well above 1 while the 5–7 ring PAHs are found in more samples outdoors than indoors, giving an I/O < 1. Lower I/O ratios for the 5–7 ring PAHs were also seen in Mexico City. This may indicate the presence of emission sources of mainly low molecular weight PAHs (vapour phase PAHs) indoors. PAH levels in the $PM_{2.5}$ fraction of Beijing urban outdoor air were lowest in summer and highest in winter (He et al., 2006).

The size distributions of ambient particles showed no significant difference between summer and winter, either outdoors or indoors. This suggested that the distributions were strongly affected also by the sources. On the other hand, ∑PAH associated with particles was predominantly (>80%) found in $PM_{2.5}$ fractions. The proportion of ∑PAH in $PM_{2.5}$ fractions was about one order of magnitude higher than in coarse particles (PM>10 and $PM_{2.5–10}$) (Ohura et al., 2004$_a$). The concentrations of indoor ∑PAH associated with $PM_{2.5}$ (∑PAH–$PM_{2.5}$) measured in his study ranged from 1.6 to 23.7 ng/m^3). In outdoors, the ∑PAH– $PM_{2.5}$ concentrations ranged from 1.1 to 29.5 ng/m^3. The concentrations of indoor ∑PAH associated with PM>10 (∑PAH–PM>10) and PM2.5–10 (∑PAH– $PM_{2.5–10}$) ranged from 0.13 to 1.8 and 0.18 to 2.05 ng/m^3, respectively. Outdoors, ∑PAH–PM>10 and ∑PAH– $PM_{2.5–10}$concentrations ranged in 0.10–1.64 and 0.17–2.48 ng/m^3, respectively. The variability of outdoor ∑PAH associated with particles was considerably larger than that of the corresponding particles. This may be due to the difference of emission sources, lifetime, and physical properties, between PAH and particles. For light molecular PAHs, concentrations were strongly affected by seasonal factors because of their high vapor pressures.

Smoking is a major emission source of indoor pollutants (Liu et al., 2001; Ohura et al., 2002). Humans spend about 90% of their daily lives indoors; thus, estimating indoor pollution is critical to ensuring a healthy environment. BaP is widely used as an indicator of human health risks and pollution levels because of its strong carcinogenicity (Finlayson-Pitts and Pitts, 2000).

In Guanzhou (China), the seasonal concentration for ∑PAH ranged from 8.11 to 106.26 ng/m^3. The average ∑PAH measured were highest in winter and lowest in summer. 5-6 ring PAHs were the abundant compounds, which accounted for 65-90% of ∑PAH and benzo[b+]fluoranthene dominated in the four seasons. Moreover, the total PAHs in Guangzhou were significantly higher than those in Hong Kong, Naples, Rom and Seoul; however, the distribution patterns of PAHs of PAHs in Guangzhou showed some similarities. In these cities high molecular weight (HMW) compounds dominated with Bbkf as maxima (except Naples) and Ant had the lowest concentration in all PAHs (Tan et al., 2005).

Fourteen polycyclic aromatic hydrocarbons were identified and quantified in PM10 rang in Guangzhou during the period from March 2002 and June 2003. The ∑PAH ranged from 8.11 to 106 ng/m^3. A strong seasonal trend of PAHs abundance was observed. The highest level was in winter and the lowest was in summer. The average ∑PAH in winter was 3.8 times higher than that in summer; the Asian monsoon had a strong influence on the seasonal variation. The distribution pattern of PAHs varied with the season. PM10-bound PAHs were dominated by high molecular weight compounds. LMW PAHs tend top have a higher contribution in the winter while HMW PAHs tend to have a higher contribution in the summer. Based on the diagnostic ratios, it is concluded that vehicular were the predominant source of airborne PAHs in Guangzhou (Tan et al., 2005).

Occurrence of PAHs in Dust

Settled house dust can be a source of human exposure to toxic PAHs through non-dietary ingestion and dermal contact. Information regarding the concentrations of various contaminants in house dust would be useful in estimating the risk associated with exposure to these compounds. PAHs in settled house dust were collected from homes in three locations: Sumgayit, Azerbaijan; Shanxi Province, China; and southern Texas, United States. The highest PAH floor surface loadings were observed in China, followed by Azerbaijan and Texas. Median concentrations of high molecular weight (four ring and larger) PAHs ranged from a low of 0.11 μg/m^2 in Texas, to 2.9 μg/m2 in Azerbaijan and 162 μg/m^2 in China as listed in Table 7. These trends in total surface loading and relative carcinogenicity indicate that the risk of health effects from exposure to PAHs in house dust is highest in the Chinese population and lowest in the Texas population (Naspinski et al., 2008).

PAHs in house dust may originate from both outdoor and indoor sources. Typical indoor sources include combustion of fossil fuels and biomass for heating and cooking; burning of tobacco products, candles and incense; and cooking fumes. Maertens et al. (2004) estimated the excess lifetime cancer risk from non-dietary ingestion of PAH in settled house dust to exceed the acceptable range for pre-school children in homes with dust PAH loading at the 95th percentile. Chuang et al. (1999) reported that ingestion of house dust and soil contributed 24% of the estimated daily dose of B2 PAH in a study of children in low-income households.

The floor dust collection procedure used in this study is quick, inexpensive, easily portable, and requires no specialized equipment. Although surface loading cannot be used directly in an exposure calculation, it is valuable for the hazard identification step of risk assessment. Carcinogenic PAHs in the house dust collected from different places was found in general, that the heavier PAHs are more toxic. All of the carcinogenic PAHs (cPAHs) contain 4 or more rings and are therefore in the HMW fraction. The percentage of HMW PAHs was highest in China and lowest in Texas (Naspinski et al., 2008).

Table 7. Concentrations of total and high molecular weight PAHs of dust samples collected from houses.

			μg/m^2
	Azerbaijan	China	Texas
Total PAHs			
Median	6.1	232	1.34
Min	0.7	42	0.24
Max	49.6	6720	35.3
High molecular weight PAHs			
Median	2.9	162	0.11
Min	0.5	30	0.03
Max	20.0	4440	2.90
EPA B2 carcinogen PAHs			
Median	1.0	53	BD
Min	0.2	12	BD
Max	5.4	1620	0.27

In Egypt, road dust PAHs was received relatively little attention to date and studies focused on heavy metals almost. Consequently, background level of PAHs in dust of Egyptian roads is not available. Recently, a study carried out by Hassanien and Abdel-Latif (2008) on the road dust collected from 17 sites over greater Cairo (Fig. 2). The concentration of PAHs was determined by using HPLC technique. Twelve common environmental PAHs were found to be distributed. The data illustrated that the total average of PAHs over the investigated sites was ranged from 0.045 mg/kg to 2.6 mg/kg. On individual scale, the highest concentrations were 1.031 and 1.028 mg/kg for pyrene and phenanthrene, meanwhile the lowest was benzo(a)pyrene with value 0.0001 mg/kg. The obtained results showed that the carcinogenic content of PAHs (benzo(a)anthracene, chrysene, benzo(a)pyrene, and benzo(k)fluoranthene) ranged from 0.2 % to 23.9% of total PAHs.

The highest average concentrations for PAH species in Cairo's road dust were for *PHE* (phenanthrene) of 0.404 mg/kg followed by *PYR* (pyrene) and *FLA* (flouoranthene) with average levels of 0.166 and 0.068 mg/kg, respectively, in consistent with those results found in Beijing (Li et la., 2006). Moreover, Bostrom et al. (2002) found *PHE* (phenanthrene) to be the most abundant of 14 different PAHs compounds monitored at the street-level site in Stockholm. Meanwhile the lowest measurable species in dust were *BaP* (benzo(a)pyrene), *ACE* (acenaphthene), *BaA* (benzo(a)anthracene), and *BbF* (benzo(b)flouranthene) with average concentrations of 0.0024, 0.0027, 0.0028, and 0.0035 mg/kg, respectively.

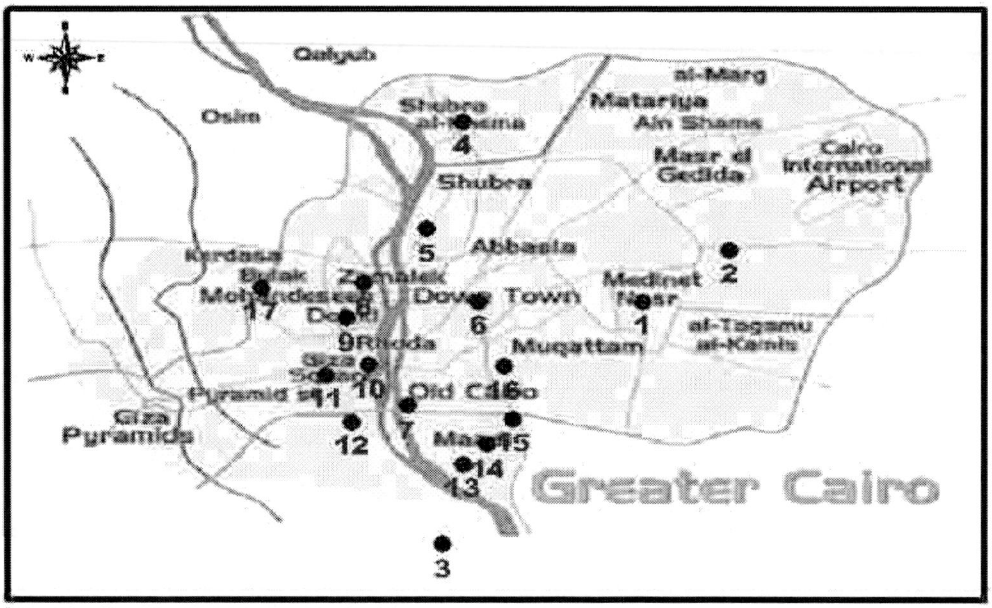

Figure 2. Map of the Greater Cairo with location of road dust sampling sites (solid circles).

The spatial variability in road dust PAHs concentration in Cairo was presented in Figure 3. It involves 17 representative locations, with site-specific information gathered at each location Distribution patterns of PAH in Cairo's road dust is obtained by processing data of species concentrations rang between 0.045 and 2.6 mg/kg, as shown in Fig. 3. Spatial distribution of PAHs highlighted sites of high concentration, and as it has revealed from results, these hot-spot-sites are distributed through out whole Cairo, i.e. sites 4 and 5 in the north, site 6 in the down town, site 11 in the west and site 13 in the south. These sites are heavily traffic areas, in addition to the industrial activities at sites 4 and 13. A moderate pollution was recorded also in sites 5, 15 and 16.

This is the first comprehensive information on PAH spatial distribution pattern in different district areas of Greater Cairo, Egypt. This work revealed the road dust contamination with PAHs residues in different district areas of Cairo. It extends our understanding of the current PAHs contamination status in this mega city. The average concentration of 12 PAHs was 0.704 mg/kg with carcinogenic content of 17.1% of total PAHs. The present study confirms the contribution of vehicular traffic, industrial emissions and the incomplete combustion of open waste burning to road dust PAHs. Plotting PAH results on a map of city can be used to advantage in investigating road dust pollution and its potential deleterious effects. This study suggested that the pollution of PAHs in Cairo should emphasize on controlling the industrial emissions and traffic exhausts.

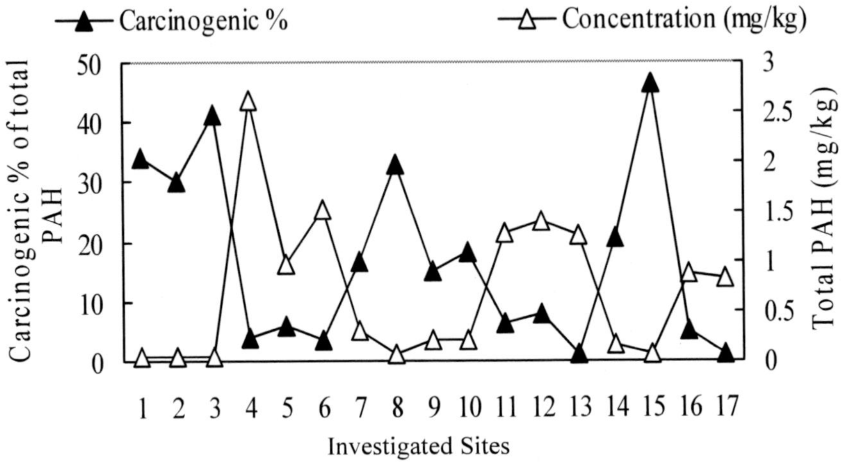

Figure 3. Total concentrations of PAH and carcinogenic percent at investigated sites.

PAHs in Food Materials

Although many PAHs are included in the "priority pollutant" listings, there are many uncertainties over the significance of PAHs in relation to their transfer from the natural environment into the human food chain. There are very few available studies concerned with PAHs in food materials in developing countries. Therefore, it should be important to see the experiences from developing countries.

Based on the premise that the plant leaf contents can provide a 'time integral' of the airborne concentration (Carberg et al., 1983, Eriksson et al., 1989, Bacci et al., 1991, Franzaring et al., 1992) the paddy leaves were selected as passive samplers to observe contemporary levels of PAHs in the atmosphere. PAHs reach the surface of plant leaves primarily from the atmosphere (Nash and Beall, 1970, Bacci and Gaggi, 1986, Trapp et al., 1990) PAHs in the atmosphere exist in either the gas or the particulate phase or both. Thus, the mechanism of transfer to plant leaves may be through gaseous dry deposition, particulate dry deposition, or wet deposition (including rain, snow and fog). Gaseous dry deposition has been shown to be the main process for many organic contaminants (Umlauf et al., 1994, McLachlan et al., 1995) PAHs with a vapor phase component in air are regarded as being subject to an air-plant exchange process, moving toward equilibrium over time. In addition, the paddy root was used for observing the adsorption of PAHs from soil and water because of root uptake, bioconcentration and translocation.

In Japan, a study was carried out to estimate the PAHs content in rice as component of food chain. The total PAHs content in rice ranged from 46 to 77 μg/kg dry weight with an arithmetic mean value of 58 μg/kg dry weight. Phenanthrene was the most abundant in most cases. For analyzed PAHs, the arithmetic mean value of acenaphthylene, fluorene, anthracene, pyrene, weight (152–202) to heavier molecular weight (228– 252) PAHs in paddy leaf (73 : 27) was similar to that in rice (70 : 30) (Liu et al., 2001). The mobility of light molecular weight PAHs through the atmosphere was greater than that of the heavy molecular weight PAHs because of the greater gas phase component. Conversely, heavy molecular weight PAHs were considered to be only particle-bound and will be subject to deposition processes that effectively remove them from the atmosphere as air masses move away from source regions (Eriksson et al., 1989, Bacci et al., 1991, Franzaring et al., 1992).

POLYCYCLIC AROMATIC HYDROCARBONS IN MARINE BIOTA

Concentrations of PAHs in marine biota indicate a high degree of contamination of certain areas of the region. Data given for the Ligurian Sea for *Mytilus* sp. range from 14 to 571 ng/g dw. In other studies, the 14 priority PAHs in mussel tissues (*Mytilus* sp.) and fish livers (*Serranus* and *Mullus* sp.) of specimens collected in the French and Spanish Mediterranean coasts and found values of 25-82, 27-87 and 15-49 ng/g dw, respectively, n values up to 390 ng/g dw were found in mussels collected inside harbours. Similar determinations carried out in deep sea fish from the NW Mediterranean basin (*Mora* sp.) showed liver concentrations of 7-16 ng/g ww (UNEP, 2002).

A large survey of mussels carried out during 1995-99 in the French Mediterranean and Atlantic coasts gave mean values of 590 and 411 ng/g dw (Σ16 PAHs), respecively. A survey carried out recently in bivalves (*Perna perna*) from 18 stations along the Atlantic coast of Morocco gave results from 3.4- 34.6 ng/g (Σ14 PAHs) with values up to 87-109 ng/g ww in the vicinities of the cities of Agadir and El Jadida. Values of 170-750 ng/g dw were found in mussels from the Thermaikos Gulf (Greece). The concentrations of 16 PAHs in mussels collected in 1999 from Izmit Bay (Turkey) were found in the range 1.25-34.0 ng/g. Blubber samples were collected in 1993 and 1996 from live specimens of fin whales *(Balenoptera physalus)* and stripped dolphins *(Stenella coeruleoalba)* from the Ligurian and Ionian Seas. In whales, the median value of total PAHs was 350 ng/g ww while median carcinogenic PAH values were 114 ng/g ww, in dolphins. The median values of total and carcinogenic PAHs were 938 and 670 ng/g ww, respectively (UNEP, 2002).

HEALTH AND ENVIRONMENTAL IMPLICATIONS

The hazards caused by toxic chemicals are devastatingly dangerous in the developing countries because:

- People and governments are unaware of the inherent dangers of pesticides and other chemicals;
- The necessary financial resources are either scarce or do not exist;
- Facilities either for containing the wastes or for its destruction are not available
- Appropriate legal measures do not exist or, if they do, are either impossible or difficult to implement;
- Environmental activists are either absent or too weak to stage public demonstration, and
- Expertise or skilled manpower is not available. However, some of the adverse health and environmental impacts, include

- Direct exposure of workers and farmers, local residents, to unsecured pesticides stocks
- Indirect exposure through contaminated food and water, thus endangering public health
- Poisoning of farm lands and other workers
- Pollution of ground and surface water sources, which impacts on fauna and flora
- Impacts on land use resulting in limiting the potential of development of the area surrounding any chemical site
- Destruction of beneficial insects
- Risks and costs of destruction and incineration.

The effects associated with exposure to PAHs are enormous. They could be classified into environmental and health effects. While the environmental effects are associated with the pollution of the air, water and soil ecosystems, the later being the most dominant is associated with disturbances and disorders ranging from the disorders of the nervous or immune system, to increases in the risk of certain cancers.

They persist in the environment, and may bioaccumulate and magnify in the food web due tof the tendency of these chemicals to accumulate in fatty tissue. There may be special windows of vulnerability in the development of fetuses when these chemicals can have long-term, irreversible effects on the reproductive and neurological systems.

Carcinogenic Assessment of PAHs

Toxic Effect of PAHs

Short-term studies showed adverse hematological effects. Systemic effects caused by long-term treatment with PAHs have been described only rarely, because the end-point of most studies has been carcinogenicity. Significant toxic effects are manifested at doses at which carcinogenic responses are also triggered. In studies of adverse effects on the skin after dermal application, non- or weakly carcinogenic PAHs were inactive, whereas carcinogenic compounds caused hyperkeratosis. Benz[*a*]anthracene, benzo[*a*]pyrene, dibenz[*ah*]anthracene, and naphthalene are embryotoxic to mice and rats. Benzo[*a*]pyrene also have teratogenic and reproductive effects (Ravindra et al., 2008).

PAHs have also been studied extensively in assays for genotoxicity and cell transformation; most of the PAHs are genotoxic or probably genotoxic. The only compounds for which negative results were found in all assays were anthracene, fluorene, and naphthalene. Owing to inconsistent results, phenanthrene and pyrene could not be reliably classified for genotoxicity. PAHs have generally been reported to have immunosuppressive effects (Ravindra et al., 2001).

The toxicity of PAHs represents a risk to human health, and there are varied risk assessment approaches to quantifying the risk posed by PAHs based on exposure routes and scenarios. PAHs are not carcinogenic until they are metabolically activated as the body attempts to break them down and forms reactive metabolites that bind to the DNA causing subsequent replication in the cells. Fundamental to assessing risk posed by PAHs is understanding metabolism of PAHs. Since exposure to PAHs is never to single PAHs, understanding what differences may occur in mixtures of PAHs gives accurate assessment of the dangers of PAHs. Understanding the dynamics of complex metabolism vis-à-vis single metabolism of PAHs and possible effects on the toxicity expression of PAHs is a necessary advancement to accurately impact and guide remediation strategies.

The bioaccumulation of PAHs and their bioamgnification along successive trophic levels is another means of exposure to human and animal life. PAHs can cause mutations, malignant and benign tumors, among a host of genotoxic effects. The link between PAHs and harmful effects in humans was made since 1775 by Sir. Percival Pot, surgeon to St. Bartholomew's Hospital in London. He observed the increased incidence of scrotal cancer in chimney sweepers originated from their exposure to soot; (Cerniglia 1980). This observation provided an important link into studying occupational related diseases and their causes. Increasingly, links were made between tumors and occupation that had PAH exposure, in workers such as miners, chemical plant workers, tar-producing industry workers.

HUMAN EXPOSURE

Air pollution is often suggested as being partly accountable for an increased incidence of lung cancer in cities. Over the past decade, an increasing body of scientific evidence has accrued associating outdoor air pollution with certain types of cancer. In many researching scientists have found out that exposure to air pollution, even at low levels, can lead to a wide range of adverse health effects including lung cancer.

An individual's total exposure to air pollution depends on indoor as well as outdoor exposures. Indoor air quality has large potential health implications because people may spend considerable amounts of time indoors. Indoor air pollution may stem from incoming outdoor air or may originate indoors from tobacco smoking, building materials, soil gases, household products, and combustion from heating and cooking. The exposure of human populations to carcinogens in outdoor air may be the result of proximity to more localized sources such as industrial facilities, small businesses, municipal facilities, or areas with high vehicular traffic.

PAHs are lipophilic compounds and can be absorbed by the lung and gastrointestinal tract tissues and by the skin. Several studies have been carried out to determine the levels of exposure of humans to PAHs (Ciganek et al., 2002). The dietary sources of PAHs are cereals and vegetables, rather than meat, excepting populations with a high consumption of meat cooked over an open flame. The recently developed biomonitoring procedures allow the assessment of human exposure to PAHs, as well as identification of diet as the major exposure source. The U.S. Environment Protection Agency published "The Exposure Factors Handbook" in which available statistical data on various factors used in assessing human exposure are summarized (http://www.epa.gov/ncea/exposfac.htm). The major steps in exposure assessment include: (1) determination of exposure pathways, (2) identification of environmental media conveying the contaminant, (3) determination of the contaminant concentration, (4) determination of exposure time, frequency, and duration, and (5) identification of the exposed population.

Exposure and Routes of Exposure

Exposure to chemical hazards may occur via all types of exposure: inhalation, oral ingestion, absorption through the skin, absorption through the eyes, and placental transfer from pregnant woman to the fetus, inoculation and direct penetration to target organs and from mother to child through breastfeeding. In the non occupational environment, ingestion of substances containing chemicals is the most common route of exposure. In the workplace, because of the nature of exposure, duration of the workday, and character of the compounds; inhalation is the most significant route of entry, followed by skin absorption and ingestion.

Health effects, of course, derive from exposure and thus it can be expected that global ill-health from air pollution is dominated by indoor exposures in developing countries as well. The precise extent of impact, unfortunately, is not known yet, but enough information is available to make reasoned estimates for India, where more data are available than in most other developing countries (Smith, 2000). The main sources of general population exposure are: vehicular traffic, residential heating, smoke from open fireplaces and cooking, tobacco smoke, contaminated food and drinking-water, and the use of PAHs contaminated products. The calculated total daily intake of PAHs from food has been estimated to be 3 μg/day (1.4 μg/day carcinogenic PAHs) per person in Italy (UNEP, 2002).

In addition to a lack of studies on the risks of exposures, there is a dearth of studies on the distribution and magnitude of exposure to PAHs in developing countries itself. Such information is needed both to estimate the total burden of disease and to identify the regions and populations where interventions need to be focused. Compared with the thousands of studies on outdoor and indoor air pollution in developed countries, only few studies have been carried out in the developing world (African countries). Although striking because of the high exposures to major pollutants that have been demonstrated, the available studies have not been conducted in a fashion that allows the results to be confidentially extrapolated to large populations.

PAH Potencies

PAH potencies are used to determine quantitative health risks posed by PAH exposure. The risks posed by a mixture of PAHs are based on an assumption of additivity of the individual risks posed by the PAHs. The IPCS monogram on PAHs (IPCS, 1998) describes three approaches used to calculate PAH potencies: 1) toxicity equivalence factors approach is based on expressing of individual potencies relative to benzo(a)pyrene, 2) comparative potency approach, which does not identify or quantify the individual compounds but determines the potency of the mixture of compounds and 3) benzo(a)pyrene surrogate approach assumes that benzo(a)pyrene is an indicator of all the PAHs.

As the toxicities (carcinogenicity and mutagencity) of individual PAHs differ considerably, toxicity assessments of PAHs are complex. A number of approaches have been developed for evaluating the potencies of the various PAHs with regard to the possible inhalation cancer risk to humans (IARC, 1983). One approach is to calculate the inhalation unit risk for excess lung cancer over the risk posed by BaP for each of its copollutant carcinogenic PAH in the polluted ambient air divided the particular PAH's risk by the risk for BaP to obtain their individual potency equivalence factors (PEFs); thus, the PEF for BaP would be, by definition, 1.00. These PEFs can be used to estimate the relative carcinogenicity of ambient air with a known distribution of PAHs (Ohura et al., $2004_{a,b}$).

The risk associated with the inhalation of particulate PAHs indoors showed that the contribution of BaP to the total carcinogenic potential was dominant in the range of 51% to 64%. Human activity patterns indicate that people tend to spend 90% or more their time indoors. In addition, indoor BaP concentrations are strongly affected by the corresponding outdoor ones. Thus, the risk assessment for inhalation of PAH could be roughly estimated by only the outdoor BaP concentrations.

PAHs half-lives in humans are in fact in the range of days/hours, and metabolism is responsible of the formation of toxicologically active (carcinogenic) metabolites. Because of the lack of persistence and to the complexity of PAH mixtures, human exposure is usually characterised in terms of "internal dose" rather than of "body-burden", by the use of biomarkers of exposure, mainly 1-OH pyrene in urine and DNA adducts in peripheral blood lymphocytes.

Risk Assessment of PAHs Risk Research Needs

Hundreds of urban developed-country epidemiological studies have been carried out in recent decades, allowing in some cases large-scale meta-analyses and reviews for important pollutants like small particulates (e.g. WHO, 1999; Kreweski et al., 2000; Samet et al., 2000). Extrapolating these relatively well-characterized results to the populations using solid fuels indoors is difficult, however, because of different pollutant mixtures, health conditions, age-distributions, exposure patterns and so on, as well as the scarcity of actual exposure measurements. There is clearly need to find ways to make these extrapolations with more confidence (McMichael and Smith, 1999). In addition, however, there are needs to conduct additional research directly within the communities of interest. Given what is known about the relative importance of the major air-pollution-related diseases in developing countries, the risks from existing studies, and the relative difficulties of measuring effects in studies of reasonable duration, the following kinds of studies would seem to have the highest priority. Although many people associate public exposures to air pollution primarily with urban outdoor settings, readers of this chapter are well aware that indoor environments can also be contaminated, both from pollution penetrating from outside and from indoor sources. General population exposure to PAHs mainly occurs from atmospheric pollution and diet, this latter accounting for about 90% of the total exposure (Hassanien et al., 2001).

The largest exposures to health-damaging air pollution probably occur in the developing world. As a result, much of the health impacts from air pollution worldwide seem to occur among the poorest and most vulnerable populations, largely women and young children who are most exposed to the indoor pollution sources of importance in poor countries.

For many PAHs, toxicological effects are still unknown or not fully clarified, and mechanistic information is lacking. Moreover, even for PAHs whose human health hazard has been characterised, no risk assessment is actually feasible due to the lack of exposure data in most countries. Food being the main exposure route, data on PAHs daily intakes are of major relevance and represent the necessary basis to develop an adequate risk assessment.

These high individual and population exposures are the result of a set of factors:

- About half the world's households use unprocessed solid fuels for cooking, ranging roughly from near zero in developed countries to more than 80% in China, India, and Sub-Saharan Africa (Holdren et al., 2000).
- The individual peak and mean exposures experienced in such settings are large by comparison with WHO guidelines and national standards.

BaP is one of the polycyclic aromatic hydrocarbons formed when gasoline, garbage or any animal or plant materials burn incompletely. It was first isolated from coal tar. People may be exposed to BaP from air, water, soil, cigarette and other plant product smoke, food and some work environments through inhalation, ingestion and skin contact.

Exposure to PAHs may cause harmful health effects. When humans are exposed to BaP at high concentrations for relatively short periods of time, BaP may cause red blood cell damage leading to anemia and suppressed immune system. Mixtures of PAHs including BaP such as coal tar were shown to be dermal carcinogens in animals as early as 1918. BaP has caused tumors in laboratory animals when administered in the diet, when applied to their skin or when inhaled for a long period of time. Humans exposed to mixtures of PAHs and other compounds at high concentrations over long periods of time can also develop cancer.

To set priorities with respect to control, an excess lifetime cancer risk, expressed in terms of the BaP concentration and based on observations in coke oven workers exposed to mixtures of PAHs, is presented here. It must be emphasized that the composition of PAHs to which coke-oven workers are exposed may not be similar to that in ambient air, although it was noted that similar risks have been derived from studies of individuals exposed to other mixtures containing PAHs. Having also taken into consideration some recent animal data from which a unit risk of the same order of magnitude can be derived, it was concluded that the occupational epidemiology data should serve as the basis for the risk estimate. Based upon epidemiological data from studies in coke-oven workers, a unit risk for BaP as an indicator in air constituent is estimated to be 8.7×10^{-5} $(ng/m^3)^{-1}$ which is the same as that established by WHO (1987). The corresponding concentrations of BaP producing excess lifetime cancer risks of 1/10 000, 1/100 000 and 1/1 000 000 are 1.2, 0.12 and 0.012 ng/m^3, respectively.

To assess health-risks associated with PAH exposures, Lin et al. (2008) reported that it is important to know the total carcinogenic potency arising from the exposures of various PAH compounds. In principle, the carcinogenic potency of a given PAH compound can be assessed according to its benzo[a]pyrene equivalent concentration (BaPeq). Calculating the BaPeq concentration for a given PAH compound requires the use of its toxic equivalent factor (TEF; using benzo[a]pyrene as a reference compound) to adjust its original concentration (Chu and Shen, 1984; Thorslund and Farrer, 1991; and Nisbet and LaGoy, 1992). Among the available TEFs lists, the one established by Nisbet and LaGoy has been demonstrated to best reflect the actual state of knowledge of the toxic potency of each individual PAH species (Nisbet and LaGoy, 1992). By using the TEF list the carcinogenic potency of total PAHs (total BaPeq) can be determined as the sum of BaPeq concentrations of the 21 selected PAH compounds.

For estimating the lung cancer risk associated with inhalatory PAH exposures, the World Health Organization (WHO) has suggested a unit risk of 8.7×10^{-2} $(\mu g/m^3)^{-1}$ for the lifetime (70 years) PAH exposure, assuming one was exposed to BaP concentration of 1 $\mu g/m^3$ (WHO, 1987). It is worth noting that the above unit risk was proposed for lifetime exposure, therefore, it has been adopted for assessing the exposure of general adults to the ambient atmospheric PAHs (Zmirou et al., 2000). For occupational exposure, Pott established a relationship between BaP exposure and lung cancer risk (Pott, 1985), based on a data bank provided by an epidemiological study conducted by Redmond et al. (1976). He suggested the unit risk of 7.0×10^{-2} $(\mu g/m^3)^{-1}$ for a 25-year occupational PAHs exposure with the averaged BaP concentration of 1 $\mu g/m^3$. By using the same data bank, the US Environmental Protection Administration (EPA, 1984), however, suggested a different unit risk of 6.4×10^{-4} $(\mu g/m^3)^{-1}$ for PAHs exposure based on its total PAH content (expressed as the benzene soluble fractions). Since another study has indicated BaP could be a better indicator than total PAH content on characterizing the carcinogenic potency of PAHs (Petry et al., 1996).

Interventions and Risk Research Needs

In spite of the clear need to conduct more research on risks and exposures, the current, if imperfect, knowledge of the health burden imposed by air pollution in developing countries argues that action is warranted now. Unfortunately, however, there also is a severe lack of good information on the interventions that might be best applied to effectively reduce the risks. Better ventilation, better stoves, better fuels, and behavioral changes would seem to encompass the range of potential interventions, but remarkably little systematic work has been performed on any of these, considering again the potential scale of the problem and consequent potential benefit.

There is research needed not only on the technologies and their acceptance at the household level, but also on the policy tools needed to promote exposure-reducing changes in household fuels and stoves.

- How to develop confounding for exposure-reduction efforts among developing sectors (health, energy, housing, and environment) in the worst affected regions.
- Conduct research on ways that different sectors can implement programs with exposure reduction potential, for example, fuel pricing in the energy sector.
- Conduct systematic program of research to evaluate the effectiveness of various policy measures

Conclusion and Future Research

The environmental pollution mixtures in major population centers should be characterized both in terms of physical and chemical constituents and in terms of sources of major constituents. If possible, retrospective characterization of levels of certain constituents could be accomplished. Although a program of risk, exposure, and intervention studies as outlined above would not be inexpensive by some measures, it is likely to be highly cost-effective as a means to address the large exposures now being experienced widely in developing countries.

Better estimates of the amount of effect will need additional epidemiologic studies. Large numbers of cases will be essential to measure the effects of environmental pollution by PAHs and to measure combined effects of pollutants and factors such as occupation, smoking and such studies will probably require pooling data from multiple locales. New studies will need to develop and apply improved methods and to compare the effect of exposure to the pollutant mixtures on lung cancer in different cities while effectively controlling confounding factors including different exposure sources. Also we need methods for the retrospective estimation of lifetime exposure to environmental pollutants.

Acknowledgment

The author would like to report that this chapter is the result of the work of many dedicated professionals who helped him in design, concept, and research. The author wishes heartily to thank his colleagues in Air Pollution Department, National Research Center for their sincere support and encouragement. He would like also to thank Prof. Alia A. Shakoiur, Dr. Mamdouh El melegy, Dr. Nasser M. Abdel-Latif and Aliaa A. Ezoo for supplying him with important data and information. Finally, the author thanks his sons for editorial assistance and their patience.

References

Asante-Duah, K. (2002). Public health risk assessment for human exposure to chemicals. Kluwer Academic Press, Boston.

Bacci, E. & Gaggi, C. (1986). Chlorinated pesticides and plant foliage: translocation experiments. *Bull Environ Contam Toxicol*, 37, 850–857.

Bacci, E., Vighi, M., Gaggi, C., Calamari, D., Focardi, S., Morosini, M. (1991). Role of plant biomass in the global environmental tobacco smoke evaluation of a portable pm-10 impactor sampler. *Environ Sci Technol*, 25, 1489–1495.

Baek, S.O., Field, R.A., Goldstone, M.E., Kirk, P.W., Lester, J.N., Rerry, R., (1991). A review of atmospheric polycyclic aromatic hydrocarbons: sources, fate and behavior. *Water, Air and Soil Pollution*, 60, 279–300.

Barakat, A.O. (2002). PAHs and petroleum biomarkers in the atmospheric environment of Alexandria City, Egypt. *Water, Air and Soil Pollution*, 139, 289–310.

Bartkow, M.E., Kennedy, K.E., Huckins, J.N., Holling, N., Komarova, T., Mu¨ ller, J.F. (2006). Photodegradation of polyaromatic hydrocarbons in passive air samplers: field testing different deployment chambers. *Environmental Pollution*, 144, 371-376.

Bjoerseth, A., & Randahl, T. (1985). Emission sources and recent progress in analytical chemistry. Handbook of polycyclic aromatic hydrocarbons, Vol. 2 Bjoeresth, A. & Ramadahl, T. eds. Marcel Dekker, New York.

Bohlin, P., Jones, K.C., Tovalin, H., Strandberg, B. (2008). Observations on persistent organic pollutants in indoor and outdoor air using passive polyurethane foam samplers. *Atmospheric Environment*, 42, 7234–7241.

Bostrom, C.E., Gerde, P., Hanberg, A., Jernstrom, B., Johansson, C., Kyrklund, T., Rannung, A., Tornqvist, M., Victorin, K.,Westerholm, R., (2002). Cancer risk assessment, indicators, and guidelines for polycyclic aromatic hydrocarbons in ambient air. *Environmental Health Perspectives*, 110, 451-489.

Carberg, G.E., Ofstad, E.B., Drangsholt, H., Steinnes, E. (1983). Atmospheric deposition of organic micro pollutants in norway studied by means of moss and lichen analysis. *Chemosphere*, 12, 341– 356.

Cerniglia, C., Dodge, H., Gibson, D.T. (1980). Studies on the fungal oxidation of polycyclic aromatic hydrocarbons. *Botany Mar*, 23, 121-124.

Cerniglia, C.E., (1984). Microbial metabolism of polycyclic aromatic hydrocarbons. *Adv Appl Microbiol*, 30, 31-64.

Chang, K.F., Fang, G.C., Chen, J.C., Wu, Y.S. (2006). Atmospheric polycyclic aromatic hydrocarbons (PAHSs) in Asia: a review from 1999 to 2004. *Environmental Pollution*, 142, 388-396.

Chiou, C.T., McGroddy, S.E, Kile, D.E. (1998). Partition characteristics of polycyclic aromatic hydrocarbons on soils and sediments. *Environ Sci Technol*, 32, 264-269.

Chu, M.M. & Chen, C.W. (1984). Evaluation and estimation of potential carcinogenic risks of polynuclear aromatic hydrocarbons; Conference Record of the 1984 Pacific Rim Risk Conference on polycyclic aromatic hydrocarbons in the workplace, Honolulu, HI.

Chuang, J.C., Callahan, P.J., Lyu, C.W., Wilson, N.K. (1999). Polycyclic aromatic hydrocarbon exposures of children in low-income families. *J Expo Anal Environ Epidemiol*, 9,85–98.

Ciganek, M.R., Ulrich, R., Neča, J., Raszyk, J. (2002). Exposure of pig fatteners and dairy cows to polycyclic aromatic hydrocarbons. *Vet Med – Czech*, 47(5), 137–142.

Coleman, P.J., Lee, R.G.M., Alcock, R.E., Jones, K.C., (1997). Observations on PAH, PCB, and PCDD/F trends in UK urban air, 1991–1995. *Environmental Science & Technology*, 31, 2120–2124.

Commins, B.T. (1958). Polycyclic aromatic hydrocarbons in rural and urban air. *Int J Air Poll*, 1, 14-17.

Conde, F.J., Ayala, J.H., Afonso, A.M., Gonza'lez, V. (2005). Emissions of polycyclic aromatic hydrocarbons from combustion of agricultural and sylvicultural debris. *Atmospheric Environment*, 39, 6654-6663.

Cupr, P., Kla'nova', J., Barto˘s, T., Flegrova', Z., Kohoutek, J., Holoubek, I. (2006). Passive air sampler as a tool for long-term air pollution monitoring: Part 2. Air genotoxic potency screening assessment. *Environmental Pollution*, 144, 406-413.

Dabestani, R., & Ivanov, I. (1999). A compilation of physical, spectroscopic and photophysical properties of poly aromatic hydrocarbons. *Photochem Photobio*, 70, 10-34.

de Souza Pereira, M., Heitmann, D., Reifenha¨user, W., Meire, R.O., Santos, L.S., Torres, J.P.M., Malm, O., Ko¨rner, W. (2007). Persistent organic pollutants in atmospheric deposition and biomonitoring with Tillandsia usneoides (L.) in an industrialized area in Rio de Janeiro state, southeast Brazil - Part II: PCB and PAH. Chemosphere, 67, 1736-1745.

EPA (1984). Health effects assessments for polycyclic aromatic hydrocarbons (PAHs). US Environmental Protection Agency EPA, Environmental Criteria and Assessment Office, EPA 549/1-86-013, Cincinnati, OH.

EPA (1997). Field Evaluation of Screening Techniques for Polycyclic Aromatic Hydrocarbons, 2,4-Diphenoxyacetic Acid, and Pentachlorophenol in Air, House Dust, Soil, and Total Diet. EPA/600/SR-97/109

Eriksson, G., Jensen, S., Kylin, H., Strachan, W. (1989). The pine needle as a monitor of atmospheric pollution. *Nature* (London), 341, 42–44.

Ezzo, A.A. (2006). Effect of industrial emission on the environment at Sadat City. MSc Thesis, Faculty of Science, Cairo University.

Fang, G.C., Wu, Y.S., Cheng, P.F., Yang, I.L., Chen, M.H. (2004). PAHs in ambient air of suburban and industrial regions of central Taiwan. *Chemosphere*, 54, 443-452.

Finlayson-Pitts, B.J. & Pitts J.R. (2000). Chemistry of the Upper and Lower Atmosphere. Academic Press, San Diego.

Franzaring, J., Bierl, R., Ruthsatz, B. (1992). Active biological monitoring polycyclic aromatic hydrocarbons using kale (Brassica oleracea) as a monitor- species. *Chemosphere*, 25, 827–834.

Grimmer, G. (1983). Environmental carcinogens: *Polycyclic aromatic hydrocarbons*. CRC Press, Boca Raton, FL.

Han, X. & Naeher, L.P. (2006). A review of traffic-related air pollution exposure assessment studies in the developing world. *Environment International*, 32, 106–120.

Harrad, S., Hazrati, S., Ibarra, C. (2006). Concentrations of polychlorinated biphenyls in indoor air and polybrominated diphenyl ethers in indoor air and dust in Birmingham, United Kingdom: implications for human exposure. *Environmental Science & Technology*, 40, 4633–4638.

Hassan, S.K. (2006). Atmospheric polycyclic aromatic hydrocarbons and some Heavy Metals in suspended particulate matter in urban, Industrial and Residential areas in Greater Cairo. Ph.D. Thesis. Faculty of Science, Cairo University, Egypt.

Hassanien, M.A. & Abdel latif, N.M. (2008). Polycyclic aromatic Hydrocarbons in road dust over Greater Cairo. *Journal of Hazardous Materials*, 151, 247-254.

Hassanien, M.A. (2006). Outdoor/indoor air polycyclic aromatic hydrocarbon concentrations in a heavy traffic area of Cairo, Egypt. *Cent Eur J Occ Environ Med*, 12(1), 23-34.

Hassanien, M.A., Horváth, A., Györgyi, V., Katalin, M. (1996). Environmental pollution in a heavy traffic related area in Hungary. *Cent Eur J Occ Environ Med* 2(3), 249-258.

Hassanien, M.A., Rieuwerts, J., Shakour, A.A., Bitto, A. (2001). Seasonal and annual variations in air concentrations of Pb, Cd and PAHs in Cairo, Egypt. *Int J Environ Health Research*, 11(1), 13-27.

He, L-Y., Hu, M., Huang, X-F., Zhang, Y-H., Tang, X-Y. (2006). Seasonal pollution characteristics of organic compounds in atmospheric fine particles in Beijing. *Sci Total Environ*, 359, 167–76.

Holdren, J.P., Smith, K.R. et al. (2000). Energy, the environment, and health. In: Goldemberg, J. et al. (eds) The World Energy Assessment: Energy and the Challenge of Sustainability, New York, United National Development Programme, 61–110.

http://cfpub.epa.gov/ncea/cfm/recordisplay.cfm?deid=38617.

Huckins, J.N., Petty, J.D., Booij, K., (2006). Monitors of Organic Chemicals in the Environment e Semi permeable Membrane Device. Springer Science, New York, USA.

IARC (1983). Monographs on the evaluation of the carcinogenic risk of chemical to humans: *Polynuclear Aromatic Hydrocarbons*. Vol. 32, WHO, Lyon, France.

IPCS (International Programme on Chemical Safety) (1998). Selected non-heterocyclic polycyclic aromatic hydrocarbons, Environmental health Criteria 202. World Health Organization, Geneva.

Juhasz, L. & Naidu, R. (2000). Bioremediation of high molecular weight polycyclic aromatic hydrocarbons: A review of the microbial degradation of benzo[a]pyrene. *Inter Biodeter Biodegr*, 45, 57-88.

Kanaly, K.A, Harayama, S. (2000). Biodegradation of high-molecular weight polycyclic aromatic hydrocarbon by bacteria. *J Bacteriol*, 182, 2059-2067.

Kipopoulou, A.M., Manoli, E., Samara, C. (1999). Bioconcentration of polycyclic aromatic hydrocarbons in vegetables grown in an industrial area. *Environ Pollution* 106, 369-380.

Kreweski, D., Burnett, R.T., Goldberg, M.S., Hoover, K., Siemiatycki, J., Abrahamowicz, M., White, W.W. (2000). Reanalysis of the Harvard Six Cities Study and the American Cancer Society Study of Particulate Air Pollution and Mortality, Cambridge, MA, Health Effects Institute.

Lehndorff, E. & Schwark, L., (2004). Biomonitoring of air quality in the Cologne conurbation using pine needles as a passive sampler-Part II: polycyclic aromatic hydrocarbons (PAH). *Atmospheric Environment*, 38, 3793-3808.

Li, X.H., Ma, L.L., Liu, X.F., Fu, S., Cheng, H.X., Xu, X.B. (2006). Polycyclic aromatic hydrocarbon in urban soil from Beijing, China. *Environ Sci,* 18, 944-950.

Lim, H.L., Harrison, R.M., Harrad, S., (1999). The contribution of traffic to atmospheric concentrations of polycyclic aromatic hydrocarbons. *Environmental Science &Technology*, 33, 3538-3542.

Lin, Y.C., Lee, W., Chend, S.J., Chang-Chien, G.P., Tsai, P.J. (2008). Characterization of PAHs exposure in workplace atmospheres of a sinter plant and health-risk assessment for sintering workers. *Journal of Hazardous Materials*, 158, 636–643

Liu, Y., Zhu, L., Shen, X., (2001). Polycyclic aromatic hydrocarbons (PAHs) in indoor and outdoor air of hangzhou, China. *Environmental Science and Technology*, 35, 840–844.

Lohmann, R., Corrigan, B.P., Howsam, M., Jones, K.C., Ockenden, W.A. (2001). Further developments in the use of semi permeable membrane devices (SPMDs) as passive air samplers for persistent organic pollutants: field application in a spatial survey of PCDD/Fs and PAHs. *Environmental Science & Technology*, 35, 2576-2582.

Maertens, R.M., Bailey, J., White, P.A. (2004). The mutagenic hazards of settled house dust: a review. *Mutat Res*, 567, 401–25.

Maliszewska-Kordybach, B. (1999). Sources, concentrations, fate and effects of polycyclic aromatic hydrocarbons (PAHs) in the environment. Part A: PAHs in air. *Polish Journal of Environmental Studies*, 8, 131–136.

Marr, L.C., Kirchstetter, T.W., Harley, R.A., (1999). Characterization of polycyclic aromatic hydrocarbons in motor vehicle fuels and exhaust emissions. *Environmental Science & Technology*, 33, 3091-3099.

Mastral, A.M. & Callen, M.S. (2000). A reviewon polycyclic aromatic hydrocarbon (PAH) emissions from energy generation. *Environmental Science and Technology*, 34, 3051–3057.

McLachlan, M.S., Tolls, J., Welsch-Pausch, K. (1995). Field validation of a model of the uptake of gaseous SOC in Lolium multiflorum (rye grass). *Environ Sci Technol*, 29, 1998–2004.

McMichael, A.J. & Smith, K.R. (1999). Air pollution and health: seeking a global perspective. *Epidemiology*, 10, 1–4.

Motelay-Massei, A., Harner, T., Shoeib, M., Diamond, M., Stern, G., Rosenberg, B. (2005). Using passive air samplers to assess urban-rural trends for persistent organic pollutants and polycyclic aromatic hydrocarbons. 2. Seasonal trends for PAHs, PCBs, and organochlorine pesticides. *Environmental Science & Technology*, 39, 5763-5773.

Nash, G.R. & Beall, M.L. (1970). Chlorinated hydrocarbon insecticides: root uptake versus vapor contamination of soybean foliage. *Science*, 168, 1109–1111.

Naspinski, C., Lingenfelter, R., Cizmas, L., Naufal, Z., He, L.Y., Islamzadeh, A., Li, Z., Li, Z., McDonald, T., Donnelly, K.C. (2008). A comparison of concentrations of polycyclic aromatic compounds detected in dust samples from various regions of the world. *Environment International*, 34, 988–993.

Nisbet, C. & LaGoy, P. (1992). Toxic equivalency factors (TEFs) for polycyclic aromatic hydrocarbons (PAHs). *Reg Tox Pharmocol*, 16, 290–300.

Ohura, T., Amagai, T., Fusaya, M., Matsushita, H. (2004_b). Spatial distributions and profiles of atmospheric polycyclic aromatic hydrocarbons in two industrial cities in Japan. *Environmental Science & Technology*, 38, 49-55.

Ohura, T., Amagai, T., Sugiyama, T., Fusaya, M., Matsushita, H. (2004_a). Characteristics of particle matter and associated polycyclic aromatic hydrocarbons in indoor and outdoor air in two cities in Shizuoka, Japan. *Atmospheric Environment*, 38, 2045–2054.

Ohura, T., Sugiyama, T., Amagai, T., Fusaya, M., Matsushita, H. (2002). Simultaneous liquid chromatographic determination of 39 polycyclic aromatic hydrocarbons in indoor and outdoor air and application to a survey on indoor air pollution in Fuji, Japan. *Journal of AOAC International*, 85, 188–202.

Petry, T., Schmid, P., Schlatter, C. (1996). The use of toxic equivalency factors in assessing occupational and environmental health risk associated with exposure to airborne mixtures of polycyclic aromatic hydrocarbons (PAHs). *Chemosphere*, 32, 639–648.

Pott, F., Pyrolyseagase, (1985). PAH und lungenkrebrisiko–daten und bewertung, Staub-Reinhaltung dedr Luft, 45, 368–379.

Ravinadra, K., Mittal, A.K., VanGrieken, R. (2001). Health risk assessment of urban suspended particulate matter with special reference to polycyclic aromatic hydrocarbons: a review. *Reviews on Environmental Health*, 16, 169-189.

Ravindra, K., Sokhi, R., Grieken, R.V. (2008). Atmospheric polycyclic aromatic hydrocarbons: source attribution, emission factors and regulation. *Atmospheric* Environment, 42, 2895-2921.

Redmond, C., Strobino, B., Cypess, R. (1976). Cancer experience among coke by-product workers. *Ann N Y Acad Sci,* 217, 102– 115.

RIVM (1989). National Institute of Public Health and Environmental Protection. Integrated criteria document PAHs. The Netherlands, Report no. 758474011; PB90-255050.

Samet, J.M., Dominici, F., Zeger, S.L., Schwartz, J., Dockery, D.W. (2000). National Morbidity, Mortality, and Air Pollution Study, Cambridge, MA, Health Effects Institute.

Smith, K.R. (2000). National burden of disease in India from indoor air pollution. *Proc Natl Acad Sci*, 97, 13286–13293.

Smith, M.E. & Frankeberg, T.T. (1975). Improvement of ambient sulfur dioxide concentrations by conversions from low to high stacks. *J Air Pollut Contr Asssoc*, 6, 595-601.

Soderstrom, H., Hajsˇlova´, J., Kocourek, V., Siegmund, B., Kocan, A., Obiedzinski, W., Tysklind, M., Bergqvist, P.A. (2005). PAHs and nitrated PAHs in air of five European countries determined using SPMDs as passive samplers. *Atmospheric Environment*, 39, 1627-1640.

Strandberg, B., Gustafson, P., Soderstrom, H., Barregard, L., Bergqvist, P.A., Sallsten, G. (2006). The use of semipermeable membrane devices as passive samplers to determine persistent organic compounds in indoor air. *Journal of Environmental Monitoring,* 8, 257–262.

Tan, J.H., Bi, Xin-H, Duan, J.C., Rahn, K.A. Sheng, G.Y., Fu, J.M. (2005). Seasonal variation of polycyclic aromatic hydrocarbons associated with PM10 in Guanzhou, china. Atm Res Doi:10.1016/J.atmosres. 2005.09.004.

Thorslund, T. & Farrer, D. (1991). Development of Relative Potency Estimated for PAHs and Hydrocarbon Combustion Product Fractions Compared to Benzo(a)pyrene and Their Use in Carcinogenic Risk Assessments, U.S. Environmental Protection Administration (EPA), Washington, DC.

Trapp, S., Scheunert, I., Matthies, M., Topp, E.M. (1990). Modeling the bioconcentration of organic chemicals in plants. *Environ Sci Technol*, 24, 1246–1252.

Umlauf, G., Hauk, H., Reissinger, M. (1994). Deposition of semivolatile organic compounds to spruce needles. *Environ Sci Pollut Res Int*, 1, 209– 222.

UNEP (2002). United Nations Environment Programme Chemicals, Mediterranean REGIONAL REPORT

Viana, M., Perez, C., Querol, T., Alastuey, A., Nickovic, S., Baldasano, J.M. (2005). Spatial and temporal variablility of PM levels and complex summer atmospheric in Barcelona (NE Spain). *Atmospheric Environment,* 39,5343-5361.

WHO (1987). Air quality guideline for Europe. WHO European Series, WHO Regional office for Europe, Copenhagen, 105–117. http://www.euro.who.int/air.

WHO (World Health Organization) (1999) Global Air Quality Guidelines, Geneva, World Health Organization.

Yang, H.H. & Chen, C.M. (2004). Emission inventory and sources of PAHs in the atmosphere at a suburban area in Taiwan. *Chemosphere*, 56, 879-887.

Yu, X.Z., Gao, Y., Wu, S.C., Zhang, H.B., Cheung, K.C., Wong, M.H. (2006). Distribution of polycyclic aromatic hydrocarbons in soils at Guiyu area of China, affected by recycling of electronic waste using primitive technologies. *Chemosphere*, 65, 1500-1509.

Zander, M. (1980). (Origins, chemical and physical properties of PAH) in: (Air pollution through polycyclic aromatic hydrocarbons: registration and evaluation). Dusseldorf, VDI-Verl eng, pp 11-21 (VDI Repot No. 358 in Germany).

Zhang, S., Zhang, W., Shen, Y., Wang, K., Hu, L., Wang, X. (2008). Dry deposition of atmospheric polycyclic aromatic hydrocarbons (PAHs) in the southeast suburb of Beijing, China. *Atmospheric Research*, 89, 138–148.

Zhu, X., Pfister, G., Henkelmann, B., Kotalik, J., Silke, B., Fiedler, S., Schramm, K. (2008). Simultaneous monitoring of profiles of polycyclic aromatic hydrocarbons in contaminated air with semipermeable membrane devices and spruce needles. *Environmental Pollution,* 156, 461-466.

Zmirou, D., Masclet, P., Boudet, C., Dor, F., Dechenaux, J. (2000). Personal exposure to atmospheric polycyclic aromatic hydrocarbons in a general adult population and lung cancer risk assessment. *J Occup Environ Health*, 42, 121–125.

In: Polycyclic Aromatic Hydrocarbons: Pollution, Health … ISBN 978-1-60741-462-9
Editors: Pierre A. Haines and Milton D. Hendrickson

Chapter 4

ADIPOSE TISSUE AS A KEY CONTRIBUTOR TO POLYCYCLIC AROMATIC HYDROCARBON-INDUCED CARCINOGENESIS

P. Irigaray[1*] and D. Belpomme[2]

[1]Cancer Research Center, Association for Research and Treatments Against Cancer (ARTAC), Paris, France

[2]Department of Medical Oncology, European Hospital Georges Pompidou (HEGP), University of Paris, Paris, France

ABSTRACT

We have previously shown that benzo[a]pyrene used at extremely low dose can cause weight gain in mice and that the increase in adipose tissue mass is due to inhibition of β-adrenergic stimulation of lipolysis. Moreover we have suggested that adipose tissue is a reservoir for many carcinogens including Polycyclic Aromatic Hydrocarbons (PAHs).In this paper we demonstrate that the two C4 PAHs pyrene and phenanthrene can accumulate into adipocytes, but at the difference of benzo[a]pyrene have no impact on lipolysis.Since benzo[a]pyrene has been proved to be mutagenic while C4 PAHs are tumor promoters, we consider that a mixture of the three PAHs may act as a complete carcinogen. Furthermore, because adipose tissue can be induced by benzo[a]pyrene, is a reservoir for many PAHs and other carcinogens and can stimulate proliferation of normal or cancer cells, through the secretion of adipokines by adipocytes, we therefore propose that adipose tissue is a key contributor to chemical carcinogenesis by acting not only upon initiation and promotion but also upon progression.

Keywords: PAH, pyrene, phenanthrene, benzo[a]pyrene, adipocytes, lipogenesis, lipolysis, obesity, cancer, mutation, promotion, progression.

---

* P. Irigaray, PhD,ARTAC ,57-59 rue,de la convention,75015 Paris, Phone: +33 (0)1 45 78 53 53,Fax: +33 (0)1 45 78 53 50,E-mail: philippei.artac@gmail.com.

Low molecular weight polycyclic aromatic hydrocarbons (*l*-PAHs) (3 to 4 rings) such as the C4 PAHs, pyrene and phenanthrene have been shown to be non genotoxic tumor promoters, while high molecular weight PAHs (*h*-PAHs) (5 to 7 rings), such as the C5 PAH benzo[a]pyrene (B[a]P) are mutagenic though DNA adduction and may act as tumor initiators and/or tumor promoters [1].

We have previously shown that in addition to its genotoxic properties, B[a]P can inhibit adipocyte lipolysis and contribute to the hypertrophy/hyperplasia of adipose tissue [2], and have consequently hypothesized that in humans, B[a]P can generate both obesity and cancer [3].

Furthermore we have proposed the concept that the adipose tissue in addition to its endocrine function acts as a reservoir for many lipophilic liposoluble exogenous carcinogens and that these carcinogens can be permanently released in the blood circulation and target peripheral tissues for carcinogenesis. [3, 4].

In this paper, we attempt to further elaborate the concept that adipose tissue act as a reservoir for chemical carcinogens by analysing the *ex vivo* incorporation of PAH into adipocytes and by testing whether the C4 PAHs pyrene and phenanthrene can inhibit lipolysis as the C5 PAH B[a]P does.

In this study, we confirm that adipocytes can incorporate *h*-PAH and *l*-PAH in a similar manner and that B[a]P in addition to its mutagenic carcinogenic properties can significantly inhibit lipolysis. So because adipocytes can incorporate both mutagenic and promoting PAHs, we conclude that PAHs cocktails may be equivalent to complete carcinogens and that the adipose tissue may play a key role in PAH-induced carcinogenesis.

Materials and Methods

Isolated Adipocyte Preparation

Adipocytes were isolated from mice adipose tissue using Rodbell's method as previously described [2].

Lipogenesis Assays

Lipogenesis assays were performed as previously described [5]. Primary adipocytes freshly isolated from murine white adipose tissue were incubated with insulin (90 nM) and HAPs. This insulin concentration permits to obtain good lipogenesis [5].

More precisely, cell suspensions were aliquoted in Eppendorf tubes containing Krebs Ringer. Bicarbonate Buffer (KRBB) supplemented with 4% (w / v) bovine serum albumin (BSA), 1mM glucose and 0,278 µCi of ^{14}C-phenanthrene (0.0066 µmol/l) or 0.296 µCi of ^{14}C-pyrene (0.007 µmol/l) or 0,272 µCi of ^{14}C-benzo[a]pyrene (0,0065 µmol/l). After different incubation time at 37°C under gentle agitation (40 r.p.m) with or without insulin (90 nM), aliquots of the media were removed for radioactivity determination by liquid scintillation (Ultima Gold) counting in a Packard Tricarb in order to determine the percent of extra cellular radioactivity.

Lipolysis Assays

Lipolysis assays were performed as previously described [2]. Primary adipocytes freshly isolated from murine white adipose tissue were incubated with increasing concentrations of HAPs and with a dose of 10μM of epinephrine. This process was followed by measurement of the Free Fatty Acids (FFA) released in the media.

Preliminary experiments using different concentrations of epinephrine (Figure 1A) and incubation times with epinephrine (Figure 1B) allowed us to determine the epinephrine concentration of 10 μM and incubation time of 45 minutes that permit obtaining good lipolysis.

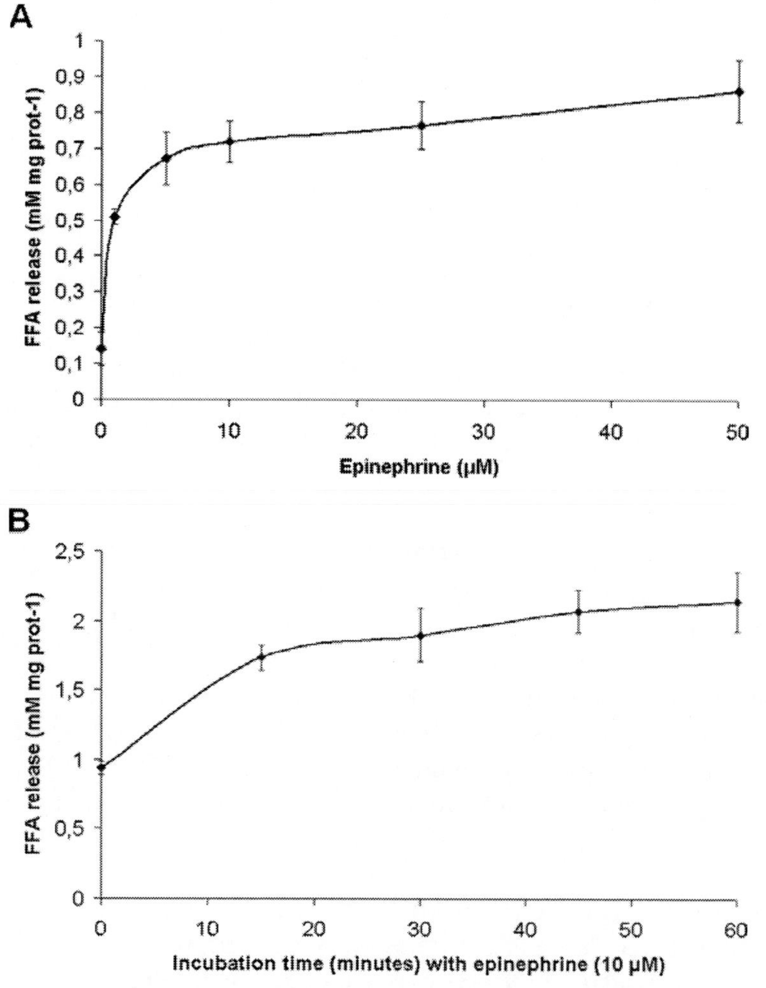

Figure 1. Effect of different concentrations and incubation times of epinephrine on the release of FFA from mice adipocytes. Freshly isolated mice adipocytes were incubated (A) for 60 min at different concentrations of epinephrine and (B) for different times with a previously determined concentration of 1.8 mg/l (10 μM) of epinephrine. Each point represents the mean ± SEM of 4 experiments during which FFA levels were measured in duplicate.

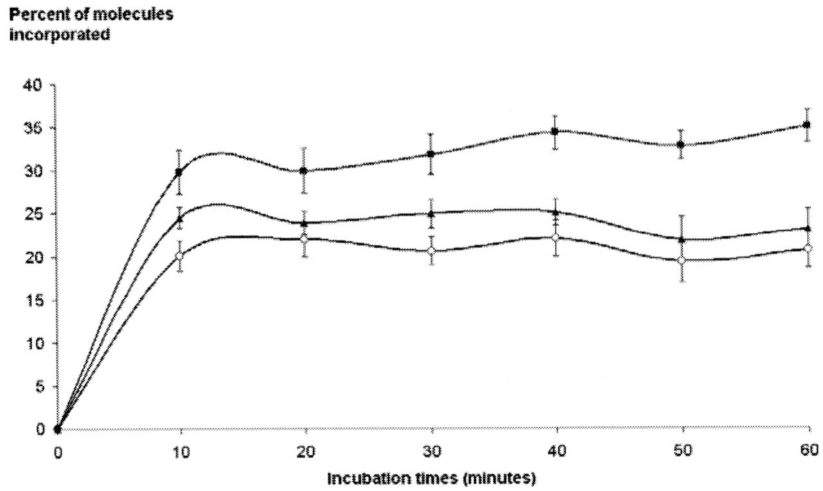

Figure 2. Kinetics of incorporation into adipocytes of B[a]P (■), pyrene (○) and phenanthrene (▲) in the presence of 90 nM of insulin. Each point represents the mean ± SEM of 4 experiments during which FFA levels were measured in duplicate. Data are expressed as the percent of molecules incorporated into mice adipocytes.

Cell suspensions were aliquoted in Eppendorf tubes containing KRBB supplemented with 4% (w/v) BSA, 5 mM glucose and one of the three PAHs, i.e. B[a]P, pyrene and phenanthrene at different concentrations during 45 min. After this preincubation period of 45 min, an epinephrine solution was added to the medium at a final concentration of 10 μM for 45 minutes, before measurements of the FFA released in the media were made.

RESULTS

1. Incoporation study

As reported in figure 2, the kinetics of incorporation was identical for the three PAHs. PAHs incorporate into adipocytes uniquely during the first ten minutes ($p = 0.001$). This rapid incorporation suggests that PAHs incorporation is not insulino-dependant [5]. In addition, as indicated in tables 1 and 2 incorporation appears to depend upon the lipophilicity of molecules.

Table 1. Incorporation rates of PAHs after 60 minutes of incubation at a concentration of 90 nM of insulin. Values indicated are the mean ± SEM of 4 experiments during which FFA levels were measured in duplicate.

Compounds	Percent of PAHs introduced in or adsorbed on adipocyte
Benzo[a]pyrene	35.04±2.28
Pyrene	20.62±2.55
Phenanthrene	23.08±2.89

Table 2. Physico-chemical properties of the three PAHs tested.

Compounds	Number of benzenoïd rings	Molecular weight (g/mol)	Water solubility (mg/l at 25°C)	Lipophilicity (log Kow)
Benzo[a]pyrene	5	252.3	3.80.10-3	6.31
Pyrene	4	202.3	1.30.10-1	4.88
Phenanthrene	3	178.2	1.21	4.6

2. Lipolysis Inhibition

A significant inhibitory effect on epinephrine-induced FFA release ($P < 0.01$) was achieved with 1 mg/l B[a]P concentration (Figure 3A). The estimated Kd value for B[a]P inhibition of lipolysis is 0.9 mg/l (3.5 μM). This value is in the same range as that measured for the known β blocker atenolol (Kd = 0.4 mg/l or 1.5 μM for β_1-adrenoreceptor and 2.3 mg/l or 8.6 μM for β_2-adrenoreceptor) [6]. The inhibitory effect of 1.8 mg/l of B[a]P on epinephrine-induced lipolysis was detectable within 5 min of preincubation (Figure 3B). This suggests that B[a]P directly affects the cellular signal transduction pathway rather than affecting gene expression. However, for the two C4 PAHs (pyrene and phenanthrene), no effect was observed on the epinephrine-induced FFA release (Figures 4A and 4B).

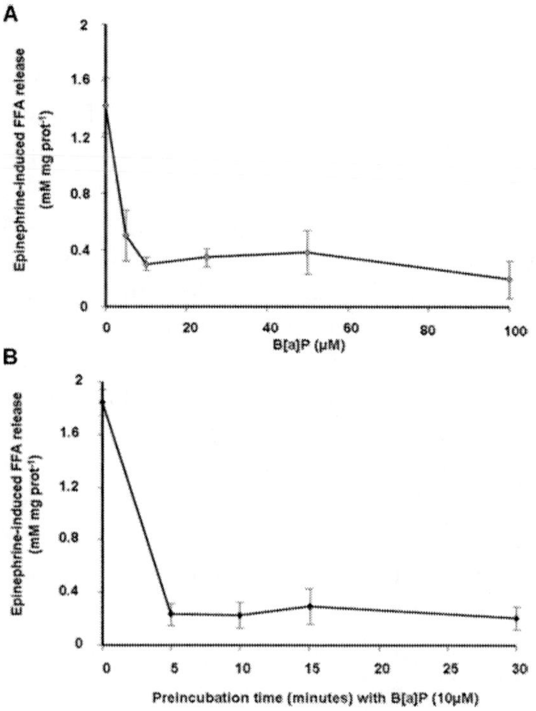

Figure 3. Effect of B[a]P on the release of FFA from mice adipocytes. Freshly isolated mice adipocytes were incubated (A) for 45 min with different concentrations of B[a]P and (B) during different preincubation periods with a predetermined concentration of 2.5 mg/l (10 μM) of B[a]P. Epinephrine (1.8 mg/l i.e. 10 μM) was then added and the incubation continued for 45 min. Each point represents the mean ± SEM of the 4 experiments during which FFA levels were measured in duplicate.

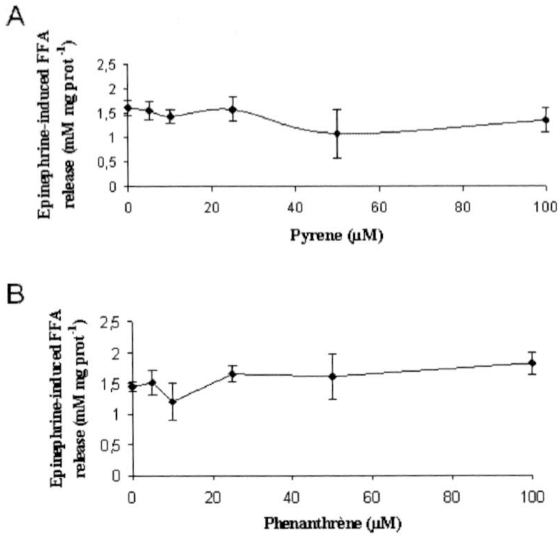

Figure 4. Effect of the two C4 PAHs, pyrene and phenanthrene on the release of FFA from mice adipocytes. Freshly isolated mice adipocytes were incubated for 45 min with different concentrations of Pyrene (A) or Phenanthrene (B). Epinephrine (1.8 mg/l i.e. 10 µM) was then added and the incubation continued for 45 min. Each point represents the mean ± SEM of 4 experiment during which FFA levels were measured in duplicate.

DISCUSSION

B[a]P is a well-known ubiquitous carcinogen and consumption of B[a]P-rich foods have been suggested to contribute to the overall cancer burden affecting human populations [7]. The carcinogenicity of B[a]P is well-documented. This pollutant is metabolized via the cytochrome P450 system into reactive dihydrodiol epoxide derivatives (e.g. B[a]P-7,8-dihydrodiol-9,10-epoxide, BPDE). These metabolites bind covalently to DNA and form BPDE adducts, that may lead to mutations and consequently to uncontrolled cell growth and tumour formation in various tissues (lung adenocarcinoma, mammary adenocarcinoma, lymphoma and hepatocellular carcinoma) [8]. B[a]P has also been shown to display immunotoxic properties that affect macrophage function [9] and increase local inflammatory response which may contribute to increased atherosclerotic lesion size [10].

In this paper, we show that in *ex vivo* experiments, B[a]P can inhibit epinephrine-induced FFA release whereas the two C4 PAHs pyrene and phenanthrene cannot. This confirms our previous observation that the administration of a single dose of B[a]P (0.5 mg/kg in IP) in male C57BL6/J mice induces a significant reduction of the epinephrine-induced lipolysis. This phenomenon occurs in less than 30 minutes with maximum inhibition 24 hours later before returning to the basal lipolytic capacity levels 72 hours after the injection. In addition, we have shown that repeated intraperitoneal injections of B[a]P (0.5 mg/kg/48h) for 14 days induce a significant body weight gain (+43%, $p>0.1$) in mice without detectable changes in food consumption [2].

We previously proposed that the inhibition of lipolysis by B[a]P most likely results from physical perturbations of the plasma membrane phospholipid bilayer. Our interpretation is that B[a]P strongly and rapidly inhibits the signalling capacity of the β1-, β2-, β3-adrenergic receptors and of the ACTH receptors. These receptors share common features: all contain seven transmembrane spanning domains and are coupled to G-proteins which are themselves anchored to the inner leaflet of the plasma membrane. In contrast atrial natriuretic peptide-induced lipolysis via the stimulation of natriuretic peptide receptor A, that contains a single transmembrane-spanning domain, was not affected by B[a]P [2]

Yet, physicochemical studies using differential scanning calorimetry, infrared spectroscopy and small-angle X-ray diffraction have shown that B[a]P incorporates into phospholipid bilayers and localizes in the most apolar region of the phospholipid matrix. This phenomenon may account for the observation of an expanded and swollen membrane [11]. We have therefore, proposed that distortion of the physiochemical properties of the adipocyte plasma membrane by B[a]P decreases the signalling capacity of G-coupled receptors intimately linked to the phospholipid bilayer, via their seven transmembrane domains.

Our data tends to show that the capacity of PAHs to incorporate into adipocytes depends upon their Coeficient Octanol Water (Kow). We therefore suggest that because the capacity of lipophilic molecules to be incorporated into the phospholipids bilayer depends on their Kow, the lack of lipolysis-inhibiting effect of the two C4 PAH pyrene and phenanthrene might be related to their low Kow.

A striking observation in our study is that the concentrations needed to achieve the B[a]P-induced lipolysis inhibition effect are 2000-fold lower than those causing carcinogenesis in rodents and 20-fold lower than those causing alterations of the epidermal growth factor receptors in cultured human uterine cells, RL95-2 [12]. Another observation is that B[a]P can directly inhibit epinephrine-induced lipolysis, whereas in order to adduct DNA and produce mutagenic and carcinogenic effects, B[a]P needs to be activated through the CYP1A1 cytochrome P450 process. It has been shown that in the ob/ob obese mouse model, CYP1A1 levels are decreased [13]. We thus assume that because B[a]P contribute to the hypertrophy/hyperplasia of the adipose tissue, it may simultaneously increase its bioaccumulation in this tissue by reducing its metabolic bioactivation. Since pyrene and phenanthrene are also activated through CYP1A1, a similar process of storage in the adipose tissue may also concern these two C4 PAHs, so B[a]P might in fact generate the storage in the adipose tissue of both types of mutagenic and promoting PAHs.

In our study, all three PAHs incorporated adipocytes in a similar manner. Our results are similar to those found by Victor, who obtained a complete incorporation of the 2,3,7,8-TCDD in hepatocytes suspension within 2 minutes [14]. These data reinforce the concept that adipose tissue may act as a reservoir for many different lipophilic exogenous chemicals including PAHs.

This concept is opposed to the hypothesis according to which adipose tissue might be a ''protective reservoir'' [15]. This assumption was based on the opinion that because adipose tissue is able to bioaccumulate chemical pollutants, it might protect the organism against their deleterious effects [15-18] Indeed, we believe that this concept needs to be revised. Following environmental exposure to low-dose chemical pollutants during many years, accumulation of carcinogens in the adipose tissue might in fact correspond to a stronger exposure, as it is observed in hot spot polluted area. On the basis of many experimental toxicological data, it is assumed that prolonged chronic exposure to low dose xenochemicals may in fact lead to carcinogenesis, as it is the case for occupational cancers during hotspot exposure [19, 20]. As a result of this new concept, it clearly appears that the carcinogenic effects of environmental xenochemicals, be they mutagens and/or promoters may be the consequence of a chronic exposure to repetitive lower doses than those at which no effect level is observed in classical rodent tests. This concept is based on the fact that many exogenous chemical carcinogens are persistent organic pollutants accumulating in fat and that they can be released in the organism from adipose tissue at doses which do not correspond to those observed in the environment. Therefore, during prolonged environmental exposure, cancer may be induced by repetitive lower doses than it is usually expected [20, 21]. Furthermore, it has been shown that mixtures of pollutants may have synergic effects and that some types of ''cocktail effects'' that occur in adipose tissue may contribute to carcinogenesis [22] and/or cancer cell proliferation [23].

Since our study shows that C4 as well as C5 PAHs can incorporate into adipocytes, we hypothesize that PAH cocktails that bioaccumulate in the adipose tissue may be equivalent to complete carcinogens (Figure 5).

As summarized in figure 5, we believe that the concept of chronic toxicity must also be totally revised in the light of the role of reservoir of the adipose tissue for lipophilic carcinogens. Even in lean people, white adipose tissue (WAT) represents about 15 to 25 % of body weight, and this percentage can increase by more than 50 % in the case of obese patients. Environmental lipophilic pollutants can enter the organism via food, water and air, bioaccumulate and often concentrate in the adipose tissue, to be permanently released in the blood circulation and elicit their effect in peripheral tissues at convenient doses according to their potential biologic effects. The amount of pollutants that bioaccumulate in WAT may depend on dietary exposure and on individual susceptibility [24]. The postnatal stock of innate preadipocytes, the volume of WAT and the capacity of the organism to assure energy homeostasis might be risk factors, inter alia. Also weight loss may result in an increased concentration of pollutants in reduced WAT [25] and in their decreased elimination rate [26].

Finally we conclude that in addition to its endocrine properties, the adipose tissue can act as a reservoir for many exogenous chemical molecules including PAHs and that B[a]P may play a pivotal role in carcinogenesis not only by inducing cancer through its mutagenic carcinogenic effects but also by increasing the bioaccumulation capacity of the adipose tissue mass.

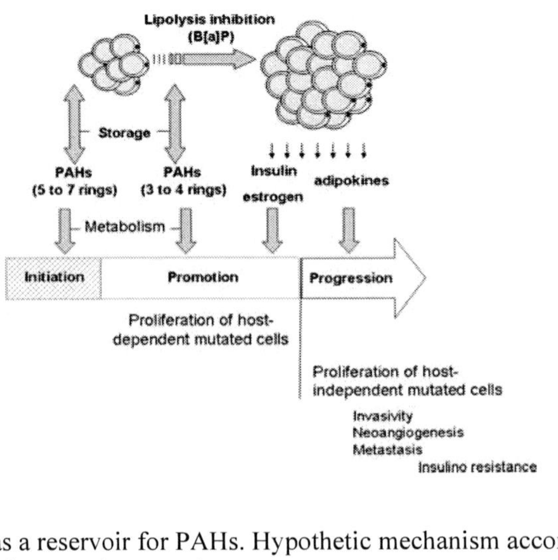

Figure 5. Adipose tissue as a reservoir for PAHs. Hypothetic mechanism according to which repeated low doses of PAHs can induce carcinogenesis.

REFERENCES

[1] Trosko, JE; Upham, BL. (2005).The emperor wears no clothes in the field of carcinogen risk assessment: ignored concepts in cancer risk assessment. *Mutagenesis*, *20*, 81–92.

[2] Irigaray, P; Ogier, V; Jacquenet, S; Notet, V; Sibille, P; Mejean, L; et al. (2006). benzo[a]pyrene impairs beta-adrenergic stimulation of adipose tissue lipolysis and causes weight gain in mice: a novel molecular mechanism of toxicity for a common food pollutant. *FEBS J*, *273*, 1362-1372.

[3] Irigaray, P; Newby, JA; Lacomme, S; Belpomme, D. (2007). Overweight/obesity and cancer genesis: more than a biological link. *Biomed Pharmacother*, *61*, 665-678.

[4] Irigaray, P; Belpomme, D. (2009). Mechanism of Environmental Carcinogenesis. In *Encyclopedia of Environmental Health*, Elsevier, *accepted for publication in 2009.*

[5] Irigaray, P; Mejean, L; Laurent, F. (2005). Behaviour of dioxin in pig adipocytes. *Food Chem Toxicol*, *43*, 457-460.

[6] Smith, C; Teitler, M. (1999). Beta-blocker selectivity at cloned human beta 1- and beta 2-adrenergic receptors. *Cardiovasc Drugs Ther*, *13*, 123–126.

[7] Kazerouni, N; Sinha, R; Hsu, CH; Greenberg, A; Rothman, N. (2001). Analysis of 200 food items for benzo[a]pyrene and estimation of its intake in an epidemiologic study. *Food Chem Toxicol*, *39*, 423–436.

[8] IARC. (1983) *IARC Monographs on the Evaluation of Carcinogenic Risk of Chemicals to Humans – Polynuclear Aromatic Compounds*, Vol. 32 (33-91), pp. 211–224. International Agency for Research into Cancer, Lyon.

[9] van Grevenynghe, J; Rion, S; Le Ferrec, E; Le Vee, M; Amiot, L; Fauchet, R; Fardel, O. (2003). Polycyclic aromatic hydrocarbons inhibit differentiation of human monocytes into macrophages. *J Immunol*; *170*, 2374–2378.

[10] Curfs, DM; Lutgens, E; Gijbels, MJ; Kockx, MM; Daemen, MJ; van Schooten, FJ. (2004). Chronic exposure to the carcinogenic compound benzo[a]pyrene induces larger and phenotypically different atherosclerotic plaques in ApoE-knockout mice. *Am J Pathol*, *164*, 101–108.

[11] Jimenez, M; Aranda, FJ; Teruel, JA; Ortiz, A. (2002). The chemical toxic benzo[a]pyrene perturbs the physical organization of phosphatidylcholine membranes. *Environ Toxicol Chem*, *21*, 787–793.

[12] McGarry, MA; Charles, GD; Medrano, T; Bubb, MR; Grant, MB; Campbell-Thompson, M; Shiverick, KT. (2002). Benzo(a)pyrene, but not 2,3,7,8-tetrachlorodibenzo-p-dioxin, alters cell adhesion proteins in human uterine RL95-2 cells. *Biochem Biophys Res Commun*, *294*, 101–107.

[13] Roe, AL; Howard, G; Blouin, R; Snawder, JE. (1999). Characterization of cytochrome P450 and glutathione S-transferase activity and expression in male and female ob / ob mice. *Int J Obes Relat Metab Disord*, *23*, 48–53.

[14] Victor, J; Wroblewski, VJ; Olson, JR. (1985). Hepatic metabolism of 2,3,7,8-Tetrachlorodibenzo-p-dioxin (2,3,7,8-TCDD) in the rat and guinea pig. *Toxicol Appl Pharmacol*, *81*, 231-240.

[15] Lassiter, RR; Hallam, TG. (1990). Survival of the fattest: implication for acute effects of lipophilic chemicals on aquatic populations. *Environ Toxicol Chem*, *9*, 585-595.

[16] Geyer, HJ; Schramm, KW; Scheunert, I; Schughart, K; Buters, J; Wurst, W; et al. (1997). Considerations on genetic and environmental factors that contribute to resistance or sensitivity of mammals including humans to toxicity of 2,3,7,8-tetrachlorodibenzo-p-dioxin (TCDD) and related compounds. Part 1: Genetic factors affecting the toxicity of TCDD. *Ecotoxicol Environ Saf*, *36*, 213-230.

[17] Geyer, HJ; Scheuntert, I; Rapp, K; Kettrup, A; Korte, F; Greim, H; et al. (1990). Correlation between acute toxicity of 2,3,7,8-tetrachlorodibenzo-p-dioxin (TCDD) and total body fat content in mammals. *Toxicology*, *65*, 97-107.

[18] Geyer, HJ; Scheunert, I; Rapp, K; Gebefugi, I; Steinberg, C; Kettrup, A. (1993). The relevance of fat content in toxicity of lipophilic chemicals to terrestrial animals with special reference to dieldrin and 2,3,7,8-tetrachlorodibenzo-p-dioxin (TCDD). *Ecotoxicol Environ Saf*, *26*, 45-60.

[19] Belpomme, D; Irigaray, P; Hardell, L; Clapp, R; Montagnier, L; Epstein, S; Sasco, AJ. (2007). The multitude and diversity of environmental carcinogens. *Environ Res*, *105*, 414-429.

[20] Howard, CV; Newby, JA. (2004). Could the increase in cancer incidence be related to recent environmental changes?. In: Nicolopoulou-Stamati P, Hens L, Howard CV, Van Larenbeke L, editors. *Cancer as an environmental disease*. Dordrecht, The Netherlands: Kluwer Academic Publishers; 2004. p. 39-56.

[21] Hutcheon, DE; Kantrowitz, J; Van Gelder, RN; Flynn, E. (1983). Factors affecting plasma benzo[a]pyrene levels in environmental studies. *Environ Res*, *32*, 104-110.

[22] Knerr, S; Schrenk, D. (2006). Carcinogenicity of ''non-dioxinlike'' polychlorinated biphenyls. *Crit Rev Toxicol*, *36*, 663-694.

[23] Payne, J; Scholze, M; Kortenkamp, A. (2001). Mixtures of four organochlorines enhance human breast cancer cell proliferation. *Environ Health Perspect*, *109*, 391-397.

[24] Smeds, A; Saukko, P. (2001). Identification and quantification of polychlorinated biphenyls (PCBs) and some endocrine disrupting pesticides in human adipose tissue from Finland. *Chemosphere*, *44*, 1463-1471.

[25] Jandacek, RJ; Anderson, N; Liu, M; Zheng, S; Yang, Q; Tso, P. (2005). Effects of yo-yo diet, caloric restriction, and olestra on tissue distribution of hexachlorobenzene. *Am J Physiol*, *288*, G292-G299.

[26] MichalekK, JE; Tripathi, RC. (1999). Pharmacokinetics of TCDD in veterans of operation ranch hand: 15-year follow-up. *J Toxicol Environ Health*, *57*, 369-378.

In: Polycyclic Aromatic Hydrocarbons: Pollution, Health … ISBN 978-1-60741-462-9
Editors: Pierre A. Haines and Milton D. Hendrickson

Chapter 5

CHLORINATE POLYCYCLIC AROMATIC HYDROCARBONS: A NEW CLASS OF ENVIRONMENTAL TOXIN?

Takeshi Ohura** and Hiroyuki Sakakibara
Institute for Environmental Sciences, University of Shizuoka, Japan

ABSTRACT

A review has provided the environmental sources and behavior of chlorinated 3- to 5-ring polycyclic aromatic hydrocarbons (ClPAHs). ClPAHs have been known to be ubiquitous contaminants, which can be found in urban air, vehicle exhaust gas, snow, tap water and sediments. The concentrations in the each environment are generally higher than those of chlorinated dioxins, although which are significantly lower than the concentrations of parent compounds, PAHs. Based on the environmental data and emission sources analysis for the ClPAHs, the dominant production process is by incineration of organic materials including chlorine like as the case of dioxins. On the other hand, treatment of PAHs with chlorine in water also produced ClPAHs. Indeed, such secondary reaction for ClPAHs generation could occur in the gaseous phase. For the biological effects, the toxicities such as mutagenicity and aryl hydrocarbon receptor activity of certain ClPAHs are emphasized than those of parent PAH, in which the toxicities of exposure to ClPAHs based on AhR activity were estimated to be approximately 30-50 times higher than that of dioxins. To investigate the sources and environmental behavior of ClPAHs is of great importance to assess human health risks.

INTRODUCTION

To date, the studies on the environmental behaviors and biological effects of PAHs have been carried out in the world, and the numbers of papers are more than 10,000. On the other

* Address correspondence to T. Ohura,Institute for Environmental Sciences, University of Shizuoka, 52-1 Yada, Shizuoka 422-8526, Japan.Telephone: +81 54 264 5789. Fax: +81 54 264 5798.E-mail: ooura@smail.u-shizuoka-ken.ac.jp

hand, those studies on chlorinated congeners, ClPAHs, are very limited [1]. One of the reasons will be due to the lack of purified, individual ClPAH analytical standard. Recently, Ohura et al. succeeded to synthesize 12 individual ClPAHs with 3 to 5 ring as analytical standards (Figure 1) [2]. It has been advanced the studies on environmental behaviors and biological effects of ClPAHs. This section outlines the latest advance of ClPAHs study.

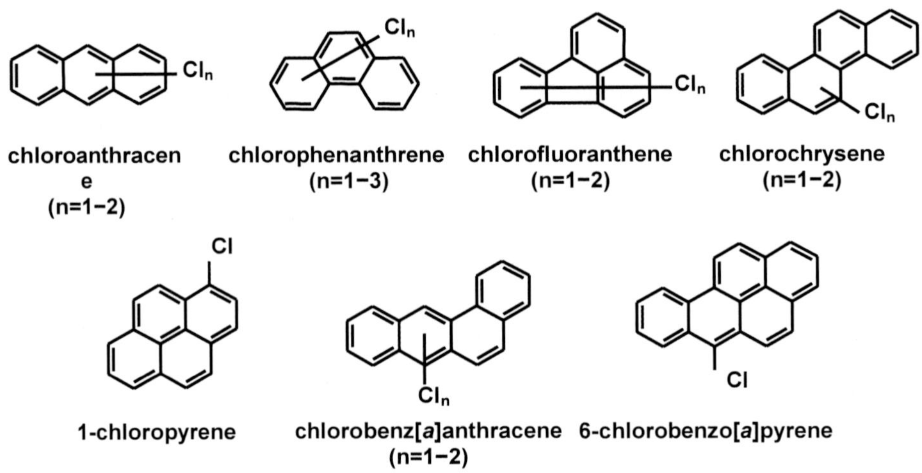

Figure 1. Structures of ClPAH derivatives.

Occurrences of Clpahs in the Environment

The first report of environmental ClPAHs will be on the survey of tap water performed by Shiraishi et al., which were performed as long ago as 1980s [3]. In this paper, relatively low molecular weight ClPAHs (< 4-ring) such as chlorophenanthrene and chloropyrene derivatives have been detected, which have been suggested to produce during chlorine treatment of water as the possible production mechanism. Indeed, Oyler et al. have confirmed to be caused production of low molecular weight ClPAHs by process of chlorine disinfection in water [4, 5]. Some time later, Nilsson et al. have reported that ClPAHs with 3 to 5-ring were present in both gas and particle phase in urban street and road tunnel air [6]. Based on the profile analysis, the authors suggested that the ClPAHs were formed in secondary reactions with PAH and leaded gasoline scavenger containing dichloroethane. In the mid-2000s, Ohura and co-workers have been carrying on the studies on atmospheric behaviors and sources of ClPAHs [2, 7-9]. In a Japanese urban air, 26 species of ClPAHs have been detected in either the gaseous or particulate phase. The ambient (gaseous + particulate) concentration of total ClPAH ranged from 18 to 330 pg/m^3 [8]. Those ClPAHs levels tended to be approximately ~1000 times lower than PAHs. A seasonal variability of both gaseous and particulate ClPAHs concentrations was observed, with higher concentrations in winter than in summer, which was different profile from PAH. Only under the study, the natural logarithm of total concentrations of particulate ClPAHs were significantly correlated ($p < 0.05$) with the reciprocal mean temperatures, but not for gaseous ClPAHs. It implies that the gaseous ClPAHs, rather than particulate ClPAHs, would be affected by various factors such as long-range transport, photodecay reactions, and sources. According to factor analysis using particulate ClPAH in addition to particulate PAH concentrations, the predominant source of particulate ClPAHs was suggested to be incineration. Indeed, Ohura and Kannan group have been vigorously carried out the analysis of possible emission sources for ClPAHs, and detected much of ClPAHs in fly ash and bottom ash from waste incinerators [10, 11]. With regard to the occurrences of ClPAHs in other environments such as soil and sediment, there is not report, yet.

Biological effects of ClPAHs

AhR activity

A majority of the adverse effects of xenobiotics such as PAHs and dioxins is thought to be mediated through AhR activation. Therefore, to investigate AhR-mediated activity of ClPAHs will be important to assess the exposure risks. Ohura et al. have investigated the AhR activities of 18 ClPAHs with 3 to 5 rings using the yeast assay system and compared the potencies to the corresponding PAHs [12]. All ClPAHs tested showed appreciable dose-dependent increases in AhR-mediated reporter gene expression. Among the ClPAHs, 3,8-dichlorofluoranthere was the most potent AhR ligand, followed by 6-chlorochrysene and 7-chlorobenz[*a*]anthracene, 3,9,10-trichlorophenanthrene, having activities that were about 6, 2, and 0.8 times higher than that of benzo[*a*]pyrene.

AhR-mediated cross-talk

In the addition of AhR activation described above, some xenobiotics such as dioxins and 3-methylcholanthrene have been recently reported to stimulate estrogen response elements (EREs)-mediated transcription without binding directly to estrogen receptors (ERs) or affecting expression levels of ERs, and consequently exert estrogen-related effects [13]. These activations were verified to be exerted via directly association of ligand-binding AhR/Arnt heterodimer with ERs, so-called AhR-ER cross-talk. One of our recent topics is whether ClPAHs affect AhR-ER cross-talk pathway or not. As reported by many researchers including our group [12, 14], phenanthrene itself have vastly weak effect on AhR activation. On the other hand, mono-, di-, and tri-chlorination increased the AhR activation and consequently activate the transcription of genes related to the drug-metabolizing enzymes such as cytochrome P-450s with the number of chlorine atom on the phenanthrene skelton [12]. However, these chlorinated phenanthrene derivatives did not affect the gene expression of cathepsin D, which is one of the target genes expressed via ER pathway, in spite of there ligand activity to AhR. These results suggest that estrogenic action mediated ER signaling through AhR activation could not necessarily occur in every AhR activated ligands. We are now under investigation on AhR-ER cross-talk activity of other chlorinated PAHs, such as chlorinated benzo[*a*]pyrene, and also other halogenated PAHs.

Risk of Exposure to Clpahs

To evaluate the risk of human exposure to ClPAHs, it is important to demonstrate the relationship between residue levels in the environment and toxicity. Most of the adverse effects of halogenated aromatic compounds are thought to be mediated through AhR activation. Therefore, the toxic potencies of these compounds can be expressed in terms of relationship between congener-specific concentrations and a relative potency (REP), which is specific to an experiment, or a TCDD equivalency factor (TEF), which is consensus-derived from many relative potency values. Ohura et al. established the REP values of each ClPAH relative to BaP (REP_{BaP}) in a yeast assay system [12]. In a similar system, the intensity of the AhR-associated toxicity of TCDD was 60-fold higher than that of BaP [15]. Therefore, the overall toxicity or toxic equivalents (TEQs) of a mixture of ClPAHs relative to TCDD can be defined as follows:

$$TEQ = \sum[C_i] \times REP_{BaPi} / 60 \tag{1}$$

where C_i represents the concentration of an individual ClPAH.

For air samples collected from a Japanese urban city, the calculated TEQ concentration of total ClPAHs based on equation 1 was 1.18 pg-TEQ/m^3, whereas the TEQ concentrations of dioxins ranged from 0.024 to 0.043 pg-TEQ/m^3 in 2005 [12]. Thus, it shows that the toxicity of exposure to ClPAHs would be approximately 30 to 50 times higher than that of dioxins.

On the other hand, the exposure risk of ClPAHs and PAHs in the air was also calculated using the individual REP_{BaP} and corresponding mean concentrations obtained from ambient particles in Japanese urban air [16]. The calculated TEQ concentrations of total ClPAHs and PAHs were 2.95 and 650 pg-TEQ/m^3, respectively, indicated that the contribution of ClPAHs to PAHs was less than 1 %.

Genotoxicity of ClPAHs

The study on genotoxicity of ClPAHs had long been performed in comparison to other biological effects as well as the environmental surveys. The first report will be to investigate the mutagenicity of ClPAHs using *Salmonella typhimurium* TA98 and TA100 in the presence or absence of S9 activation enzyme system [17]. In the study, some ClPAHs such as 1-chloropyrene showed the direct mutagenicity, which was apparently different pattern from parent PAH, pyrene. The reason of such toxic activation occurs remain unclear. Fu et al. have investigated the tumogenicity of ClPAH, especially focused on 7-chlorobenz[*a*]anthracene (7-ClBaA), indicated that 7-ClBaA are potent carcinogen and that bay-region diol epoxides are the ultimate metabolites that lead to DNA adduct formation and tumor initiation [18, 19]. Conversely, 6-chlorobenzo[a]pyrene (6-ClBaP) reduced the level of mammary epithelial tumors of female Sprague-Dawley rats, compared to parent PAH, BaP [20]. It was suggested that the weak activity of 6-ClBaP could be due to poor substrate for metabolism by cytochrome P-450.

Conclusion

The risk assessments based on relative potencies of each AhR activity showed that ambient ClPAHs was extremely low contributors, as well as order of the ambient concentrations. In general, the significant contributions of PAHs to AhR activities in the environmental samples have been often observed in the assessment of persistent organics from sediments, which were 5~50 times order higher than that of polychlorinated dibenzo-*p*-dioxins and dibenzofurans (PCDD/Fs). Similarly, the contribution of PCBs to the total TEQ in sediment was less than 1%, PAHs occupied the considerable portion, whereas PCBs accounted for a greater proportion (> 95%) of total TEQs in coastal and tidal flatfishes. PCDD/Fs and PCBs are extremely persistent and bioaccumulative compared to PAHs. Although metabolic characteristics of halogenated PAHs remain unknown, they are expected to be more lipophilicity and be difficult to metabolize rather than origin PAHs. Thus, such ClPAHs may be accumulated at high level in biological tissues, and exert the toxic action. In order to reveal the toxicities of ClPAHs, further studies concerning the metabolic dispositions will be needed.

References

[1] Ohura, T. (2007) Environmental behavior, sources, and effects of chlorinated polycyclic aromatic hydrocarbons. *ScientificWorldJournal.* 7, 372-380.

[2] Ohura, T.; Kitazawa, A.; Amagai, T. & Makino, M. (2005) Occurrence, profiles, and photostabilities of chlorinated polycyclic aromatic hydrocarbons associated with particulates in urban air. *Environ. Sci. Technol.* 39, 85-91.

[3] Shiraishi, H.; Pilkington, NH.; Otsuki, A. & Fuwa, K. (1985) Occurrence of chlorinated polynuclear aromatic hydrocarbons in tap water. *Environ. Sci. Technol.* 19, 585-590.

[4] Oyler, AR.; Liukkonen, RJ.; Lukasewycz, MK.; Cox, DA.;Peake, DA. & Carlson, RM. (1982) Implications of treating water containing polynuclear aromatic hydrocarbons with chlorine: A gas chromatographic-mass spectrometric study. *Environ. Health Perspect.* 46, 73-86.

[5] Oyler, AR.; Liukkonen, RJ.; Lukasewycz, MT.; Helkklla, KE.; Cox, DA. & Carison, RM. (1983) Chlorine "disinfection" chemistry of aromatic compounds. Polynuclear aromatic hydrocarbons: Rates, Products, and Mechanisms. *Environ. Sci. Technol.* 17, 334-342.

[6] Nilsson, UL. & Ostman, CE. (1993) Chlorinated polycyclic aromatic hydrocarbons: Method of analysis and their occurrence in urban air. *Environ. Sci. Technol.* 27, 1826-1831.

[7] Kitazawa, A.; Amagai, T. & Ohura, T. (2006) Temporal trends and relationships of particulate chlorinated polycyclic aromatic hydrocarbons and their parent compounds in urban air. *Environ. Sci. Technol.* 40, 4592-4598.

[8] Ohura, T.; Fujima, S.; Amagai, T. & Shinomiya, M. (2008) Chlorinated Polycyclic Aromatic Hydrocarbons in the Atmosphere: Seasonal Levels, Gas-Particle Partitioning, and Origin. *Environ. Sci. Technol.* 42, 3296-3302.

[9] Ohura, T.; Kitazawa, A. & Amagai, T. (2004) Seasonal variability of 1-chloropyrene on atmospheric particles and photostability in toluene. *Chemosphere*. 57, 831-837.

[10] Horii, Y.; Ok, G.; Ohura, T. & Kannan, K. (2008) Occurrence and Profiles of Chlorinated and Brominated Polycyclic Aromatic Hydrocarbons in Waste Incinerators. *Environ. Sci. Technol.* 42, 1904-1909.

[11] Ohura, T.; Kitazawa, A.; Amagai, T. & Shinimiya, M. (2007) Relationships between chlorinated polycyclic aromatic hydrocarbons and dioxins in urban air and incinerators. *Organohalogen Compd.* 69, 2902-2905.

[12] Ohura, T.; Morita, M.; Makino, M.; Amagai, T. & Shimoi, K. (2007) Aryl hydrocarbon receptor-mediated effects of chlorinated polycyclic aromatic hydrocarbons. *Chem. Res. Toxicol.* 20, 1237-1241.

[13] Ohtake, F.; Takeyama, K.; Matsumoto, T.; Kitagawa, H.; Yamamoto, Y.; Nohara, K.; Tohyama, C.; Krust, A.; Mimura, J.; Chambon, P.; Yanagisawa, J.; Fujii-Kuriyama, Y. & Kato, S. (2003) Modulation of oestrogen receptor signalling by association with the activated dioxin receptor. *Nature*. 423, 545-550.

[14] Lampen, A.; Ebert, B.; Stumkat, L.; Jacob, J. & Seidel, A. (2004) Induction of gene expression of xenobiotic metabolism enzymes and ABC-transport proteins by PAH and a reconstituted PAH mixture in human Caco-2 cells. *Biochim. Biophys. Acta*. 1681, 38-46.

[15] Kawanishi, M.; Sakamoto, M.; Ito, A.; Kishi, K. & Yagi, T. (2003) Construction of reporter yeasts for mouse aryl hydrocarbon receptor ligand activity. *Mutat. Res.* 540, 99-105.

[16] Ohura, T.; Sawada, K.; Amagai, T. & Shinomiya, M. (2009) Atmospheric behaviors and exposure risks of novel halogenated polycyclic aromatic hydrocarbons associated with particulates. submitted.

[17] Colmsjo, A.; Rannug, A. & Rannug, U. (1984) Some chloro derivatives of polynuclear aromatic hydrocarbons are potent mutagens in *Salmonella typhimurium*. *Mutat. Res.* 135, 21-29.

[18] Fu, PP.; Von Tungeln, LS.; Zhan, D-J. & Bucci, T. (1996) Potent tumorigenicity of 7-chlorobenz[*a*]anthracene and 7-bromobenz[*a*]anthracene in the neonatal B6C3F$_1$ male mouse. *Cancer Lett.* 101, 37-42.

[19] Xia, Q.; Yi, P.; Zhan, D-J.; Von Tungeln, LS.; Hart, RW.; Heflich, RH. & Fu, PP. (1998) Liver tumors induced in B6C3F$_1$ mice by 7-chlorobenz[*a*]anthracene and 7-bromobenz[*a*]anthracene contain *K*-ras protooncogene mutations. *Cancer Lett.* 123, 21-25.

[20] Cavalieri, E.; Rogan, E.; Cremonesi, P.; Higginbotham, S. & Salmasi, S. (1988) Tumorigenicity of 6-haloganated derivatives of benzo[*a*]pyrene in mouse skin and rat mammary gland. *J. Cancer Res. Clin. Oncol.* 114, 10-15.

In: Polycyclic Aromatic Hydrocarbons: Pollution, Health ... ISBN 978-1-60741-462-9
Editors: Pierre A. Haines and Milton D. Hendrickson

Chapter 6

RECENT TRENDS IN ANALYTICAL METHODS TO CHEMICALLY MONITOR PAH POLLUTION

Verónica Pino
Department of Analytical Chemistry, Nutrition and Food Science,
University of La Laguna, Spain

ABSTRACT

Polycyclic aromatic hydrocarbons (PAHs) are a large group of organic compounds with two or more fused aromatic rings. They have received increased attention in recent years in pollution studies because some of these compounds are highly carcinogenic or mutagenic. Some, while not carcinogenic, may act as synergists. These compounds are widespread in the environment. Their occurrence is largely a result of anthropogenic and natural emissions such as fossil fuel burning, motor vehicles, waste incinerators, oil refining, coke and asphalt production, aluminum production, etc. PAHs have relatively low solubility in water but are highly lipophilic. Most of the PAHs with low vapor pressure in the air are adsorbed on particles. When dissolved in water or adsorbed on particulate matter, PAHs can undergo photodecomposition when exposed to UV light. In the atmosphere, PAHs can react with many other pollutants such as ozone, nitrogen oxides and sulfur dioxide, yielding diones, nitro- and dinitro-PAHs, and sulfonic acids, respectively. PAHs may also be degraded by some microorganisms in soils. PAHs in soils/sediments can be taken up by biota, and some bio-magnification of these toxic compounds may occur as they pass through the food chain. The US Environmental Protection Agency (EPA) has identified 16 un-substituted PAHs as priority pollutants. The European Food Security Authority (EFSA) has classified 15 PAHs by their food toxicity. Thus, potential exposure to PAHs must be assessed by testing contaminated soils, sediments, air, water and foods, and estimating the degree of intake of each of these media into the human body, as well as the overall toxicity.There are several main problems associated with the determination of PAHs in environmental and food samples: their low concentrations, the complexity of the samples, and the necessity for fast and reliable analytical methods. To achieve successful results, many extraction-separation techniques have been developed depending on the nature of the sample. An in-depth description of the most commonly-used techniques, together with recent trends in analytical methods to monitor PAHs levels in different kind of samples, will be the main focus of this chapter. Among those that will be cited are Soxhlet extraction, liquid-liquid extraction (LLE), microwave-assisted extraction (MAE), ultrasound-assisted extraction (UAE), pressurized fluid extraction (PFE), supercritical fluid extraction (SFE), solid-phase extraction (SPE), and solid-phase microextraction (SPME), mainly in combination with chromatographic techniques (HPLC and GC). Moreover, the utilization of active or passive samplers in pollution monitoring, as well as the increased utilization of semi-permeable membrane passive samplers (SPMDs), will be described.

1. Introduction

Polycyclic aromatic hydrocarbons (PAHs) refers to a group of a few hundred organic compounds consisting of two or more condensed benzene rings that can incorporate rings with an odd number of carbon atoms, sharing at least two atoms in each pair of contiguous rings. Table 1 shows several of the most significant PAHs.

In general, PAHs are originated during the incomplete combustion of organic matter in natural processes (i.e., forest fires, volcanoes, and hydrothermal processes) [1,2] and in anthropogenic activities (fossil fuel burning, motor vehicles, waste incinerators, oil refining, coke and asphalt production, aluminum production, heat, food processing, etc.) [3-5].

The formation mechanism of PAHs involves several steps, not well established, but probably related with pyrolysis and pyrosynthesis reactions. It has been established that the rupture of organic matter at high temperatures in combustion processes forms radicals of low molecular weight, which later will recombine by pyrosynthesis to form PAHs [6].

In 1973, Crittenden et al. [7] proposed a PAH formation mechanism in which the acetylene molecules react to form unsaturated radicals C4, C6, C8 and C2n. These species can interact to form polyacetylenes, or can suffer a ring closure reaction to form radicals constituted by an aromatic ring attached to a 2-carbon chain (phenyl-acetylene or styrene). PAHs are then formed by reactions involving the radicals phenyl-acetylene and styrene.

In 1983, Bockhorn et al. [8] established a similar mechanism, but proposed that the phenyl-acetylene is formed by the reaction between acetylene and 1,3-butadiene, as can be observed from Figure 1. The authors described in depth the different formation reactions of PAHs starting with phenyl-acetylene. Other authors have suggested the formation of PAHs starting with cyclopentadienyl radicals [9,10].

In spite of the difficulty in establishing the formation mechanism of PAHs [3], it can be ensured that, in general, the first PAH to be formed is naphthalene, mainly due to its high volatility and stability. Heavier PAHs, especially the ones with higher thermodynamic stability, are formed by addition of acetylene or by intra-molecular ring closure of lighter PAHs when the combustion temperature is increased [9-12].

Table 1. Names and structures of several significant PAHs.

PAH (Abbreviation)	Structure	PAH (Abbreviation)	Structure	PAH (Abbreviation)	Structure
Naphthalene (N)	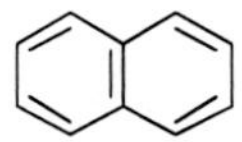	Anthracene (A)	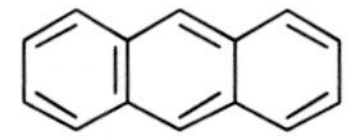	5-Methylchrysene (5-MeChy)	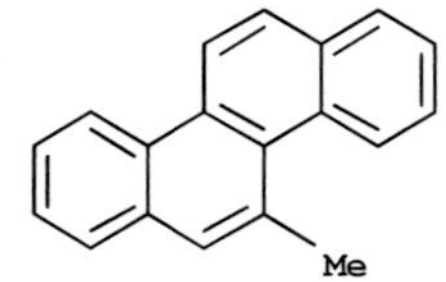
Acenaphthene (Ace)	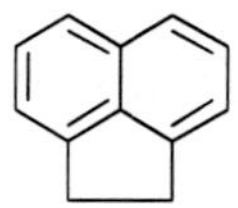	Fluoranthene (Ft)	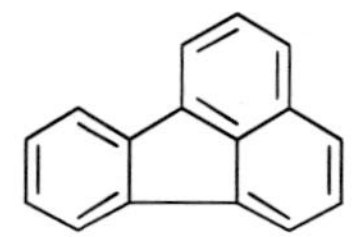	Benzo(b)fluoranthene (B(b)Ft)	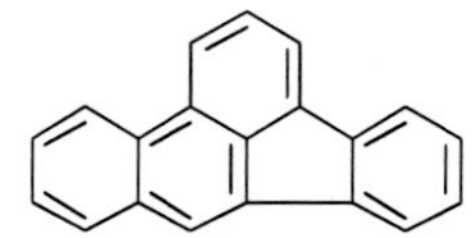
Acenaphthylene (Acy)	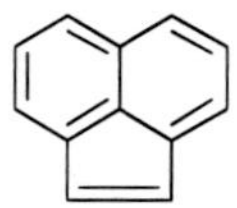	Pyrene (Py)	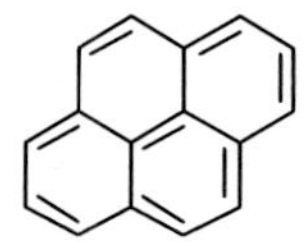	Benzo(k)fluoranthene (B(k)Ft)	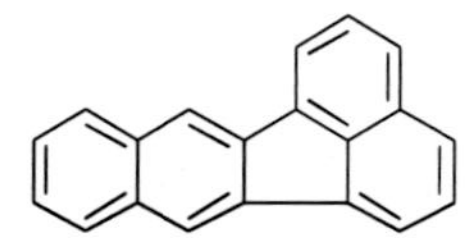
Fluorene (Fl)	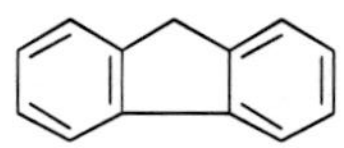	Benzo(a)anthracene (B(a)A)	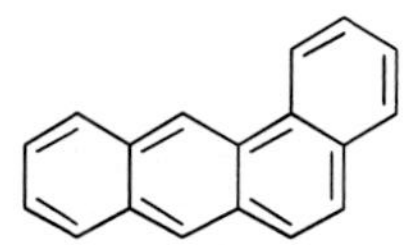	Benzo(j)fluoranthene (B(j)Ft)	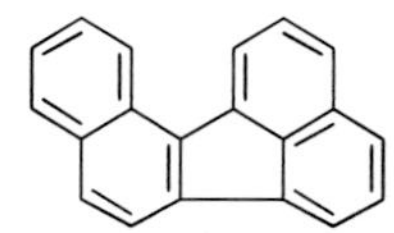
Phenanthrene (Phe)	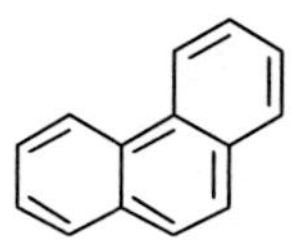	Chrysene (Chy)	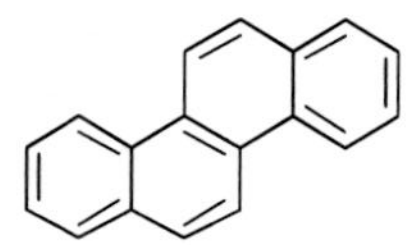	Perylene (Per)	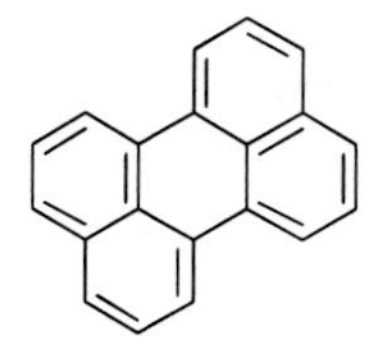

Table 1 (Continued) Names and structures of several significant PAHs.

PAH (Abbreviation)	Structure	PAH (Abbreviation)	Structure	PAH (Abbreviation)	Structure
Benzo(e)pyrene (B(e)Py)	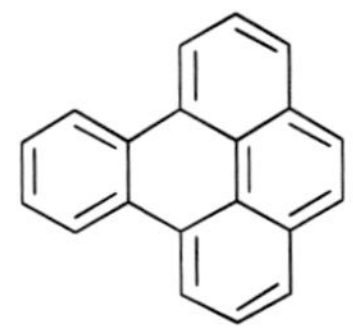	Dibenzo(a,c)anthracene (diB(a,c)A)	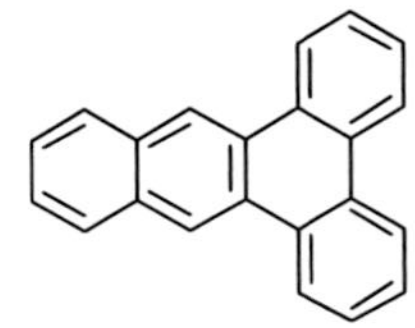	Dibenzo(a,l)pyrene (diB(a,l)Py)	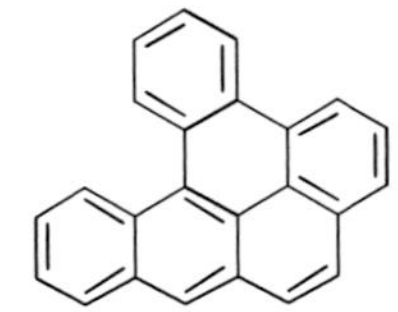
Benzo(a)pyrene (B(a)Py)	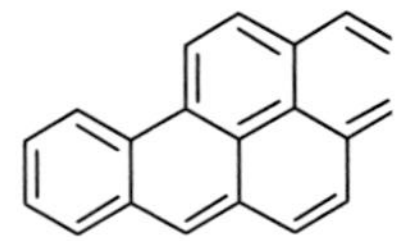	Dibenzo(a,e)pyrene (diB(a,e)Py)	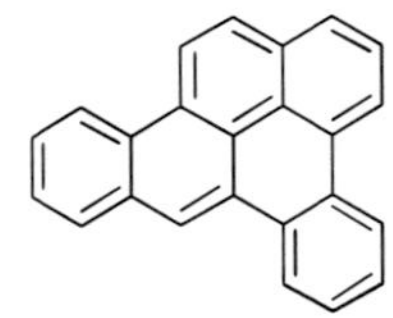	Dibenzo(a,h)anthracene (diB(a,h)A)	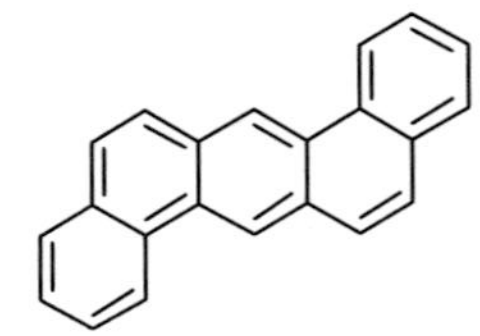
Cyclopenta(c,d)pyrene (cyPent(c,d)Py)	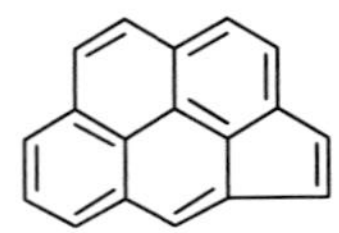	Dibenzo(a,h)pyrene (diB(a,h)Py)	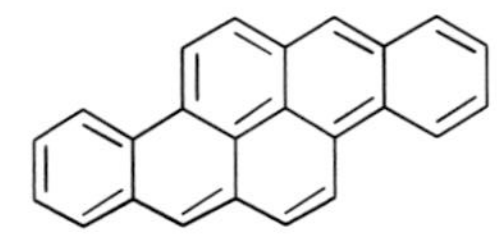	Indeno(1,2,3-cd)pyrene (I(1,2,3-cd)Py)	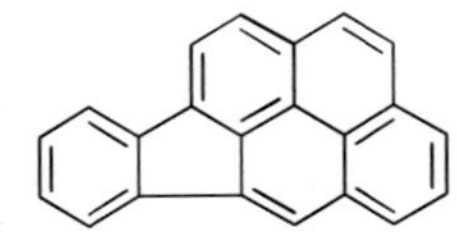
Benzo(ghi)perylene (B(ghi)Per)	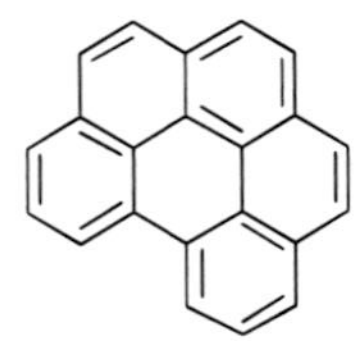	Dibenzo(a,i)pyrene (diB(a,i)Py)	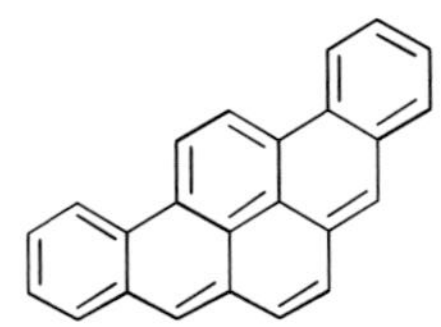	1-Hydroxypyrene (1-OHPy)	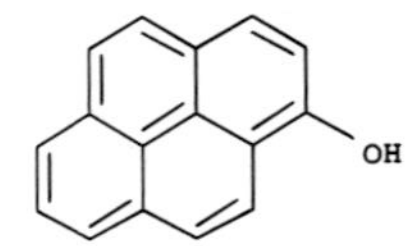

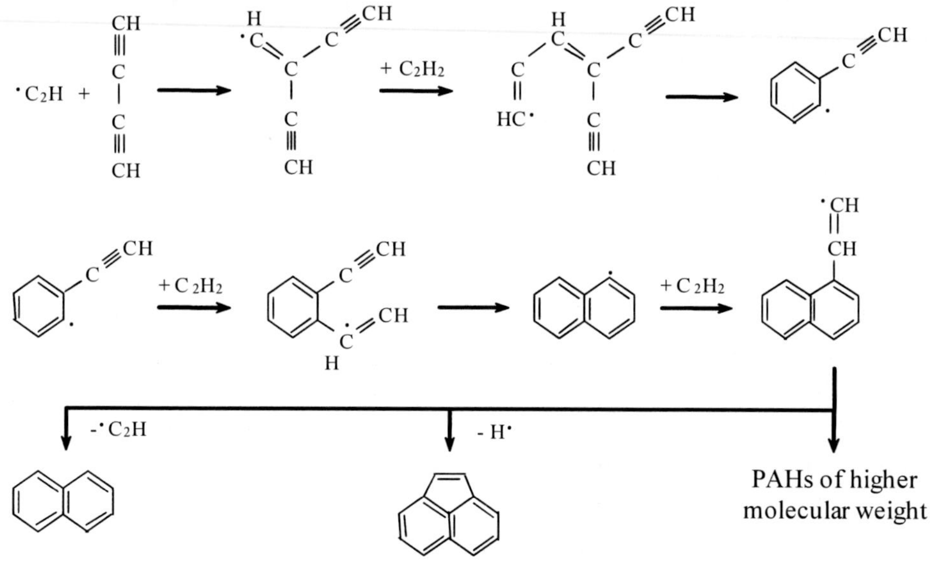

Figure 1. Formation mechanism of PAHs proposed by Bockhorn et al. [8].

1.1. Physicochemical Properties

PAHs are compounds of high chemical stability, due to the conjugated systems of π orbital. Table 2 shows some physical properties of several significant PAHs. PAHs are in general solids at room temperature. Many of them can be classified as semivolatile organic compounds. They are soluble in many different organic solvents, but practically nonsoluble, or slightly soluble, in water. Nevertheless, the aqueous solubility can be increased by the presence of detergents, salts, fatty acids, or organic solvents [13]. In the same manner, the aqueous solubility can be three times higher when increasing the temperature. Their physicochemical properties, especially nonpolarity and hydrophobicity, are responsible for their persistence in the environment. Their lipophilicity justifies the bioaccumulation of these compounds in the lipid tissues of living organisms.

Table 2. Physicochemical properties of several PAHs

PAHs	Vapor pressure (Pa)	Boiling point (°C)	Solubility ($mg \cdot L^{-1}$)
Naphthalene	10.4	218.0	31.0
Acenaphthylene	0.90	270.0	16.1
Acenaphthene	0.30	277.5	3.80
Fluorene	0.09	295.0	1.90
Phenanthrene	0.02	339.0	4.57
Anthracene	$1.0 \cdot 10^{-3}$	340.0	0.05
Fluoranthene	$1.2 \cdot 10^{-3}$	375.0	0.26

Table 2 (Continued) Physicochemical properties of several PAHs

PAHs	Vapor pressure (Pa)	Boiling point (°C)	Solubility ($mg{\cdot}L^{-1}$)
Pyrene	$6.0{\cdot}10^{-4}$	360.0	0.13
Benz(a)anthracene	$2.8{\cdot}10^{-5}$	435.0	0.01
Chrysene	$5.7{\cdot}10^{-7}$	407.0	$9.6{\cdot}10^{-3}$
Benzo(b)fluoranthene	$6.7{\cdot}10^{-8}$	481.0	$1.5{\cdot}10^{-3}$
Benzo(k)fluoranthene	$5.2{\cdot}10^{-8}$	481.0	$8.0{\cdot}10^{-4}$
Benzo(a)pyrene	$7.0{\cdot}10^{-7}$	495.0	$3.8{\cdot}10^{-3}$
Indeno(1,2,3-cd)pyrene	$1.4\ 10^{-8}$	536.0	$1.9{\cdot}10^{-3}$
Dibenzo(a,h)anthracene	$3.7{\cdot}10^{-10}$	524.0	$6.0{\cdot}10^{-4}$
Benzo(g,h,i)perylene	$1.3{\cdot}10^{-11}$	525.0	$2.6{\cdot}10^{-4}$
Benzo(j)fluoranthene	$2.4{\cdot}10^{-6}$	480.0	$5.0{\cdot}10^{-3}$
Cyclopenta(c,d)pyrene	$2.4{\cdot}10^{-5}$	438.3	0.014
Dibenzo(a,e)pyrene	$1.5{\cdot}10^{-9}$	552.3	$2.7{\cdot}10^{-4}$
Dibenzo(a,h)pyrene	$1.5{\cdot}10^{-9}$	552.1	$2.7{\cdot}10^{-4}$
Dibenzo(a,i)pyrene	$1.5{\cdot}10^{-9}$	552.1	$2.7{\cdot}10^{-4}$
Dibenzo(a,l)pyrene	$1.5{\cdot}10^{-9}$	552.1	$2.7{\cdot}10^{-4}$
5-Methylchrysene	$1.0{\cdot}10^{-5}$	449.4	$9.7{\cdot}10^{-3}$

1.2. Carcinogenic Effects

The first evidence of the carcinogenic effect of PAHs was reported in 1775, when the British medical doctor Percival Pott (1714–1788) pointed out the relationship between the scrotum cancer in chimney sweeps and their long-term exposure to tar and soot. In 1930 it was demonstrated that the PAH benzo(a)pyrene, contained in the soot, was carcinogenic [14]. Up to the present, several studies have evaluated the risk for cancer development in humans exposed to PAHs via the lungs or dermis [15-17]. Recent epidemiological studies suggest that PAH-DNA adduct formation in blood and breast tissue may play an important role in breast cancer development [18].

Given their carcinogenic characteristics, several agencies have included PAHs in their list of priority pollutants. The US Environmental Protection Agency (EPA) has selected 16 PAHs as contaminants of high priority. In the same fashion, the Food Scientific Committee of the European Union (FSC), now named the European Food Security Authority (EFSA), has selected 15 PAHs by their potential toxicity in relation to foods. In 2002, the former European Commission's Scientific Committee on Food (SCF) considered the study of the 33 PAHs included in an Environmental Health Criteria document on PAHs [19].

Not all PAHs have been demonstrated to be carcinogenic. In many instances, the carcinogenic effect is in fact due to the presence of more than one member of the PAH family, as well as several of their derivatives. Hence, some PAHs, while not carcinogenic, may act as synergists. Thus, the implication of human exposure to mixtures of PAHs, rather than to individual substances, is important. The toxicity equivalent factor (TEF) is a useful tool to quantitatively estimate the overall carcinogenic effect of an environmental sample containing several PAHs. Based on toxicological studies, a toxicity value is given to each PAH by comparison with a well-known carcinogenic PAH, benzo(a)pyrene, which has a reference TEF value of 1. The overall carcinogenic effect of a sample is then determined considering both the amount of each PAH in the sample and their respective TEF values. Table 3 shows several TEF values for a group of PAHs, proposed by different authors and/or agencies. It can be observed that hydrocarbons with a high number of condensed rings in the structures (between 5 and 6) are the ones with higher TEF values. The potential toxicity of a sample, expressed as equivalents of benzo(a)pyrene (B(a)Py), depends on the TEF values selected for each hydrocarbon.

The carcinogenic and mutagenic activities of PAHs are closely related with their chemical structures. In general, planar PAHs are more stable and less toxic than non-planar PAHs. In addition to this, the introduction of an alkyl group into a PAH moiety generates a structural distortion of the hydrocarbon molecule, which can provoke activation or deactivation effects [20]. PAHs are apparently not carcinogenic by themselves, but they are converted into active carcinogens when metabolized by a living organism. The induction process of cancer in mammals involves a group of enzymes able to convert PAHs in water soluble compounds [21]. These oxydase enzymes of multiple functions belong to the group of cytochrome P450. As shown in Figure 2, PAHs are transformed into diol-epoxide adducts, followed by a rapid transformation in carbocations, able to covalently bind with DNA macromolecules, and therefore modifying the genetic message during the formation of new cells [22,23]. Similar reactions have also been observed with other macromolecules such as albumin and hemoglobin [20]. The presence of PAHs-DNA adducts in human tissues has been conclusively established [24]. A recent study has shown that chromosomal aberrations in cord blood are associated with prenatal exposure to carcinogenic polycyclic aromatic hydrocarbons [25].

There are various industrial workplaces which present a significant risk of certain cancer diseases. This may be attributed to an unusually high exposure to PAH [17]. Furthermore, as highly exposed occupational groups, chimney sweeps, roadmen (pavement-tarring) and roofers (roof-tarring) are also under increased risk [23].

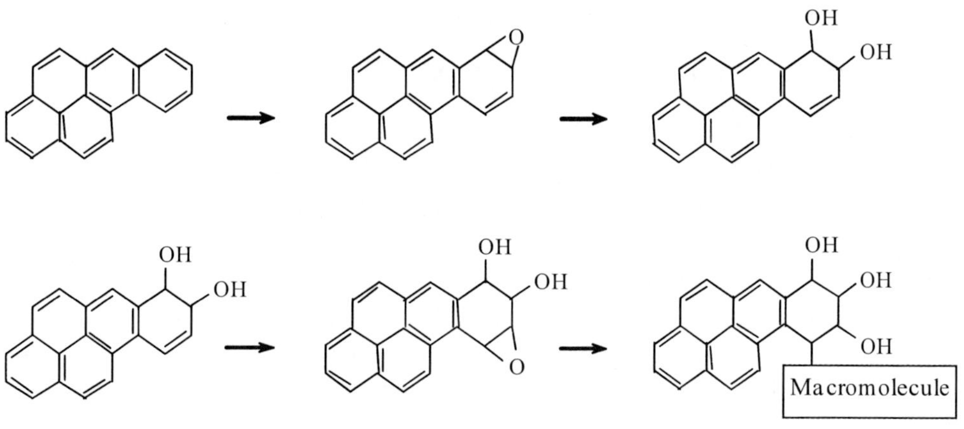

Figure 2. Schematic representation of the metabolic activation of benzo(a)pyrene.

Table 3. Toxicity equivalent factors proposed by different authors for the 16 PAHs listed by the EPA (numbered from 1 to 16) and for the 15 PAHs listed by the EFSA (numbered from 9 to 23).

PAHs	Nisbet [26]	EPA [27]	Kalberlah [28]	Cecinato [29]	Larsen [30]
1. Naphthalene	0.001	—	—	—	—
2. Acenaphthylene	0.001	—	0.01	—	—
3. Acenaphthene	0.001	0	0.001	—	—
4. Fluorene	0.001	—	0	—	—
5. Phenanthrene	0.001	—	0	—	0.0005
6. Anthracene	0.01	—	0.01	—	0.0005
7. Fluoranthene	0.001	—	0.01	—	0.05
8. Pyrene	0.001	—	0.001	—	0.001
9. Benz(a)anthracene	0.1	0.1	0.1	0.06	0.005
10. Chrysene	0.01	0.001	0.01	—	0.03
11. Benzo(b)fluoranthene	0.1	0.1	0.1	0.07	0.05
12. Benzo(k)fluoranthene	0.1	0.01	0.1	0.07	0.1
13. Benzo(a)pyrene	1	1	1	1	1
14. Indeno(1,2,3-cd)pyrene	0.1	0.1	0.1	0.08	0.1
15. Dibenzo(a,h)anthracene	1	1	1	0.6	1.1
16. Benzo(g,h,i)perylene	0.01	—	0.01	—	0.02
17. Benzo(j)fluoranthene	—	—	0.1	—	0.05
18. Cyclopenta(c,d)pyrene	0.1	—	—	—	0.02
19. Dibenzo(a,e)pyrene	—	—	—	—	0.2
20. Dibenzo(a,h)pyrene	—	—	—	—	1
21. Dibenzo(a,i)pyrene	—	—	—	—	0.1
22. Dibenzo(a,l)pyrene	—	—	—	—	1
23. 5-Methylchrysene	—	—	—	—	—

1.3. Environmental Distribution of PAHs

Once PAHs are formed in the atmosphere, many of them can suffer photo-oxidation reactions with nitrogen oxides, ozone, and hydroxyl radicals, with reaction rates varying between hours and days [31].

PAHs, as semivolatile organic compounds, exist in both the gaseous and the particulate phase in air. The distribution of PAHs between the gas phase and the particulate matter (PM) is a complex process which depends on the temperature, the relative humidity, the composition of the particulate matter, and the nature of the PAH, mainly, the molecular weight and vapor pressure. In general, it has been demonstrated that PAHs of lower molecular weight tend to stay in the gas phase whereas PAHs of higher molecular weight tend to exist associated to the particulate matter, especially with particulate matter having diameters lower than 10 μm (PM10) [32,33]. In general terms, heavy PAHs are associated with the particulate matters by adsorption and absorption mechanism [34]. Independently on the sorption process that is taken place, the distribution constant is inversely dependent on the vapor pressure of the hydrocarbon.

The presence of PAHs in the aquatic ecosystems is mainly as a result of anthropogenic inputs [35]. PAHs can be transported by direct effluents, lixiviation of soils, dry atmospheric deposition, and, mainly, humid atmospheric deposition (rain). For some watersheds, atmospheric deposition represents a significant input to the total surface water PAH burden, and may be independent of adjacent land use [36].

Once PAHs are in the aqueous medium, they can associate or bind to colloids, and so being transported in water. This justify the presence of PAHs in living organisms that are located in areas far away of an intensive industrial activity, and justify the fact that sediments are much richer in PAHs than the surrounding aqueous media [37-39]. Atmospherically derived PAHs in marine sediments have been linked to industrial and athropogenic sources [40,41]. The degradation of PAHs in sediments is in general a very slow process [42,43]. PAHs can accumulate in aquatic living organisms [44], entering into their biologic processes and affecting physiological lethal and sub-lethal levels [44,45]. Evidently, this effect is not isolated, because everything belongs to the ecological equilibrium. PAHs are metabolized by fishes [46] and into a lesser extent by mollusks or crustaceans, generating polar metabolites that can be excreted. In addition to this, the presence of PAHs may change the mechanisms of metal uptake by several plants, thus influencing kinetics and extent of metal phytoextraction [47]. It results clear that the biomagnification of PAHs into higher trophic levels across the food chains results into dispersal of pollutants far from their point of origin and from areas of significant human activity. Figure 3 illustrates a general scheme of the environmental distribution of PAHs.

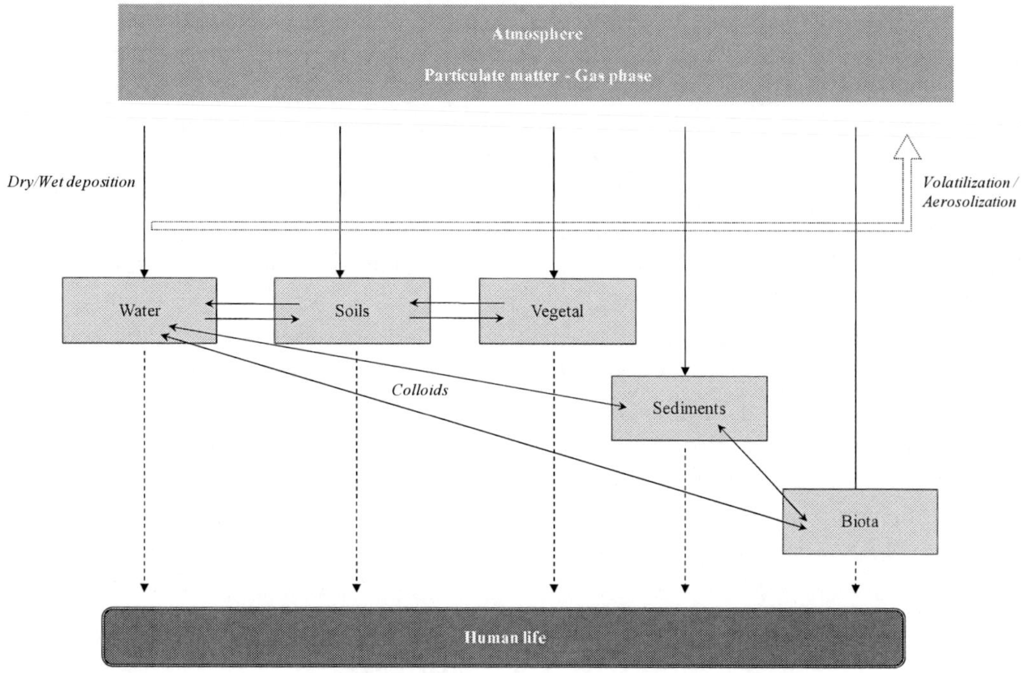

Figure 3. Environmental distribution of PAHs.

1.4. Monitoring PAHs

Given all these precedents, the wide environmental distribution and the carcinogenic risk associated to PAHs, it results necessary to monitor these contaminants in all kind of environmental samples, as well as to establish their impact in living organisms [48]. They must therefore be measured in the atmosphere for air quality assessment, in biological tissues for health-effects monitoring, in waters, sediments and mollusks/fishes for environmental monitoring, and in foodstuffs for safety reasons. To understand the extent of human exposure to PAHs, reliable sampling and analytical methods are necessary.

The monitoring of PAHs in environmental samples presents several problems. One of the main limitations is related with the low concentration levels of PAHs to be monitored in the samples, with the consequent difficulty for being detected and quantified. Another limitation is related with the inadvertent errors during the sampling. In addition to this, the complexity of the samples, the undoubtedly presence of interfering substances, and the necessity of accurate and fast analysis, do the monitoring of PAHs an analytical challenge.

In the analysis of PAHs, chromatographic techniques, mainly gas chromatography (GC) and liquid chromatography (LC), are often the preferred approach for separation, identification and quantification. Spectrofluorimetric techniques have also been used with screening purposes. In general, capillary GC is preferred over LC for the determination of PAHs mainly because of favorable combination of greater selectivity and resolution [49-51]. On the other hand, HPLC with fluorescence detection (FD) presents the best detection limits for PAHs [15].

Nevertheless, to reach the adequate sensitivity limits in gas chromatography or liquid chromatography, independently on the detection technique used (mass-spectrometry or fluorescence, respectively), it results necessary to carry out preconcentration procedures as well as clean-up steps [52]. The utilization of clean-up and preconcentration procedures introduces more sources of error in the analytical method, as well as increases the analysis time. The most common extraction-preconcentration procedures are Soxhlet extraction, liquid-liquid extraction (LLE), ultrasound-assisted extraction (UAE), microwave-assisted extraction (MAE), pressurized fluid extraction (PFE), dialysis, solid-phase extraction (SPE), and solid-phase microextraction (SPME).

The sampling is a critical step that must be carried out carefully, with the appropriate sampling design and sampling device. The increased utilization of semi-permeable membrane passive samplers (SPMDs) during the sampling stage of PAHs should be highlighted.

The main aim of this chapter is to describe the most common procedures to monitor PAHs in environmental samples, with emphasis to the analytical methods used in extraction and determination procedures. This chapter is divided in analytical monitoring depending on the nature of the sample, though some methods turn out to be common for different samples.

2. Air Monitoring

The highest concentrations of atmospheric PAHs can be found in the urban environment, due to the traffic and the small dispersion of the atmospheric pollutants [15,53]. Hence, the cancer risk development in humans is much higher in the cities, considering the population density [54]. Sources of PAHs in urban atmosphere also include re-suspended soils, refineries and power plants [55-57]. Lighter PAHs, which have weaker carcinogenic effects by themselves, are the most abundant in the urban atmosphere; however, their presence is quite risky because they can react with other air pollutants to form more toxic derivatives [58], i.e., they can react with ozone, nitrogen oxides and sulfur dioxide, yielding diones, nitro- and dinitro-PAHs, and sulfonic acids, respectively [31].

Despite the efforts conducted to reduce urban particulate pollution in cities, the traffic still contributes to increase urban particulate pollution. Besides, particles produced by cars are much smaller than coal particles and can be found in the breathable size fraction [59]. Heavier PAHs, with higher carcinogenic effects, are usually associated to particulate matter in the atmosphere. Particles of concern to human health are those known as inhalable particles (PM2.5, PM with aerodynamic diameter less than 2.5 μm). Most particulate-phase PAH are adsorbed onto fine particles, which are deposited slowly and, depending on atmospheric conditions and chemical reactivity, may be transported over long distances, being placed even in remote areas. The European Union requires continuous air-quality monitoring with long-term assessment of benzo(a)pyrene (B(a)Py) and other relevant PAHs [60]. The Council Directive 2004/107/EC establishes a target value for B(a)Py of 1 $ng \cdot m^{-3}$, and a long-term objective of 0.1 $ng \cdot m^{-3}$, applying to the B(a)Py content in the particulate matter fraction with an aerodynamic diameter lower than 10 μm (PM10) [60].

Other sources of PAHs are the incomplete biomass combustion which can take place in both natural processes (as fires or volcanic eruptions) and anthropogenic processes (as elimination of residues, cooking or smoking of foods). Smokes coming from biomass combustion not only increases the amount of PAHs in the atmosphere but also constitute an important source of particulate matter in the atmosphere [61], contributing to increase the risk of sorption of PAHs.

The composition of PAHs in association with airborne particles depends significantly on the emission sources. PAH concentration ratios can be used to identify possible emission sources: combustion of vegetal biomass or combustion of fossil fuels. [62]. These ratios are usually restricted to PAHs with equal molecular weights, which are normally present in combustion processes due to their chemical stability [63-65]. Table 4 shows several isomeric ratios of PAHs used to distinguish among combustion sources.

Table 4. Isomeric ratios of PAHs used to distinguish among different combustion sources.

Isomeric ratio	Combustion source	
	Petroleum	Biomass
Ft / (Ft + Py) = Ft/202	0.40-0.50	> 0.50
A / (A + Phe) = A/178	< 0.10	> 0.10
1,7-diMePhe / (1,7-diMePhe + 2,6-diMePhe) = 1,7-diMePhe/206	< 0.45	> 0.70
I(1,2,3-cd)Py / (I(1,2,3-cd)Py + B(ghi)Per) = I(1,2,3-cd)Py/276	0.20–0.50	> 0.50

Indoor air quality is also of great interest in the world, especially if considering that most people spend approximately 90% of their time indoors (in their residences, at work, or elsewhere), where scarce ventilation usually helps in increasing PAHs contents. Indoor air in houses is polluted by PAHs that come from not only outdoor air but also indoor emission sources such as cooking [66] and gas home appliances [67,68]. Additional contributions to ambient air levels arise from tobacco smoking [69-71].

To monitor air quality, it results therefore important not only to determine the PAHs contents in the gas phase but also the content associated to the particulate matter. In addition to this, the monitoring methods for airborne PAHs maybe classified in those concerned with either direct occupational exposures or with ambient exposures (outdoor) [72]. Occupational environment is usually monitored by active sampling, whereas ambient exposure is usually monitored by passive sampling, but not exclusively.

Active sampling methods require pumping the air stream through a medium (e.g., filter, sorbent, impinger, cryogenic trap) which will retain the PAHs. The analysis of PAHs from air samples in active sampling is followed by an extraction step, clean-up procedures and/or exchange of solvents, and then chromatographic separation and quantification. Sample preparation is considered to be the bottleneck of most analytical methodologies, as it constitutes the principal source of error in the analytical process and remains one of the most time-consuming steps [73]. Figure 4 shows a general scheme of the usual procedure for the monitoring of PAHs in air. The active sampling methods commonly used to determine and quantify PAHs in air are combination of fiber filters, to retain the particle-bound PAH fraction, whereas impregnated filters, impingers, solid sorbents like polyurethane foam plugs (PUF) and XAD resins, and cryogenic traps are used to trap the semivolatile and volatile PAH fractions, respectively [74-76]. After sampling, the most common extraction and preconcentration methods are Soxhlet [77] or sonication [78] of the systems used to trap the PAHs. These routine methods are time and solvent consuming. Apart of these common extraction procedures, many times other clean-up techniques are necessary given the complexity of the samples, such as a cleaning using a chromatographic column (filled with silica gel, Florisil or alumina, as sorbent phases), or a solid-phase extraction (SPE) step.

Passive sampling techniques rely on the controlled transport of the analyte along a concentration gradient towards a surface which acts as a sink for the PAHs. Passive air samplers are preferred to conventional active air samplers since they do not require electricity or maintenance and are less expensive. The main disadvantage is the long sampling time required due to the low PAH levels and limited analytical sensitivity [79-81]. Followed passive sampling, extraction-preconcentration, clean-up, and chromatographic procedures, identical to the ones used in active sampling approaches, are conducted.

2.1. Sampling Systems

The most critical step to ensure an accurate monitoring of PAHs in air is the choice of the sampling strategy. The main problems associated to the sampling systems of PAHs in air are related to insufficient sorption on the sampling devices, the re-evaporation of already retained PAHs, and even the chemical destruction of the PAHs already retained. Then, PAHs losses

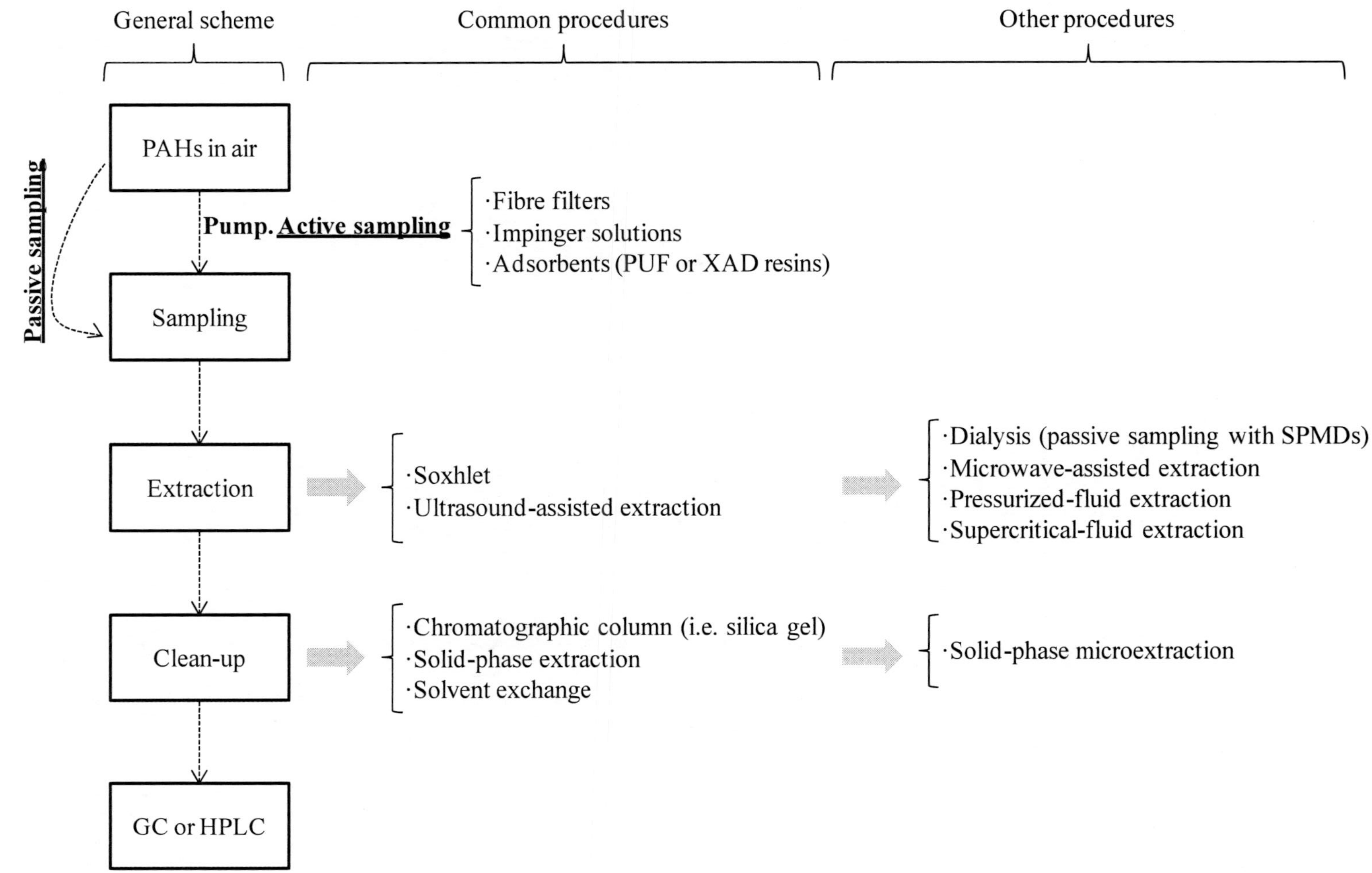

Figure 4. Common steps of the analytical method to monitor PAHs in air.

are practically unavoidable due to their chemical reactivity and volatility. The importance of the sampling is such that, indeed, the posterior utilization of any extraction and/or chromatographic method does not turn out to be so critical in the analysis of PAHs for air monitoring.

2.1.1. Active Sampling

While diffusive passive sampling offers advantages in terms of cost and simplicity, active-pumped sampling remains the most versatile option for simultaneous monitoring of multiple target PAHs. The major drawbacks of active sampling are the high cost of the equipment and requirement of power and training; these restrict its application to widespread locations especially to remote regions. Given that large air volumes (>100 L) may be drawn through the tubes in many cases, pumped active sampling also offers better detection limits than other air monitoring techniques. There are multiple pump models on the market, from a wide variety of brands, designed to fulfill different needs depending on the specific sampling. High/low volume pumps designed for active sampling typically operate at flow rates ranging from 0.5-1400 L min^{-1}. The sampling time is normally a few hours up to one day.

Sampling of airborne PAHs requires special equipment since PAHs with a lower molecular weight are found mainly as vapor, while those with a higher molecular weight are to be found in the condensed phase. Consequently, an adequate sampling method usually combines a filter for the collection of particulate, and one or more sorbents such as XAD-2 and/or polyurethane foam (PUF) for the collection of the vapor phase.

Table 5. Sampling systems recommended by standard methods.

Standard method	Application	Sampling system
ASTM D 6209-98 (2004)	Standard test method for determination of gaseous and particulate PAHs in ambient air	Fine particulate filter followed by a vapor trap containing PUF or resin XAD-2.
ISO 11338-2 (2003)	Stationary source emissions—determination of gas and particle-phase PAHs—Part 2: sample preparation, clean-up, and determination	Filters to collect the particulate phase whereas gas phase is trapped on an adsorbent (XAD-2 or PUF).
ISO 12884 (2000)	Ambient air—Determination of total (gas and particle phase) PAHs—Collection on sorbent-backed filters with gas chromatographic–mass spectrometric analyses	Fine-particle filter followed by a vapor trap containing PUF or XAD-2.
ISO 16362 (2005)	Ambient air—Determination of particle-PAHs by HPLC	Fine-particle filters.

Historically, two types of filters, glass fiber (GFF) and silver membrane have been utilized most often in the sampling of airborne PAHs in both active and passive sampling. Glass filters have mainly been used in active sampling whereas silver membranes have mainly been utilized in passive sampling [72].

New materials have recently been developed for filters [82], like silica quartz (QFF), teflon-coated glass (TGFF), reagents-coated glass [83], PTFE membrane bonded to polyethylene net, and PTFE member supported by PTFE fibers, among others.

Improvements in sampling methods designed to minimize the losses of PAHs volatized during sampling have also relied on addition of solid sorbent back-up cartridges and cryogenic traps. For the sampling of PAHs, the sorbents most commonly used include Tenax, XAD-2, polyurethane foam (PUF), Florisil, Chromosorb 102, Chromosorb 106 and SepPak-C_{18} on Porasil [84].

The procedures UNE-77250 and EPA TO-13A described a sampling system to determine PAHs in ambient air based on the utilization of a particulate fiber filter followed by a PUF cartridge or a XAD-2 resin [85,86]. Table 5 shows several sampling devices utilized in standard methods.

The most common sampling devices utilized in recent years for the trapping of PAHs in air monitoring are XAD-2 resins [87], glass-quartz filters [88-90], combinations of filters and PUF [15,91], combination of glass-quartz filters, PUF, and XAD-2 [92], combination of PTFE filters and XAD-2 [93], or combination of denuders and filter pack (XAD-4 impregnated quartz) [94].

With respect to the PAHs determination in biomass smoke, Table 6 includes the most common sampling systems utilized in recent years. It can also be observed that glass-quartz filters, PUF and XAD-2 are the preferred trapping systems.

Table 6. Sampling procedures to determine PAHs in biomass smoke.

Combustion source	Sampling procedure	Ref.
Wood, barley, corn, rice and wheat	G-Q* filter, XAD-2 and impingers	[95]
Woods	G-Q* filter and XAD-2	[5]
Woods	G-Q* filter, XAD-2 and impingers	[96]
Woods residues	Impingers (glass wool and CH_2Cl_2)	[97]
Pines wood and needle	G-Q* filter and PUF/XAD-2/PUF	[98]
Agricultural and sylvicultural debris	G-Q* filter and PUF/XAD-2/PUF	[99]
Cork residues	G-Q* filter and XAD-2	[100]
Cattle residues	G-Q* filter and cartridge XAD-2/PUF	[101]
Urban residues	G-Q* filter and PUF	[102]
Medical residues	G-Q* filter and cartridge XAD-16/PUF	[4]
Pine and coal	G-Q* filter, XAD-2 and active coal	[74]
Coal	XAD-2 impregnated with CH_2Cl_2	[103]
Coal	G-Q* filter and PUF	[104]

*G-Q: glass or quartz

A recent method has been reported for the ambient air sampling of the 16 EPA PAHs plus B(e)Py. The method is based on active sampling on sorption tubes consisting of polydimethylsiloxane (PDMS) foam, PDMS particles and a TENAX TA bed [105]. It was demonstrated that both the high molecular weight PAHs and the more volatile PAHs, such as naphthalene, are efficiently trapped and quantitatively desorbed, without breakthrough and chemical reaction artifacts. An important result is that the total toxicity equivalence value

(TEQ) for PAHs was ca. two times higher compared to the conventional sampling method, illustrating that the concentrations of PAHs in ambient air have been underestimated until now [105]. This important conclusion points out the need to develop accurate and quantitative methods of sampling in order to report authentic values of PAHs contents in the air.

In any case, the best exposure and risk assessments for PAHs for the general population and the working population can be done using personal monitors. Ideally, this should consider a sampling time relevant to exposure scenarios, and hence be representative for possible health effects. These personal samplers are usually equipped with a battery to facilitate the pumping, and are available for a variety of brands. The personal sampler is characteristics for being portable and should be carried by a person as close to the breathing zone as possible [16,17,93,106]. In addition to this, people do not need to be excessively trained on the utilization of these systems. These personal samplers are equipped with filters and/or sorbents, as any sampling device above described.

2.1.2. Passive Air Samplers (PAS)

Diffusive samplers, also called diffusive monitors, passive samplers or passive monitors, are utilized for sampling without the need for an air mover, that is, without a pump. Man-made diffusive sampling operates by allowing gas or vapor molecules to diffuse through a defined volume of still air or through a polymer membrane, until they reach a sorbent bed. The principles of uptake are to consider that the passive sampling medium is uniform and "porous" and that it traps PAHs from the atmosphere by gaseous diffusion, and sorption. The mass collected is a function of the external concentration and the diffusion coefficient of the molecules. The diffusion coefficient varies in a known manner with temperature and pressure, and so the result can be corrected for these parameters [107].

Passive samplers provide information about long-term contamination. They can be used in a screening method for semi-quantitative comparison of different sites contamination with the advantage of low sensitivity to accidental short-time changes in concentration of pollutants. Other advantages of PAS are the simplicity and easiness of handle. In addition to this, they are cheaper because they do not need a pump. The main disadvantage is the long sampling time required due to the low PAH levels and limited analytical sensitivity.

Most of the passive air sampling measurements have been performed using semi-permeable membrane devices (SPMDs) [80,108,109], polyurethane foam disks (PUFs) [79,110-112] and XAD resins [113], which can be exposed over the period of several weeks or months.

SPMDs were initially developed by Huckins et al. [114]. The importance of SPMDs for air monitoring is expected to increase in the near future. SPMDs were initially designed to mimic the behavior of aquatic animal tissues in bioconcentrating contaminants, that is, the purpose was that only the bioavailable fraction of contaminants is absorbed and not compounds bonded to macromolecules [80]. As pointed out by Esteve-Turrillas et al. [80], semi-permeable membrane devices comprise a low-density polyethylene (LDPE) tube, 70-90 μm wall thickness, filled with triolein (1,2,3-tris-cis-9-octadecenoyl glycerol). Standard, these devices are 106 cm long, 2.54 cm wide, and contain 1 mL of triolein. Such devices are usually located in chambers to avoid external contamination. Due to both their high membrane surface area and their relatively large volume of receiving medium, SPMDs proved to be most effective in their capacity to accumulate lipophilic contaminants such as

PAHs. The main disadvantage of the SPMD technique is the complex sample preparation procedure required to recover the accumulated pollutants from the collecting phase (triolein). This is achieved by dialysis using considerable amounts of organic solvents, followed by concentration of the extracts and an expensive cleanup before the chromatographic analysis. SPMDs composed of some other materials for the membrane and for the internal solvent have also been described in recent years [80]. Photo-degradation of PAHs has been described in several SPMDs [115], so recent studies have focused in developing the best sorption chamber to avoid such photo-degradation [116]. SPMDs have been used not only to monitor PAHs in air [117], but also to monitor PAHs and nitro-PAHs in a wide survey which includes five European countries [118]. SPMDs must be calibrated in the laboratory, and the equations to relate the extracted concentrations with environmental concentrations must be established.

Some other devices for low-flow sampling are the denuders, which consist of sections of capillary gas chromatographic columns in which the vapor-phase of the semi-volatile organic compounds are sorbed by diffusion into the stationary phase, while the particulates pass through and are collected on a filter. Advantages of the method are the more accurate measurement of the vapor-particle partitioning and the more quantitative sampling of the most volatile compounds. Disadvantages are the low sampling flows resulting in very long sampling times (several days) [94,119].

Many studies utilize biomaterials as pollutant sorbents, such as plant samples, e.g. bark [120-123], lichen or pine needles [117,123-125], to monitor ambient air concentrations assuming that their spatial pollution distribution in the plants and the atmosphere will be closely related. However, the concentration capacity and sampling rate of plant tissues vary with the plant species and age, location and season. Therefore, these kinds of plant data are still subject to several uncertainties, which complicate their interpretation and limit the potential of plants as monitoring tools. For this reason, man-made passive air samplers are preferred up to date, despite their additional costs and that they have to be deployed, since the variability of samplers at different locations is low, and the sampling period can be controlled.

2.2. Extraction Procedures

After sampling, convenient transport and storage of PAHs must be conducted, to avoid losses by volatilization, sorption on recipients, or photo-degradation. In general, the sampling devices are protected from light until they reach the laboratory. It is also advisable to transport the filters and adsorbents that contain the PAHs in a refrigerated bag or icebox [94,99]. Once in the laboratory, samples are conserved at $\leq 4°C$, in absence of light, during at period not longer than 7 days prior to extraction. The most common extraction methods when monitoring PAHs in air are Soxhlet and ultrasound-assisted extraction (Figure 4).

2.2.1. Soxhlet Extraction

Soxhlet has been a standard technique during more than one century and it is still the main reference procedure to which the performance of other analytical methods is compared. The Soxhlet extraction is based on the apparatus invented in 1879 by Franz von Soxhlet, and is mainly used when extracting solid and complex samples. Figure 5 shows a typical scheme of a Soxhlet apparatus. The typical Soxhlet procedure starts by placing the sample in a cellulose thimble, which is loaded into the main chamber of the extractor. The Soxhlet extractor is then placed onto a distillation flask containing the organic extraction solvent, being acetone, toluene, and methylene chloride the most common organic solvents used. Once the organic solvent is heated to reflux, the solvent vapor travels up to the distillation arm and floods into the chamber housing the thimble. When the Soxhlet chamber is almost full, the chamber is automatically emptied by a siphon side arm, with the solvent running back down to the distillation flask. This cycle is repeated several times, over hours or days. It is obvious that during each cycle a portion of the PAHs dissolves in the solvent. After many cycles the PAHs are then concentrated in the extracting solvent. After extraction the solvent is removed, typically by means of a rotary evaporator, to decrease the amount of organic solvent and so to preconcentrate the analytes. The non-soluble portion of the extracted sample remains in the thimble, and is usually discarded.

Soxhlet extraction is still today the preferred extraction technique to ensure quantitative extraction of complex samples. Table 7 shows significant examples of works which utilize Soxhlet extraction in the determination of PAHs for air monitoring. Soxhlet is applied in the extraction of the solid systems used to trap PAHs in air monitoring. Table 7 also includes the most common organic solvents used, and the extraction time required to complete the extraction step. It should be noted the high extraction times (around 24 h following official procedures) and the high volumes of toxic organic solvents required in this traditional extraction approach, which are not only expensive to dispose off but which cause additional environmental issues. These are undoubtedly the main drawbacks associated with the Soxhlet extraction.

Official procedures also recommend the utilization of Soxhlet extraction in PAHs air monitoring, in the methods ISO 11338-2 (2003) "Stationary source emissions — determination of gas and particle-phase PAHs—Part 2: sample preparation, clean-up, and determination"; and ISO 12884 (2000) "Ambient air—Determination of total (gas and particle phase) PAHs—Collection on sorbent-backed filters with gas chromatographic–mass spectrometric analyses". The ASTM D 6209-98 (2004) "Standard test method for determination of gaseous and particulate PAHs in ambient air" also recommends Soxhlet extraction. It is also recommended by EPA TO-13A (toxic compounds in ambient air). Protection from light is always recommended during Soxhlet extraction performance, to avoid photo-degradation of PAHs.

Given the above mentioned disadvantages of Soxhlet, several authors have tried to improve the conventional Soxhlet device, either by modifying its design or the operational procedure [126]. The main updates of this "old-fashioned" technique have been related to the miniaturization (and therefore diminution of the organic solvent consumption), automation, and particularly, the combination with microwaves [127]. Main applications have been found when combining focused microwave-assisted extraction with Soxhlet, which is based on the same principles as conventional Soxhlet extraction but using microwaves as auxiliary energy to accelerate the process. However, up to date there are no reported applications of the focused microwave-assisted-Soxhlet in the analysis of PAHs in air monitoring. There are few works which utilize devices based on automatic Soxhlet in air-monitoring of PAHs [79,112]. It is therefore necessary that the recent developments in analytical procedures are followed by an implementation in monitoring techniques.

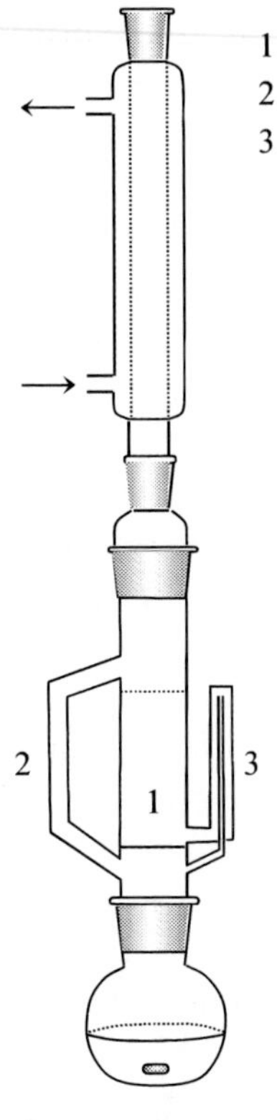

Figure 5. Typical scheme of a Soxhlet extractor.

Table 7. Recent works which utilize Soxhlet for the monitoring of PAHs in air.

Extraction time (hours)	Organic solvent	Organic solvent volume (mL)	Reference
20	hexane:acetone (1:1) or toluene	150	[77]
18	diethyl ether:hexane (1:9)	200	[90]
13	hexane:methylene chloride (1:1)	90	[121]
8 (each)	methylene chloride	1st 250 and 2nd 150	[94]
8	methylene chloride	—	[15]
24	methylene chloride	200	[122]
16	methylene chloride	—	[16]
72	methylene chloride	—	[104]
18	diethylether:hexane (1:9)	300	[99]

2.2.2. Ultrasound-Assisted Extraction (UAE)

Any sonication waves with a frequency higher than the ones detected by the human hear are called ultrasounds (US), that is, ultrasounds have wave frequencies higher than 16 kHz. The production of ultrasounds is possible due to the discovery of Marie and Pierre Curie in 1880 of the piezoelectric effect and its inverse. Nevertheless, the first commercial equipment of ultrasounds was not developed until 1917. Such equipment was curiously proposed in a contest of ideas in 1912, with the purpose of detecting icebergs right after the Titanic shipwreck. The utilization of ultrasounds in analytical chemistry laboratories did not take place until 1970, which appears to be illogical considering that it is obvious that the propagation of the sonication waves in any medium is going to increase the mobility of the molecules present in such medium. The most common ultrasounds devices in an analytical laboratory are the sonication bath and the sonication probe. The ultrasonic bath is so common in the chemical laboratories that it does not need presentation. It is widely used due to their economical feasibility, ensuring at the same time a relatively good energy distribution in the reactor placed in the bath. The sonication probe has a higher power, which can be modulated. The main applications of the ultrasounds in analytical chemistry are the ultrasound-assisted extraction and the ultrasound-assisted digestion, among others [128-130]. The most common procedure to carry out an ultrasound-assisted extraction is to place the sample in a recipient containing the extractant organic solvent, and to activate the ultrasounds during a fixed period of time. Protection from light of recipients-containing PAHs during the sonication extraction is always recommended.

EPA Method 3550 C describes the utilization of sonication for solid samples, which can be extended to several samples coming from air monitoring. Recent applications of sonication in the extraction of PAHs from the sampling devices used in air monitoring are summarized in Table 8. The extraction times are lower compared to traditional Soxhlet, and also the amounts of organic solvents used, but higher enough to be an environmental issue. Sonication probes or the most-recent US-devices have not been used up to date in the analysis of PAHs in air monitoring.

A paradoxical situation has emerged as most of the analytical methods employed to investigate environmental problems also generate chemical wastes, like high amounts of organic solvents, resulting in an environmental impact. In some circumstances, the chemicals employed are even more toxic than the species being monitored. In this sense, the work of some analytical chemists has focused on the development of methods less harmful to humans and to environment. Great concern over the disposal of the toxic organic solvents employed in the described method and their effect to the environment has led to move towards cleaner extraction methods such as microwave-assisted extraction and pressurized fluid extraction, among others. Improvements in clean-up methods also include solid-phase extraction (SPE) and solid-phase microextraction (SPME).

Table 8. Recent works reporting the utilization of ultrasound-assisted extraction to remove PAHs from sampling devices in air monitoring

Sampling device/sample	Extractant	Extraction time (min)	Ref.
Filters	5 mL cyclohexane	30 min	[93]
XAD-2 sorbents	1 mL acetonitrile	30 min	[93]
Pine needles/bark	30 mL hexane:CH_2Cl_2 1:1	10 min (2 times)	[123]
Glass filter	cyclohexane	30 min	[88]
SRM[a] 1648 or 1649a	20 mL hexane:acetone 1:1	30 min (2 times)	[77]
SRM[a] 1648 or 1649a	20 mL toluene	30 min (2 times)	[77]
Aerosol	25 mL CH_2Cl_2	10 min (4 times)	[89]
Filters	150 mL CH_2Cl_2	30 min (3 times)	[104]

[a]Standard reference material (certified)

2.2.3. Microwave-Assisted Extraction (MAE)

Microwaves are non-ionizing radiation with a frequency between 300 and 300000 MHz. Molecules under microwaves are moved by ion migration and dipole rotation. Commercial microwaves operate at 2450 MHz, which generate the alignment of molecules under the radiation followed by return to disorder $4.9 \cdot 10^9$ times per second. This quick movement-relax is consequently followed by a fast heating. The only condition for a molecule to suffer such heating is to be a dielectric compound. Fastest heating are obtained with analytes of high dielectric constants.

The first analytical application of microwaves was carried out by Abu-Samra in 1975 [131]. The utilization of microwaves to extract organic compounds from a number of samples (soil, seeds, foods and feeds) was carried out at the end of the 80s utilizing a domestic microwave [132]. The analytical applications in extraction procedures are based on the heating of the extractant by microwaves. The extractant is usually an organic solvent characteristic for being a dielectric compound. The use of solvent mixtures with and without dipoles opens up a variety of potential solvent mixtures. The extractant is placed in a reactor altogether with the sample, followed by heating under microwaves during a certain period of time. It should be highlighted that microwave heat the sample and the extractant without the need of heating the reactor, therefore reaching high temperatures with lower extraction times. After the application of the microwaves, the analytes originally bound to the sample are dissolved in the extractant solvent. The efficiency of the microwave extraction process depends on the nature of the sample and the extractant.

While domestic microwaves are still used in several analytical laboratories, commercial microwaves specifically-designed for laboratory use are preferred, especially attending to safety reasons. Most analytical applications have been carried out using cavity-diffuse or, to a lesser extent, focused microwave systems. Usually they are referred to as closed- or open-vessel systems, respectively, but this terminology is not correct as the so-called open vessels in focused systems are not completely open, and as it is also possible to operate closed vessels in a focused microwave [133]. The utilization of diffuse microwaves needs of a cavity (an oven) where samples are placed in a rotor to be moved for ensuring reproducibility in microwave irradiation, because the electric field is not uniform inside the cavity. Several samples can be placed simultaneously in a cavity-microwave. Focused microwaves are more efficient in terms of homogeneous heating. Figure 6 shows a scheme with the differences between diffuse and focused microwaves devices.

The main variables to be optimized in an analytical extraction by microwaves are: the microwave power, the extraction temperature and/or extraction pressure, the extraction time, the nature and volume of the extractant, and the characteristics of the matrix such as amount of water or presence of interferences. The optimization of these variables is time consuming unless an experimental design is used [134].

Microwave-assisted extraction has successfully been used in the extraction of sampling devices utilized for monitoring of PAHs in air, as it can be observed in Table 9. It can be seen that shorter extraction times and lower amounts of organic solvents are needed, especially when compared to Soxhlet extraction. In spite of the advantages of the microwave-assisted extraction method, including high extraction efficiencies, its utilization in samples coming from air monitoring is not widespread. EPA Method 3546 utilizes microwaves to extract solid samples, which can be extended to several samples coming from air monitoring.

Table 9. Microwave-assisted extraction procedures used in the monitoring of PAHs in air.

Sample	Extractant	Power-Time	Microwave	Ref.
SRM1648 SRM1649a	15 mL hexane:acetone (1:1)	400 W – 20 min	commercial	[77]
SRM1649a	15 mL toluene (10% water) or 15 mL CH_2Cl_2	400 W – 20 min	commercial	[77]
XAD-2 resins	10 mL ethyleneglycol:NaCl 1*M* (7:3)	120 W – 40 min	domestic[a]	[87]
bark / needles	90 mL hexane:CH_2Cl_2 (1:1)	513 W – 30 min	domestic[a]	[123]

[a]modified domestic oven

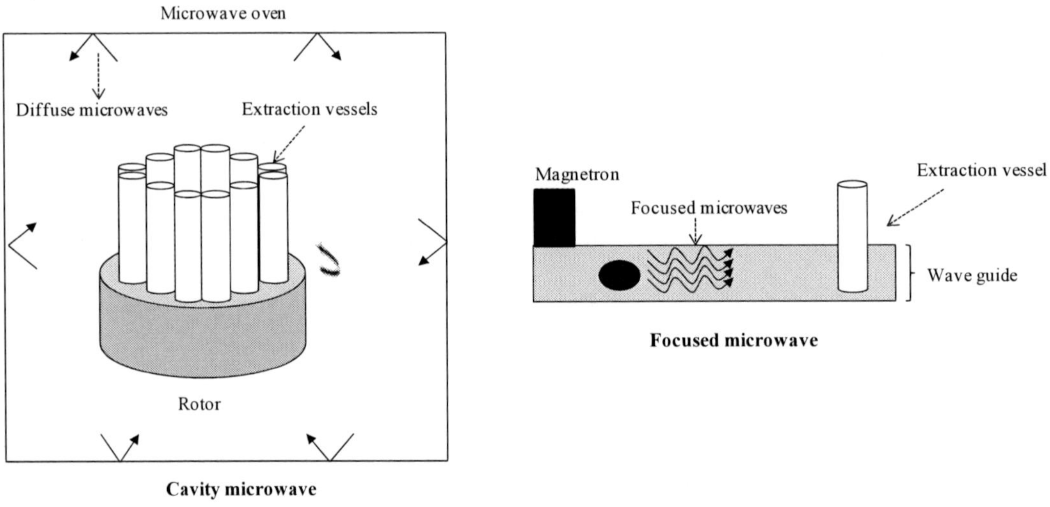

Figure 6. Differences between cavity and focused microwave devices.

2.2.4. Pressurized Liquid Extraction (PLE)

The technique pressurized liquid extraction (PLE), also named pressurized fluid extraction (PFE), was originally launched by Dionex Inc. in 1996 under the name accelerated solvent extraction (ASE[TM]) [135,136]. PLE is a solid-liquid extraction process performed with organic solvents in stainless steel closed-vessels into which solvent is pumped and brought to relatively elevated temperatures, between 80 and 200°C, and elevated pressures, between 10 and 20MPa [137]. The sample must be placed in the extraction vessel typically dispersed in a drying or inert sorbent such as sodium sulfate, Hydromatrix (Isco, Lincoln, NE, USA), or diatomaceous earth. During the heating cycle, solvent is pumped in and out of the cell to maintain the pressure and to perform the number of static cycles indicated by the user. In PLE, the organic solvents used as extractant are still below their critical conditions but with enhanced solvation power and with lower viscosities and surface tensions. The combination of high temperature and pressure results in better extraction efficiency, so minimizing solvent use and expediting the extraction process. PLE is like Soxhlet extraction, the only difference being the use of solvents that are raised to the near-supercritical region, where they show better extraction properties.

Water can also be employed as the extraction solvent in PLE, and so the technique is called pressurized hot water extraction, subcritical water extraction (SWE), super-heated water extraction, or high temperature water extraction [138].

The pressure in PLE is not a significant variable to optimize, because its role is simply to maintain the solvent in its liquid state. The main variables to optimize are extraction time and temperature, which considerably simplifies the experimental design. The main advantages of PLE are: fastness (around 15 min), low amounts of toxic organic solvents are required (between 15 and 40 mL), and several samples can be extracted simultaneously (up to 24). However, the instrumentation is much more expensive than Soxhlet or even microwaves. EPA method 3545 uses PLE in the determination of semivolatile organic compounds from solid samples [139], which can be extended to solid devices used to trap PAHs in air monitoring. Furthermore, PLE has proven to be an efficient method to extract analytes from complex samples, with efficiencies comparable to Soxhlet extraction [140].

Nevertheless, few applications of PLE have been reported for the extraction of sampling devices when monitoring PAHs in air. Table 10 reports several examples of reported applications in the literature for this kind of samples.

Table 10. Utilization of PLE in the determination of PAHs for air monitoring.

Sample	Pretreatment	Cell	Extractant	Conditions	Ref.
PM10 filters SRM1649a*	diatomaceous earth and cellulose filter	22 mL	hexane : acetone (1:1)	125°C - 1500 psi 2 static cycles 5 min hold	[90]
pine needles	dried	—	hexane : acetone (1:1)	100°C - 1500 psi 2 cycles 20 min hold	[117]
XAD-2 PUF	dried portions of 30 g	100 mL	dichloromethane	100°C - 1500 psi 1 cycle 15 min hold	[92]

*Method optimized by experimental design: extraction efficiencies between 96-103%

2.2.5. Dialysis

Separation processes brought about differences in concentration through a semi-permeable membrane are generically known as dialysis. Dialysis is a membrane barrier separation process in which differential concentration forces one or more analytes to transfer from the sample solution to another fluid through a semi-permeable membrane, which allows contact between the fluids without mixing. The extraction of PAHs from SPMDs filled with triolein, utilized in passive air monitoring, is widely performed by dialysis using organic solvents such as hexane [80]. In dialysis of SPMDs, PAHs diffuse from the inner triolein solution, where they were initially retained, to the hexane outer side of the LDPE membrane as a result of a concentration gradient (see Figure 7). The extraction is normally carried out in 48 h [107]. The main disadvantages of dialysis are the long times required to perform the extraction, and the high volumes of hexane required (usually between 100 and 200 mL). The main advantage is the fact that there are practically no losses of triolein during the extraction step, and so the SPMDs devices can be reutilized. The conventional dialysis procedure is quite simple because it only requires placing the SPMDs device in a recipient containing hexane, all protected from UV light to avoid PAHs' photodegradation. It has been suggested the improvement of the conventional dialysis by utilizing ASE [141] in order to decrease both, extraction time and organic solvent consumption. Other extraction techniques have been proposed for desorption of SPMDs [80], however, they have not been applied yet to the analysis of PAHs in air monitoring.

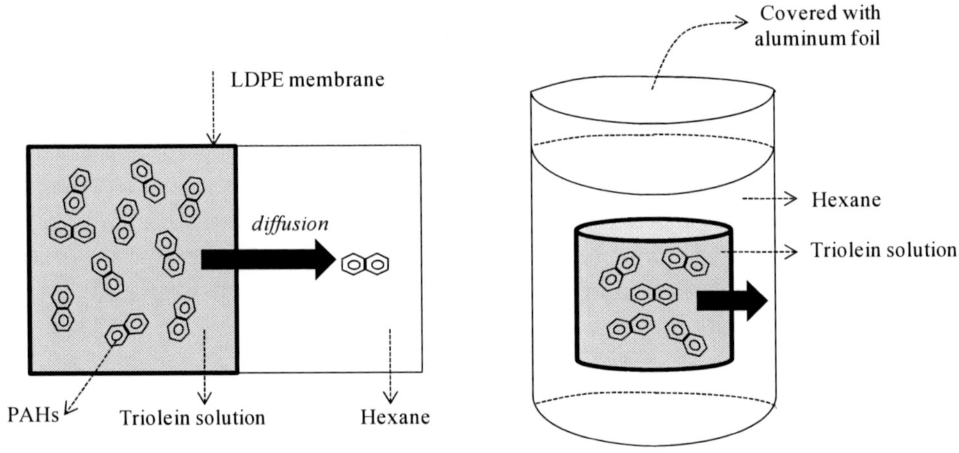

Figure 7. Dialysis extraction of PAHs contained in SPMDs devices.

Table 11 summarizes the main applications of conventional dialysis in the extraction of SPMDs devices used in air monitoring of PAHs. It can be observed that, in all cases, extraction times of 48 h, and organic solvent volumes higher than 200 mL, are required.

Table 11. Dialysis extraction reports of SPMD used in the air monitoring of PAHs.

Organic solvent	Volume of extractant	Time	Reference
Hexane	—	2 × 24 h	[117]
Hexane	180 mL × 2	2 × 24 h	[115]
Hexane	180 mL × 2	2 × 24 h	[116]
95:5 cyclopentane:CH_2Cl_2	1st 100 mL 2nd 180 mL	2 × 24 h	[118]
Hexane	130 mL × 2	2 × 24 h	[141]

2.2.6. Supercritical Fluid Extraction (SFE)

Supercritical fluid extraction (SFE) is an extraction procedure which utilizes solvents in their supercritical state as extractive agents. These supercritical fluids have similar densities to liquids, but lower viscosities and so higher solvation power. Supercritical fluids constitute an acceptable alternative to conventional liquid solvents in the analytical extraction of environmental samples [137]. Supercritical fluids are obtained in commercially available instruments, applying high temperatures and pressures to the solvent to ensure conditions above the critical point. The sample is located in an inert extraction cell, where the supercritical fluid is pumped.

The most common fluid to date has been supercritical CO_2 because of its reasonable critical properties (relatively low critical pressure and temperature, i.e., 7 MPa and 31.3°C, respectively), low toxicity, and chemical inertness. In addition, supercritical CO_2 has Lewis base characteristics, induced dipole interactions, and quadrupole interactions that allow it to solvate different analytes ranging in polarity from nonpolar to moderately polar. While CO_2 is a relatively good solvent, these solvating interactions are weak and it has been shown that pure CO_2 often cannot extract environmentally persistent pollutants [142]. The problem is usually overcome using modifiers to adjust the overall polarity, such as acetone, methanol and dichloromethane, among others.

Extraction can be performed in static, dynamic or recirculation mode. In static mode, the cell containing the sample is filled with the supercritical fluid, pressurized and allowed to equilibrate. In dynamic mode, the supercritical fluid is continuously passed through the extraction cell. In the recirculation mode, the same supercritical fluid is repeatedly pumped through the sample until it is finally pumped out to the collection system.

After depressurization, the extracted analytes can be collected in a small volume of organic solvent or on a solid-phase cartridge (adsorbent trap) [137].

An important advantage of SFE is the fact that the extracts are relatively quite clean, and only moderate additional clean-up is required. In fact, combined with solid adsorbent traps, the extracts are practically ready for the chromatographic determination.

Several parameters must be optimized in order to achieve successful extraction efficiency, such as supercritical fluid nature, working temperature and pressure, extraction time, shape and size of the extraction cell, sample particle size, moisture content of the matrix and the analyte collection system. Due to the high number of parameters affecting the overall performance, the optimization in SFE is often tedious and difficult, which has avoided a wider applicability of the technique. Other disadvantages of the SFE technique include: limited sample size and high cost of the equipment.

Since 1980s, several applications of SFE in the analysis of environmental samples have been reported [143-145]. In addition, SFE has been adopted by the EPA as a method for extracting PAHs (Method 3561) from solid environmental matrices, which can be extended to several samples coming from air monitoring. Table 12 lists several reported applications of SFE in the analysis of PAHs for air monitoring. The reported optimum extraction conditions are also listed.

In aerosol analysis, the most abundant compounds are nonpolar mixture of branched, cyclic and unsaturated hydrocarbons, generally characterized as an unsolved complex mixture (UCM) [149]. The UCM generates a large hump in the gas chromatogram, obscuring the peaks of other trace compounds. Several works have suggested the online coupling of SFE-LC-GC-MS to determine PAHs in aerosols and so to avoid the UCM problems [146,148]. The coupling of multidimensional techniques to sample extraction methods further decreases the manual sample pretreatment steps, which results convenient.

Table 12. Applications of SFE for the monitoring of PAHs in air.

Sample-PAHs	Technique	Optimum SFE conditions	Reference
SRM 1649 urban air PM (0.3 g)	off-line SFE GC-MS	CO_2 + different modifiers. 400 atm. 2.5 mL extraction cell 1) 5 min static 2) 10 min dynamic Elution on CH_2Cl_2	[142]
GFF 16 EPA PAHs	on-line SFE LC-GC-MS	CO_2. 400 atm. 150°C 45 min Elution with n-pentane:ethyl acetate 97:3 v/v	[146]
GFF 6 nitro-PAHs 7 oxy-PAHs	off-line SFE GC-ECD	CO_2 + toluene* 95:5 v/v. 350 atm. 90°C 5 mL extraction cell. 1) 5 min static 2) 30 min dynamic 1 $mL{\cdot}min^{-1}$ Elution: a) 2 mL hexane for aliphatic, b) 3 mL hexane:CH_2Cl_2 4:1 for PAHs, and c) 6 mL CH_2Cl_2 for nitro- and oxy-PAHs	[147]
QFF 17 PAHs	on-line SFE LC-GC-MS	CO_2 + CH_2Cl_2 (400 μL). 400 atm. 150°C 3 mL extraction cell. 1) 10 min static 2) 45 min dynamic 1 $mL{\cdot}min^{-1}$ Solid phase trap, Elution with n-hexane:ethyl acetate 95:5 v/v	[148]

*A first step with pure CO_2 (150 atm, 45°C) is carried out to remove other components.

2.3. Clean-up Procedures

After the sampling and the extraction procedure, independently on the technique used for extraction, various matrix components are frequently present in the extract altogether with the PAHs extracted. These non-desirable components must be eliminated to ensure an accurate identification and quantification of PAHs, which are present at low levels of concentration in the extract. In addition, several matrix components present in the extract can later damage the chromatographic system. Clean-up steps are therefore necessary to remove not only the bulk of the co-extracted material, but also those compounds closely related to the analytes that could potentially interfere in the final determination (i.e., aliphatic hydrocarbons when trying to determine PAHs). Furthermore, many times a solvent exchange is necessary, right after the extraction step, to allow the injection of the extracted PAHs in a chromatographic system.

2.3.1. Liquid adsorption chromatography in columns

Several procedures have been developed to achieve a successful clean-up. Among them, classical liquid adsorption chromatography in columns is still the dominant technique for purification of extracts. This technique involves passing the extract through an adsorbent column prepared in the laboratory. The interferences remain retained in the sorbent whereas the analytes are eluted from the sorbent using a convenient organic solvent. The operating procedure is simple, and once the sample is loaded into the glass column already filled with the sorbent, an adequate organic solvent will elute PAHs from the sorbent. The flow of the eluent must be carefully controlled to avoid insufficient elution of PAHs or elution of matrix interferences. Flows between 1 and 2 $mL \cdot min^{-1}$ are usually employed for sorbent amounts of 10 g. The most common sorbents used in these columns are alumina, silica gel, and Florisil, with different mesh sizes. These sorbents must be carefully cleaned and conditioned before being used, being Soxhlet the most common pre-cleaning step [85]. In many occasions, the columns are also loaded with 1 cm of anhydrous Na_2SO_4, located into the upper side of the sorbent, to remove any water content of the samples, especially if a GC procedure is going to be used. EPA methods 3630C, 3610B and 3620B, which use silica gel, alumina and Florisil, respectively, have been approved for the purification of organic extracts from solid environmental samples. The official method UNE-77250 and the method EPA TO-13A also recommend the utilization of a silica gel column as a clean-up step for PAHs [85,86]. Table 13 includes common clean-up procedures with sorbents-columns for PAHs samples in air monitoring. It can be observed the complexity of the step and the timing required to carry out a successful clean-up. Several reports in air monitoring of PAHs also utilize silica-gel columns [79,91,95,104,112,117,118] or alumina columns [88,95], but without specifying the conditions in which the clean-up procedure is conducted.

2.3.2. Solid Phase Extraction (SPE)

Solid-phase extraction (SPE) is a widely used sorbent extraction technique which involves the use of sorbent materials to trap analytes and separate them from the bulk solution. As the sample solution passes through the activated sorbent bed of the device, analytes concentrates together with some other matrix components. To remove the analytes of interest, an adequate organic solvent is employed in the final elution step. The most widely used trapping mechanism for SPE is hydrophobic interaction. The most common sorbent material is alkyl-bonded silica (C18 silica, C8 silica, and C2 silica), because it is reactive enough to permit its surface to be chemically modified and yet stable enough to allow its use with a wide range of solutions. Copolymer sorbents such as cross-linked polystyrene divinylbenzene and hydrophilic lipophilic balanced polymers are very popular too [150]. Most recently, molecularly imprinted polymers (MIPs), which are synthetic materials with artificially generated recognition sites able to specifically rebind a target molecule in preference to other closely related compounds, start to be quite common [151]. Today, there is a wide variety of SPE sorbent materials commercially available, which are valid for PAHs clean-up [152]. The size of the sorbent particles usually ranges from 10 to 60 μm.

The stationary phase in SPE comes in the form of a packed syringe-shaped cartridge, a 96-well plate [153], or a flat disk. These devices can be mounted on commercially available extraction manifold which allow many samples to be processed simultaneously. Up to 24 cartridges can be placed in a typical cartridge SPE manifold, while 6 disks can be placed in a typical disk SPE manifold. These SPE manifolds are usually equipped with a vacuum port to speed up the extraction process. The analytes are collected in sample tubes inside or below the manifold after they pass through the sorbent.

Table 13. Specific conditions for a successful clean-up of PAHs samples coming from air monitoring utilizing sorbent columns.

Sorbent	Pre-conditioning	Conditioning	Interferences elimination	Elution of PAHs	Reference
Silica gel Alumina	130°C (silica) 350°C (alumina)	Deactivated silica. Soxhlet – alumina: 1) CH_2Cl_2:MeOH 2:1 (12 h) 2) CH_2Cl_2:hexane 30:70 (12 h)	Hexane (aliphatic)	CH_2Cl_2:hexane (30:70)	[77]
Silica gel 8 g glass column 1 cm i.d. × 30 cm	130°C 16 h	Cleaning with 50 mL of hexane prior to sample loading	20 mL hexane (aliphatic)	30 mL CH_2Cl_2:hexane (3:7)	[121]
Alumina 1 g Silica 2 g Na_2SO_4 1 cm	180°C (alumina) 140°C (silica) 450°C (Na_2SO_4)	Deactivated silica	Hexane	25 mL CH_2Cl_2:hexane (1:1)	[116]
Silica gel 10 g	—	Cleaning with 25 mL of hexane, 2 $mL \cdot min^{-1}$, prior to sample loading	25 mL n-pentane	25 mL pentane:CH_2Cl_2 (2:3). 1 mL sample	[16]
Silica gel 10 g	16 h 150°C	Soxhlet 6 h. Cleaning with 40 mL of pentane prior to sample loading	25 mL pentane (aliphatic)	25 mL pentane:CH_2Cl_2 (6:4). 1.5 $mL \cdot min^{-1}$	[5,98,99]
Alumina 5 g 8 mm i.d.	12 h 160°C	Cleaning with cyclohexane	20 mL cyclohexane	30 mL toluene	[74]

The SPE extraction is commonly performed in four steps: conditioning-wetting (that is, a solvation of the functional groups of the sorbent bed to make them interact with the sample), retention-loading (that is, the sample is loaded into the cartridge to make the analytes bound to the sorbent), selective washing (that is, the elution of undesired matrix components by adequate choosing of the eluting agent), and elution (the analytes are desorbed using an adequate solvent and collected to be further analyzed). Sometimes, column regeneration is accomplished. Figure 8 shows a scheme with the common steps followed in SPE. Method development in SPE is accomplished optimizing pH, ionic strength, polarity and flow-rate of the eluent, and physicochemical characteristics of the sorbent bed. The main advantages of this clean-up technique are the lower amount of organic solvents required and the shorter times needed. In addition to this, the technique is cheap when compared to other extraction and/or clean-up procedures, and quite efficient in terms of extraction recoveries.

Several attempts have been conducted to improve SPE devices in order to decrease the consumption of organic solvent or to minimize the human handling. The main step forward has been the introduction of on-line (and sometimes automated) SPE–LC [154,155] or SPE-GC [156]. The SPE procedures are essentially the same as off-line ones, the major difference being that the elution volumes are usually smaller and often different adsorbent materials are used for the extraction.

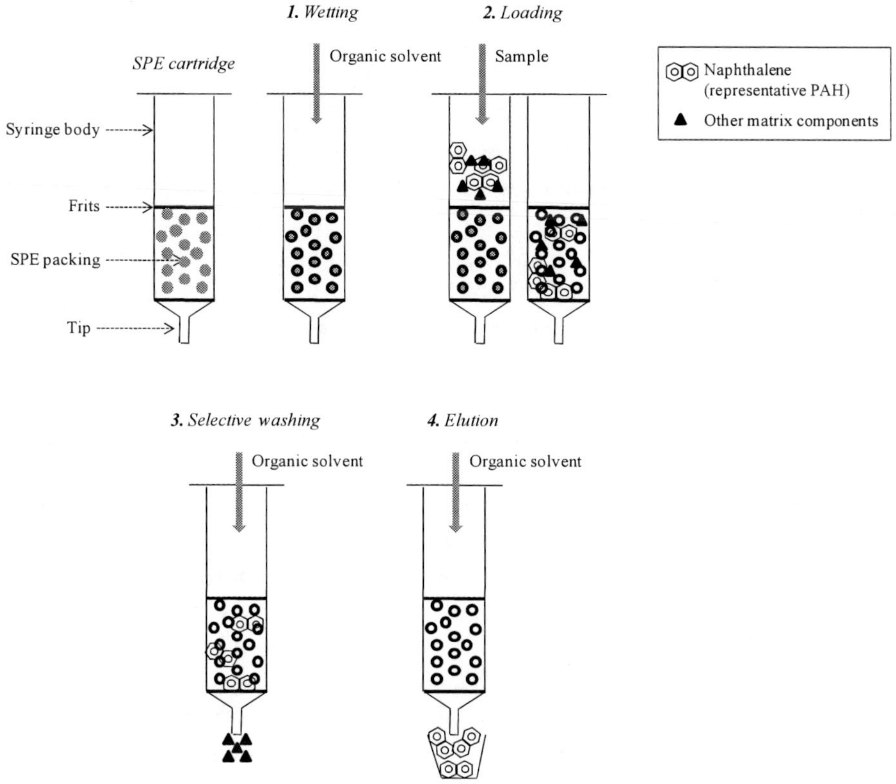

Figure 8. Common steps to carry out a SPE clean-up procedure.

Recent applications of SPE in the clean-up of PAHs extracts obtained in air monitoring have been carried out by off-line SPE. Ratola et al. [123] used SPE cartridges which were conditioned with 50 mL of hexane, and eluted with 50 mL of hexane:dichloromethane (1:1) followed by 50 mL of dichloromethane. Xie et al. utilized a 10 port vacuum SPE manifold to carry out the clean-up procedure. The authors paid special attention to the drying steps of the procedure. The sample was successfully eluted with 3 mL of a solution composed of 20% of dichloromethane in hexane (v/v) at a flow rate of 1 mL·min^{-1} [89].

2.3.3. Solid Phase Microextraction (SPME)

Solid-phase microextraction (SPME) is a relatively novel extraction technique developed by Pawliszyn in the nineties [157,158]. The working procedure of this non-exhaustive extraction technique always involves two steps. In the first step, the analytes contained in an aqueous matrix are extracted by contact during a certain period of time with a fiber coated by an active phase. This fiber coating is linked to a syringe to facilitate its manipulation. This sorption step can be achieved by direct immersion or in the head space of the matrix solution. The second step implies desorption of the analytes retained in the fiber into an analytical system. Figure 9 shows a general scheme of the SPME process. Desorption in a GC system is simple, because it only requires thermal desorption in the hot injection port of the GC. On-line desorption in CE or HPLC needs a specific interface device to accomplish such step [159]. The most common commercial SPME fiber coatings are the polyacrylate (PA) and the polydimethylsiloxane (PDMS). Nevertheless, many new polymeric sorbents are currently being developed with specific purposes [151,160]. The main advantage of SPME is the fact that is a solvent-free technique, with undoubtedly environmental interest. Furthermore, the technique is simple, fast, easy to automate, portable, not expensive, and require low amounts of sample (sample volumes lower than 20 mL). The main disadvantage of the technique when used in direct immersion is the fact that analytes must be dissolved in water, which undoubtedly limits its utilization to polar analytes and/or to water samples. If intending to extract solid samples, headspace mode or utilization of hollow-fiber membranes is required, but the latter is not always quite successful with hydrophobic analytes such as PAHs. In addition, the technique is limited to volatile compounds when used in the headspace mode. One important feature that must be kept in mind is the fact that SPME is a non-exhaustive extraction procedure. In fact, only a small fraction of analytes are extracted from the sample. This non-exhaustive characteristic is usually compensated by the high-preconcentration factors achieved, by the avoiding of co-extracting matrix components, and by the utilization of an external calibration, with standards subjected to the whole SPME step [161].

The main parameters to optimize in a SPME extraction procedure are the selection of the optimum fiber coating and width (wider coatings are preferred to ensure higher extraction efficiencies), extraction time (equilibrium conditions are usually preferred), extraction temperature (mainly for headspace mode, it should be noted however that the sorption process on the fiber coating is not favored at high temperatures), selection of modifiers to increase the extraction efficiency (such as small amounts of organic solvents or high contents of salts), pH, stirring, and sample volume.

The main application of SPME as clean-up procedure for PAHs monitored in air has been its headspace (HS) mode combined with microwaves for the analysis of XAD-2 resins [87]. Three different SPME fibers were used in the study: 65 μm PDMS-DVB, 85 μm PA, and 100 μm PDMS. The microwave power was 120 W during 40 min, while the HS-SPME mode was simultaneously carried out.

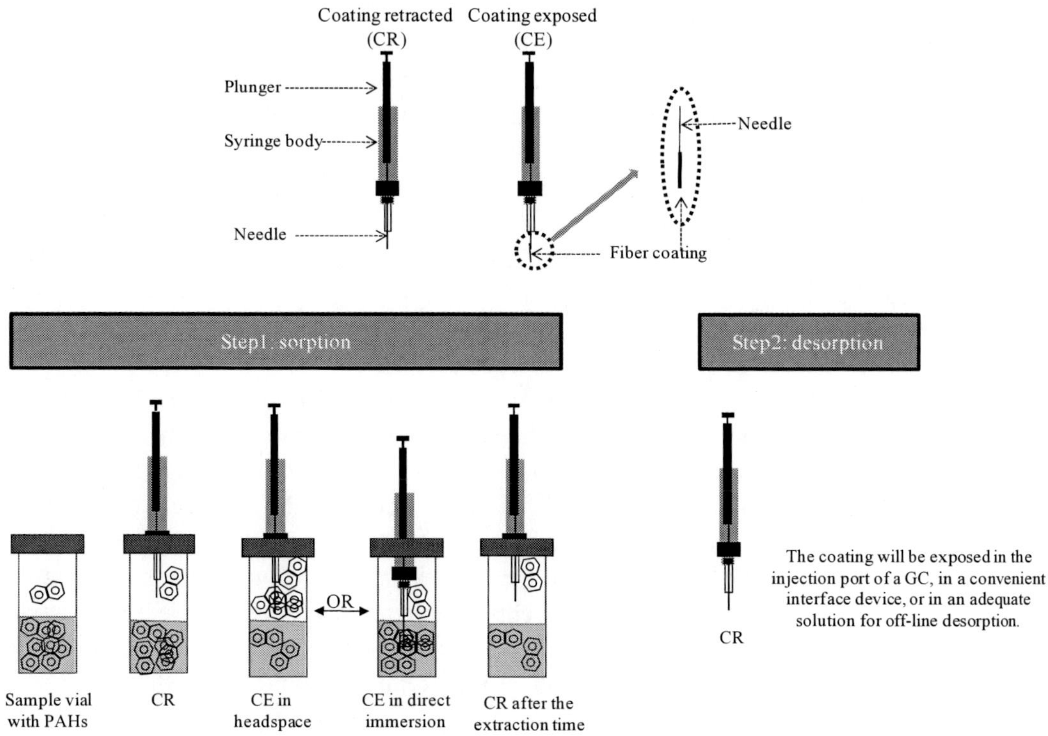

Figure 9. General scheme of the SPME procedure.

Both, SPE and SPME can be used as extraction techniques. However, when dealing with complex samples such as the ones obtained in air PAHs monitoring, the techniques are not totally successful as extraction procedures, and therefore their utilization is in general restricted to clean-up steps, or used in combination with other extraction procedures.

2.3.4. Gel-permeation chromatography (GPC)

Gel-permeation chromatography (GPC) is a technique also referred to as size exclusion chromatography (SEC), and based on molecular size separation. This size exclusion cleanup procedure utilizes hydrophobic gels in the separation of macromolecules and organic solvents to elute retained analytes. The packing gel is porous and is characterized by the range or uniformity (exclusion range) of that pore size. The GPC column retains molecules that are small enough to enter the pores of the gel. Larger molecules are un-retained and therefore eluted from the column first. In the choice of gels, the exclusion range must be larger than the molecular size of the molecules to be separated. The most common sorbent is a cross-linked divinylbenzene-styrene copolymer (SX-3 Bio Beads or equivalent). The main advantage of GPC is its more applicability than adsorption chromatography to unknown contaminants isolation when little information on the sample is available. Furthermore, it can be fully automated, with pumps, detector and fractions collector. Another advantage is the lifetime of the GPC column, up to several months. An EPA method (Method 3640A GPC Cleanup) has been approved for the purification of organic extracts from solid environmental samples. The main disadvantage of GPC is that, mainly for quite complex samples, is necessary to combine it with other clean-up procedures, such as classical liquid adsorption chromatography in columns. Nevertheless, it should be highlighted herein that many of the already described clean-up procedures are not always used isolated. In fact, it is very common to combine some of them in order to ensure an exhaustive purification of the PAHs' extract.

Several works of PAHs monitoring in air utilize GPC as a clean-up step [90,116,118]. These reported applications utilize an automated GPC instrument, eluting PAHs with dichloromethane at flow rates between 4.5 and 5 $mL{\cdot}min^{-1}$, and collecting the hydrocarbons fractions at around 20 min.

2.3.5. Other clean-up and preconcentration steps

Clean-up steps must be always followed by a preconcentration procedure. The injection of the final extract into a chromatographic system requires the reduction of the final amount of organic solvent volume to reach the adequate limits of sensitivity. In addition to this, many times is also necessary to exchange the solvent, because the solvent used in the extraction/clean-up step is not compatible with the chromatograph. Such exchange is preceded by a reduction and/or elimination of the original solvent volume. The most common systems to carry out such reduction of the solvents volume are the rotary evaporator (rotavap) and the TurboVap® evaporation system. A rotary evaporation is in general used to separate solvents such as n-hexane or ethyl acetate (with relatively low boiling points) from compounds which are solid at room temperature and pressure. A rotavap consists of a motor unit which rotates an evaporation flask containing the sample extract. The flask is partially submerged in a water bath, with a temperature not exceeding 35 - 40°C. A condenser is also connected to the system to produce the condensation of the solvent that is gathered in a collector. The whole system is under vacuum, to reduce the pressure within the evaporator system. It is well-known that lowering the pressure above a bulk liquid lowers the boiling points of the component liquids, and this is the basis of the rotavap. The TurboVap® is an evaporation system which uses a patented gas vortex shearing action and optical sensors to provide fast and efficient evaporation of organic solvents. The time required to carry out this solvent reduction depends on the amount of solvent to evaporate. The main advantage of the rotavap is that there are no losses of analytes by evaporation, whereas the TurboVap® needs to be optimized to avoid such losses. On the other hand, only one sample can be placed in the rotavap whereas several samples can be placed in a TurboVap®. Table 14 includes a summary of the most common clean-up procedures utilized in the monitoring of PAHs in air. In all cases, it can be observed the complexity of the overall procedure, and the long time and human handling required. Sample preparation is still the bottleneck of any analytical method.

Table 14. Clean-up procedures most frequently used in the monitoring of PAHs in air.

Sample	Clean-up steps right after the extraction procedure	Reference
SRM1648 SRM1649a	1) Filtration-Centrifugation. 2) Redissolve in hexane. 3) Silica-gel:alumina column. 4) Rotavap (up to 0.2 mL or dryness). 5) Redissolve in hexane (1 mL) for GC and in acetonitrile (0.5 mL) for HPLC.	[77]
PM10 (filters) SRM1649a	1) N_2. 2) Redissolve in 700 μL of CH_2Cl_2. 3) GPC (elution with CH_2Cl_2). 4) TurboVap®-N_2. 5) Redissolve in hexane (0.2 mL) for GC.	[90]
GFF and PUF	1) Rotavap. 2) Silica-gel:alumina column. 3) Rotavap (up to 0.1-0.5 mL).	[104]
Bark	1) Rotavap. 2) Redissolve in 1 mL of hexane. 3) Sillica-gel column. 4) Rotavap. 5) Redissolve in hexane (1 mL).	[121]
QFF and XAD	1) Rotavap (up to 1 mL). 2) Ultrapure N_2 (up to 0.5 mL).	[94]
GFF and PUF	1) Rotavap. 2) N_2 (up to dryness). 3) Redissolve in acetonitrile (2 mL) for HPLC.	[15]
Pinus bark	1) Glass wool. 2) Anhydrous Na_2SO_4. 3) Rotavap (up to 10 mL). 4) N_2 (up to 0.4 mL).	[122]
SPMDs	1) Solvent exchange to CH_2CL_2. 2) GPC (elution with CH_2Cl_2). 3) Silica-gel:alumina column + anhydrous Na_2SO_4 (eluted with hexane:CH_2Cl_2 (1:1)). 4) N_2 (up to 50 μL).	[116]
SPMDs + pine needles	1) Silica-gel:alumina column. 2) GPC.	[117]
SPMDs	1) Rotavap (up to 0.5 mL). 2) GPC. 3) Silica-gel:alumina column. 4) Rotavap. 5) N_2 (up to 100 μL).	[118]
Biomass	1) Solvent exchange to cyclohexane. 2) Silica-gel:alumina column. 3) Rotavap (up to 1 mL). 4) 1 mL acetonitrile.	[16]
Pinus needles and bark	1) Rotavap. 2) SPE. 3) N_2 at room temperature. 4) Solvent exchange to 1 mL hexane.	[123]

2.3.6. Mechanisms control during sampling

Given the complexity and labor-intense of the extraction and clean-up procedures, it is clearly necessary to utilize mechanisms to control and quantify possible losses of PAHs during these steps. In this sense, it results quite useful the utilization of internal standards (IS) and recovery standards (RS). An internal standard, by definition, is a compound, different from the analyte to be measured, which is added with a perfectly known concentration to the unknown sample (or to the standards). The calibrations plots are then constructed by plotting the ratio of the signal from the analyte to the signal from the internal standard *versus* the analyte concentration. The internal standard must be added right before conducting the quantification, that is, right after the last clean-up step. It is quite difficult to select the most adequate internal standard, which must be a compound that matches as closely, but not completely, the analytes to be monitored. The utilization of deuterated standards is quite common if a GC procedure is going to be accomplished, because a deuterated analyte is not going to be present in the sample, and at the same time present quite similar characteristics with the analytes to be monitored. Recovery standards take into account the whole extraction/clean-up procedure, and so they are added to the sample since the first beginning of the process, in order to quantify the efficiency of the extraction-preconcentration procedure developed in the laboratory. Decafluorobiphenyl (DFB) as recovery standard for PAHs is recommended in the regulations UNE-77250 and US EPA-TO13 [85,86].

The performance of SPMDs, which implies longer exposition and so higher risk of losses of PAHs during sampling, needs to be evaluated too. It results advisable to use reference compounds to validate them, the so-called performance reference compounds (PRCs) for SPMDs. These standards act like RS, but are only referred to the sampling stage.

Table 15 shows the most common IS, RS and PRCs, employed in several analytical methods used to monitor PAHs in air.

Table 15. IS, RS, and PRCs used in analytical methods to monitor PAHs in air.

IS	RS	PRCs	Ref.
—	Ace-d_{10}, Fl-d_{10}, Phe-d_{10}, Ft-d_{10}, B(a)A-d_{12}, Chy-d_{12}, B(b)Ft-d_{12}, B(a)Py-d_{12}, B(ghi)Per-d_{12}, I(1,2,3-cd)Py-d_{12}	A-d_{10}, Py-d_{10}	[115]
N-d_8	—		[95]
Me_6Benzene	—		[89]
A-d_{10}	N-d_8, Ft-d_{10}, Py-d_{10}, Acy-d_{10}, Fl-d_{10}, Phe-d_{10}, 2-MeN-d_{10}, Biphenyl-d_{10}, Ace-d_{10}, B(a)A-d_{12}, Chry-d_{12}, B(b)Ft-d_{12}, B(k)Ft-d_{12}, B(a)Py-d_{12}, B(e)Py-d_{12}, Per-d_{12}, I(12,3-cd)Py-d_{12}, diB(a,h)A-d_{12}, B(ghi)Per-d_{12}		[92]
Chy-d_{12}, Per-d_{12}	Py-d_{10}, B(k)Ft-d_{10}		[90]
A-d_{10}, Py-d_{10}	—		[162]
Me_6Benzene	N-d_8, Phe-d_{10}, Ace-d_{10}, Chy-d_{12}, Per-d_{12}		[104]
—	N-d_8, Phe-d_{10}, Ace-d_{10}, Py-d_{10}, Chy-d_{12}, B(a)Py-d_{12}		[121]
p-terphenyl-d	N-d_8, Ace-d_{10}, Py-d_{10}, B(a)Py-d_{12}, B(ghi)Per-d_{12}		[94]
Ace-d_{10}, Phe-d_{10}, Chy-d_{12}, Per-d_{12}	—		[122]
B(e)Py-d_{12}	Fl-d_{10}, Phe-d_{10}, Ace-d_{10}, Phe-d_{10}, B(a)A-d_{12}, Chry-d_{12}, B(b)Ft-d_{12}, B(a)Py-d_{12}, I(12,3-cd)Py-d_{12}, B(ghi)Per-d_{12}	A-d_{12}, Py-d_{12}	[116]
terfenyl	N-d_8, Phe-d_{10}, Per-d_{12}		[79]
Ace-d_{10}, Phe-d_{10}, Chy-d_{12}, Per-d_{12}	DFB		[98]
—	N-d_8, Phe-d_{10}, Ace-d_{10}, Py-d_{10}, Chy-d_{12}, Per-d_{12}		[117]

2.4. Chromatographic Procedures

Gas chromatography and high-performance liquid chromatography are the preferred techniques to carry out the final determination and quantification of PAHs. It is not the purpose of this book chapter to deeply describe and revise the principles and applications of both well-known chromatographic procedures [49,163]. Furthermore, several reviews have focused on the utilization of GC [51] and HPLC [164] for PAHs.

Traditional PAHs separations by reverse phase (RP)-HPLC is conducted using gradient elution, mainly with acetonitrile:water mixtures and sometimes with methanol:water mixtures. RP-HPLC on chemically bonded C_{18} phases has been widely used in the environmental analysis of PAHs. Up to date, a wide variety of brands of HPLC columns is available for PAHs' determination. Packed columns are rarely used in contemporary GC methods for PAHs, being practically 100% of applications carried out with capillary columns. Many commercial brands of GC and HPLC include in their web pages information related to best columns to work with in a specific application, advices to optimize the best gradient elution, and even suggestions for troubleshooting.

The most common detectors used in HPLC for the determination of PAHs are ultraviolet (UV), fluorescence (FD), diode-array (DAD) and electrochemical (ECD). In GC the most utilized detector is flame ionization (FID). Enhancements to GC (large-volume injection, thermal desorption and fast GC) have also been applied in the analysis of PAHs [51]. Mass-spectrometry (MS) is the preferred detection system GC to identify and determine PAHs.

Advantages of HPLC are good selectivity and extremely good sensitivity when fluorescence detection is used. On the other hand, GC is the method with the highest resolution power when a PAH mixture analysis is needed, and with good selectivity. Both, GC and HPLC are good choices for the determination of PAHs in air monitoring. It must be taken into account that the main problems in the monitoring of PAHs are encountered during the sampling, and also during the extraction and preconcentration of PAHs, and so the final choice of the chromatographic procedure is not that critical. Nevertheless, there is an increased tendency to use GC over LC in recent years [51].

The coupling of liquid chromatography with gas chromatography (LC-GC) results in a powerful technique that combines the best characteristics of both techniques. To perform LC-GC, the whole sample is injected into the LC and then subsequently analyzed by GC. Christensen et al. have reported a LC-GC/MS method for the determination of PAHs in particulate matter from ambient air [165].

Several attempts have also been conducted to develop portable GC for PAHs monitoring. Several sorbents can also be used as selective concentrators for sample air as part of the analysis protocol for air taken from canisters or air sampled directly by automated gas chromatographs (autoGCs), which is essentially equivalent to ambient air monitoring. However, these autoGCs have not been quite successful up to date for PAHs in air monitoring.

The utilization of GC-MS is recommended by EPA TO-13A. The official procedures ISO 12884 (2000) and ISO 16362 (2005) include the utilization of GC-MS and HPLC-FD (or – DAD), respectively, for the determination of PAHs in air.

Table 16. Extraction techniques combined with chromatographic procedures for the monitoring of PAHs in air.

Sample / sampling	PAHs determined	Extraction technique* — Chromatographic procedure	Extraction efficiency— Detection limits (DLs)	Reference
SPMDs	16 EPA, B(e)Py, Per	dialysis-GC-MS	45-105% (PRCs)	[115]
GFF and XAD-2		Soxhlet-GC-MS	93-122%	
GFF and XAD-2	16 EPA, B(e)Py	UAE-HPLC-UV	DLs around 0.3 ng	[5]
FF and XAD-2	16 EPA, 2-MeN, B(e)Py, Per	Soxhlet-GC-MS	DLs: 2-50 ng	[95]
GFF	16 EPA	UAE-SPE-GC-MS	83-97% from aerosols	[89]
QFF, PUF+XAD-2	16 EPA, 1-MeN, 2-MeN, 2,6-DMeN, Biphenyl, 2,3,5-TMeN, 1-MePhe, B(j)Ft, B(e)Py, Per	ASE-GC-MS	DLs around 0.1 $ng \cdot m^{-3}$	[92]
SRM 1649a	Ft, B(a)A, B(b)Ft, B(j)Ft, B(k)Ft, B(a)Py, I(1,2,3-cd)Py, diB(ah)A	ASE-GC-MS	DLs: 0.26-0.30 $pg \cdot m^{-3}$	[90]
PUF	16 EPA, B(e)Py, Coronene	Soxhlet or UAE –GC-MS	74-98%	[104]
bark	16 EPA	Soxhlet-GC-MS	DLs: 0.03-0.5 $ng \cdot g^{-1}$	[121]
pine	16 EPA	MAE or UAE –GC-MS	70-130%. LODs around 0.2 $ng \cdot g^{-1}$	[123]
QFF and PUF + XAD-2	16 EPA	Soxhlet-GC-MS	45-105%	[99]
QFF and XAD-4	16 EPA	Soxhlet-GC-MS	DLs down to 0.002 $ng \cdot m^{-3}$	[94]
FF + PUF	16 EPA except N, B(e)Py, diB(a,e)Py, diB(a,c)A, diB(a,e)Py, diB(a,e)Ft, diB(a,h)Py, diB(a,i)Py	Soxhlet-HPLC-UV or FD	83–110%	[15]
XAD-2	N, Ace, Acy, Fl, Phe, A, Ft, Py	MW-HS-SPME-GC-MS	80%. DLs: 0.02-1.0 ng	[87]
Mixed bed	16 EPA	GC-MS	DLs: 13 $pg \cdot m^{-3}$	[105]
Pinus bark	16 EPA except N, 2-MeA, 9-MeA, 1-MePy, Per	Soxhlet-GC-MS	88%	[122]
SPMDs	16 EPA, B(e)Py, Per	dialysis-GC-MS	32–68% (PRCs)	[116]
SPMDs	16 EPA	dialysis-GC-MS	95–120%	[117]
GFF + PUF		Soxhlet-GC-MS	72–112%	
pine needles		ASE-GC-MS	80–110%	
SPMDs	16 EPA	dialysis-GC-MS	—	[118]
GFF	Ft, B(a)A, Py, B(a)Py, B(b)Ft	UAE-GC-FID	—	[88]
GFF	B(a)A, Chy, B(b)Ft, B(k)Ft, B(a)Py, diB(a,h)A, I(1,2,3-cd)Py	Soxhlet-HPLC-UV	82-94% DLs: 0.5-5 $ng \cdot mL^{-1}$	[16]

*Clean-up procedures are not listed

Table 16 summarizes the most recent extraction techniques combined with GC or HPLC to determine PAHs for air monitoring.

3. Water Monitoring

Waters are important receptors for PAHs coming from several sources. The contamination of watercourses is clearly undesirable if considering that rivers, lakes and ponds are frequently used for potable water supply [166]. Waters can be divided in drinking waters, surface waters, groundwater and coastal waters, each one with different implications with regards to human health.

PAHs enter surface waters mainly via atmospheric fallout (wet and dry deposition), urban run-off, municipal effluents, industrial effluents and oil spillage or leakage. Urban run-off contains PAHs deposited on surfaces, as well as mobile-related PAHs from gasoline and oil drips or spills, exhaust products, and bitumen from road surfaces [167]. Although specific sources are known to be responsible for the presence of PAHs in surface waters, their occurrence cannot always be related to a particular source [166].

PAHs in groundwater may originate from polluted surface waters, agricultural irrigation with effluents, leaching from solid wastes disposal sites or contaminated soils. Groundwater is naturally filtered as it flows through various soil matrices, and PAHs adsorb well to organic soil [166].

Coastal waters are mainly characterized for its salinity and high particulate organic matter content. The quality of coastal waters has social, economic and environment impacts [168]. It is important to consider that PAH concentrations at off-shore sites are usually at low or undetectable levels, while higher concentrations are observed in many coastal and estuary sites [169]. According to the World Health Organization study in 1997, the concentration of individual PAHs in surface and coastal waters is generally around 0.05 $\mu g \cdot L^{-1}$ and concentration above this point indicates some contamination.

The presence of PAHs in drinking water may be due to the contamination of surface or groundwater used as raw water sources [166]. The chlorination of drinking water which contains PAHs has been related to the formation of oxygenated and chlorinated PAHs, compounds that are even more toxic.

The Water Framework Directive (WFD) 2000/60/EC [170] intends to achieve and preserve the good ecological status of the waters of the European Union member states until 2015. For achieving such purposes, it is required to monitor the waters quality. The new European Drinking Water Guideline (98/83/EC) [171] gives special attention to benzo(a)pyrene by setting a threshold limit of 10 $ng \cdot L^{-1}$ for this single compound in drinking waters. The guideline also sets a limit of 0.10 $\mu g\ L^{-1}$ for the sum of benzo(b)fluoranthene, benzo(k)fluoranthene, benzo(ghi)pyrene, and indeno(1,2,3-cd)pyrene. Moreover, the European Union has fixed very restrictive limits for these compounds in different kinds of superficial waters in the proposal for a new Directive 2006/0129EC [172], which was approved in 2007/C 97/02 [173]. The limits for benzo(a)pyrene were fixed on 0.05 $\mu g\ L^{-1}$ as annual medium value and 0.1 $\mu g\ L^{-1}$ as maximum admissible concentration for different kinds of superficial waters. The limits for the sum of benzo(b)fluoranthene-benzo(k)fluoranthene and the sum of benzo(ghi)perylene-indeno(1,2,3-cd)pyrene were fixed to 0.03 and 0.002 $\mu g \cdot L^{-1}$, respectively, as annual medium value and as maximum admissible concentration. Naphthalene, fluoranthene, and anthracene were also included in this directive. All the above limits must be achieved in 2015. Furthermore, analytical methods used for the assessment of compliance with the requirements of the WFD should demonstrably meet stated minimum performance criteria and be implemented taking into account approved guidelines on quality assurance and quality control (QA/QC) [174]. Some other countries in the world have fixed low admissible limits for benzo(a)pyrene in driking waters, i.e. the maximum contaminant level (MCL) is 25 $ng \cdot L^{-1}$ in Japan [175]. Health Canada and the Ontario Ministry of the Environment (MOE) have established a maximum acceptable concentration (MAC) of 10 $ng \cdot L^{-1}$ for benzo(a)pyrene in drinking waters [176], whereas the US-EPA establish a MCL of 200 $ng \cdot L^{-1}$ for benoz(a)pyrene. The World Health Organization has recommended the maximum permissible concentration of six representative PAHs: fluoranthene, benzo(b)fluoranthene, benzo(k)fluoranthene, benzo(a)pyrene, benzo(g,h,i)perylene and indeno(1,2,3-c,d)pyrene, not to collectively exceed 200 $ng \cdot L^{-1}$ in drinking waters [who2].

The solubility of PAHs in water is low and decreases with increasing molecular weight, which is mainly due to their hydrophobic nature (log $K_{o/w}$ = 3-8). Nowadays, the analysis of natural waters for PAHs, given the direct implications with human health, has been developed to high standards. Nevertheless, the low concentration levels of PAHs in waters continue to promote research interest, often directed towards achieving more convenient, cost-effective, and environmental-friendly methods.

Main problems with water samples are found during the sampling stage. In addition to this, PAHs in water, due to their low solubility, present adsorption problems in containers. Methanol, acetonitrile, 2-propanol or surfactants are added during sampling and storage to avoid these adsorption problems [178,179]. For the extraction of water samples, most methods use liquid-liquid extraction [180] and solid-phase extraction [178]. Other extraction methods, such as supercritical fluid extraction (SFE) [181], solid-phase microextraction (SPME) [182], stir-bar sorptive extraction (SBSE) [183], single-drop microextraction (SDME) [184], and cloud-point extraction (CPE) [185], have been developed more recently. Figure 10 shows a typical scheme of the steps usually involved in the determination of PAHs for water monitoring. It should be highlighted the important decrease in the clean-up steps when dealing with water samples, mainly with drinking water samples, due to the lower complexity of the matrixes, especially when compared to samples obtained for air monitoring.

3.1. Sampling Strategies

The nature of the water sample has an important influence on the sampling strategy. Drinking waters generally have low suspended solids content, and are therefore relatively straightforward to sample. In contrast, river, lake, coastal waters and groundwater exhibit much greater temporal and spatial heterogeneity [186].

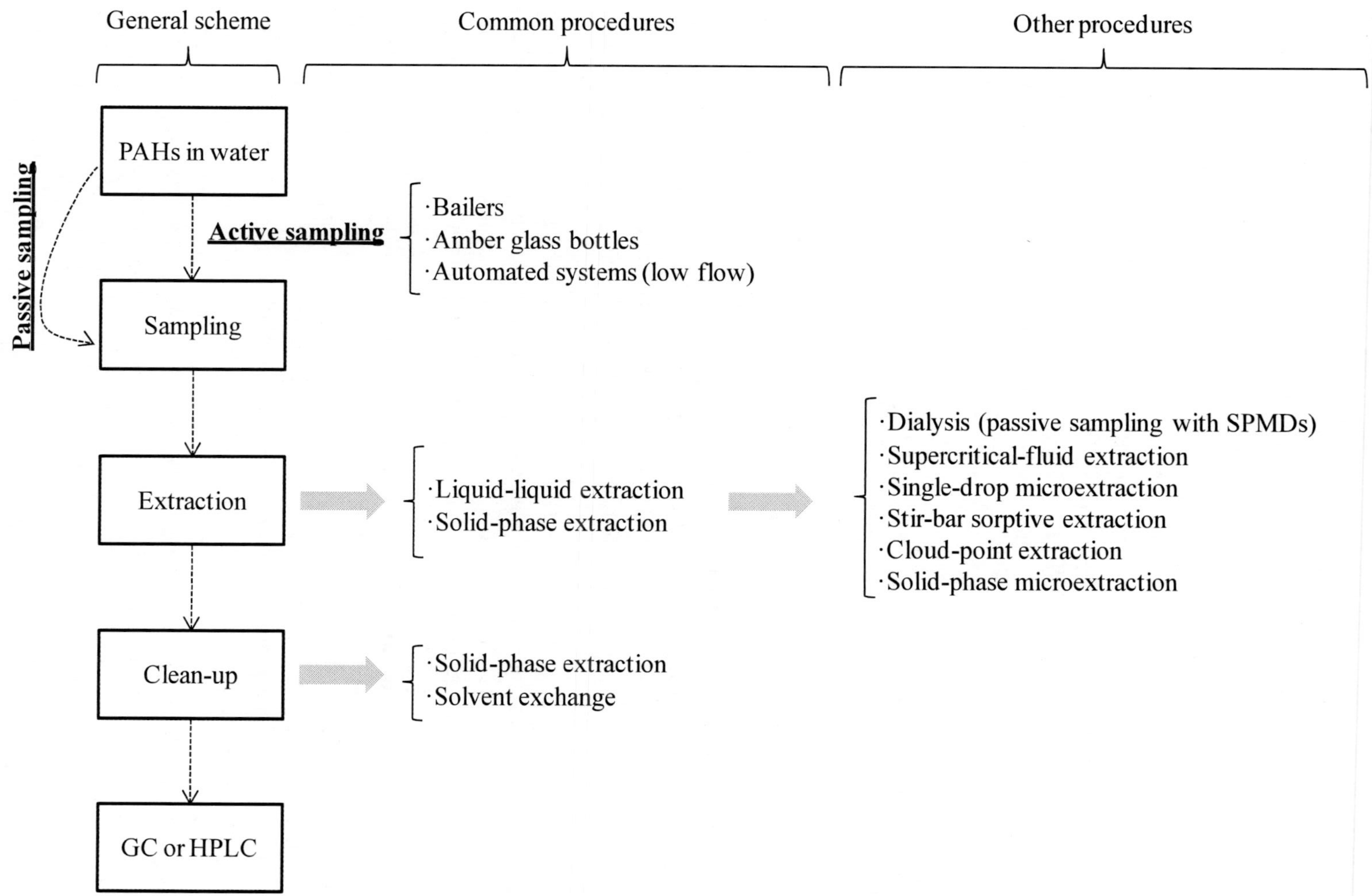

Figure 10. Scheme of the analytical steps employed in the determination of PAHs for water monitoring.

Disturbances of water composition during sampling as well as alterations during transport and storage, all can irreversibly affect the outcome and subsequent interpretation of water analysis results, and so the sampling strategy must be carefully carried out. The sampling strategy should also reflect the known or expected variability of the system. The changes in concentration with sampling position at a site are important where point source inputs are found. Moreover, when samples are needed to measure discharge loads in rivers or fluxes in lakes, it is important to plan the sampling so as to cover periods when the concentrations are likely to differ from baseline values, e.g. during spates or storm periods [186].

The higher molecular-weight PAHs, due to their hydrophobic nature, tend to be particle- or colloid associated, not truly dissolved in water [187,188]. These species have limited potential for off-site transport unless through facilitated transport, i.e., while associated with mobile colloids [187,189]. Thus, if sampling methods tend to increase the amount of suspended solids in water samples; reported PAH levels may be biased upward.

3.1.1. Active Sampling

The conventional sampling technique is the active removal, so-called snap-shot sampling, of a defined volume of water at a given time mainly using amber glass bottles [190], bailers or pumps [191].

Amber borosilicate glass containers are used for *in-situ* measurements, normally capped with PTFE-lined tops. They must be conveniently rinsed prior to sampling. Usually, apart from washing of the glassware with detergent and water, it includes rinsing with an organic solvent [192]. Alternatively, non-volumetric borosilicate glassware may be heated to 400°C for 1 h [186]. The glass containers are usually completely filled with water during sampling, to avoid an excessive amount of dissolved oxygen in the aqueous sample. Ballasted bottles, which are opened at the desired water depth to carry out the sampling, are also commonly used.

Bailers are sampling systems composed of two caps, which are closed at the desired water depth to carry out the sampling. They are commonly used when sampling deep waters (rivers, seawaters, ponds, etc.). Hand bailers are commonly used to collect samples due to their simplicity, low-cost, and universality but can impact chemical constituents due to the turbulence associated with their use.

The automatic water samplers have two variants. They can be time-dependent, that is, the automatic samplers take sub-aliquots of waters every fixed period of time, without accounting for flow variations, whereas other automatic samplers are caudal dependent. These automated system usually accomplish the process using low flow pumping (<0.2 L·min^{-1}). Sampling devices such as low-flow peristaltic pumps remove water slowly and smoothly from wells to obtain representative samples but possess significant potential for sorption of PAHs to long segments of tubing and losses of volatile compounds through volatilization [193].

The limitations of the active sampling technique are well known: PAHs can suffer sorption onto tubes and containers [194], and can also be degraded during transport. Disturbed flow regimes due to pumping, particularly when sampling groundwater, may lead to the pulling in of clean or contaminated water from zones not intended for sampling [194] In fact, measurements of PAH concentrations in groundwater are highly prone to bias stemming from the choice of sampling techniques [187]. Pumping can also cause losses of volatile compounds [195].

3.1.2. Passive Sampling

In passive sampling, PAHs are trapped into appropriate sampling devices along a diffusion gradient without the use of pumps or bailers [196]. The diffusion gradient is established and maintained by means of a receiving sorbent with a high affinity to the hydrocarbons. If the diffusion gradient can be maintained over extended periods by providing sufficient sorption capacity, PAHs can be accumulated continuously over time and time-weighted average (TWA) concentrations determined. Thus, in addition to providing undisturbed water sampling, passive approaches can be used to integrate fluctuations in contaminant concentrations over time without the need for excessive snap-shot sampling. A variety of passive sampling devices are available for water sampling [107].

Semipermeable membrane devices (SPMDs) filled with triolein, identical to the ones used in air sampling, are also commonly employed in water monitoring. Gustavson and Harkin [191] reported a comparison between SPMDs and typical active samplers (bailers and automated systems) for the monitoring of PAHs in groundwater. Among other important results, the authors concluded that SPMDs only sample truly dissolved PAHs. Extraction efficiencies for PAHs when using SPMDs are around 40% for naphthalene and between 57 and 94% for the rest of EPA PAHs [197,198], which results acceptable if considering the difficulties for trapping naphthalene. Recent modifications of SPMDs include the substitution of the inner triolein solution for other reagents. It has been proposed the utilization of an ionic liquid (IL) as adequate substituent of triolein, specifically, 1-buthyl-3-methylimidazolium hexafluorophosphate (BMIM-PF_6), with comparable results [199]. ILs are non-molecular solvents, characterized for presenting low melting points and negligible vapor pressure, which have gained significant attention recently as a newer class of potential green solvents, mainly due to the recent trend to substitute conventional organic solvents with less harmful reagents [200]. The advantages of using ILs arise in the avoiding of the further dialysis step, because 50 μL of the IL-containing the extracted PAHs can be directly injected in a HPLC-FD [199].

Another integrative passive sampling system has been described, and consists of a solid poly(dimethylsiloxane) (PDMS) material (tube or rod), acting as hydrophobic organic receiving phase, enclosed in a water-filled or an air-filled low-density polyethylene (LDPE) membrane tubing [201]. The sampler was proved to be valid for light PAHs (with detection limits of 3 $pg \cdot L^{-1}$ for fluorene), but not so adequate for heavy PAHs.

Solid-phase microextraction (SPME) devices have successfully been employed in water samples not only in extraction or clean-up procedures, but also as passive samplers. Field trials demonstrated that three types of SPME samplers (fiber retracted, PDMS-rod and PDMS-membrane) were suitable for the long term monitoring of PAHs (acenahpthene, fluorene, phenanthrene, anthracene, fluoranthene and pyrene) in water [202]. These samplers possessed all of the advantages of SPME: solvent-free, sampling, extraction and concentration were combined into one step, followed by direct injection into a gas chromatograph (GC). These samplers also addressed additional advantages of passive samplers: economical, easy to deploy, and the time-weight average (TWA) concentrations of target analytes could be obtained with one sampler. Moreover, the mass uptake of these samplers was independent of the face velocity, or the effect could be calibrated, which is desirable when the convection conditions of the sampling environment are difficult to measure and calibrate. Among the three types of SPME samplers that were tested, the PDMS-membrane possessed the highest surface-to-volume ratio, which resulted in the highest sensitivity and mass uptake and the lowest detection level [202].

The ceramic dosimeter is a time-integrative passive sampling device which is based on a ceramic tube as diffusion-limiting barrier enclosing a receiving phase that consists of solid sorbent beads. It has been successfully applied to the extraction of PAHs from groundwater using Amberlite as sorbent bead [194].

It has also been described the utilization of polymer films as passive sampler for the determination of pyrene and anthracene in clear and turbid aquatic environments, followed by monitoring the laser induced fluorescence, emitted from the film, via optical fibers [203]. Detections limits around 10 $ng \cdot L^{-1}$ are achieved with these samplers.

3.2. Extraction Procedures

Likewise in air monitoring, after the sampling it is necessary to conveniently carry out the transport and storage of PAHs, to avoid losses by volatilization, sorption on recipients, or photo-degradation. In general, the sampling devices are protected from light until they reach the laboratory, and transported in a refrigerated bag or icebox. Regarding storage, and independently on the sampling strategy used, samples are normally stored at 4°C in absence of light for no more than 7 days. The most common extraction methods when monitoring PAHs in water are liquid-liquid extraction and solid-phase extraction (Figure 10).

3.2.1. Liquid-Liquid Extraction (LLE)

Liquid-liquid extraction (LLE) is probably the oldest method of extraction. It allows the separation of two or more analytes due to their unequal solubility in two immiscible liquid phases. A feed aqueous solution initially containing the analytes is thoroughly mixed with an immiscible solvent having a different density. The mixing provides a large interfacial area so that efficient mass transfer of the desired analytes can occur between the two immiscible liquids. The liquid-liquid dispersion created during the mixing is then separated by gravity or by centrifugal force depending upon the type of extractor selected. Both, the mixing step and the separation step constitute one stage of extraction.

The extraction solvent is chosen to selectively extract specific analytes from the feed aqueous solution. The extraction solvent is usually selected for its extraction efficiency and selectivity, its inertness and its boiling point. Other factors which are considered are the toxicity of the extracting solvent, relative densities of the two phases and their tendency to form emulsions. A typical extraction solvent for aqueous extraction in LLE is dichloromethane. Depending upon the selectivity of the solvent and the amount of mass transfer required to achieve the desired analyte recovery, several stages of extraction may be required.

LLE has the advantages of low equipment costs and quantitative recoveries, but there are a number of disadvantages. Among them, it should be cited the relatively large volumes of frequently toxic and flammable organic solvents used in the extraction process (around 100 mL for 0.25-1 L of sample), such as disposal and resulting environmental concern. Other problems are emulsion formation, contamination from glassware and analyte losses owing to sorption on laboratory ware.

EPA Method 610 reports the utilization of LLE with dichloromethane (3 × 60 mL) for the extraction of PAHs in the analysis of municipal and industrial wastewaters. EPA Methods 3510C and 3520C report the utilization of a separatory funnel in liquid-liquid extraction and a continuous liquid-liquid extraction process, respectively. Nowadays, LLE is one of the most common and routinely extraction technique employed in the monitoring of PAHs in water.

Table 17 reports several examples which use LLE in the monitoring of PAHs from water samples, or methods which use LLE as a standard method to compare with. It can be observed the high amounts of organic solvents needed in the extraction step, which is the main disadvantage of the technique.

Table 17. Analytical applications of LLE to monitor PAHs in water samples.

Sample	PAHs	LLE conditions	Performance	Ref.
Tap water, seawater (500 mL)	16 EPA PAHs	3 × 60 mL CH_2Cl_2 (sep. funnel 1 L)	68.6-84.5% (tap) 50.3-74.3% (sea)	[180]
Groundwater	16 EPA PAHs	3 × 60 mL CH_2Cl_2 (sep. funnel 2 L)	Comparison with SPMDs	[191]
Amberlite of ceramic dosimeter	16 EPA PAHs + 1-MeN + 2-MeN	3 × 10 mL acetone	DLs: 30-80 $ng \cdot L^{-1}$	[194]

3.2.2. Solid-Phase Extraction (SPE)

Solid-phase extraction (SPE) can be used as an extraction technique, rather than a clean-up technique, when dealing with water samples not highly contaminated. In fact, its utilization with drinking waters is quite successful. SPE can be conducted in both, off-line and on-line procedures.

A drawback of off-line SPE procedures is that they can be time consuming and labor intensive, often requiring many steps before reaching a concentrated extract suitable for instrumental analysis, of which only a small portion is actually injected onto the chromatographic column [204].

The utilization of on-line SPE offers a series of advantage if compared to off-line SPE. The use of on-line SPE techniques has made possible the development of faster methods by reducing the sample preparation time and thus increasing the sample throughput [154]. Conditioning, washing and elution steps can be performed automatically and some systems also permit to extract one sample while another one is being analyzed by LC [204]. Other important advantages of on-line coupling are decreased risk of contamination of the sample or sample extract, elimination of analyte losses by evaporation or by degradation during sample preconcentration, and improved precision and accuracy. Higher sensitivity is also achieved in on-line configurations due to the transfer and analysis of the totality of the extracted species to the analytical system. Furthermore, on-line SPE has lower solvent consumption requirements thereby decreasing the costs for organic solvents waste disposal. SPE coupled on-line to LC has also the advantage that it is not necessary to remove all residual water from cartridges because elution solvents are compatible with the LC separation methods [155].

Although the reduction of solvent usage can be accomplished in SPE, it still requires organic solvent volumes of some milliliters and has a limiting sample loading volume. These are the two main disadvantages of SPE.

Table 18. Utilization of SPE as extraction procedure for monitoring PAHs in waters.

Sample	PAHs	Technique	SPE procedure	Performance	Ref.
Coastal waters (500 mL)	16 EPA	off-line SPE (C18) -GC-MS	*condit.*: 1) 3 mL CH_2Cl_2 2) 2 × 3 mL methanol 3) 3 mL HPLC water. *elut.*: 8 mL CH_2Cl_2 ↓ vol. (N_2) up to 0.3 mL	DLs: ≈ 2 $ng·L^{-1}$	[190]
Groundwater (200 mL with 10% MeOH v/v)	16 EPA	off-line SPE (C18) -GC-MS	Filtration of samples *condit.*: 1) 5 mL ethyl acetate 2) 5 mL MeOH 3) 5 mL water with MeOH 2% v/v (all at 0.5 $mL·m^{-1}$) and rinsing with 5 mL HPLC water. *elut.*: 5 × 5 mL ethyl acetate, 10 $mL·m^{-1}$ ↓ vol. (N_2) - reconstitute 250 µL hexane	DLs: 0.3–15 $ng·L^{-1}$ Recovery: 35–113%	[46]
Tap waters and seawaters (500 mL with 10 mL MeOH)	16 EPA	off-line SPE (C18) -MAE-GC-MS	Filtration of samples *condit.*: 1) 10 mL ethyl acetate 2) 10 mL MeOH 3) 10 mL pure water. Vacuum. *elut.*: microwaves: 50%, 10 mL acetone or CH_2Cl_2, 100°C, 7 min ↓ vol. 1 mL	≈ 70% recovery	[180]
Rain water (180 mL, 25% 2-propanol v/v)	16 EPA	off-line SPE (C18) -HPLC-UV or FD	*condit.*: 1) 5 mL CH_2Cl_2 2) 4 mL 2-propanol. *elut.*: 2 mL THF or CH_2Cl_2 ↓ vol. (N_2) up to 0.3 mL – reconstitute MeOH up to 1 mL	DLs: 0.03-36 $ng·L^{-1}$ Rec.: 10–52% N to Fl 69-98% rest of PAHs	[178]

Table 18 (Continued). Utilization of SPE as extraction procedure for monitoring PAHs in waters.

Sample	PAHs	Technique	SPE procedure	Performance	Ref.
Drinking waters (1 L). 100 mL for SPE	16 EPA	off-line SPE -GC-MS	*condit.*: 1) 5 mL MeOH 2) 5 mL HPLC water, 1.5 $mL·min^{-1}$. Vacuum. *elut.*: 1 mL CH_2Cl_2. reconstitute up to 1 mL	Recovery: 65–96%, some exceptions	[206]
Purified waters, drinking waters, surface waters (100 mL). 50 mL for SPE	16 EPA	on-line SPE (C18) -HPLC-FD	Filtration of samples. *condit.*: 1) 3 mL acetonitrile 2) 7 mL HPLC water. *load* and rinsing with 3 mL of water and increasing T (thermally assisted desorption). *elut.*: gradient of analysis.	DLs $< 100\ ng·L^{-1}$	[207]
Surface waters, drinking waters, groundwater (800 mL)	17 PAHs	off-line SPE (C18) -GC-MS	*condit.*: 1) 5 mL MeOH 2) 5 mL water (2 $mL·m^{-1}$). *load*: 20 $mL·m^{-1}$. *elut.*: 2 × 2 mL toluene. ↓ vol. (TurboVap): 1 mL	MDLs $\approx 1\ ng·L^{-1}$ Recovery: 65–114%	[208]

US EPA has suggested the use of SPE for the recovery of PAHs from drinking water. The cartridges are proposed to be packed with silica, whose surface has been modified by chemically bonded octadecyl (C18) groups, while methylene chloride is proposed as eluting solvent [205].

Table 18 includes several reports on the utilization of SPE in the monitoring of PAHs from water samples.

It should be highlighted that, although SPE (used as extraction technique) works well with "clean" samples, its applicability is limited by the presence of particulate matter. This often requires preliminary filtration of the sample, and analytes may be lost in this step owing to adsorption on the solid particulate matter. SPE used as an extraction technique is not that convenient when dealing with seawaters or wastewater.

3.2.3. Supercritical-Fluid Extraction (SFE)

SFE is an extraction procedure mainly applied to complex samples, like the ones coming from air or soils-sediments monitoring. Nevertheless, several successful applications have been reported for the monitoring of PAHs in water. Glazkov et al. have reported the extraction of Ace, Acy, Fl, Phe, A, Ft, Py, B(a)A and Chy from 10 mL water samples (in 15 mL extraction cells) using off line SFE in combination with GC-FID or GC-MS, with detection limits ranging from 10 to 100 $ng \cdot L^{-1}$ [181]. Sargenti and McNair have also reported successful extraction of the 16 EPA PAHs from drinking water (5 mL) using CO_2+CH_2Cl_2 (80:20 v/v) in off line SFE-GC-MS [206].

3.2.4. Solid-Phase Microextraction (SPME)

In addition to the the solid-phase microextraction (SPME) applications in clean-up steps or in passive sampling, it has also been proposed the use of SPME in extraction schemes for PAHs, mainly contained in drinking waters. Table 19 summarizes several applications of SPME in water samples, as well as the optimum extraction conditions to reach successful limits of detection. The applications are always carried out in direct immersion mode.

The main disadvantage of SPME is due to the low amount of sorbent coated on the SPME fiber. This, altogether with the fact that SPME is a non-exhaustive extraction technique, usually brings sensitivity problems. SPME has also limiting lifetime, fragility, and the fibers and assembly holder are relatively expensive. In addition, SPME is not convenient when dealing with contaminated waters, waters with high salinity or with a high content of particulate matter. Nevertheless, and as it has been mentioned, it results quite convenient for drinking waters, being able to follow the European Union water directive 2006/0129 EC [182].

EPA method 8272 describes the determination of hydrocarbons in sediment pore water using SPME followed by GC-MS.

Table 19. Reported applications of SPME in the extraction step of PAHs from water samples

Sample	PAHs	SPME fiber	Technique	Extraction conditions	DLs	Ref.
Drinking waters, surface waters. 18 mL	16 EPA + B(e)Py + B(b)Fl + 6 alkyl-PAHs	65 μm PDMS[a]-DVB[b]	GC-MS	60ºC, 60 min, stirring	0.07–0.76 ng·L^{-1}	[182]
Sediment porewaters. 30 mL	16 EPA	100 μm PDMS[a]	GC-MS	Room T, 45 min, stirring-sonication	1–29 ng·L^{-1}	[209]
Estuary waters (filtered). 30 mL	16 EPA (+ 6 PCBs + phthalate esters)	30 μm PDMS[a]	GC-MS	50ºC, 50 min, stirring optimization by experimental design	30–120 ng·L^{-1}	[210]
Lake waters (filtered). Sampling at 1.5 depth 10 mL	Py, B(k)Ft, B(a)Py, B(ghi)Per, diB(a,e)Py	30 μm PDMS[a]	HPLC-FD	80ºC, 35 min, stirring (900 rpm) pH 6	2–6 ng·L^{-1}	[192]

[a]PDMS: polydimethylsyloxane
[b]DVB: divinylbenzene

3.2.5. Stir-Bar Sorptive Extraction (SBSE)

The stir bar sorptive extraction (SBSE) is a relatively new preconcentration and solventless extraction method that was first described in 1999 by Baltussen et al. [211], and further developed. SBSE is based on the same principles than the solid-phase microextraction (SPME) and relies in the partitioning of the analytes between a sample and a polymeric sorbent, followed by desorption of the analytes retained in the sorbent into an appropriate analytical system.

In SBSE, an iron stir bar (or a magnetic rod encapsulated in a glass jacket) is coated by a polydimethylsyloxane (PDMS) sorbent. The main difference between SBSE and SPME is the amount of sorbent used in both applications. In SBSE, the amount of PDMS coating oscillates between 55 and 300 μL, depending on the size of the stir bar, whereas in SPME the PDMS fiber coating is 0.5 μL for a 100 μm fiber thickness. The stir bars are commercially available under the name Twister™ with two different sizes: 10 mm x 3.2. e.d. (for sample volumes between 1-50 mL), and 40 mm x 3.2 mm e.d. (for samples volumes between 100-250 mL). Due to the larger amount of PDMS relative to the SPME fiber, this technique increases the recovery of analytes and therefore enhances sensitivity. SBSE can further extend detection limits to part-per-quadrillion due to the high capacity of the high film thickness of PDMS. Typically, the life-time of a single stir bar is 20 to more than 50 extractions, depending on the matrix.

Nowadays, it is only possible to acquire commercially stir bars with PDMS coatings, whereas in SPME there is a huge range of polymeric sorbents available. Hence, the application extent of this technique is limited to nonpolar compounds or polar compounds if they are conveniently derivatized. Nevertheless, it is expected that the number of polymeric sorbents in the stir bar is increased in the near future, and thus the number of applications. Other disadvantages of SBSE are high bleed, injection artifacts and carry over [212,213].

The extraction performance in SBSE takes place by introducing the stir bar in the sample solution or in its headspace, and allowing the equilibration analyte-coating stir bar during a certain period of time. Afterwards, the stir bar is removed, gently washed with water, dried with a paper cloth, and introduced in a desorption unit connected on line with a gas chromatograph, or desorbed in an organic solvent followed by injection of the organic extract in HPLC or GC. The thermal desorption is the majority option when employing SBSE. It has been commercially developed a thermal desorption unit to accomplish such desorption step coupled on-line to a gas chromatograph. The overall device is commercialized by Gerstel. The system is basically formed by a combination of two vaporization injectors with programmable temperatures (PTV) connected in series. The first PTV posses a glass tube to place the stir bar, and the thermal desorption is accomplished in it. The desorbed analytes are collected in the second PTV because this second injector is maintained at very low temperatures (around -150°C) to trap the compounds. After completion of desorption, the temperature of the second PTV is increased to introduce the analytes in the column by the carrier gas flow. The main advantage of this device is the possibility of coupling on-line desorption step and GC analysis. Off-line procedures imply an organic solvent extraction (100-1000 μL of organic solvent) of the analytes retained in the stir bar, followed by direct injection of the organic extracts in a HPLC or in a GC that uses large-volumes injection. The use of HPLC increases the applicability of this technique to high-molecular or thermal-labile compounds.

It has been demonstrated [214] that loaded stir bars can be stored at 4°C for 1 week without loss of solutes. This opens interesting prospects for on-site sampling and extraction, because loaded stir bars can be sent to the laboratories for analysis, not the samples.

The variables that affect SBSE extraction performance are similar to the ones affecting SPME, namely, extraction time, extraction temperature (mainly for headspace mode, it should be noted however that the sorption process on the PDMS coating is not favored at high temperatures), selection of modifiers to increase the extraction efficiency (such as small amounts of organic solvents or high contents of salts), pH, stirring, and sample volume.

Several applications of the SBSE extraction technique to the determination of PAHs in water samples are reported in Table 20, together with several analytical features.

Table 20. SBSE applications of PAHs in water monitoring.

Sample	PAHs	Extraction conditions	DLs (ng·L^{-1})	Ref.
Seawaters Puddle waters (10 mL)	11 OH-PAHs	24 μL PDMS. SBSE-GC-MS. *wash*: MeOH:CH_2CL_2 1:1 v/v 5 min US, 100% acetonitrile *derivat*: 100 mg $NaHCO_3$ + 20 μL acetic anhydride *extr*: 1000 rpm, 360 min	0.27–25	[218]
Water (10 mL)	16 EPA, 1-MeN, 2 Me-N	24 μL PDMS. SBSE-GC-MS. *extr*: 500 rpm, room T, hyamine 10 μg·L^{-1}, 210 min	0.1–2	[219]
Seawaters Estuary waters (filtered) (20 mL)	16 EPA and other analytes	24 μL PDMS. SBSE-GC-MS. *extr*: 20% MeOH v/v, 20% NaCl, 12 h	0.1–14.7	[183]

3.2.6. Single-Drop Microextraction (SDME)

In 1979, Murray developed a type of liquid microextraction technique for gas chromatography, in which 200 μL of organic solvent was used to extract analytes from moderate amounts of aqueous matrices [215]. In 1996, two independent works of Jeannot and Cantwell [216] and Liu and Dasgupta [217] introduced a novel and miniaturized technique termed solvent microextraction (SME) or liquid-phase microextraction (LPME). Such microextraction procedure was accomplished between a small amount of a water-immiscible solvent and an aqueous phase containing the analytes of interest. The small volume of extractant phase (in the micro-liters or submicro-liters range) could be directly immersed in or suspended above the aqueous sample (for headspace extraction). Thus, high enrichment factors were obtained due to the high sample volume-to-acceptor phase volume ratio.

Since the extraction medium is in the form of a single drop, this type of LPME has also been termed as single-drop microextraction (SDME). A microsyringe is commonly used in SDME and has two functions: as a microseparatory funnel in the extraction step, and as device to further inject the analytes into GC or HPLC. The stability of the small drop is the key to achieve successful extraction recoveries: the higher the size of the drop during long extraction times, the lower the obtained detection limits.

The variables that affect SDME extraction performance are quite similar to the ones affecting SPME and SBSE, namely, extraction time, extraction temperature (mainly for headspace mode), selection of modifiers to increase the extraction efficiency (such as small amounts of organic solvents or high contents of salts), pH, stirring, nature of the extractant agent used in the drop, and sample volume.

Single-drop microextraction (SDME) has become a very popular liquid-phase microextraction technique because it is inexpensive, easy to operate and nearly solvent-free [220]. SDME overcomes many of the problems associated to the solid-phase microextraction: the limited lifetime of the SPME fiber coatings, the fiber coatings are expensive, they tend to degrade with the use (with subsequent loose of efficiency), and the carry-over (especially in HPLC applications). The main disadvantage of the SDME technique is related with the difficulty in maintaining the stability of the drop, especially with high stirring rates of the aqueous sample solution, and during long periods of time. This is inconvenient because the velocity of the stirring process is extremely important to ensure optimum extraction efficiency. Furthermore, SDME is not yet suitable as a routinely applicable online preconcentration procedure.

Few applications of SDME have been reported for the monitoring of PAHs in water samples. Two of them have been carried out in headspace mode (HSDME) [184,221], thus being limited to volatile PAHs. The direct immersion mode SDME for the determination of PAHs in water has been reported using several ILs as the extractant phases [222]. Table 21 shows these reported applications.

Table 21. Determination of PAHs in water samples using SDME.

Sample	PAHs	Technique	Extraction conditions	DLs	Ref.
Lake waters (filtered) 10 mL	N, A, Phe, Ft, Py	HSDME-HPLC-FD	10 μL drop β-cyclodextrin 0.2 g·mL^{-1} NaCl content 40°C, 1000 rpm, 10 min	1.5–28 μg·L^{-1}	[184]
Tap waters, wastewater 6 mL	N, Ace, A, Phe, Ft, Py	HSDME-GC-FID	3 μL drop 1-butanol needle -6°C, 2.3 *M* NaCl 40°C, 400 rpm, 12 min	4–41 μg·L^{-1}	[221]
Water 10–15 mL	16 EPA	SDME-HPLC-FD	3 μL drop ILs: BMIm-PF$_6$[a], HMIm-PF$_6$[b], OMIm-PF$_6$[c] 15% 2-propanol v/v 20°C, 30 min	—	[222]

[a]BMIm-PF$_6$: 1-buthyl-3-methylimidazolium hexafluorophosphate
[b]HMIm-PF$_6$: 1-hexyl-3-methylimidazolium hexafluorophosphate
[c]OMIm-PF$_6$: 1-octyl-3-methylimidazolium hexafluorophosphate

3.2.7. Cloud-Point Extraction (CPE)

The cloud-point extraction (CPE) was first reported by Watanabe [223]. CPE is based on the temperature-induced phase separation of aqueous micellar solutions of polyoxyethylene-type nonionic surfactants [224,225]. Micellar solutions refer to any surfactant solution containing aggregates, thus with a surfactant concentration above the critical micelle concentration (CMC). Upon heating, aqueous micellar solutions of many nonionic surfactants become turbid at a temperature known as the cloud point, above which there is a separation of the solution into two phases. One phase is water-rich and the other one is surfactant-rich. The surfactant-rich phase volume (V_s) is quite smaller than the aqueous phase volume. Any analyte originally present in the aqueous solution that bind to the micellar aggregate can thus be extracted and concentrated in the small volume of the surfactant-rich phase; with elimination of the hydrophilic matrix components to the bulk aqueous phase once CPE is conducted. V_s can be further separated by centrifugation or gravity settling, and removed with a syringe from the remaining aqueous solution. This allows a great enhancement in the sensitivity of further chromatographic analysis. Depending upon the density of the surfactant-rich phase in relation to that one on the aqueous phase, the surfactant-rich phase can be either the bottom or the top layer. Figure 11 shows a scheme with the common steps involved in the CPE procedure.

The main parameters to be optimized in order to get successful CPE extraction-preconcentration are: surfactant nature, surfactant concentration, extraction time, extraction temperature, modifiers in solution (salts, organic solvents) and pH. The preconcentration factors (C_F) depend on the volume of the surfactant-rich phase, which at the same time varies with the surfactant concentration in solution. A compromise must be struck between the surfactant concentration required to achieve a maximum C_F and at the same time provide an adequate volume V_s for subsequent volume manipulation and extraction efficiency [226].

Important advantages of CPE are safety and cost benefits, only very small amounts of relatively nonflammable and nonvolatile surfactants are required, and its compatibility with the micellar or hydroorganic mobile phases commonly used in high-performance liquid chromatography (HPLC). In fact, the surfactant-rich phase containing the analytes of interest can be many times directly injected into a HPLC without the need for subsequent solvent exchange [227]. Furthermore, the obtained V_s containing the extracted analytes can be kept for more than a week without losing efficiency [228,229]. CPE represents a good alternative for extracting and preconcentrating hydrophobic analytes from liquid samples in the same extraction step.

One disadvantage of using some nonionic surfactants in the preconcentration step of PAHs is their tendency to undergo clouding and liquid–liquid separation at elevated temperatures [226]. Sometimes, the high temperatures used in the extraction provoked losses of PAHs with low molecular weight [227]. It has been observed that the observed cloud-point temperature of a mixture of two nonionic surfactants is intermediate between the individual cloud-point temperatures of the nonionic surfactants involved [230]. Therefore, it is often very convenient to use mixtures of surfactants in order to modify and control the cloud-point temperature for a given use [185].

The coacervation extraction has been employed for the phase separation of ionic surfactants, and it is induced by other conditions such as the presence of HCl (pH induced coacervation) for anionic surfactants [231] or the presence of saturated sodium chloride and 1-octanol for cationic surfactants [232]. The separation of anionic surfactants into two isotropic phases in an acid medium takes place at room temperature. The main advantages of this approach are the absence of UV chromophore groups in alkylsulfate or alkylsulfonate molecules of ionic surfactants, the lack of time and temperature dependence in the extraction-preconcentration step, the speed of extraction, and the ability to extract thermally labile and polar compounds. An important disadvantage is the utilization of quite acidic conditions, which must be controlled before injection in HPLC. More recently, the coacervation extraction has also been conducted with reverse micelles [228], and its operating procedure is quite simple: addition of the extractant colloids followed by centrifugation.

Main applications of CPE for the analysis of PAHs in water samples are included in Table 22, together with the operational conditions and the method performance (recoveries and detection limits).

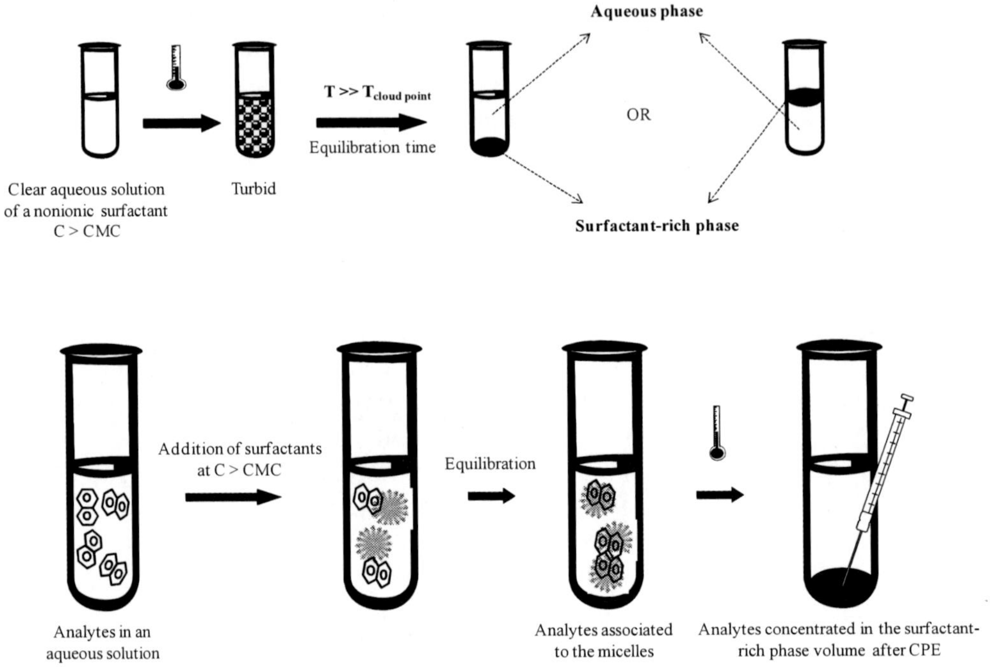

Figure 11. Steps involved in the CPE phenomena.

3.2.8. Dialysis

The utilization of SPMDs (filled with triolein) in water monitoring is commonly followed by dialysis to carry out desorption of retained PAHs [107,191,197,198]. This process is in general accomplished using hexane for 48 h, as described in air monitoring.

3.2.9. Capillary Microextraction (CME) and Solid-Phase Nanoextraction (SPNE)

Two novel extraction approaches have been proposed for the monitoring of PAHs in water. Capillary microextraction (CME) uses a chromatographic capillary column as both extractor and preconcentrator to enrich analytes from aqueous samples. The sorption of the analytes on the thin stationary phase PDMS film of a gas chromatographic capillary column helps to extract and preconcentrate the analytes. The analytes were then desorbed with either suitable solvent or thermal desorption. The sorption area in the capillary microextractor is much higher than SPME, resulting in high recovery and high sample capacity. Peeraprasompong et al. have reported the application of a CME-GC-MS procedure to determine B(a)Py, B(b)Ft and diB(a,h)A from tap waters and raw waters, with average recoveries of 90.5% and detection limits in the $ng \cdot L^{-1}$ region using 7 mL of sample [233].

Table 22. Application of CPE to the extraction of PAHs from water samples.

Sample	PAHs	Surfactant	Extraction	Clean-up	Technique	Performance	Ref.
Seawaters (filtered) 3 mL	14 EPA	polyoxyethylene-10-lauryl-ether ($C_{16}E_{10}$) 1% w/w	90 min, 95°C V_s = 170 µL	No 25 µL directly	HPLC-FD	1–150 ng·L^{-1} N: 38% rest: 62-107%	[227]
Seawaters (filtered) 3 mL	15 EPA	Mixture $C_{16}E_{10}$ + Brij 30, total surf. 1% (w/w), weight fraction Brij 30 = 0.74	60 min, 78°C V_s = 270 µL	No 20 µL directly	HPLC-FD	23–200 ng·L^{-1} 72–99%	[185]
Seawaters (filtered) 3 mL	B(a)Py, B(k)Ft	Mixture $C_{16}E_{10}$ + Brij 30, total surf. 1% (w/w), weight fraction Brij 30 = 0.74	60 min, 78°C V_s = 270 µL	No V_s in the well	Fluorescence (screening)	Cut-off values: µg·L^{-1} region	[229]
Wastewater (filtered) 36 mL	B(a)A, B(a)Py, Chy, diB(a,h)A, I(1,2,3-cd)Py	70 mg decanoic acid in 4 mL THF, stirring 5 min, 700 rpm pH = 2	Centrif. 10 min V_s = 115 µL	No V_s in the well	HPLC-FD	0.1–1 ng·L^{-1} 88–95%	[228]
Tap, river, groundwater (filtered) 35 mL	15 EPA	0.05 g sodium dodecane sulfonic acid + 15 mL HCl conc. 5 min stirring	15 mL centrifuge tubes 10 min V_s = 300 µL	50 µL acetonitrile to 100 µL of V_s: 20 µL directly	HPLC-FD	0.1–8 ng·L^{-1} 63–106%	[235]
Tap water 25 mL	15 EPA	1% Triton X-114 (w/w)	40 min, 40°C	V_s + cyclohexane, silica gel column	HPLC-FD (20 µL)	0.3–12 ng·L^{-1} N: 0% Ace: 30%, rest: 61–109%	[236]
River, tap (filtered) 15 mL	16 EPA minus N, Aci, Chy, Ace, Fl and Phe	0.5% Triton X-114 (w/w)	5 min, 40°C Centrif. 5 min	V_s + 100 µL acetonitrile: 10 µL directly	HPLC-FD	2–120 ng·L^{-1} 62–110%	[237]
Tap water 3 mL	Py, Phe, N, Chy, B(a)A, B(a)Py	0.26 g SDS + 3 mL HCl 12 *M*	Centrif. 5 min	No	Fluorescence	67-95%	[231]

Another novel alternative is an approach named solid-phase nanoextraction (SPNE), which takes advantage of the strong affinity that exists between PAH and gold nanoparticles [234]. The novel extraction method has been applied to the determination of PAHs in waters. Among other features, it must be highlighted the small volume of water sample needed (500 μL). Furthermore, the entire extraction procedure consumes less than 100 μL of organic solvents per sample, which makes it environmentally friendly. The limits of detection ranged from 0.9 (anthracene) to 58 $ng \cdot L^{-1}$ (fluorene), and the analytical recoveries from tap water samples of the six regulated PAH varied from 83.3 to 95.7% [234].

3.3. Clean-up Procedures

The lower complexity of the water samples compared to that of samples coming from air monitoring, especially when dealing with drinking waters, justify the lower number of clean-up procedures and the high number of applications that can be carried out on-line.

Nevertheless, for contaminated samples it is still necessary to carefully carry out a number of clean up procedures. Among them it should be cited classical liquid adsorption chromatography in columns, solid-phase extraction, gel-permeation chromatography, reduction of solvent volume and solvent exchange, which imply an important timing of the process. Their characteristics and operational procedure have been described in air monitoring, and will not be further detailed here.

The utilization of internal standards (IS), recovery standards (RS), and performance reference compounds (PRCs) for SPMDs, described in air monitoring, is also applicable to water samples.

3.4. Chromatographic Procedures

The selection of GC and HPLC to carry out the final determination of PAHs has been described in air monitoring. Along this section dedicated to water monitoring, the utilization of GC or HPLC in each extraction technique has been commented (Tables 18 to 22) and will not be further repeated here.

4. Soils and Sediments Monitoring

Soils are important long-term repositories for PAHs and so they are used as an indicator of the environmental pollution. Soils are continuously subjected to pollution because of their open-system nature. Together with their capability to accumulate various pollutants, soil serves as a medium to integrate historical and recent pollution status indirectly. Accumulation of PAHs in soils may lead to further potential contamination of vegetables and food chains [238], and then cause direct or indirect exposure to humans. Moreover, leaching, evaporation and migration are possible PAHs sources of atmospheric or groundwater contamination. Therefore, knowledge of soil contamination with PAHs is needed to avoid any food production risk and to reduce the effect of these contaminants. The fate and transport of PAHs in the environment are controlled by sorption and degradation processes [239]. Sorption also plays a key role in controlling transport of PAHs during leaching, as well as in transformation and bio-accumulation processes [240].

Yunker et al. have reported that most PAHs found in soils are derived from oil seeps, bitumen, coal, plant debris, and forest fires [241]. Lichtfouse et al. have reported that soil PAHs are mainly of pyrolytic origin, with a small contribution of uncombusted fossil fuels [242]. With regard to PAHs contents, Wilcke and Amelung reported that the total of 20 PAHs in the soils from North America Great Plain ranges from 60 to 180 $\mu g \cdot kg^{-1}$ [243]. PAHs in vegetable soils in China are in higher contents [244], especially when irrigated with wastewaters [245]. Agricultural soils close to highways also present higher PAHs levels [246]. Higher concentrations have also been found in urban soils, with amounts ranging from 400 to 600 $\mu g \cdot kg^{-1}$ [247], or up to 3700-3900 $\mu g \cdot kg^{-1}$ [248,249]. Average contents for the 16 EPA PAHs of 140 $\mu g \cdot kg^{-1}$ have been reported in 138 soils in Hong Kong, with hotspots contents of 19,500 $\mu g \cdot kg^{-1}$ in one urban park [250].

In the same fashion than soils, sediments always belong to any control program to locate contamination sources due to their preferential sequester of contaminants and accumulation [251]. Higher weight PAHs deposited in sediments are quite resistant to bacterial degradation and so the determination of sediment concentrations yields valuable information about the PAH load and possible adverse impacts on the marine environment. Sediment retention capacity may be related to physicochemical properties such as grain size and organic matter. PAHs in sediments are mainly associated with organic carbon, and therefore the PAH concentrations in sediments also tend to be related to the particle size distribution. Sediments with a smaller particle size accumulate PAHs to a greater degree than do coarse, sandy sediments [252]. The presence of PAHs in sedimentary systems can have a large (negative) impact on sedimentary microbial communities and dynamics [253]. PAHs have been monitored in the marine environment since the 1960s and this continues today in local, regional, and national-scale programs [254,255]. Major long-term monitoring programs, for example the Arctic Monitoring and Assessment Program (AMAP), the Baltic Marine Environment Protection Commission (Helsinki Commission), and the National Status and Trends Program (US), focus on PAHs in marine biota and sediment as part of their strategies to assess marine environmental quality [255]. Sediment contamination evidently affects the surrounding media [256]. Heavy pollutions of halophytic plants, especially *Salicornia* species, have been reported further to the significant impacts of sediments by fuel oil [257].

With respect to PAHs content is sediments, Sprovieri et al. have reported PAHs contents in highly contaminated surface sediments in Naples harbor varying from 9 to 318,000 $\mu g \cdot kg^{-1}$ [258], whereas total content values up to 3766 $\mu g \cdot kg^{-1}$ have been found in the Mersey Estuary (UK) [259]. Slater et al. have reported significant temporal variability in PAH deposition to surface sediments both proximate and distal to the industrialized shoreline [253].

The monitoring of PAHs in sewage sludge results also necessary. Sewage sludge samples have been used as fertilizers for agricultural lands over a long time [260]. This is part of a sustainable practice utilizing and recycling the macronutrients back to land but, on the other hand, the negative effect due to the possible presence of contaminants in sludge has to be controlled [261,262]. PAHs concentrations about 10 times higher in urban than in rural sludge samples have also been reported [261]. A draft directive of the European Union limits to 6000 μg/kg the maximum allowable concentration of 10 PAHs in sewage sludge used in agriculture [263].

PAH distributions and concentration ratios can also be used to distinguish PAHs of petrogenic and pyrolytic origin in contaminated soils and sediments, as in air monitoring [264,265].

The monitoring of PAHs in soil and sediment samples utilizes, in general, tedious analytical schemes, due to the complexity of the matrix. Several official methods include procedures to determine PAHs in soil/sediment samples. ISO standard 13877:1998 describes a method for the determination of PAHs in soil by HPLC, whereas the recently published standard ISO 18287:2006 specifies a method for the determination of PAHs in soil by GC-MS.

Several EPA methods report the utilization of different extraction procedures for solid samples. As examples, the EPA method 8275 A describes the determination of semivolatile organic compounds (PAHs and PCBs) in soils/sludges and solid waste using thermal extraction (TE) followed by GC-MS. EPA method 3540 describes the utilization of Soxhlet extraction for solid samples. EPA method 3545 is related to the utilization of PLE in the determination of semivolatile organic compounds from solid samples [139]. EPA method 3561 adopts the utilization of SFE for the extraction of PAHs from solid environmental matrices, and EPA method 3550 C recommends the utilization of sonication for solid samples.

Most common extraction methods for PAHs with soils and sediments are Soxhlet extraction, ultrasound-assisted extraction, microwave-assisted extraction, pressurized liquid extraction and supercritical fluid extraction. Figure 12 shows a scheme with the usual steps involved in the monitoring of PAHs in soil and sediment samples.

On-line combination of extraction system suitable for sample pre-treatment of solid samples is much more complex than for water samples. The main reason for this is that more exhaustive methods are required for efficient extraction of target analytes from a solid matrix, and this in practice means that the extracts typically contain high amount of co-extracted matrix components, that should not be transferred to a chromatographic system. Moreover, relatively high volumes of organic solvents are required for efficient extraction, making the on-line combination more challenging [156].

4.1. Sampling Strategies

Sampling of solids is more difficult than liquids or gases and larger sampling errors may easily arise due to the lack of uniformity of the material. Indeed, literature indicates that up to half of the variability between similar soils may occur within a distance of one meter. The sampling of solid samples is quite complicated, and usually requires an important amount of time to design an adequate sampling plan [264]. Sample spacing, sample collection depth, and exposure period, among others, are key factors for both passive and active approaches. The investigator must decide the number of samples to be taken, the field sample size, and the cost of sampling and analysis. Furthermore, the investigator will have to determine if gravel, cobbles, and other non-soil fractions will have to be sampled and how these are to be handled if they are sampled.

The minimum amount of sample that should be taken related to total size of the bulk material is around 2.5 kg per 0.5 tons of soil if the bulk material is not excessively heterogeneous. Very heterogeneous materials may need to be sampled in great bulk, amounting to 1 percent or more of the total. The resulting sample then needs to be reduced in size by some means that preserves its representative character. "Coning and quartering" is one approach. The original sample is formed into a cone-shaped pile and then flattened into a disk. The disk is divided into four quadrants. Two opposite quadrants are shoveled into a second pile, mixed together, and then coned and quartered again. This sequence continues until the selected material has been reduced to a size small enough for a useful laboratory sample.

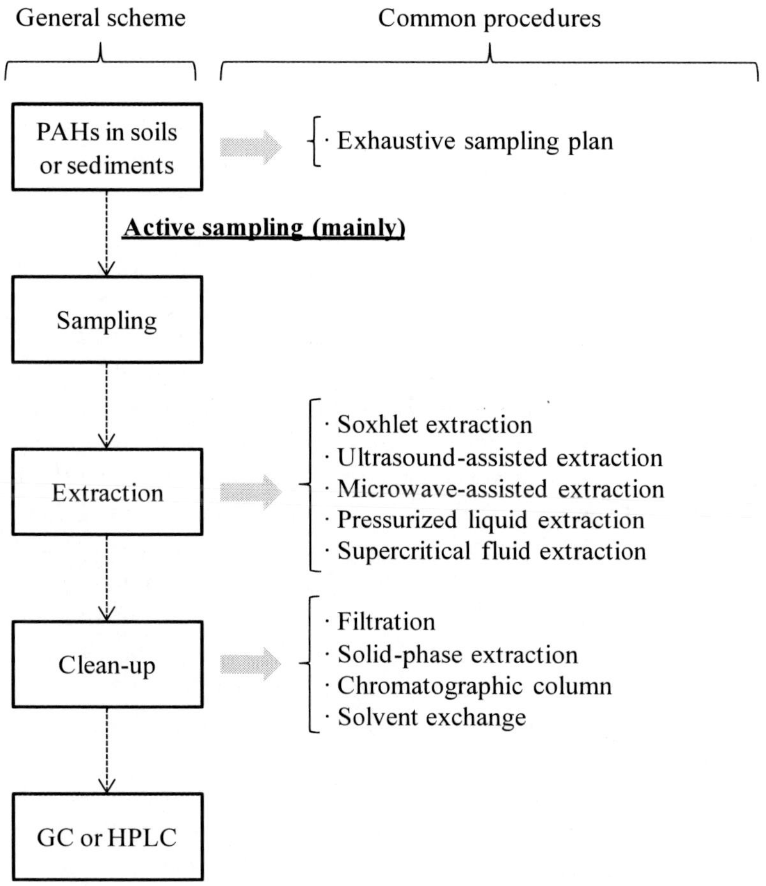

Figure 12. Common steps followed in the analytical procedure for the monitoring of PAHs in soil, sediment or sludge samples.

According to the EPA, there are essentially five major types of sampling situations that the environmental investigator is likely to encounter. These are:

- Large area studies, where the pollution is in the surface or shallow layers, e.g., in support of an ambient monitoring effort.
- Large area studies, where pollution has moved down into the soil profile, e.g., assessing the impact from a major industrial complex.
- Localized area studies, where pollution is in the surface layers, e.g., sampling around a recent hazardous chemical spill or capacitor rupture.
- Localized plume studies, where the major source of contamination is below the surface at some depth, e.g., sampling near a leaking hazardous waste disposal site.
- And special research studies conducted for litigation, scientific understanding, method development, or source identification.

There are two portions of the soil that are important to the environmental investigator. The surface layer (0–15 cm) reflects the deposition of airborne pollutants, especially recently deposited pollutants and also pollutants that do not move downward because of attachment to soil particles. On the other hand, pollutants which have been deposited by liquid spills, by long-term deposition of water soluble materials, or by burial may be found at considerable depth. The methods of sampling each of these are slightly different, but all make use of basic techniques [266].

The most common sampling approaches are: simple random sampling, stratified sampling and composite sampling. Simple random sampling is carried out locating the sampling points referenced to coordinate axis fixed in the sampling area, and selecting the sampling coordinates points with random numbers. Divisions of 1/100 are considered sufficient. Classically, the sampling is achieved through random sampling schemes. However, such simple random sampling schemes can be very demanding of laboratory effort to analyze the large numbers of samples that are collected, and can therefore be costly to carry out [264].

One of the tools of sampling that can be used to reduce the variability of the sample is stratified sampling. Strata are identified as regions of the site that are expected to be uniform in character. The variance within the strata should be smaller than the variance between strata. Sampling points within the strata are selected systematically or by some random process. In the soil environment, strata are often associated with soil types or as areas of known pollution versus areas where pollutants are not expected to be present.

Composite sampling is standard practice in the analysis of soils, and is mainly used to estimate mean concentrations for an area at a reduced cost. Composite sampling involves the physical combination of homogenized samples to form a set of new samples (i.e., composite samples). The chemical analyses are then performed on aliquots of the composite samples. Because the compositing physically averages the individual samples, averaging the analytical results of the smaller number of composite samples can theoretically produce an estimated mean that is more precise than, or at least as precise as, one based on many more individual sample results [264]. The main anticipated benefits of composite sampling are that it may achieve better precision of an estimated mean at less cost, and data analysis is usually straightforward. The limitation of composite sampling is that it yields a reduced amount of information on variability; for example, composite sampling loses information on individual samples and on spatial (or temporal) patterns.

After the convenient sampling plan, active sampling strategies are mainly conducted [267].

4.1.1. Active Sampling

Active sampling relies on taking the soil (or sediment, or sludge) sample, to further carry out the analytical extraction of PAHs in the laboratory. Samples can either be collected with some form of core sampling or auger device, or they may be collected by use of excavations or trenches. In the latter case, the samples are cut from the soil mass with spades or short punches. Drags are usually employed for the sampling of the surface layers, and are composed of two spades (or shovels) with opposite movements. Tubular probes are often used when intending a study of temporal evolution. A tubular probe is formed by a stainless steel cylinder containing a plastic tube, with a valve which closes the whole system when the desired depth is reached. Glass tubes are also used for surface layer sampling. It is obvious that care must be taken to ensure that no fragments of the container contaminate the sample [267].

Active sampling is the main sampling approach when intending to monitor PAHs in soil, sediment or sludge samples.

4.1.2. Passive Sampling

Passive sampling relies on trapping the PAHs directly from the soil sample, without the necessity of sampling the soil itself. Passive sampling from solid matrices is in essence a soil gas or soil vapor sampling methodology, and so it is mainly limited to volatile hydrocarbons. In addition to this, the concentration of the volatile hydrocarbons in the vapor phase is dependent on many variables, such as soil particle size and mineralogy, organic and moisture contents, temperature and overall heterogeneity of particle size and composition, leading to difficulties in converting the amount of analyte collected by the sampler to its concentration in the sampled medium. As a result, most passive solid sampling applications are restricted to qualitative or screening purposes only [268]. Typically, soil gas sampling involves collection of vapor occupying the pore space of soils. The vapor is composed of air and water vapor, naturally occurring organic compounds, anthropogenic organic compounds, organic compounds migrating from deep fossil fuel reserves, and other constituents capable of partitioning from liquid and solid materials into the vapor phase under ambient conditions [107].

Solid-phase microextraction (SPME) devices can be employed with solid samples as passive samplers, but this approach must be conducted in headspace mode and so it is limited to the more volatile PAHs [269]. Furthermore, it appears that SPME is a better approach to extract primarily PAHs present in sediment porewater rather than in the colloidal phase [270]. EPA method 8272 describes the determination of parent and alkyl polycyclic aromatics in sediment pore water by SPME-GC-MS. In a similar way, SPMDs have been mainly used for the determination of PAHs contained in sediment porewater [271].

4.2. Extraction Procedures

Chemical preservation of solids is generally not recommended and therefore cooling to 4°C is usually the best approach to preserve solid samples. Wide mouth glass containers with Teflon lined caps are mainly utilized during transport and storage of the samples. If composite sampling techniques or multiple grabs are employed, equal portions of solid sample from each location are deposited into a stainless steel, plastic, or other appropriate composition (e.g., Teflon) containers [266]. Once in the laboratory, water content must be removed from solids samples, mainly by lyophilization (also called freeze-drying) [46,259,261,264,265]. It is a relatively complex and expensive form of drying, and it usually takes between 24 to 48 h the complete dryness of a sediment sample. Important advantages of the technique are the avoiding of changes in the structural characteristics of soils and sediment samples during the drying process and the avoiding of losses of the most volatile PAHs. Other procedures for water removing are heating [253] and air drying [250,272-274].

PAHs contained in soils and sediments are mainly extracted using Soxhlet extraction [46,250,259,272,275], ultrasound-assisted extraction [46,264,265,270], microwave-assisted extraction [253,276,277], pressurized liquid extraction [258,272,275,278-280] and supercritical fluid extraction [280,281]. These techniques have been widely described in air and water monitoring. Focused-microwave assisted Soxhlet (FMAS), which up to date have not had any applications in air monitoring, is however widely employed with soils and sediments for PAHs monitoring [282-284]. FMAS have demonstrated successful quantitative results, being characteristic for the necessity of short extraction times if compared to Soxhlet. PAHs contained in sludge are mainly extracted by Soxhlet extraction [285], but more recently some alternative techniques have been employed, including ultrasonic-assisted extraction [286], microwave-assisted extraction [287] and pressurized liquid extraction [288]. Table 23 includes main extraction techniques used in the monitoring of PAHs in soil, sediment and sludge samples in recent years, altogether with some other analytical features.

An interesting environmental improvement to traditional extraction methods for solid samples, mainly with microwave-assisted extraction and ultrasound-assisted extraction, has been conducted in recent years. The environmental modification implies the avoiding of organic solvents in the extraction step, and the successful substitution of such organic solvents by aqueous micellar media [289-291], as can be observed in Figure 13. The aqueous micellar extract containing PAHs can be further injected in a HPLC after filtration or centrifugation. The micellar microwave-assisted extraction and the micellar ultrasound-assisted extraction procedures have been validated using certified reference material [292]. IL-micelles have been also employed in extraction processes of PAHs bounded to sediments [293]. An important disadvantage is that surfactants usually bind to the C18 stationary phases commonly used in HPLC, and thus their co-elution with analytes can be problematic if both, analyte and surfactant, have similar absorbance (or fluorescence) spectra. Another disadvantage of micellar assisted extraction with microwaves or ultrasounds is the fact that applications are limited to the further utilization of HPLC. The combination of the micellar extracts with GC is not possible, unless complicated clean-up steps are carried out to remove the surfactants, and so organic solvents are then again involved in the procedure. Nevertheless, it has recently been proposed the utilization of solid-phase microextraction to link micellar extracts with GC. The technique has been named as micellar solid-phase microextraction (MSPME) [294]. MSPME combines the high extraction efficiency and analyte preconcentration advantages of micellar assisted extraction with the speed and high separation efficiency of GC. MSPME has been applied to the determination of PAHs in sediment samples using traditional surfactant micelles [294] and IL-micelles [295], and the procedures have been validated using certified materials. Several reports have also described the utilization of cloud-point extraction as a preconcentration technique to be conducted after a micellar microwave- or ultrasound-assisted extraction of PAHs from soil/sediment samples [237,296], therefore improving the whole detection limits of the technique. Such preconcentration procedures can be carried out on-line [297]. Several applications which utilize micellar media in microwave-assisted extraction, ultrasound-assisted extraction, MSPME, and micellar assisted extraction followed by CPE, for PAHs contained in solid samples, are included in Table 23, together with other conventional extraction techniques.

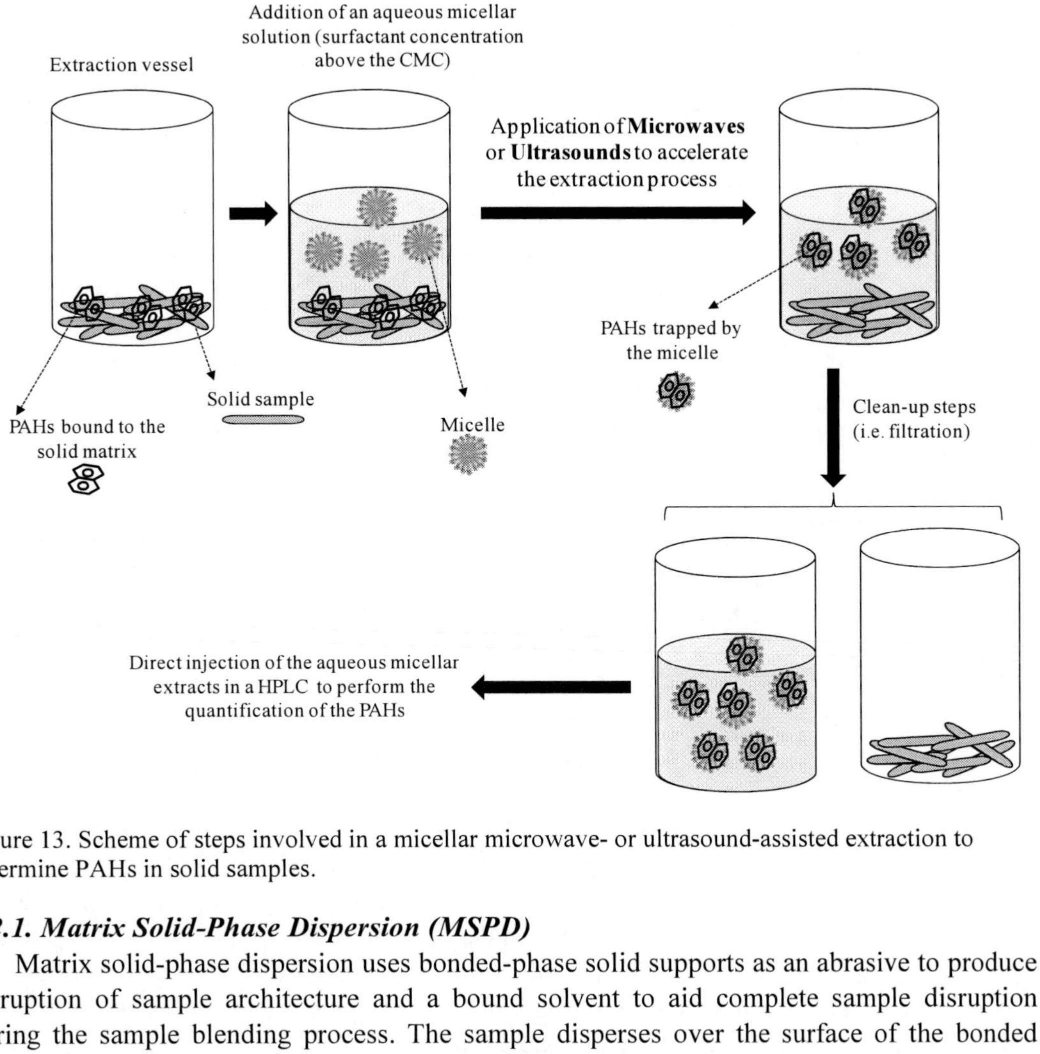

Figure 13. Scheme of steps involved in a micellar microwave- or ultrasound-assisted extraction to determine PAHs in solid samples.

4.2.1. Matrix Solid-Phase Dispersion (MSPD)

Matrix solid-phase dispersion uses bonded-phase solid supports as an abrasive to produce disruption of sample architecture and a bound solvent to aid complete sample disruption during the sample blending process. The sample disperses over the surface of the bonded phase-support material to provide a new mixed phase for isolating analytes from various sample matrices [298]. The solid support is commonly a silica-based material such as derivatized silica, silica gel, sand and florisil, and sometimes graphitic fibers and alumina [52]. The blended mixture should then be packed into a column and a sequential elution is conducted with solvents to collect the analytes by fractionation. The eluate may be directly used for further instrumental analysis, but additional SPE is suggested to remove the co-eluted interferences or to clean-up the analytes by further fractionation. MSPD is characteristic for a significant reduction in solvent consumption compared to traditional extraction techniques, without requiring special expensive instrumentation [261]. The method is however fairly labor intensive. Applications of MSPD for the determination of PAHs in soil, sediment or sludge samples are included in Table 23.

4.3. Clean-up Procedures

In general, clean-up procedures are necessary for solid samples, due to the common high level of interfering materials (i.e., aliphatic hydrocarbons) present in the extracts. Filtration [253,276,289,290,297], SPE [46,152,261], and classical liquid adsorption chromatography in columns [250,253,265,270,276,278] are the most common clean-up procedures conducted with soil, sediment or sludge samples. Solvent exchange and reduction of solvent volume before injection in the chromatographic system are often conducted [250,253,264,270,274,275]. These procedures have been described in the section of air monitoring in this chapter. The utilization of internal standards (IS) and recovery standards (RS) is a common quality control practice in soil/sediment monitoring, as in air or water monitoring.

4.4. Chromatographic Procedures

GC and HPLC (widely described in previous sections of this book chapter) are the common chromatographic techniques to carry out the final determination of PAHs in soil monitoring. Its utilization in combination with each specific extraction procedure is included in Table 23.

5. Food Monitoring

Food appears to be the main source of PAH intake for humans not occupationally exposed to PAH [48]. Vyskocil et al. have pointed out that food consumption represents the main source of PAH exposure for the children even in big cities that are heavily polluted by PAH from industrial sources [303]. PAH contamination of food arises from two sources, environment and food-processing technique. The PAHs present in water and rivers are available to fishes and other biota via different uptake routes, such as diet and/or through direct transport across external membranes [304], and so to human beings by direct consumption. Consumers can also be exposed to PAHs by eating contaminated milk, cereals, flour, bread and vegetables [305,306]. PAHs can enter plant tissues by partitioning from contaminated soil to the roots and translocation into the shoot. However, the extent of PAH uptake by roots is still debated in the literature [305,307,308]. It seems that the main pathway of plant contamination is atmospheric exposure. Heat processing of meat and dairy products, such as charcoal grilling [309], roasting, and smoking [310-312], constitute important sources for contamination with PAHs. Several cooking practices, common round the globe, such as

Table 23. Analytical techniques most commonly used in the monitoring of PAHs in soil, sediment, and sludge samples.

Sample	PAHs	Analytical procedure	Analytical features	Ref.
sandy soil, industrial soil CRM-524	Py, B(a)A, B(a)Py, B(k)Ft, B(ghi)Per	PLE (hot water + SDS)-on-line-FD	Quantitative extraction, short extraction time	[278,279]
19 sludge samples and one CRM	27 PAHs	MSPD-US-SPE-GC-MS	DLs: 0.03-0.45 $\mu g \cdot kg^{-1}$	[261]
clay soil	Py, B(a)A, B(a)Py, B(e)Py, B(k)Ft	FMAS (CH_2Cl_2)-GC-ECD	Recoveries > 94%, overall time: 1 h	[282]
clay soil	Py, B(a)A, B(a)Py, B(e)Py, B(k)Ft, B(ghi)Per	FMAS (acetonitrile)-on line-FD	Recoveries > 98%	[283]
clay soil	nitro-PAHs	FMAS (CH_2Cl_2)-GC-ECD	Recoveries > 89.7%, DLs: 0.5-1.5 $\mu g \cdot kg^{-1}$	[284]
organic and sandy soils	16 EPA PAHs	miniaturized PLE-LVI[a]-GC-MS	Recoveries > 83% except N (59%), DLs: 0.8-30 $\mu g \cdot kg^{-1}$, comparison with Soxhlet	[272]
sea sediment	16 EPA PAHs minus Ace, Aci and diB(a,h)A	PLE (hot water)-LC-GC	QLs[b]: 0.001-0.01 $\mu g \cdot g^{-1}$, comparison with SFE-GC-MS	[280]
sediments, soils and one CRM	16 EPA PAHs	PLE (hot water)-DTS[c]-on line-GC-MS	Successful recoveries, extraction times: 35–60 min, comparison with Soxhlet	[275]
sediments	16 EPA PAHs	UAE (CH_2Cl_2)-GC-MS	DLs: 1-4 $\mu g \cdot kg^{-1}$	[270]
sediments	16 EPA PAHs	UAE (CH_2Cl_2)-GC-MS	Recoveries > 80%	[264]
2 CRM sediments	N, Ace, Aci, Fl, A, Ft, Py	HSPME-GC-FID	DLs: 0.3-3 $pg \cdot g^{-1}$	[269]
sediments and CRM	16 EPA PAHs	MAE (hexane:acetone 1:1)-GC-MS	Recoveries > 80%	[253]
sediments and CRM	16 EPA PAHs	MAE (CH_2Cl_2)-GC-MS	—	[276]
sediments	16 EPA	UAE (CH_2Cl_2:methanol 1:1)-GC-MS	QUASIMEME[d]	[265]
soils	Ace, Fl, Phe, A, Ft, Py	HS-LPME (hollow fiber)-GC-MS	drop 3 μL octanol, DLs: 5.9-76 $ng \cdot g^{-1}$	[273]
soils	16 EPA	UAE or PLE (hexane)-GC-MS	DLs: 0.02-0.75 $\mu g \cdot kg^{-1}$, recoveries: 59-110% except N (in PLE)	[274]
138 soils and CRM	16 EPA	Soxhlet (acetone:CH_2Cl_2 1:1)-GC-MS	DLs: 5-9 $\mu g \cdot kg^{-1}$, recoveries > 78%	[250]
clay soil	16 EPA	Soxhlet or UAE (acetone:CH_2Cl_2 1:1)-GC-FID	DLs: 0.5-1 $mg \cdot kg^{-1}$	[299]
sediments	N, Ace, Aci, Fl, Phe, A, Ft, Py	SPME (100 μm PDMS)-GC-MS	Comparison with LLE (CH_2Cl_2)	[270]
soils	Phe, Py, Chy, B(k)Ft, B(a)Py	off-line micellar MAE or UAE (1% tergitol)-on-line CPE-HPLC-FD	Overall recoveries > 88.7%	[297]

[a]LVI : large volume injection [b]QLs: limits of quantification
[c]DTS: different trapping systems: solid-phase extraction, micro-porous-membrane liquid extraction (hollow fiber or flat sheet). LLE is off-line.
[d]QUASIMEME: quality assurance of information for marine environmental monitoring in Europe

Table 23 (Continued).

Sample	PAHs	Analytical procedure	Analytical features	Ref.
soils, sediments, sludges (CRMs)	Py, B(a)A, B(b)Ft, B(k)Ft, B(a)Py, B(e)Py, I(1,2,3-cd)Py	micellar extraction-CPE (2% SDocS[e])-off-line HPLC-FD	Recoveries: 71-98%, comparison with Soxhlet	[300]
soils, sediment (CRM)	16 EPA minus Aci, Ace, Fl, diB(a,h)A and B(ghi)Per	SFE-GC-MS	Recoveries > 88%, comparison with Soxhlet	[281]
sediments and CRM	16 EPA PAHs + B(e)Py	PLE/ASE (hexane:acetone 80:20 v/v)-GC-MS	Recoveries: 94-107%, DLs ~ 1 $ng{\cdot}g^{-1}$	[258]
sediments and CRM	16 EPA PAHs minus Aci	Soxhlet (CH_2Cl_2)-HPLC-FD	—	[259]
sediments	16 EPA	Soxhlet or UAE (CH_2Cl_2:hexane 1:1)-GC-MS	Recoveries: 47–122% except N	[46]
soils and CRMs	16 EPA PAHs minus Aci	Soxhlet (THF) or UAE (THF or acetonitrile)-HPLC-FD	Comparison with ELISA. Recoveries: 60–102%	[301]
CRM sediment	16 EPA	PLE/ASE (CH_2Cl_2)-GC-MS	Quantitative recoveries	[302]
marine sediments	16 EPA minus A, Ace, Aci	off-line micellar MAE (1% POLE[f])-HPLC-UV	Recoveries > 87%	[289]
CRM sediment	Py, B(a)A, B(b)Ft, B(k)Ft, B(a)Py, B(e)Py, I(1,2,3-cd)Py	off-line MAE (45m*M* IL HDBIm-Br[g])-HPLC-FD	Average recoveries 91.1%, DLs: 0.8–53 $ng{\cdot}g^{-1}$	[293]
CRM sediment	Phe, Fl, Py, B(a)A, Chy, B(b)Ft, B(a)Py, B(e)Py, Per, diB(a,h)A, B(ghi)Per, I(1,2,3-cd)Py	focused MW (CH_2Cl_2)-GC-MS	Quantitative recoveries	[277]
CRM sediment	Fl, Phe, A, Ft, Py, B(a)A, Chy, B(b)Ft, B(k)Ft, B(a)Py	off-line MAE (POLE)-SPME-GC-MS	R: 58.6-111.5%, DLs: 20-547 $ng{\cdot}g^{-1}$	[294]
sludge (30)	Ace, Fl, Phe, Ft, Py, B(b)Ft, B(j)Ft, B(k)Ft, B(a)Py, B(ghi)Per, I(1,2,3-cd)Py	UAE (CH_2Cl_2)-GC-MS	—	[286]
marine sediments	16 EPA minus N, Aci, A and B(k)Ft	micellar UAE or micellar MAE (with POLE)-CPE-HPLC-UV	R: 58-118%, DLs: 50–1300 $ng{\cdot}g^{-1}$	[296]
sludge	16 EPA	PLE/ASE (CH_2Cl_2:hexane 1:1)-GC-MS	R: 60.6-93.4%, comparison with Soxhlet	[288]
soils	16 EPA	ELS[h] (CH_2Cl_2:acetone)-HPLC-FD	—	[245]
soils and one CRM	16 EPA	UAE (acetone:petroleum ether 1:1)-GC-MS	DLs: 2-42 $ng{\cdot}g^{-1}$, R: 75–120%	[249]
soils	16 EPA	Soxhlet (hexane)-GC-MS	R: 32-96%, QLs ~ 0.1–0.2 $ng{\cdot}g^{-1}$	[247]

[e]SDocS: sodium dodecane sulphonate
[f]POLE: polyoxyethylene-10-lauryl ether
[g]HDBIm-Br: 1-hexadecyl-3-methylimidazolium bromide
[h]ELS: liquid-solid extraction with mechanical shaking

repeated frying of vegetarian and non-vegetarian foods in edible oil, are quite risky. Pandey et al. have studied the cytotoxicity of polycyclic aromatic hydrocarbon residues arising through repeated fish fried oils in humans [313]. In addition, PAHs could arrive in food products through contaminated packaging material [314]. *In vivo* studies have suggested a transfer in human intestinal epithelium by diffusion, which appears extensively governed by the physicochemical properties of PAHs, particularly lipophilicity [21,315,316].

The PAHs content have been assessed in several kinds of foods. The purpose of this section is not to deeply summarize all the works conducted to estimate PAHs levels in foods rather but to give an overview of the common levels found in different kind of foods. These levels will give an idea about the sensitivity required for the analytical methods to be used in food monitoring. Kishikawa et al. have reported that the average concentrations of total PAHs in commercial milk, infant formula and human milk were 0.99, 2.01 and 0.75 $\mu g{\cdot}kg^{-1}$, respectively [317]. The total PAHs content have also been examined in green tea leaves, with average values ranging from 497 to 517 $\mu g{\cdot}kg^{-1}$ whereas brick tea leaves presented total contents 1,048 to 1,162 $\mu g{\cdot}kg^{-1}$ [318]. The PAHs content in fats and oils has been reviewed by Moret et al. [319]. The determination of PAHs in smoked meat products and smoke flavouring food additives has recently been reviewed by Simko [310], whereas the determination of PAHs is smoked fish has recently been reviewed by Stolyhwo et al. [320]. The highest contents of B(a)Py have been found in smoked fish [310] if compared to other smoked foods. With respect to marine foods, the bioaccumulation of PAHs by various marine organisms has been extensively reviewed by Meador et al. [321]. Boehm et al. have reported total PAHs contents in mussels ranging from 3 to 1200 $ng{\cdot}g^{-1}$ (dry weight) [322]. Average contents for the 16 EPA PAHs around 5400 $ng{\cdot}g^{-1}$ (wet weight) have been reported in mussels of the Red Sea coast by Nemr et al. [323]. Kong et al. have examined PAHs content in freshwater fishes in Hong Kong [324]. The contents found in fishes purchased from the markets supplied from the mainland showed higher values, around 76 $ng{\cdot}g^{-1}$ (wet weight), than those collected from fish ponds, with values around 60.1 $ng{\cdot}g^{-1}$. Total PAHs contents ranging from 4 to 69 $ng{\cdot}g^{-1}$ (wet weight) in *Lipophrys pholis* fishes have been reported by Lima et al. [325]. Perugini et al. have reported total PAHs content of 44.1-63.3 $ng{\cdot}g^{-1}$ in fishes taken at pools in the Adriatic Sea [326]. Cheung et al. have reported differences in PAHs content depending on the fish studied (river or marine fishes), but the differences were not that accused among marine fishes [327]. Nevertheless, with respect to the uptake of PAHs by fishes, a major fraction of the PAHs absorbed is rapidly metabolized to hydroxylated derivatives (phase 1 metabolism), then conjugated to polar groups (phase 2 metabolism), and stored in the gall bladder to be excreted. The analysis of PAH metabolite levels in fish bile can be used to assess the actual PAH uptake, rather than the analysis of the non-hydroxylated PAHs content [328]. This aspect will be further discussed in the section of biomarkers within this book chapter.

There have been several regulations with respect to PAHs levels in food samples, mainly focused on B(a)Py contents, with maximum contents around 1 $\mu g{\cdot}kg^{-1}$. In 2002, the Scientific Committee on Food of the European Union assessed 33 PAHs and identified 15 PAHs that possess both genotoxic and carcinogenic properties, and concluded that B(a)Py could be used as a marker [329]. The Committee also recommended monitoring B(a)A, B(b)Ft, B(j)Ft, B(k)Ft, B(ghi)Per, Chy, cyPent(c,d)Py, diB(a,h)A, diB(a,e)Py, diB(a,h)Py, diB(a,i)Py, diB(a,l)Py, I(1,2,3-cd)Py, and 5-MeChy in food to enable long-term exposure assessments. In 2005, the European Commission (EC) asked the EU Member States for further investigations on the 15 EU priority PAHs (Table 3) together with B(c)Fl [330], highlighted by the Joint FAO/WHO Expert Committee on Food Additives (JECFA) in 2005 [331]. The European union has set maximum levels for PAHs, specifically for B(a)Py, in certain food products. The maximum allowed concentration for B(a)Py in various food products is in the range 1-10 $\mu g \cdot kg^{-1}$ [332], and for B(a)Py and B(a)A in liquid smoke flavoring primary products is 10 $\mu g \cdot kg^{-1}$ and 20 $\mu g \cdot kg^{-1}$, respectively [333]. The PAH on-line analytical database is available on the EFSA web site [334]. Wenzl et al. have recently reviewed the current European Union (EU) legislation on PAHs for official control of food samples [335].

The foods samples most commonly analyzed for the determination of PAHs are marine foods (due to the risk of PAHs content by water contamination), smoked foods (due to the risk of PAHs formed during the smoking process), oils, and in general foods that could be potentially exposed to PAHs formation during their preparation and manufacturing.

Foods are quite complex matrixes for any kind of monitoring. Due to the multiplicity of food matrices, it is not possible to use one sample preparation technique, so many methods have been proposed for meeting all the requirements. There are few national or international standard methods related to B(a)Py in food samples, but also little information about collaborative trials has been published so far [335,336]. Furthermore, the majority of the reports of analytical methods to monitor hydrocarbons in foods have been focused on the determination of B(a)Py exclusively or on the 16 EPA PAHs. Several works are starting to show up which includes the determination of the 15 PAHs recommended by the EU in food samples [336-338].

Fidalgo-Used et al. [137] have recently reviewed the main extraction and clean-up procedures, published in the last 5 years, applied to the analytical determination of both conventional and emerging persistent organic pollutants (POPs), not exclusively PAHs, in environmental biota (vegetal and animal) samples. Buldini et al. [339] have also reviewed recent sample preparation methods for food analysis.

For the analysis of foods, the choice of the sampling strategy depends on the nature of the food sample and on the purpose of the analytical measurement. In general, foods are acquired in supermarkets, or in production lines when performing quality control steps. In any case, main cautions are taken with the recipients used to transport the samples to the laboratory to avoid contamination. The utilization of clean scissors, knifes or any other cutting device when dealing with complex samples such as mussels or fishes is also mandatory. In many occasions, freeze-drying of the samples is also advisable to ensure a better conservation.

An important analytical step which is widely conducted with food samples, especially with greasy foods, and has not been described yet in this book chapter, is the saponification [276,309,310,317,323,326,336,340-342]. Saponification is carried out in the analytical laboratory as a clean-up step to break the fatty acid esters (usually triglycerides) of vegetable oils and animal fats, and so to digest the sample releasing the PAHs. The process is conducted by reflux using a strong alkali, normally sodium or potassium hydroxide, in an aqueous alcoholic solution (methanol or ethanol). Traditional saponification procedures require 40 min up to several hours. After saponification, liquid-liquid extraction is usually conducted [276,309,317,326,336,340-342]. The overall procedure results labor intensive and time-consuming.

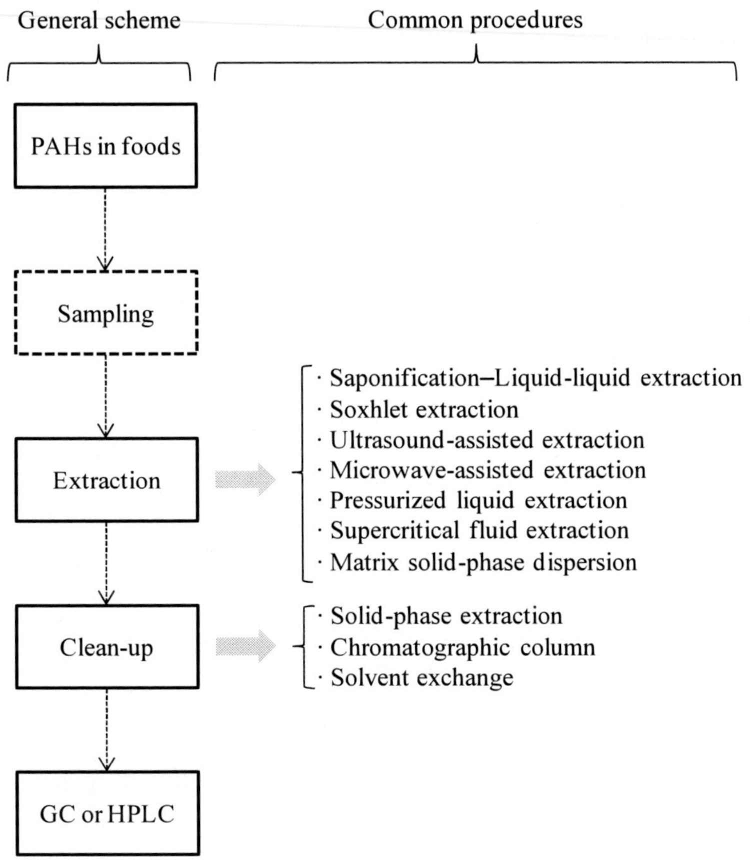

Figure 14. Common analytical steps carried out in the monitoring of PAHs in food samples.

Other common extraction techniques for the monitoring of polycyclic aromatic hydrocarbons in food samples are Soxhlet extraction [305,318,323,324,327,343], ultrasound-assisted extraction [312,338,344], supercritical fluid extraction (SFE) [345-347], microwave-assisted extraction (MAE) [340,348], pressurized liquid extraction (PLE) [46,302,337,349,350], and matrix solid-phase dispersion (MSPD) [351,352]. In some cases, the alkali saponification step is also conducted as a clean-up step right after these extraction techniques (or simultaneously) to ensure an adequate recovery of PAHs from food samples with high fat content [46,312,323,348]. In some cases, the digestion of complex samples such as fish bile requires an enzymatic hydrolysis, in general utilizing β-glucuronidase/sulfatase [328]. Other extraction techniques, such as headspace extraction techniques, have also proven to be successful when dealing with olive oil samples using a triple-quadrupole MS detector (QqQ) [353]. All these extraction techniques have been described in other sections of this book chapter and will not be further discussed here.

Table 24. List of analytical methods recently used in the extraction of PAHs from a number of food samples.

Food	PAHs	Analytical method	Analytical performance	Ref.
SRM 2974[a]	16 EPA, 2-MeN, 1-MeN, Biphenyl, 1-MePhe, B(b)Ft, Triphenylene, Per, diB(a,c)A, diB(a,j)A, B(b)Chy	PLE (CH_2Cl_2)-GC-MS	Quantitative extractions, comparison with Soxhlet	[302]
trout and sardines	Py, B(a)A, B(k)Ft, B(a)Py, B(ghi)Per	PLE (hot water)-HPLC-FD	R > 97.6%	[278]
mussels tissues, SRM 2977[a]	16 EPA	Soxhlet (acetone:hexane 1:1)-GC-MS	DLs: 0.2–1 $ng\cdot g^{-1}$	[325]
oils, SRM 2974[b]	15 EU foods + Phe, A, Ft, Py	PLE (acetone:hexane 1:1)-GC-MS-MS	DLs: 0.008–0.15 $\mu g\ kg^{-1}$	[337]
smoked and fresh fishes, rape-oils	B(a)A, B(a)Py, B(b)Ft, B(k)Ft, B(ghi)Per, I(1,2,3-cd)Py	hydrolysis (digestion)-LLE (CH_2Cl_2:hexane 1:3)-GC-MS	DLs ~ 0.3 $\mu g\cdot L^{-1}$ (oil), ~ 0.25 $\mu g\cdot g^{-1}$ (fish), R ~ 75%	[354]
mussels, SRM 2974[b]	16 EPA PAHs	Soxhlet (methanol)-saponification-LLE (hexane)-GC-MS	R: 86–108%	[323]
olive oils	Ft, Py, B(a)A, Chy, B(b)Ft, B(k)Ft, B(a)Py, I(1,2,3-cd)Py, diB(a,h)A, B(ghi)Per	HS-GC-MS-MS (QqQ)	R: 96–99% DLs: 0.02–0.06 $\mu g\cdot kg^{-1}$	[353]
pumpkin seed oils, CRM 458	16 EPA	MAE-saponification-LLE (hexane)-GC-MS	DLs: 0.01–0.12 $ng\cdot g^{-1}$ R > 83.6%	[340]
smoked meat (beef)	Phe, A, Ft, Py, B(a), Chy, B(b)Ft, B(e)Py, B(a)Py, Per	SFE-simultaneously SPE (C18)-GC-MS	DLs: 2.5–4.1 $ng\cdot g^{-1}$	[346]
mussels, tellins, SRM 1974b[c]	A, Ft, B(a)A, B(b)Ft, B(k)Ft, B(a)Py, diB(a,h)A, B(ghi)Per	miniaturized MSPD—in-tube in-valve SPME-HPLC-FD	DLs ~ 0.6 $ng\cdot g^{-1}$	[352]
mussels, SRM 1947a[d]	triphenylene, B(j)Ft,B(a)Ft, Per, B(e)Py, 16 EPA minus N, Aci, Ace, Fl	saponification-LLE (pentane)-GC-MS	R > 60% DLs ~ 0.05 $ng\cdot g^{-1}$	[276]
mussels	16 EPA	PLE (hexane:acetone 1:1 / hexane:CH_2Cl_2 1:1)-saponification-GC-MS	DLs ~ 0.5-8 $\mu g\cdot kg^{-1}$ R: 64-121%	[46]
bilvalves, fish, cephalopods, crustaceans	16 EPA minus N, Aci, Py	saponification-LLE (cyclohexane)-HPLC-FD	R: 63-100% DLs: 0.05-0.25 $\mu g\cdot kg^{-1}$	[326]
fish	16 EPA	Soxhlet (acetone:CH_2Cl_2 1:1)-GC-MS	DLs ~ 0.05 $ng\cdot g^{-1}$	[327]
tea	16 EPA	Soxhlet (acetone:CH_2Cl_2 1:1)-GC-MS	-	[318]
milk (commercial, infant, human)	16 EPA minus N, Aci, Ace, Fl	saponification-LLE (hexane)-HPLC-FD	R: 90-105% DL: 1.3-76 $ng\cdot kg^{-1}$	[317]
cabbage, carrot, lettuce, leek, endive	B(e)Py, 16 EPA minus Aci	Soxhlet (CH_2Cl_2)-HPLC-FD	R: 71-92% except N	[305]
fish	16 EPA	Soxhlet (acetone:CH_2Cl_2 1:1)- GC-MS	DLs ~ 20 $ng\cdot g^{-1}$	[324]

[a]SRM 2977: organics in freeze-dried mussels tissue
[b]SRM 2974: mussel tissue
[c]SRM 1974b: mussel tissue
[d]SRM 1947a: mussel tissue

Table 24 (Continued)

Food	PAHs	Analytical method	Analytical performance	Ref.
smoked cheese	B(c)Phe, B(j)Ft, B(e)Py, Per, 16 EPA, diB(a,l)Py, diB(a,i)Py, diB(a,h)Py, 23 alkyl-PAHs	UAE (cyclohexane)-saponification-LLE (cyclohexane)-GC-MS	use of IS	[312]
fish	16 EPA	Soxhlet (CH_2Cl_2)-GC-MS	R: 45-84% DLs: 1–87 $pg \cdot g^{-1}$	[343]
mussels	total PAHs content	UAE (1st CH_2Cl_2, 2nd hexane)-spectrofluorimetry	R: 98-10% DLs (total PAHs): 21 $ng \cdot g^{-1}$	[344]
fish, SRM 2977[a]	B(a)A, B(b)Ft, B(k)Ft, B(a)Py, diB(a,h)A, I(1,2,3-cd)Py	MAE (hexane)-saponification-HPLC-FD	R ~ 90% QLs: 0.07–0.53 $ng \cdot g^{-1}$	[348]
fish	B(e)Py, 16 EPA, 6 alkyl-PAHs	PLE (hexane:acetone 1:1)-HPLC-FD.	Comparison with Soxhlet and UAE	[349]
mussel, salmon, fish oil, fish feed, SRM 2977[a]	24 PAHs	PLE/ASE (CH_2Cl_2)-GC-MS	DLs for B(a)Py: 1.5–5.8 $pg \cdot g^{-1}$	[350]
fish, SRM 2977[a]	B(a)A, B(b)Ft, B(k)Ft, B(a)Py, diB(a,h)A, I(1,2,3-cd)Py	MSPD-HPLC-FD	R ~ 80% DLs: 0.04–0.32 $ng \cdot g^{-1}$	[351]
trout, sardine	Py, B(a)A, B(a)Py, B(k)Ft, B(ghi)Per	PLE (hot water + SDS)-on-line-FD	Quantitative extraction, short extraction time	[279]
smoked fish, bovine milk	Phe, A, Ft, Py, B(a), Chy, B(b)Ft, B(e)Py, B(a)Py, Per	SFE-simultaneously SPE (C18)-GC-MS	R > 97% DLs: 10-14 $ng \cdot g^{-1}$ (fish)	[345]
barbecued meat sausages, CRMs 458[e] – 459[e]	16 EPA	saponification-LLE (cyclohexane)-GC-MS or HPLC-UV/FD	DLs ~ 0.06 $\mu g \cdot kg^{-1}$	[309]
toasted bread	15 EU foods minus cyPent(c,d)Py, diB(a,e)Py, diB(a,h)Py, diB(a,i)Py	UAE (hexane)-HPLC-FD	R > 82% DLs: 0.01-0.70 $\mu g \cdot kg^{-1}$	[338]
strawberries, apples, tomatoes, lettuce, parsley, potatoes, kale	16 EPA	PLE/ASE (hexane:acetone 1:1)-HPLC-FD	—	[355]
sunflower oils in containers	N, 2-MeN, 1-MeN, DiMeNs, 9-H-Fl	HS-SPME-GC-MS	use of IS	[356]
commercial liquid smoke flavorings	34 PAHs	saponification-LLE (cyclohexane)-GC-MS	R > 84%	[341]
smoke flavorings	15 EU foods	saponification-LLE (cyclohexane)-GC-MS	R > 69% DLs: 0.1–1.3 $\mu g \cdot kg^{-1}$	[336]
marine plant biota	16 EPA	saponification-LLE (hexane)-GC-MS	R: 88-116% DLs: 3.5–18 $\mu g \cdot kg^{-1}$	[342]
vegetable oils	B(a)A, B(e)Py, B(a)Py, B(b)Ft, B(k)Ft, diB(a,h)A, B(ghi)Per	SFE-HPLC-FD	DLs ~ 1.55 $\mu g \cdot kg^{-1}$	[347]

CRMs 458 and 459: coconut oil reference materials containing six PAHs

The complexity of the food samples limits the utilization of on-line procedures in combination with chromatographic techniques. This problem is normally related to any complex solid matrix.

Clean-up steps such as SPE, liquid adsorption chromatography in columns, size exclusion chromatography, SPME, solvent exchange, reduction of solvent volume… are absolutely necessary when dealing with food samples. Given the complexity of the food samples, as well as the necessity of a number of clean-up steps to purify the extract, the utilization of IS and/or RS, as well as adequate certified materials, becomes an inherent necessity. These clean-up steps have been described in other sections of this book chapter.

Figure 14 shows a scheme of the typical steps used in the monitoring of PAHs in food samples. Table 24 includes a list of several analytical procedures utilized in recent years for the determination of polycyclic aromatic hydrocarbons in several food samples as a summary of the state of the art.

6. Biomarkers

Several definitions have been given for the term biomarker, which is generally used in a broad sense to include almost any measurement reflecting an interaction between a biological system and a potential hazard, which may be chemical, physical or biological [357]. Van Gestel et al. considered biomarker as any biological response to an environmental chemical at the subindividual level, measured inside an organism or in its products (urine, faeces, hair, feathers, etc.), indicating a deviation from the normal status that cannot be detected in the intact organism [358,359]. A bioindicator is defined as an organism giving information on the environmental conditions of its habitat by its presence or absence or by its behavior. Biomonitoring provides an assessment of the integrated uptake through all exposure routes. This is important since only the bioavailable chemical is likely to be assimilated-concentrated in tissues, body fluids, and excreta.

Among several classes of biomarkers, the biomarkers of exposure cover the detection and measurement of an exogenous substance or its metabolite that is measured in a compartment within an organism. Therefore, biomarkers of exposure will be the focus of this section, and more specifically, bioaccumulation markers [360]. The bioaccumulation monitoring can be defined as an exposure assessment by measuring contaminant levels in biota.

This section will specifically focus on the chemical methods used in the monitoring of PAHs and their metabolites in biota, tissues, body fluids and excreta, as a measure of exposure contamination.

Fish bioaccumulation and biomarkers in environmental risk assessment have been reviewed by Oost et al. [360]. Fish bioaccumulation markers may be applied in order to elucidate the aquatic behavior of environmental contaminants and to assess exposure of aquatic organisms. The feasibility of PAH tissue concentrations in marine species as a monitoring parameter for PAH exposure depends on their uptake, biotransformation and excretion rates. Since it remains hard to accurately predict bioaccumulation in marine species, even with highly sophisticated models, analyses of tissue levels are required. The main problem is that PAHs do not tend to accumulate in fish tissues in quantities that reflect the exposure. The analysis of PAH metabolite levels in fish bile can be used to assess the actual

PAH uptake, rather than the analysis of the non-hydroxylated PAHs content [328,361]. A number of sentinel fish species have been proposed to asses pollution by PAHs [325,326], as well as several mussels [322,323,326,352]. Several studies have also correlated the high levels of 1-OHPy and B(a)Py metabolites found in the bile of cat-shark with contamination sources such as boat traffic and combustion-based industries present in the sampling area [362].

Various plants such as moss, bryophytes, lichen, leaves, and tree bark have also been employed as bioindicators to monitor airborne PAHs pollution levels, assuming that their spatial pollution distribution in the plants and the atmosphere will be closely related [117,122,123,363,364]. This aspect has already been revised in this chapter within the section passive air samplers.

Several studies achieved in environmental conditions have shown that PAHs can be excreted in the milk of ruminants when feeded with PAHs [306]. The authors concluded there was an excretion of PAHs into milk under native forms when rumiants were fed with low PAHs content. However, a slight increase in concentration could induce the metabolism, which should lead to an increase in the excretion of metabolites into the milk. In spite of the absence of a significant transfer of parent PAHs to milk, the appearance of metabolites in milk raises questions of their impact on human health.

Urinary 1-hydroxypyrene (1-OHPy) is commonly tested as biomarker for the assessment of low level PAHs exposure of general population. The determination of 1-OHPy in urine as a biomarker was first suggested by Jongeneelen [365]. Several reports have suggested that levels of 1-OHPy reflect the levels of environmental PAH exposure from different sources such as air, food and the environment of the home life [303,366,367] or in certain working activities [368]. Controlled studies using human volunteers have shown a strong association between urinary excretion of 1-OHPy and consumption of PAH contaminated diet [21,303]. Other studies have reported that the urinary excretion of 1-OHPy was not correlated with ingested dose of pyrene under normal feeding conditions [369], and so it was not considered a reliable bioindicator of ingested pyrene. It is also well-known that cigarette smoke contains a variety of potentially toxic components including PAHs. Py from smoking is presumably the main source of 1-OHPy detected in the cigarette smokers' urine, particularly when compared with non-cigarette smokers [370,371]. However, the background urinary 1-OHPy is linked to a number of sources such as diets, ambient air, and second hand smoke. It should be also considered that the concentration of urinary 1-OHPy indicates only recent exposure because the halflife for its urinary excretion ranges from 6 to 35 h. Although 1-OHPy has been utilized as a reasonable surrogate biomarker for exposure to a wide variety of PAHs in general, determination of the urinary profile of PAH metabolites would provide more information on estimation of human exposure, internal doses, and individual variability in metabolism of a wide range of PAHs [372]. Hecht et al. have proposed *r*-1,*t*-2,3,*c*-4-tetrahydroxy-1,2,3,4-tetrahydrophenanthrene (*trans*, *anti*-PheT) as a better biomarker for assessing PAHs metabolic activations [373].

The analytical methods commonly used to biomonitor PAHs or their metabolites in marine species or plants have been described in other sections of this chapter, in food monitoring and in air monitoring, respectively, and will not be repeated here.

The analysis of PAHs metabolite in urine samples is quite complex and so intricate analytical protocols must be used for analyses. The analytical methods used are a challenge to biomonitoring studies since most PAH compounds and their metabolites are predominantly excreted in the faeces rather than urine, and so extremely low concentrations are expected in the latter. Jongeneelen et al. proposed an analytical method in which the urine sample is hydrolyzed by β-glucuronidase, purified by solid-phase extraction (SPE) and analyzed by high-performance liquid chromatography (HPLC) with fluorescence detection [374]. It is therefore quite common to carry out an enzymatic hydrolysis of the urine samples followed by HPLC-FD using off-line sample cleanup by SPE [368,369,372,373,375]. However, because of strong matrix interferences, the isolation of more polar metabolites (e.g., hydroxyphenanthrenes) with conventional SPE techniques is difficult and is subjected to error. In this sense, numerous variants have been proposed to improve the sensitivity, specificity, and speed of the method. Schedl et al. have proposed the utilization of immunosorbent of anti-pyrene antibodies which were encapsulated in a sol-gel glass (SGG) matrix in the extraction of urine samples to monitor hydroxyl PAHs [376]. It has also been proposed the utilization of SPME [377,378] or LLE [371,379] instead of SPE. When determining PAHs or its metabolites in urine, several reports suggest that correction for urinary creatinine can be useful, mainly for substances that are eliminated primarily by glomerular filtration and are not reabsorbed by the tubules; however, such corrections may be inappropriate insituations involving concentration-dependent passive elimination or reabsorption [380]. For other biological samples an alkali digestion is also commonly conducted [381]. Another problem often encountered is the small amount of sample available, which can be the limiting factor of the analysis. Table 25 summarizes the most common analytical methods employed in the determination of PAHs or its metabolites in urine, hair or blood samples.

HPLC-FD [368,369,375,376,381] and also HPLC-MS [372,382] are quite helpful chromatographic techniques for the determination of hydroxyl-PAHs. The GC method for hydroxy-PAHs requires the utilization of a time-consuming derivatization step in order to decrease the polarity of the analytes and so reduce the peak-tailing in the chromatograms [371,373,377,378]. The derivatization reaction is commonly a silylation reaction, where active hydrogens of hydroxyl PAHs are replaced by trimethylsilyl (TMS) groups. The most common derivatizating reagents are N-O-bis(trimethylsilyl)trifluoroacetamide (BSTFA) and N-methyl-N-trimethylsilyltrifluoroacetamide (MSTFA). The derivatization steps, in spite of being quite common in laboratories when dealing with polar analytes in GC, add time to the overall analytical procedure and are labor intensive.With respect to GC methods, the successful utilization of GC×GC approaches have also been reported for the analysis of hydroxyl-PAHs in urine [379].

Table 25. Analytical methods commonly employed in the determination of PAHs or its metabolites in urine, hair or blood samples.

Sample	PAHs	Analytical method and performance	Ref.
human urine	1-OHPy	enzimatic hydrolysis (β-glucuronidase/arylsulphatase, 37°C, 16 h)-SPE-HPLC-FD. Comparison with PAHs air levels.	[368]
human hair	16 EPA minus Aci	NaOH digestion-LLE (hexane)-HPLC-FD. Comparison with PAHs air levels.	[381]
human urine	1-OHPy; 2-,3-,4-,9-OHPhe; 3-OHB(a)Py	enzymatic hydrolysis (β-glucuronidase/sulphatase, 37°C, 16 h)-Inmuno extraction cartridge-HPLC-UV/FD. DLs: 1-16 $ng \cdot L^{-1}$	[376]
seabird blood	15 *Prestige* oil PAHs	MAE (hexane:acetone 1:1)-HPLC-FD. DLs: 0.01-0.05 $ng \cdot g^{-1}$	[383]
human urine	1-OHPy	enzimatic hydrolysis (β-glucuronidase/arylsulphatase, 37°C, overnight)-SPE-HPLC-FD. Comparison with PAHs in foods.	[369]
human urine	1-OHPy	enzimatic hydrolysis (glucuronidase/arylphosphatase, 37°C, 16 h)-SPE-HPLC-MS/MS. R ~ 77%. QL ~ 0.1 $\mu g \cdot L^{-1}$. Accuracy: 85-115%.	[382]
human urine	1-, 2-naphthol; 2-OHFl; 1-OHPy; 3-OHFt; 1-OHB(a)A; 3-, 9-phenanthrol; 2-OHB(c)Phe; 3-, 6-OHChy; 3-, 9-OHB(a)Py	enzimatic hydrolysis (glucuronidase/sulphatase, 37°C, overnight)-SPE-HPLC-MS/MS	[372]
human urine	1-OHPy	enzimatic hydrolysis (β-glucuronidase/sulphatase, 37°C, 1 h)-SPE-HPLC-FD. Comparison with PAHs air levels	[375]
human urine	1-, 2-OHN; 2-, 9-OHFl; 9-OHPhe; 3-OHB(a)A; 1-OHChy; 3-, 7-, 10-OHB(a)Py	enzimatic hydrolysis (β-glucuronidase, 37°C, overnight ~ 18 h)-LLE (pentane)-Derivatization (MSTFA) by microwaves-GC×GC-FID. R ~ 76%. DLs: 0.03-0.18 $\mu g \cdot L^{-1}$	[379]
crab urine	1-OHPy, 9-OHPhe	enzyme-linked immunosorbent assay (ELISA)-spectrofluorimetry	[384]
human urine	*r*-1,*t*-2,3,*c*-4-tetrahydroxy-1,2,3,4-tetraOHPhe; 1-OHPy	enzimatic hydrolysis (β-glucuronidase/arylsulphatase, 37°C, overnight)-SPE-HPLC-FD. HPLC eluates were derivatizated (MSTFA)-GC-MS.	[373]
human urine	unmetabolized N, Aci, Ace, Fl, Phe, A, Ft, Py	HS-SPME-GC-MS. Comparison with PAHs air levels.	[385]
human urine	1-, 2-OHN; 2-, 3-OHFl; 1-, 2-, 3-, 4-, 9-OHPhe; 3-OHFt; 1-OHPy; 1-, 2-, 3-OHB(c)Phe; 1-, 3-B(a)A; 3-, 6-OHChy	enzimatic hydrolysis (β-glucuronidase/arylsulfatase, 37°C, 2 h)-SPME-on-fiber derivatization (MSTFA)-GC-MS. DLs: 0.78-15.8 $ng \cdot L^{-1}$	[377]
human urine	1-OHPy	enzimatic hydrolysis-SPE-HPLC-FD	[303]
human urine	1-, 2-, 4-, 9-OHFl; 2-OH-5,6,7,8-tetrahydroPhe	enzimatic hydrolysis (β-glucuronidase/arylsulphatase, 37°C, overnight)-SPME-on-fiber derivatization (MSTFA or BSTFA)-GC-MS	[378]
human urine	1-OHPy; 1-, 2-OHN; 3-, 9-OHPhe	enzimatic hydrolysis (β-glucuronidase/arylsulphatase, 37°C, 17 h)-LLE (ethyl acetate)-derivatization (BSTFA)-GC-MS. R > 93%. DLs: 0.28-2.64 $\mu g \cdot L^{-1}$	[371]

CONCLUSIONS

This chapter has intended to give an overview of the analytical steps more commonly carried out in laboratories when determining PAHs (or their metabolites) in environmental, food or biological samples in monitoring studies. The challenges in the analytical methods are significant, because the samples to be analyzed are quite different and complex, and the PAHs are present at low levels altogether with many interfering substances. In addition to this, the developed methods are required to be fast, environmental-friendly (solventless if possible), accurate and reproducible. There are no universal approaches to carry out an analytical determination for the monitoring of PAHs, and the utilization of a number of clean-up steps is, in general, unavoidable. In spite of the fact that traditional methods to monitor PAHs successfully reach the detection limits imposed by several regulations, the overall procedures are generally time-consuming and not environment-friendly. Today, sampling of PAHs is the most critical step in overall analysis, and efforts must be made to improve the efficiency of this step in both active and passive sampling approaches.

ACKNOWLEDGMENTS

Dr. Francisco J. Conde is gratefully acknowledged for his helpful comments and discussion.

REFERENCES

[1] Yamanaka, T.; Mizota, C.; Murae, T.; Hashimoto, J. *Geochem. J.* 1999, 33, 355-367.
[2] Wilcke, W. J. Plant. *Nutr. Soil Sci.* 2000, 163, 229-248.
[3] Richter, H.; Howard, J. B. Prog. *Energy Combust. Sci.* 2000, 26, 565-608.
[4] Lee, W. J.; Liow, M. C.; Tsai, P. J.; Hsieh, L. T. *Atmos. Environ.* 2002, 36, 781-190.
[5] Oanh, N. T. K.; Albina, D. O.; Ping, L.; Wang, X. *Biomass Bioenerg.* 2005, 28, 579-590.
[6] Simoneit, B. R. T. *Appl. Geochem.* 2002, 17, 129-162.
[7] Crittenden, B. D.; Long, R. *Combust. Flame* 1973, 20, 359-368.
[8] Bockhorn, H.; Fetting, F.; Wenz, H. W. *Ber. Bunsenges. Phys. Chem.* 1983, 87, 1067-1073.
[9] Dean, A. M. J. *Phys. Chem.* 1990, 94, 1432-1439.
[10] Marinov, N. M.; Ptiz, W. J.; Westbrook, C. K.; Castaldi, M. J.; Senkan, S. M. *Combust. Sci. Technol.* 1996, 116, 211-287.
[11] Stein, S.E. J. *Phys. Chem.* 1986, 82, 566-571.
[12] Sullivan, R. F.; Boduszynski, M. M.; Fetzer, J. C. *Energy Fuels* 1989, 3, 603-612.
[13] Cerniglia, C. E.; Heitkamp, M. A. In Metabolism of PAHs in the Aquatic Environment. Varanasi, U. (Ed.) CRC Press: Boca Raton, 1989
[14] Kennaway, E. L.; Hieger, I. *Br. Med. J.* 1930, 1044-1046.
[15] Kameda, Y.; Shirai, J.; Komai, T.; Nakanishi, J.; Masunaga, S. *Sci. Total Environ.* 2005, 340, 71-80.

[16] Bhargava, A.; Khanna, R. N.; Bhargava, S. K.; Kumar, S. *Atmos. Environ.* 2004, 38, 4761-4767.

[17] Tsai, P. J.; Shieh, H. Y.; Lee, W. J.; Lai, S. O. *Sci. Total Environ.* 2001, 278, 137-150.

[18] Gammon, M. D.; Santella, R. M.; Neugut, A. I.; Eng, S. M.; Teitelbaum, S. L.; Paykin, A.; Levin, B.; Terry, M. B.; Young, T. L.; Wang, L. W.; Wang, Q.; Britton, J. A.; Wolff, M. S.; Stellman, S. D.; Hatch, M.; Kabat, G. C.; Senie, R.; Garbowski, G.; Maffeo, C.; Montalvan, P.; Berkowitz, G.; Kemeny, M.; Citron, M.; Schnabel, F.; Schuss, A.; Hajdu, S.; Vinceguerra, V. *Cancer Epidemiol. Biomarkers Prev.* 2002; 11, 677-685.

[19] EC (European Commission), Opinion of the Scientific Committee on Food on the Risks to Human Health of Polycyclic Aromatic Hydrocarbons in Food, 4 December 2002, Scientific Committee on Food.

[20] Dabestani, R.; Ivanov, I.N. *Photochem. Photobiol.* 1999, 70, 10-34.

[21] Ramesh, A.; Walker, S. A.; Hood, D. B.; Guillén, M. D.; Schneider, K.; Weyand, E. H. *Int. J. Toxicol.* 2004, 23, 301-333.

[22] Farmer, P. B.; Singh, R.; Kaur, B.; Sram, R. J.; Binkova, B.; Kalina, I.; Popov, T. A.; Garte, S.; Taioli, E.; Gabelova, A.; Cebulska-Wasilewska, A. *Mut. Res.* 2003, 544, 397-402.

[23] Jacob, J.; Seidel, A. *J. Chromatogr. B* 2002, 778, 31-48.

[24] Boysen, G.; Hecht, S. S. *Mutat. Res.* 2003, 543, 17-30.

[25] Bocskay, K. A.; Tang, D.; Orjuela, M. A.; Liu, X.; Warburton, D. P.; Perera, F. P. *Cancer Epidemiol. Biomarkers Prev.* 2005, 14, 506-511.

[26] Nisbet, L. C. T.; LaGoy, P. K. *Regul. Toxicol. Pharmacol.* 1992, 16, 290-300.

[27] U.S. EPA, Provisional guidance for quantitative risk assessment of polycyclic aromatiec hydrocarbons, Cincinnati, 1993. EPA/600/R-93/089.

[28] Kalberlah, F.; Frijus-Plessen, N.; Hassauer, M. *Altlasten-Spektrum* 1995, 5, 231-237.

[29] Cecinato, A. *Ann. Chim.* 1997, 87, 483-496.

[30] Larsen, J.; Larsen, P. Chemical Carcinogens, Hester, R. Harrison (Eds), Cambridge, 1998, pp. 33-56.

[31] Vione, D.; Barra, S.; DeGennaro, G.; DeRienzo, M.; Gilardoni, S.; Perrone, M. G.; Pozzoli, L. *Ann. Chim.* 2004, 94, 257-268.

[32] Sitaras, I. E.; Bakeas, E. B.; Siskos, P. A. *Sci. Total Environ.* 2004, 327, 249-264.

[33] Nauomova, Y. Y.; Offenberg, J. H.; Eisenreich, S. J.; Meng, Q.; Polidori, A.; Turpin, B. J.; Weisel, C. P.; Morando, M. T.; Colome, S. D.; Stock, T. H.; Winer, A. M.; Alimokhtari, S.; Kwon, J.; Maberti, S.; Shendell, D.; Jones, J.; Farrar, C. *Atmos. Environ.* 2003, 37, 703-719.

[34] Pankow, J. F.; Bidleman, T. F. *Atmos. Environ.* 1991, 25, 2241-2249.

[35] Law, R. J.; Biscaya, J. L. *Mar. Pollut. Bull.* 1994, 29, 235-241.

[36] Rose, N. L.; Rippey, B. *Environ. Pollut.* 2002, 117, 121–132.

[37] Tolosa, I.; de Mora, S.; Sheikholeslami, M. R.; Villeneuve, J. –P.; Bartocci, J. *Mar. Pollut. Bull.* 2004, 48, 44-60.

[38] Geffard, O.; Geffard, A.; His, E.; Budzinski, H. *Marine Poll. Bull.* 2003, 46, 481-490.

[39] McGarthy, L. H.; Williams, T. G.; Stephens, G. R.; Peddle, J.; Robertson, K.; Gregor, D. *J. Sci. Tot. Environ.* 1997, 197, 21-53.

[40] Soclo, H. H.; Garrigues, P. H.; Ewald, M. *Mar. Pollut. Bull.* 2000, 40, 387-396.
[41] Van Metre, P.; Mahler, B.; Furlong, E. *Environ. Sci. Technol.* 2000, 34, 4064-4070.
[42] McDow, S. R.; Sun, Q. R.; Vartiainen, M.; Hong, Y. S.; Yao, Y. L.; Fister, T.; Yao, R. Q.; Kamens, R. M. *Environ. Sci. Technol.* 1994, 28, 2147-2153.
[43] Kira, S.; Itoh, T.; Hayatsu, H.; Taketa, K.; Zheng, Y.; Li, R.; Holliday, T. L.; Giam, C. S. *Bull. Environ. Contam. Toxicol.* 1994, 53, 285-291.
[44] Nendza, M.; Herbst, T.; Kussatz, C.; Gies, A. *Chemosphere* 1997, 35, 1875-1885.
[45] Knezovich, J. P.; Harrison, F. L.; Wilhelm, R. G. *Water Air Soil Pollut.* 1987, 32, 233-245.
[46] Martinez, E.; Gros, M.; Lacorte, S.; Barceló, D. *J. Chromatogr. A* 2004, 1047, 181-188.
[47] Almeida, C. M. R.; Mucha, A. P.; Delgado, M. F. C.; CaÇador, M. I.; Bordalo, A. A.; Vasconcelos, M. T. S. D. *Mar. Environ. Res.* 2008, 66, 311-318.
[48] Srogi, K. *Environ. Chem. Lett.* 2007, 5, 169-195.
[49] Lee, H. K. In: Environmental analysis: handbook of analytical separations. Kleiböhmer, W. (ed), Elsevier Science, Amsterdam, The Netherlands: 2001, pp 39-74.
[50] Santos, F. J.; Galcerán, M. T. *Trends Anal. Chem.* 2002, 21, 672-685.
[51] Poster, D. L.; Schantz, M. M.; Sander, L. C.; Wise, S. A. *Anal. Bioanal. Chem.* 2006, 386, 859-881.
[52] Chen, Y.; Guo, Z.; Wang, X.; Qiu, C. *J. Chromatogr. A* 2008, 1184, 191-219.
[53] Rockens, E.; Dumollin, J.; Matheeussen, C. *Adv. Air Pollut.* 2000, 8, 699-707.
[54] Caricchia, A. M.; Chiavarini, S.; Pezza, M. *Atmos. Environ.* 1999, 33, 3731-3738.
[55] Omar, N. Y. M. J.; Abas, M. R. B.; Ketuly, K. A.; Tahir, N. M. *Atmos. Environ.* 2002, 36, 247-254.
[56] Yang, H. H.; Lai, S. O.; Hsieh, L. T.; Hsueh, H. J.; Chi, T. W. *Chemosphere* 2002, 48, 1061-1074.
[57] Dyke, P. H.; Foan, C.; Fiedler, H. *Chemosphere* 2003, 50, 469-480.
[58] Park, S. S.; Kim, Y. J.; Kang, C. H. *Atmos. Environ.* 2002, 36, 2917-2924.
[59] Manoli, E.; Voutsa, D.; Samara, C. *Atmos. Envrion.* 2002, 36, 949-961.
[60] Council Directive 2004/107/CE, OJ No L23, 26.1.2005.
[61] Areskoug, H.; Camner, P.; Dahlén, S.; Lastbom, L.; Nyberg, F.; Pershagen, G. E.; Sydbom, A. *Scand. J. Work Environ. Health* 2000, 26, 1-96.
[62] Venkataraman, C.; Lyons, J. M.; Fiedlander, S. *Environ. Sci. Technol.* 1994, 28, 535-562.
[63] Tsapakis, M.; Stephanou, E. G. *Environ. Pollut.* 2005, 133, 147-156.
[64] Vasilakos, C.; Levi, N.; Maggos, T.; Hatzianestis, J.; Michopoulos, J.; Helmis, C. *J. Hazard. Mater.* 2007, 140, 45-51.
[65] Kalaitzoglou, M.; Terzi, E.; Samara, C. *Atmos. Environ.* 2004, 38, 2545-2560.
[66] Bhargava, A.; Khanna, R. N.; Bhargava, S. K.; Kumar, S. *Atmos. Environ.* 2004, 38, 4761-4767.
[67] WHO, Polynuclear aromatic hydrocarbons (PAHs). In: Air quality guidelines for Europe. Copenhagen, World Health Organization Regional Office for Europe: 1987, pp 105-117.
[68] Liu, Y. J.; Zhu, L. Z.; Shen, X. Y. *Environ. Sci. Technol.* 2001, 35, 840-844.
[69] Sakai, R.; Siegmann, H. C.; Sato, H.; Voorhees, A. S. *Environ. Res. Sec. A* 2002, 89, 66-71.

[70] Wang, S. –F.; Liu, B. –Z.; Sunb, K. –J.; Su, Q. –D. *J. Chromatogr. A* 2004, 1025, 255-261.

[71] Culea, M.; Cozar, O.; Culea, E. *Indoor Built. Environ.* 2005, 14, 283-292.

[72] Peltonen, K.; Kuljukka, T. *J. Chromatogr. A* 1995, 710, 93-108.

[73] Bièvre, P. D. *Accredit. Qual. Assur.* 2002, 7, 181-181.

[74] Ross, A. B.; Jones, J. M.; Chaiklangmuang, S.; Pourkashanian, M.; Williams, A.; Kubica, K.; Andersson, J. T.; Kerst, M.; Danielka, P.; Bartle, K. D. *Fuel* 2002, 81, 571-582.

[75] Schauer, J. J.; Kleeman, M. J.; Cass, G. R.; Simoneit, B. R. T. *Environ. Sci. Technol.* 2001, 35, 1716-1728.

[76] Tsai, P. J.; Shieh, H. Y.; Hsieh, L. T.; Lee, W. J. *Atmos. Environ.* 2001, 35, 3495-3501.

[77] Piñeiro-Iglesias, M.; Grueiro-Noche, G.; López-Mahía, P.; Muniategui-Lorenzo, S.; Prada-Rodríguez, D. *Sci. Tot. Environ.* 2004, 334-335, 377-384.

[78] Manoli, E.; Kouras, A.; Samara, C. *Chemosphere* 2004, 56, 867-878.

[79] Klánová, J.; Kohoutek, J.; Hamplová, L.; Urbanová, P.; Holoubek, I. *Environ. Pollut.* 2006, 144, 393-405.

[80] Esteve-Turrillas, F. A.; Pastor, A.; Yusà, V. de la Guardia, M. *Trends Anal. Chem.* 2007, 26, 703-712.

[81] Bohlin, P.; Jones, K. C.; Strandberg, B. *J. Environ. Monitor.* 2007, 9, 501-509.

[82] Arp, H. P. H.; Schwarzenbach, R. P.; Goss, K. U. *Atmos. Environ.* 2007, 41, 8241-8252.

[83] Tucker, S. P. *J. Environ. Monitor.* 2007, 9, 1122-1130.

[84] Gjølstad, M.; Bergemalm-Rynell, K.; Ljungkvist, G.; Thorud, S.; Molander, P. *J. Sep. Sci.* 2004, 27, 1531-1539.

[85] AENOR Determinación de hidrocarburos aromáticos policíclicos totales (fase gaseosa y partículas). Captación en filtros de adsorción y análisis por cromatografía de gases-espectrometría de masas. UNE-77250, Madrid, 2001.

[86] U.S. EPA, Compendium methods for determination of toxic organic compounds in ambient air. Determination of polycyclic aromatic hydrocarbons (PAHs in ambient air using gas chromatography-mass spectrometry, GC-MS). Method TO-13A, Cincinnati, 1999, EPA/625/R-96/010b.

[87] Wei, M. C.; Chang, W. T.; Jen, J. F. *Anal. Bioanal. Chem.* 2007, 387, 999-1005.

[88] Padma, B.; Rao, S.; Faiyaz Ansari, M.; Pipalatkar, P.; Kumar, A.; Nema, P.; Devotta, S. *Bull. Environ. Contam. Toxicol.* 2007, 79, 197-201.

[89] Xie, M. X.; Xie, F.; Deng, Z. W.; Zhuang, G. S. *Talanta* 2003, 60, 1245-1257.

[90] Yusà, V.; Quintas, G.; Pardo, O.; Pastor, A.; de la Guardia, M. Talanta 2006, 69, 807-815.

[91] Blanchard, P.; Kallweit, D.; Brice, K. A.; Froude, F. A.; Chan, C. H.; Neilson, M.; Holz, J.; Millat, H. *J. Environ. Monitor.* 2006, 8, 465-471.

[92] Iavicoli, I.; Chiarotti, M.; Bergamaschi, A.; Marsili, R.; Carelli, G. *J. Chromatogr. A* 2007, 1150, 226-235.

[93] Kuusimäki, L.; Peltonen, K.; Mutanen, P.; Savela, K. *Ann. Occup. Hyg.* 2003, 47, 389-398.

[94] Poor, N.; Tremblay, R.; Kay, H.; Bhethanabotla, V.; Swartz, E.; Luther, M.; Campbell, S. *Atmos. Environ.* 2004, 38, 6005-6015.
[95] Jenkins, B. M.; Jones, A. D.; Turn, S. Q.; Williams, R. B. *Environ. Sci. Technol.* 1996, 30, 2462-2469.
[96] Bhargava, A.; Dlugogorski, B. Z.; Kennedy, E. M. *Fire Saf. J.* 2002, 37, 659-672.
[97] Khalfi, A.; Trouvé, G.; Delobel, R.; Delfosse, L. *J. Anal. Appl. Pyrol.* 2000, 56, 243-262.
[98] Conde, F. J.; Ayala, J. H.; Afonso, A. M.; González, V. *Anal. Chim. Acta* 2004, 524, 287-294.
[99] Conde, F. J.; Ayala, J. H.; Afonso, A. M.; González, V. *Atmos. Environ.* 2005, 39, 6654-6663.
[100] Cabañas, A.; Sáez, F.; González, A.; Escalada, R. *J. Agric. Eng. Res.* 2000, 77, 349-354.
[101] Chen, S. J.; Hsieh, L. T.; Chiu, S. C. *Environ. Int.* 2003, 28, 659-668.
[102] Sidhu, S.; Gullet, B.; Striebich, R.; Klosterman, J.; Contreras, J.; DeVito, M. *Atmos. Environ.* 2005, 39, 801-811.
[103] Yan, R.; Zhu, H.; Zheng, C.; Xu, M. *Energ.* 2002, 27, 485-503.
[104] Chen, Y.; Bi, X.; Mai, B.; Sheng, G.; Fu, J. *Fuel* 2004, 83, 781-790.
[105] Wauters, E.; Van Caeter, P.; Desmet, G.; David, F.; Devos, C.; Sandra, P. *J. Chromatogr. A* 2008, 1190, 286-293.
[106] Kuo, C. Y.; Hsu, Y. W.; Lee, H. S. *Arch. Environ. Contam. Toxicol.* 2003, 44, 454-459.
[107] Namiesnik, J.; Zabiegala, B.; Kot-Waski, A.; Partyka, M.; Wasik, A. *Anal. Bioanal. Chem.* 2005, 381, 279-301.
[108] Lohmann, R.; Corrigan, B. P.; Howsam, M.; Jones, K. C.; Ockenden, W. A. *Environ. Sci. Technol.* 2001, 35, 2576-2582.
[109] Bartkow, M. E.; Hawker, D. W.; Kennedy, K. E.; Muller, J. F. *Environ. Sci. Technol.* 2004, 38, 2701-2706.
[110] Wilford, B. H.; Harner, T.; Zhu, J. P.; Shoeib, M.; Jones, K. C. *Environ. Sci. Technol.* 2004, 38, 5312-5318.
[111] Gouin, T.; Harner, T.; Blanchard, P.; Mackay, D. *Environ. Sci. Technol.* 2005, 39, 9115-9122.
[112] Cupr, P.; Klánová, J.; Bartos, T.; Flegrová, Z.; Kohoutek, J.; Holoubek, I. *Environ. Pollut.* 2006, 144, 406-413.
[113] Wania, F.; Shen, L.; Lei, Y. D.; Teixeira, C.; Muir, D. C. G. *Environ. Sci. Technol.* 2003, 37, 1352-1359.
[114] Huckins, J. N.; Manuweera, G. K.; Petty, J. D.; Mackay, D.; Lebo, J. A. *Environ. Sci. Technol.* 1993, 27, 2489-2496.
[115] Bartkow, M. E.; Huckins, J. N.; Müller, J. F. *Atmos. Environ.* 2004, 38, 5983-5990.
[116] Bartkow, M. E.; Kennedy, K. E.; Huckins, J. N.; Holling, N.; Komarova, T.; Müller, J. F. *Environ. Pollut.* 2006, 144, 371-376.
[117] Liu, G.; Zhang, G.; Li, J.; Li, X.; Peng, X.; Qi, S. *Atmos. Environ.* 2006, 40, 3134-3143.
[118] Söderström, H.; Hajslová, J.; Kocourek, V.; Siegmund, B.; Kocan, A.; Obiedzinski, M. W.; Tysklind, M.; Bergqvist, P. A. *Atmos. Environ.* 2005, 39, 1627-1640.
[119] Tobias, D. E.; Perlinger, J. A.; Morrowa, P. S.; Doskey, P. V.; Perram, D. L. *J. Chromatogr. A* 2007, 1140, 1-12.
[120] Simonich, S. L.; Hites, R. A. *Science* 1995, 269, 1851-1854.

[121] Wang, Q.; Zhao, Y.; Yang, D.; Yang, L.; Li, Z.; Huang, B. *Environ. Sci. Technol.* 2004, 38, 4739-4744.

[122] Orecchio, S.; Gianguzza, A.; Culotta, L. *Environ. Res.* 2008, 107, 371-379.

[123] Ratola, N.; Lacorte, S.; Barceló, D.; Alves, A. *Talanta* 2009, 77, 1120-1128.

[124] Kylin, H.; Sjodin, A. *Environ. Sci. Technol.* 2003, 37, 2350-2355.

[125] Hellström, A.; Kylin, H.; Strachan, W. M. J.; Jensen, S. *Environ. Pollut.* 2004, 128, 29-48.

[126] Luque de Castro, M. D.; García-Ayuso, L. E. *Anal. Chim. Acta* 1998, 369, 1-10.

[127] Luque-García, J. L.; Luque de Castro, M. D. *Talanta* 2004, 64, 571-577.

[128] Santos, H. M.; Capelo, J. L. *Talanta* 2007, 73, 795-802.

[129] Priego-Capote, F.; Luque de Castro, M. D. *J. Biochem. Biophys. Methods* 2007, 70, 299-310.

[130] Priego-Capote, F.; Luque de Castro, M. D. *Anal. Bioanal. Chem.* 2007, 387, 249-257.

[131] Abu-Samra, A.; Morris, J. S.; Koirtyohann, S. R. *Anal. Chem.* 1975, 47, 1475-1477.

[132] Ganzler, K.; Salgo, A.; Valko, K. *J. Chromatogr.* 1986, 371, 299-306.

[133] Nóbrega, J. A.; Trevizan, L. E.; Araújo, G. C. L.; Nogueira, A. R. A. *Spectroc. Acta Pt. B-Atom. Spectr.* 2002, 57, 1855-1876.

[134] Montgomery, C. in: *Design and Analysis of Experiments,* 3rd Edition, Wiley, New York: 1991, Chapter 11.

[135] Richter, B. E.; Jones, B. A.; Ezzell, J. L.; Porter, N. L.; Avdalovic, N.; Pohl, C. *Anal. Chem.* 1996, 68, 1033-1039.

[136] Fitzpatrick, L. J.; Zuloaga, O.; Etxebarria, N.; Dean, J. R. *Rev. Anal. Chem.* 2000, 19, 75-122.

[137] Fidalgo-Used, N.; Blanco-González, E.; Sanz-Medel, A. *Anal. Chim. Acta* 2007, 590, 1-16.

[138] Smith, R. M. *J. Chromatogr. A* 2002, 975, 31-46.

[139] EPA Method 3545 (July 1995) Pressurised fluid extraction, test methods for evaluating solid waste, 3rd edn. Update III; EPA SW-846:US GPO, Washington DC.

[140] Mendiola, J. A.; Herrero, M.; Cifuentes, A.; Ibáñez, E. *J. Chromatogr. A* 2007, 1152, 234-246.

[141] Wenzel, K. –D.; Vrana, B.; Hubert, A.; Schuurmann, G. *Anal. Chem.* 2004, 76, 5503-5509.

[142] Langenfeld, J. J.; Hawthorne, S. B.; Miller, D. J.; Pawliszyn, J. *Anal. Chem.* 1994, 66, 909-916.

[143] Smith, R. M. *J. Chromatogr. A* 2003, 1000, 3-27.

[144] Librando, V.; Tomaselli, G.; Tringali, G. *Ann. Chim.* 2005, 95, 211-216.

[145] Talebi, S. M.; Taebi, A. *Indian J. Chem. Technol.* 2004, 11, 531-534.

[146] Shimmo, M.; Adler, H.; Hyötyläinen, T.; Hartonen, K.; Kulmala, M.; Riekkola, M. –L. *Atmos. Environ.* 2002, 36, 2985-2995.

[147] Castells, P.; Santos, F. J.; Galceran, M. T. *J. Chromatogr. A* 2003, 1010, 141-151.

[148] Shimmo, M.; Anttila, P.; Hartonen, K.; Hyötyläinen, T.; Paatero, J.; Kulmala, M.; Riekkola, M. –L. *J. Chromatogr. A* 2004, 1022, 151-159.

[149] Tsapakis, M.; Lagoudaki, E.; Stephanou, E. G.; Kavouras, I. G.; Koutrakis, P.; Oyola, P.; von Baer, D. *Atmos. Environ.* 2002, 36, 3851-3863.

[150] Huck, C. W.; Bonn, G. K. *J. Chromatog. A* 2000, 885, 51-72.
[151] Tamayo, F. G.; Turiel, E.; Martín-Esteban, A. *J. Chromatogr. A* 2007, 1152, 32-40.
[152] Marcé, R. M.; Borrull, F. *J. Chromatogr. A* 2000, 885, 273-290.
[153] Venn, R. F.; Merson, J.; Cole, S.; Macrae, P. *J. Chromatogr. B* 2005, 817, 77-80.
[154] Pyrzynska, K.; Pobozy, E. *Crit. Rev. Anal. Chem.* 2002, 32, 227-243.
[155] Rodriguez-Mozaz, S.; Lopez de Alda, M. J.; Barceló, D. *J. Chromatogr. A* 2007, 1152, 97-115.
[156] Hyötyläinen, T. *J. Chromatogr. A* 2008, 1186, 39-50.
[157] Arthur, C. L.; Pawliszyn, J. *Anal. Chem.* 1990, 62, 2145-2148.
[158] Lord, H.; Pawliszyn, J. *J. Chromatogr. A* 2000, 885, 153-193.
[159] Lord, H. L. *J. Chromatogr. A* 2007, 1152, 2-13.
[160] Mullett, W. M.; Pawliszyn, J. *J. Sep. Sci.* 2003, 26, 251-260.
[161] Ouyang, G.; Pawliszyn, J. *J. Chromatogr. A* 2008, 627, 184-197.
[162] Pleil, J. D.; Vette, A. F.; Rappaport, S. M. *J. Chromatogr. A* 2004, 1033, 9-17.
[163] Grant, D.W. Capillary gas chromatography, John Wiley and Sons, Chichester, New York, Brisbane: 1996.
[164] Lee, H. K. *J. Chromatogr. A* 1995, 710, 79-92.
[165] Christensen, A.; Östman, C.; Westerholm, R. *Anal. Bioanal. Chem.* 2005, 381, 1206-1216.
[166] Manoli, E.; Samara, C. *Trends Anal. Chem.* 1999, 18, 417-428.
[167] Bomboi, M. T.; Hernandez, A. *Water Res.* 1991, 25, 557-565.
[168] Filipkowska, A.; Lubecki, L.; Kowalewske, G. *Anal. Chim. Acta* 2005, 547, 243-254.
[169] Dachs, J.; Bayona, J. M.; Raoux, C.; Albaigés, J. *Environ. Sci. Technol.* 1997, 31, 682-688.
[170] Directive 2000/60/EC of the European Parliament and of the Council of 23 October 2000 Establishing a Framework for Community Action in the Field of Water Policy Official J. L 327, 22 December 2000, pp. 0001–0073 (www.europa.eu.int.).
[171] Council Directive 98/83/EC, on the quality of water intended for human consumption, *European Commission*, Brussels, 1998.
[172] Directive of the European Parliament and of the Council on environmental quality standards in the field of water policy and amending Directive 2000/60/EC, *European Commission,* Brussels, 2006. 2006/0129.
[173] Opinion of the European Economic and Social Committee on the Proposal for a Directive of the European Parliament and of the Council on environmental quality standards in the field of water policy and amending Directive 2000/60/EC, Off. *J. Eur. Union* 2007, C97, 3. 2007/C 97/02.
[174] Lepom, P.; Brown, B.; Hanke, G.; Loos, R.; Quevauviller, P.; Wollgast, J. *J. Chromatogr. A* 2009, 1216, 302-315.
[175] Environmental Agency of Japan (1998) Manual of analytical method for the endocrine disrupting chemicals. *Environmental Agency of Japan, Tokyo.*
[176] Regulation 459/00-Drinking Water Protection, made under the Ontario Water Resources Act, in Ontario Gazette, August 26, Queen's Printer of Ontario, Toronto, 2000.
[177] World Health Organization, Guidelines for Drinking Water Quality-Recommendations, *World Health Organization*, Geneva, 1984, vol. 1, p. 66.
[178] Kiss, G.; Varga-Puchony, Z.; Hlavay, J. *J. Chromatogr. A* 1996, 725, 261-272.

[179] Brouwer, E. R.; Hermans, A. N. J.; Lingeman, H.; Brinkman, U. A. T. *J. Chromatogr. A* 1994, 669, 45-57.
[180] Chee, K. K.; Wong, M. K.; Lee, H. K. *Anal. Chim. Acta* 1996, 330, 217-227.
[181] Glazkov, I. N.; Revelsky, I. A.; Efimov, I. P.; Zolotov, Y. A. *Chromatographia* 2000, 52, 495-498.
[182] Fernández-González, V.; Concha-Graña, E.; Muniategui-Lorenzo, S.; López-Mahía, P.; Prada-Rodríguez, D. *J. Chromatogr. A* 2007, 1176, 48-56.
[183] Prieto, A.; Telleria, O.; Etxebarria, N.; Fernández, L. A.; Usobiaga, A.; Zuloaga, O. *J. Chromatogr. A* 2008, 1214, 1-10.
[184] Wu, Y.; Xia, L.; Chen, R.; Hu, B. *Talanta* 2008, 74, 470-477.
[185] Delgado, B.; Pino, V.; Ayala, J. H. González, V.; Afonso, A. M. *Anal. Chim. Acta* 2004, 518, 165-172.
[186] House, W. A. *Int. J. Environ. Anal. Chem.* 1994, 57, 207-214.
[187] Backhus, D. A.; Ryan, J. N.; Groher, D. M.; MacFarlane, J. K.; Gschwend, P. M. *Ground Water* 1993, 31, 466-479.
[188] Villholth, K. G. *Environ. Sci. Technol.* 1999, 33, 691-699.
[189] Puls, R. W.; Powell, R. M. *Environ. Sci. Technol.* 1992, 26, 614-621.
[190] Kot-Wasik, A.; Debska, J.; Namiesnik, J. *Mar. Pollut. Bull.* 2004, 49, 264-276.
[191] Gustavson, K. E.; Harkin, J. M. *Environ. Sci. Technol.* 2000, 34, 4445-4451.
[192] Chen, H. –W. *Anal. Sci.* 2004, 20, 1383-1388.
[193] Parker, L. V. *Ground Water Monit. Rev.* 1994, 14, 130-141.
[194] Bopp, S.; Weiβ, H.; Schirmer, K. *J. Chromatogr. A* 2005, 1072, 137-147.
[195] Powell, R. M.; Puls, R. W. *Pollut. Eng.* 1997, 29, 50-54.
[196] Kot, A.; Zabiegala, B.; Namiesnik, *J. Trends Anal. Chem.* 2000, 19, 446-459.
[197] Petty, J. D.; Poulton, B. C.; Charbonneau, C. S.; Huckins, J. N.; Jones, S. B.; Cameron, J. T.; Prest, H. F. *Environ. Sci. Technol.* 1998, 32, 837-842.
[198] Luellen, D. R.; Shea, D. *Environ. Sci. Technol.* 2002, 36, 1791-1797.
[199] Zhao, W.; Han, M.; Dai, S.; Xu, J.; Wang, P. *Chemosphere* 2006, 62, 1623-1629.
[200] Davis, J. H. *Chem. Lett.* 2004, 33, 1072-1077.
[201] Wennrich, L.; Vrana, B.; Popp, P.; Lorenz, W. *J. Environ. Monit.* 2003, 5, 813-822.
[202] Ouyang, G.; Zhao, W.; Bragg, L.; Qin, Z.; Alaee, M.; Pawliszyn, J. *Environ. Sci. Technol.* 2007, 41, 4026-4031.
[203] Sluszny, C.; Gridin, V. V.; Bulatov, V.; Schechter, I. *Anal. Chim. Acta* 2004, 522, 145-152.
[204] Bones, J.; Thomas, K.; Nesterenko, P. N.; Paull, B. *Talanta* 2006, 70, 1117-1128.
[205] US EPA, Method 525 in: Methods for the Determination of Organic Compounds in Drinking Water, EPA/600/4- 88/ 039. Office of Research and Development, Washington, DC, 1988.
[206] Sargenti, S. R.; McNair, H. M. *J. Microcolumn Separations* 1998, 10, 125-131.
[207] Renner, T.; Baumgarten, D.; Unger, K. K. *Chromatographia* 1997, 45, 199-205.
[208] Crozier, P. W.; Plomley, J. B.; Matchuk, L. *Analyst*, 2001, 126, 1974-1979.
[209] King, A. J.; Readman, J. W.; Zhou, J. L. *Anal. Chim. Acta* 2004, 523, 259-267.
[210] Cortazar, E.; Zuloaga, O.; Sanz, J.; Raposo, J. C.; Etxebarria, N.; Fernández, L. A. *J. Chromatogr. A* 2002, 978, 165-175.

[211] Baltussen, E.; Sandra, P.; David, F.; Cramers, C. *J. Microcol. Sep.* 1999, 11, 737-747.
[212] David, S.; Sandra, P. *J. Chromatogr. A* 2007, 1152, 54-69.
[213] Baltussen, E.; Cramers, C. A.; Sandra, P. *Anal. Bioanal. Chem.* 2002, 373, 3-22.
[214] Benanou, D.; Acobas, F.; de Roubin, M. R.; David, F.; Sandra, P. *Anal. Bioanal. Chem.* 2003, 376, 69-77.
[215] Murray, D. A. J. *J. Chromatogr.* 1979, 177, 135-140.
[216] Jeannot, M. A.; Cantwell, F. F. *Anal. Chem.* 1996, 68, 2236-2240.
[217] Liu, H.; Dasgupta, P. K. *Anal. Chem.* 1996, 68, 1817-1821.
[218] Itoh, N.; Tao, H.; Ibusuki, T. *Anal. Chim. Acta* 2005, 535, 243-250.
[219] Kolahgar, B.; Hoffmann, A.; Heiden, A. C. *J. Chromatogr. A.* 2002, 963, 225-230.
[220] Xu, L.; Basheer, C.; Lee, H. K. *J. Chromatogr. A* 2007, 1152, 184-192.
[221] Shariati-Feizabadi, S.; Yamini, Y.; Bahramifar, N. *Anal. Chim. Acta* 2003, 489, 21-31.
[222] Liu, J.; Jian, G.; Chi, Y.; Cai, Y.; Zhou, Q.; Hu, J. –T. *Anal. Chem.* 2003, 75, 5870-5876.
[223] Watanabe, H.; Tanaka, H. *Talanta* 1978, 25, 585-589.
[224] Quina, F. H.; Hinze, W. L. *Ind. Eng. Chem. Res.* 1999, 38, 4150-4168.
[225] Paleologos, E. K.; Giokas, D. L.; Karayannis, M. I. *Trends Anal. Chem.* 2005, 24, 426-436.
[226] Hinze, W. L.; Pramauro, E. *Crit. Rev. Anal. Chem.* 1993, 24, 133-177.
[227] Pino, V.; Ayala, J. H.; Afonso, A. M.; González, V. *J. Chromatogr. A* 2002, 949, 291-299.
[228] Ballesteros-Gómez, A.; Rubio, S.; Pérez-Bendito, D. *J. Chromatogr. A* 2008, 1203, 168-176.
[229] Delgado, B.; Pino, V.; Ayala, J. H. González, V.; Afonso, A. M. *Analyst* 2005, 130, 571-577.
[230] Inoue, T.; Ohmura, H.; Murata, D. *J. Colloid Interf. Sci.* 2003, 258, 374-382.
[231] Goryacheva, I.Y.; Shtykov, S. N.; Loginov, A. S.; Panteleeva, I. V. *Anal. Bioanal. Chem.* 2005, 382, 1413-1418.
[232] Jin, X.; Zhu, M.; Conte, E. D. *Anal. Chem.* 1999, 71, 514-517.
[233] Peeraprasompong, P.; Thavarungkul, P.; Kanatharana, P. *J. Environ. Monit.* 2007, 9, 174-181.
[234] Wang, H.; Campiglia, A. D. *Anal. Chem.* 2008, 80, 8202-8209.
[235] Sicilia, D.; Rubio, S.; Pérez-Bendito, D.; Maniasso, N.; Zagatto, E. A. G. *Anal. Chim. Acta* 1999, 392, 29-38.
[236] Ferrer, R.; Beltrán, J. L.; Guiteras, J. *Anal. Chim. Acta* 1996, 330, 199-206.
[237] García-Pinto, C.; Pérez-Pavón, J. L.; Moreno Cordero, B. *Anal. Chem.* 1994, 66, 874-881.
[238] Mueller, K. E.; Shann, J. R. *Chemosphere* 2006, 64, 1006-1014.
[239] Magee, B. R.; Lion, L. W.; Lemley, A. T. *Environ. Sci. Technol.* 1991, 25, 323-331.
[240] Reeves, W. R.; McDonald, T. J.; Cizmas, L.; Donnelly, K. C. *Sci. Total Environ.* 2004, 332, 183-192.
[241] Yunker, M. B.; Macdonald, R. W.; Vingarzan, R.; Mitchell, R.H.; Goyette, D. *Org. Geochem.* 2002, 33, 489-515.
[242] Lichtfouse, E.; Budzinski, H.; Garrigues, P.; Eglinton, T. I. *Org. Geochem.* 1997, 26, 353-359.
[243] Wilcke, W.; Amelung, W. *Soil Sci. Soc. Am. J.* 2000, 64, 2140-2148.

[244] Chen, L.; Ran, Y.; Xing, B.; Mai, B.; He, J.; Wei, X.; Fu, J.; Sheng, G. *Chemosphere* 2005, 60, 879-890.

[245] Song, Y. F.; Wilke, B. –M.; Song, Y. X.; Gong, P.; Zhou, Q. X.; Yang, G. F. *Chemosphere* 2006, 65, 1859-1868.

[246] Crépineau, C.; Rychen, G.; Feidt, C.; Le Roux, Y.; Lichtfouse, É.; Laurent, F. *J. Agric. Food Chem.* 2003, 51, 4841-4845.

[247] Bucheli, T. D.; Blum, F.; Desaules, A.; Gustafsson, Ö. *Chemosphere* 2004, 56, 1061-1076.

[248] Mielke, H. W.; Wang, G.; Gonzales, C. R.; Le, B.; Quach, V. N.; Mielke, P. W. *Sci. Tot. Environ.* 2001, 281, 217-227.

[249] Ma, L. L.; Chu, S. G.; Wang, X. T.; Cheng, H. X.; Liu, X. F.; Xu, X. B. *Chemosphere* 2005, 58, 1355-1363.

[250] Chung, M. K.; Hu, R.; Cheung, K. C.; Wong, M. H. *Chemosphere* 2007, 67, 464-473.

[251] Anderson, B.; Hunt, J.; Phillips, B.; Thompson, B.; Lowe, S.; Taberski, K.; Carr, R. S. *Environ. Res.* 2007, 105, 145-155.

[252] Webster, L.; Fryer, R. J.; Dalgarno, E. J.; Megginson, C.; Moffat, C. F. *J. Environ. Monit.* 2001, 3, 591-601.

[253] Slater, G. F.; Cowie, B. R.; Harper, N.; Droppo, I. G. *Environ. Pollut.* 2008, 153, 60-70.

[254] de Boer, J.; Law, R. J. *J. Chromatogr. A* 2003, 1000, 223-251.

[255] Roose, P.; Brinkman, U. A. T. *Trends Anal. Chem.* 2005, 24, 897–926.

[256] Morales-Caselles, C.; Riba, I.; Sarasquete, C.; DelValls, T. A. *Environ. Int.* 2008, 34, 514-523.

[257] Meudec, A.; Dussauze, J.; Deslandes, E.; Poupart, N. *Chemosphere* 2006, 65, 474-481.

[258] Sprovieri, M.; Feo, M. L.; Prevedello, L.; Manta, D. S.; Sammartino, S.; Tamburrino, S.; Marsella, E. *Chemosphere* 2007, 67, 998-1009.

[259] Vane, C. H.; Harrison, I.; Kim, A. W. *Sci. Total Environ.* 2007, 374, 112-126.

[260] Laturnus, F.; Von Arnold, K.; Gron, C. *Environ. Sci. Pollut. Res.* 2007, 14, 53-60.

[261] Sánchez-Brunete, C.; Miguel, E.; Tadeo, J. L. *J. Chromatogr. A* 2007 1148, 219-227.

[262] Eriksson, E.; Christensen, N.; Schmidt, J. E.; Ledin, A. *Desalination* 2008, 226, 371-388.

[263] Council of the European Community, Working Document on Sludge, 3rd Draft, Brussels, 27 April 2000, p. 1.

[264] Ahmed, A. S.; Webster, L.; Pollard, P.; Davies, I. M.; Moffat, C. F. *J. Environ. Monitor.* 2007, 9, 730-739.

[265] Webster, L.; Twigg, M.; Megginson, C.; Walsham, P.; Packer, G.; Moffat, C. *J. Environ. Monitor.* 2003, 5, 395-403.

[266] Barth, D. S.; Mason, B. J. *Soil Sampling Quality Assurance User's Guide*. 1984 EPA-600/4-84-043.

[267] Popek, E. Sampling & analysis of environmental chemical pollutants. A complete guide. Academic Press, Amsterdam: 2003

[268] Seethapathy, S.; Górecki, T.; Li, X. *J. Chromatogr. A* 2008, 1184, 234-253.

[269] Ghiasvand, A. R.; Hosseinzadeh, S.; Pawliszyn, J. *J. Chromatogr. A* 2006, 1124, 35-42.

[270] King, A. J.; Readman, J. W.; Zhou, J. L. *Int. J. Environ. Anal. Chem.* 2007, 87, 211-225.

[271] Williamson, K. S.; Petty, J. D.; Huckins, J. N.; Lebo, J. A.; Kaiser, E. M. *Chemosphere* 2002, 49, 717-729.

[272] Ramos, L.; Vreuls, J. J.; Brinkman, U. A. Th. *J. Chromatogr. A* 2000, 891, 275-286.

[273] Jiang, X.; Basheer, C.; Zhang, J.; Lee, H. K. *J. Chromatogr. A* 2005, 1087, 289-294.

[274] Barco-Bonilla, N.; Martínez-Vidal, J. L.; Garrido-Frenich, A.; Romero-González, R. *Talanta* 2009, 78, 156-164.

[275] Lüthje, K.; Hyötyläinen, T.; Riekkola, M. –L. *J. Chromatogr. A* 2004, 1025, 41-49.

[276] Baumard, P.; Budzinski, H.; Garrigues, P.; Narbonne, J. F.; Burgeot, T.; Michel, X.; Bellocq, J. *Mar. Environ. Res.* 1999, 47, 415-439.

[277] Letellier, M.; Budzinski, H. *Analyst* 1999, 124, 5-14.

[278] Morales-Muñoz, S.; Luque-García, J. L.; Luque de Castro, M. D. *Anal. Chem.* 2002, 74, 4213-4219.

[279] Morales-Muñoz, S.; Luque-García, J. L.; Luque de Castro, M. D. *J. Chromatogr. A* 2002, 978, 49-57.

[280] Hyötyläinen, T.; Andersson, T.; Hartonen, K.; Kuosmanen, K.; Riekkola, M. –L. *Anal. Chem.* 2000, 72, 3070-3076.

[281] Librando, V.; Hutzinger, O.; Tringali, G.; Aresta, *M. Chemosphere* 2004, 54, 1189-1197.

[282] García-Ayuso, L. E.; Sánchez, M.; Fernández de Alba, A.; Luque de Castro, M. D. *Anal. Chem.* 1998, 70, 2426-2431.

[283] García-Ayuso, L. E.; Luque-García, J. L.; Luque de Castro, M. D. *Anal. Chem.* 2000, 72, 3627-3634.

[284] Priego-Capote, F.; Luque-García, J. L.; Luque de Castro, M. D. *J. Chromatogr. A* 2003, 994, 159-167.

[285] Blanchard, M.; Teil, M. J.; Ollivon, D.; Legenti, L.; Chevreuil, M. *Environ.* Res. 2004, 95, 184-197.

[286] Abad, E.; Martínez, K.; Planas, C.; Palacios, O.; Caixach, J.; Rivera, J. *Chemosphere* 2005, 61, 1358-1369.

[287] Villar, P.; Callejón, M.; Alonso, E.; Jiménez, J. C.; Guiraúm, A. *Anal. Chim. Acta* 2004, 524, 295-304.

[288] Helaleh, M. I. H.; Al-Omair, A.; Nisar, A.; Gevao, B. *J. Chromatogr. A* 2005, 1083, 153-160.

[289] Pino, V.; Ayala, J. H.; Afonso, A. M.; González, V. *J. Chromatogr. A* 2000, 869, 515-522.

[290] Pino, V.; Ayala, J. H.; Afonso, A. M.; González, V. *Talanta* 2001, 54, 15-23.

[291] Bianco-Prevot, A.; Gulmini, M.; Zelano, V.; Pramauro, E. *Anal. Chem.* 2001, 73, 3790-3795.

[292] Pino, V.; Ayala, J. H.; Afonso, A. M.; González, V. *Int. J. Environ. Anal. Chem.* 2001, 81, 281-294.

[293] Pino, V.; Anderson, J. L.; Ayala, J. H.; González, V.; Afonso, A. M. *J. Chromatogr. A* 2008, 1182, 145-152.

[294] Pino, V.; Ayala, J. H.; Afonso, A. M.; González, V. *Anal. Chim. Acta* 2003, 477, 81-91.

[295] Guerra-Abreu, L.; Pino, V.; Anderson, J. L.; Afonso, A. M. *J. Chromatogr. A* 2008, 1214, 23-29.

[296] Pino, V.; Ayala, J. H.; Afonso, A. M.; González, V. *Fresenius J. Anal. Chem.* 2001, 371, 526-531.

[297] Li, C. F.; Wong, J. W. C.; Huie, C. W.; Choi, M. M. F. *J. Chromatogr. A* 2008, 1214, 11-16.
[298] Barker, S. A. *LC GC Eur.* 1998, 47, S37-S40.
[299] Guerin, T. F. *J. Environ. Monitor.* 1999, 1, 63-67.
[300] Merino, F.; Rubio, S.; Pérez-Bendito, D. *J. Chromatogr. A* 2002, 962, 1-8.
[301] Knopp, D.; Seifert, M.; Väänänen, V.; Niessner, R. *Environ. Sci. Technol.* 2000, 34, 2035-2041.
[302] Schantz, M. M.; Nichols, J. J.; Wise, S. A. *Anal. Chem.* 1997, 69, 4210-4219.
[303] Vyskocil, A.; Fiala, Z.; Chénier, V.; Krajak, V.; Ettlerova, E.; Bukac, J.; Viau, C.; Emminger, S. *Environ. Toxicol. Pharm.* 2000, 8, 111-118.
[304] Richardson, D. M.; Gubbins, M. J.; Davies, I. M.; Moffat, C. F.; Pollard, P. M. *Environ. Toxicol. Pharm.* 2004, 17, 79-85.
[305] Kipopoulou, A. M.; Manoli, E.; Samara, C. *Environ. Pollut.* 1999, 106, 369-380.
[306] Grova, N.; Rychen, G.; Monteau, F.; LeBizec, B.; Feidit, C. *Agron. Sustain. Dev.* 2006, 26, 195-199.
[307] Lodovici, M.; Akpan, V.; Casalini, C.; Zappa, C.; Dolara, P. *Chemosphere* 1998, 6, 1703-1712.
[308] Fismes, J.; Perrin-Ganier, C.; Empereur-Bissonnet, P.; Morel, J. L. *J. Environ. Qual.* 2002, 31, 1649-1656.
[309] Mottier, P.; Parisod, V.; Turesky, R. J. *J. Agric. Food Chem.* 2000, 48, 1160-1166.
[310] Simko, P. *J. Chromatogr. B* 2002, 770, 3-18.
[311] Simon, R.; de la Calle, B.; Palme, S.; Meier, D.; Anklam, E. *J. Sep. Sci.* 2005, 28, 871-882.
[312] Guillén, M. D.; Palencia, G.; Sopelana, P.; Ibargoitia, M. L. *J. Dairy Sci.* 2007, 90, 2717-2725.
[313] Pandey, M. K.; Pant, A. B.; Das, M. *Toxicol. Vitro* 2006, 20, 308-316.
[314] Simko, P.; Khunova, V.; Simon, P.; Hruba, M. *Int. J. Food Sci. Technol.* 1995, 30, 807-812.
[315] Cavret, S.; Laurent, C.; Feidt, C.; Laurent, F.; Rychen, G. *Reprod. Nutr. Dev.* 2003, 43, 145-154.
[316] Laurent, C.; Feidt, C.; Lichtfouse, E.; Grova, N.; Laurent, F.; Rychen, G. *J. Agric. Food. Chem.* 2001, 49, 2493-2496.
[317] Kishikawa, N.; Wada, M.; Kuroda, N.; Akiyama, S.; Nakashima, N. *J. Chromatogr. B* 2003, 789, 257-264.
[318] Fiedler, H.; Cheung, C. K.; Wong, M. H. *Chemosphere* 2002, 46, 1429-1433.
[319] Moret, S.; Conte, L. S. *J. Chromatogr. A* 2000, 882, 245-253.
[320] Stolyhwo, A.; Sikorski, Z. E. *Food Chem.* 2005, 91, 303-311.
[321] Meador, J. P.; Stein, J. E.; Reichert, W. L.; Varanasi, U. *Rev. Environ. Contam. Toxicol.* 1995, 143, 79-165.
[322] Boehm, P. D.; Page, D. S.; Brown, J. S.; Neff, J. M.; Bence, A. E. *Mar. Pollut. Bull.* 2005, 50, 740-750.
[323] El-Nemr, A.; El-Sikaily, A.; Khaled, A.; Said, T. O.; Abd-Alla, A. M. A. *Environ. Monit. Assess.* 2004, 96, 251-261.

[324] Kong, K. Y.; Cheung, K. C.; Wong, C. K. C.; Wong, M. H. *J. Environ. Sci. Heal. A* 2005, 40, 2105-2115.
[325] Lima, D.; Santos, M. M.; Ferreira, A. M.; Micaelo, C.; Reis-Henriques, M. A. *Environ. Int.* 2008, 34, 94-101.
[326] Perugini, M.; Visciano, P.; Giammarino, A.; Manera, M.; Di Nardo, W.; Amorena, M. *Chemosphere* 2007, 66, 1904-1910.
[327] Cheung, K. C.; Leung, H. M.; Kong, K. Y.; Wong, M. H. *Chemosphere* 2007, 66, 460-468.
[328] Ariese, F.; Beyer, J.; Wells, D. *J. Environ. Monit.* 2005, 7, 869-876.
[329] European Commission, Opinion of the Scientific Committee on Food, 2002, (http://europa.eu.int/comm/food/fs/sc/scf/out153_en.pdf).
[330] European Union, Commission Recommendation 2005/108/EC, Off. *J. Eur. Comm.* L34 (2005) 43.
[331] Joint FAO/WHO Expert Committee on Food Additives (http://www.who.int/ipcs/food/jecfa/en/).
[332] European Union, Commission Regulation 2005/208/EC, Off. *J. Eur. Comm.* L34 (2005) 3.
[333] European Commission, Regulation 2003/2065/EC, *Off. J. Eur. Comm.* L309 (2003) 1.
[334] European Food Safety Authority (EFSA) (http://www.efsa.eu.int/).
[335] Wenzl, T.; Simon, R.; Kleiner, J.; Anklam, E. *Trends Anal. Chem.* 2006, 25, 716-725.
[336] Simon, R.; Palme, S.; Anklam, E. *J. Chromatogr. A* 2006, 1103, 307-313.
[337] Veyrand, B.; Brosseaud, A.; Sarcher, L.; Varlet, V.; Monteau, F.; Marchand, P.; Andre, F.; Le Bizec, B. *J. Chromatogr. A* 2007, 1149, 333-344.
[338] Rey-Salgueiro, L.; García-Falcón, M. S.; Martínez-Carballo, E.; Simal-Gándara, J. *Food Chem.* 2008, 108, 607-615.
[339] Buldini, P. L.; Ricci, L.; Sharma, J. L. *J. Chromatogr. A* 2002, 975, 47-70.
[340] Gfrerer, M.; Lankmayr, E. *J. Sep. Sci.* 2003, 26, 1230-1236.
[341] Guillén, M. D.; Sopelana, P.; Partearroyo, M. A. *J. Agric. Food Chem.* 2000, 48, 126-131.
[342] Meudec, A.; Dussauze, J.; Jourdin, M.; Deslandes, E.; Poupart, N. *J. Chromatogr. A* 2006, 1108, 240-247.
[343] Pointet, K.; Milliet, A. *Chemosphere* 2000, 40, 293-299.
[344] Rodríguez-Sanmartín, P.; Moreda-Piñeiro, A.; Bermejo-Barrera, A.; Bermejo-Barrera, P. *Talanta* 2005, 66, 683-690.
[345] Ali, M. Y.; Cole, R. B. *Anal. Bioanal. Chem.* 2002, 374, 923-931.
[346] Ali, M. Y.; Cole, R. B. *J. Agric. Food Chem.* 2001, 49, 4192-4198.
[347] Lage-Yusty, M. A.; Cortizo-Daviña, J. L. *Food Control* 2005, 16, 59-64.
[348] Pena, T.; Pensado, L.; Casais, C.; Mejuto, C.; Phan-Tan-Luu, R.; Cela, R. *J. Chromatogr. A* 2006, 1121, 163-169.
[349] Jánská, M.; Tomaniová, M.; Hajslová, J.; Kocourek, *V. Anal. Chim. Acta* 2004, 520, 93-103.
[350] Liguori, L.; Heggstad, K.; Hove, H. T.; Julshamn, K. *Anal. Chim. Acta* 2006, 573–574, 181-188.
[351] Pensado, L.; Casais, M. C.; Mejuto, M. C.; Cela, R. *J. Chromatogr. A* 2005, 1077, 103-109.

[352] Campíns-Falcó, P.; Verdú-Andrés, J.; Sevillano-Cabeza, A.; Molins-Legua, C.; Herráez-Hernández, R. *J. Chromatogr. A* 2008, 1211, 13-21.

[353] Arrebola, F. J.; Frenich, A. G.; Rodríguez, M. J. G.; Bolaños, P. P.; Vidal, J. L. M. *J. Mass Spectrom.* 2006, 41, 822-829.

[354] Yurchenko, S.; Mölder, U. *J. Food. Compos. Anal.* 2005, 18, 857-869.

[355] Wennrich, L.; Popp, P.; Zeibig, M. *Int. J. Environ. Anal. Chem.* 2002, 82, 677-690.

[356] Guillén, M. D.; Goicoechea, E.; Palencia, G.; Cosmes, N. *J. Agric. Food Chem.* 2008, 56, 2028-2033.

[357] WHO International Programme on Chemical Safety (IPCS), 1993. Biomarkers and risk assessment: concepts and principles. *Environmental Health Criteria* 155, World Health Organization, Geneva.

[358] Van Gestel, C. A. M.; Van Brummelen, T. C. *Ecotoxicology* 1996, 5, 217-225.

[359] Sarkar, A.; Ray, D.; Shrivastava, A. N.; Sarker, S. *Ecotoxicology* 2006, 15, 333-340.

[360] van der Oost, R.; Beyer, J.; Vermeulen, N. P. E. *Environ. Toxicol. Pharmacol.* 2003, 13, 57-149.

[361] Neves, R. L. S.; Oliveira, T. F.; Ziolli, R. L. *Mar. Pollut. Bull.* 2007, 54, 1813-1838.

[362] Fuentes-Rios, D.; Orrego, R.; Rudolph, A.; Mendoza, G.; Gavilán, J. F.; Barra, R *Chemosphere* 2005, 61, 192-199.

[363] Krommer, V.; Zechmeister, H. G.; Roder, I.; Scharf, S.; Hanus-Illnar, A. *Chemosphere* 2007, 67, 1956-1966.

[364] Domeño, C.; Blasco, M.; Sánchez, C.; Nerín, C. *Anal. Chim. Acta* 2006, 569, 103-112.

[365] Jongeneelen, F. J.; Anzion, R. B. M.; Leijdekkers, Ch. M.; Bos, R. P.; Henderson, P. Th. *Int. Arch. Occup. Environ. Health* 1985, 57, 47-55.

[366] Kanoh, T.; Fukuda, M.; Onozuka, H.; Kinouchi, T.; Ohnishi, Y. *Environ. Res.* 1993, 62, 230-245.

[367] Jongeneelen, F. J. *Toxicol. Lett.* 1994, 72, 205-212.

[368] Kim, M. K.; Oh, S.; Lee, J. H.; Im, H.; Ryu, Y. M.; Oh, E.; Lee, J.; Lee, E.; Sul, D. *Exp. Mol. Med.* 2004, 36, 396-410.

[369] Viau, C.; Diakité, A.; Ruzgyté, A.; Tuchweber, B.; Blais, C.; Bouchard, M.; Vyskocil, A. *J. Chromatogr. B* 2002, 778, 165-177.

[370] Li, H.; Krieger, R. I.; Li, Q. X. *Sci. Total Environ.* 2000, 257, 147-153.

[371] Serdar, B.; Waidyanatha, S.; Zheng, Y.; Rappaport, S. M. *Biomarkers* 2003, 8, 93-109.

[372] Xu, X.; Zhang, J.; Zhang, L.; Liu, W.; Weisel, C. P. *Rapid Commun. Mass Spectrom.* 2004, 18, 2299-2308.

[373] Hecht, S. S.; Chen, M.; Yagi, H.; Jerina, D. M.; Carmella, S. G. *Cancer Epidemiol. Biomarkers Prev.* 2003, 12, 1501-1508.

[374] Jongeneelen, F. J.; Anzion, R. B. M.; Henderson, P. Th. *J. Chromatogr. B* 1987, 413, 227-232.

[375] Huang, W.; Smith, T. J.; Ngo, L.; Wang, T.; Chen, H.; Wu, F.; Herrick, R. F.; Christiani, D. C.; Ding, H. *Environ. Sci. Technol.* 2007, 41, 2711-2716.

[376] Schedl, M.; Wilharm, G.; Achatz, S.; Kettrup, A.; Niessner, R.; Knopp, D. *Anal. Chem.* 2001, 73, 5669-5676.

[377] Smith, C. J.; Walcott, C. J.; Huang, W.; Maggio, V.; Grainger, J.; Patterson Jr, D. G. *J. Chromatogr. B* 2002, 778, 157-164.

[378] Gmeiner, G.; Gärtner, P.; Krassnig, C.; Tausch, H. *J. Chromatogr. B* 2002, 766, 209-218.

[379] Amorim, L. C. A.; Dimandja, J. –M.; Cardeal, Z. L. *J. Chromatogr. A* 2009, 1216, 2900-2904.

[380] Waidyanatha, S.; Rothman, N.; Fustinoni, S.; Smith, M. T.; Hayes, R. B.; Bechtold, W.; Dösemeci, M.; Guilan, M.; Yin, S.; Rappaport, S. M. *Carcinogenesis* 2001, 22, 279-286.

[381] Toriba, A.; Kuramae, Y.; Chetiyanukornkul, T.; Kizu, R.; Makino, T.; Nakazawa, H.; Hayakawa, K. *Biomed. Chromatogr.* 2003, 17, 126-132.

[382] Pigini, D.; Cialdella, A. M.; Faranda, P.; Tranfo, G. *Rapid Commun. Mass Spectrom.* 2006, 20, 1013-1018.

[383] Pérez, C.; Velando, A.; Munilla, A.; López-Alonso, M.; Oro, D. *Environ. Sci. Technol.* 2008, 42, 707-713.

[384] Fillmann, G.; Watson, G. M.; Howsam, M.; Francioni, E.; Depledge, M. H.; Readman, J. W. *Environ. Sci. Technol.* 2004, 38, 2649-2656.

[385] Campo, L.; Addario, L.; Buratti, M.; Scibetta, L.; Longhi, O.; Valla, C.; Cirla, P. E.; Martinotti, I.; Foà, V.; Fustinoni, S. *Toxicol. Lett.* 2006, 162, 132-138.

In: Polycyclic Aromatic Hydrocarbons: Pollution, Health... ISBN 978-1-60741-462-9
Editors: Pierre A. Haines and Milton D. Hendrickson

Chapter 7

NEW TENDENCIES IN LIQUID CHROMATOGRAPHY FOR CONTROLLING PAHS POLLUTION

R. Herráez-Hernández, P. Campíns-Falcó*, J. Verdú-Andrés, A. Sevillano-Cabeza and L. Ferrando-Climent
Department of Analytical Chemistry, University of Valencia, Valencia, Spain

ABSTRACT

Different configurations were studied for the screening analysis of common polycyclic aromatic hydrocarbons (PAHs) in environmental waters by liquid chromatography (LC). Analyte enrichment on a traping microcolumn connected by column switching to the analytical column, and in-valve in-tube solid-phase microextraction (IT-SPME) were tested as alternatives to conventional solid-phase extraction (SPE) with disposable cartridges. For separation conventional LC, capillary LC and LC with a monolithic column were tested and compared, whereas UV-diode array detection (DAD) or fluorescence detection were used for signal monitoring. The compounds assayed were naphthalene, anthracene, fluoranthene, benzo[*a*]anthracene, benzo[*b*]fluoranthene, benzo[*k*]fluoranthene, benzo[*a*]pyrene, dibenzo[*a,h*]anthracene and benzo[*ghi*]perylene. On the basis of the results obtained, new methods are presented for the on-line screening analysis of PAHs in several water matrices. The main advantages of the proposed procedures are: (i) the minimum off-line sample manipulation as the samples only need to be filtered; (ii) the rapidity, as the total analysis time can be reduced to less than 20 min, (iii) the improved detection limits, especially for the in-valve IT-SPME configurations, as they are compatible with the injection of very large sample volumes (up to 4.0 mL), and (iv) the total elimination of organic solvents in the sample treatment. Therefore, the proposed methods are environmentally friendly and cost-effective alternatives for the routine monitoring of PAHs in waters. The occurrence of the tested PAHs in waters of The Comunidad Valenciana (Spain) was also tested.

* Corresponding author: Tel.: + 34 963543002; Fax: + 34 963544436; e-mail address: pilar.campins@uv.es.

1. Introduction

To date, many analytical procedures have been developed for the analysis of PAHs in water, soil, air and food matrices, most of them using LC or gas chromatography (GC). The conventional approach for the analysis of these pollutants in water samples involves the previous preconcentration of the analytes by liquid-liquid extraction (LLE) or SPE, often combined with solvent evaporation [2-8]. Recent research efforts are oriented towards the development of simplified and miniaturized sample treatment methods which reduce the time of analysis, the consumption of chemicals and the generation of wastes. In this respect, different alternatives have been proposed for PAHs based on the miniaturization of LLE [9, 10] or SPE [11, 12], but the most popular approach is solid-phase microextraction (SPME). SPME permits the detection and quantification of PAHs at low to sub ppb levels using for analyte enrichment either fibres [13-18] or stir bars [19, 20] coated with an extractive phase. The long adsorption times required to extract the analytes, up to several hours in some of the reported assays [19], is the most serious limitation of SPME for routine monitoring of PAHs.

Ideally, an analytical method for monitoring of PAHs in waters should be simple, rapid, and sensitive. Sample conditioning should be integrated in the chromatographic system. However, owing to the low concentrations of PAHs expected, on-line sample conditioning would be only possible if the chromatographic system is compatible with the injection of large sample volumes. Another approach to improve the sensitivity in LC is the reduction of the column diameter dimensions. For example, Titato and Lancas demonstrated that the employment of capillary LC substantially increases the mass sensitivity in the analysis of PAHs with respect to conventional LC [21]. However, the sample volumes that can be injected in capillary chromatography are rather low and thus, some kind of analyte enrichment is still necessary. The assays reported for the analysis of PAHs by capillary LC generally involve conventional LLE or SPE [21, 22].

In this chapter we have evaluated different strategies for the sensitive analysis of PAHs in waters which combine the injection of large sample volumes and modern forms of LC. For analyte enrichment two approaches have been tested and compared: SPE into a trapping column connected to the analytical column by means of a switching arrangement and in-valve in-tube-SPME (IT-SPME). Analyte enrichment by column switching (CS) is a technique of routine use in conventional LC [7, 8, 23]. However, the employment of this technique in capillary LC is still rather limited [24]. As regards IT-SPME, different publications have demonstrated that this methodology is especially well-suited for the sensitive analysis of organic pollutants such as triazines, organophosphorous pesticides or phenols [25-27]. Recently, IT-SPME has been employed to preconcentrate and purify the extracts obtained after applying matrix solid-phase dispersion (MSPD) to extract PAHs from bivalve organisms [28]. However, the potential of IT-SPME for the sensitive analysis of PAHs in waters has not been fully exploited. For separation of the analytes capillary LC and LC with a monolithic column [29] have been tested and compared, whereas for signal monitoring UV-DAD and fluorescence detectors have been used. The reliability of the proposed methods has been tested by comparing the results obtained with those achieved by a classical approach based on the employment of SPE with disposable cartridges and conventional LC. Advantages, disadvantages and possible applications are discussed.

EXPERIMENTAL

Reagents and Solutions

All the reagents were of analytical grade. Naphthalene, anthracene, fluoranthene, benzo[*b*]fluoranthene, benzo[*k*]fluorantene, benzo[*a*]pyrene, dibenzo[*a,h*]anthracene and benzo[*ghi*]perylene (10.0 μg/mL in acetonitrile) were obtained from Dr. Ehrenstorfer (Augsburg, Germany). Working solutions of these compounds were prepared by dilution of the stock solutions with water.

Acetonitrile of HPLC grade was obtained from J. T. Baker (Deventer, The Netherlands); tetrahydrofuran (THF) was purchased from Fluka (Buchs, Switzerland).

Water was deionised and filtered through 0.45 μm nylon membranes (Teknokroma, Barcelona, Spain). All solutions were stored in the dark at 2°C.

APPARATUS AND ANALYTICAL CONDITIONS

SPE-Conventional LC-FLD

The chromatographic system consisted of a quaternary pump, an automatic sample injector (Hewlett-Packard 1050 Series, Palo Alto, CA, USA) and a programmable fluorimetric detector (Hewlett-Packard, 1046 Series). The detector was linked to a data system (Hewlett-Packard, HPLC ChemStation) for data acquisition and calculation. According to the results of preliminary experiments obtained by injecting directly into the analytical column standard solutions of PAHs, the detector was programmed to measure the fluorescence intensity at the excitation/emission wavelengths pairs listed in Table 1.

A LiChrospher 100 RP_{18}, 125 mm × 4 mm i.d. (Merck, Darmstadt, Germany) column was used for separation. The mobile-phase was a mixture of acetonitrile-water in gradient elution mode (see also Table 1). The solvents were filtered through 0.45 μm nylon membranes (Teknokroma,) and degassed with helium before use.

For SPE Envirolut PAH 1 g/6 mL cartridges (Variant, Harbor City, CA, USA) were used. Cartridges were conditioned by drawing successively 2 mL of acetonitrile and 5 mL of nanopure water. Then the samples (25 mL) were passed through the cartridges. All these operations were carried out by means of a vacuum station (IST VacMaster, Hengoed, UK). After sample loading the remaining water was manually eliminated from the cartridges by drawing the minimum volume of air with a plastic syringe. Finally, the analytes were desorbed from the cartridges with acetonitrile or THF, and 20 μL of the collected extracts were injected into the LC equipment. All the experiments were done in triplicate at ambient temperature.

CS-Capillary LC-DAD

A scheme of the system used for on-line enrichment by CS is shown in Figure 1a. The system consisted of a capillary pump (Agilent 1100 Series, Waldbronn, Germany) equipped with a Rheodyne model 7725 high-pressure six port injection valve (valve 1) with a internal loop (2 μL) or with an external loop (50, 100 or 200 μL). Analyte enrichment was effected into a trapping column containing a Zorbax SB C_{18} phase (35 mm x 0.5 mm i.d., 5 μm particle diameter) from Agilent. For separation of the analytes a Zorbax SB C_{18}, 150 mm x 0.5 mm i.d., 5 μm particle diameter column (Agilent) was employed. Both columns were connected by means an automatic switching microvalve (valve 2) (Agilent 1100 Series). The UV-DAD detector (Hewlett-Packard, 1100 Series) was coupled to a data system (Agilent, HPLC ChemStation) for data acquisition and calculation. The UV chromatograms were registered from 200 to 450 nm.

After filling the injection loop with the sample, valves 1 and 2 were rotated and the chromatographic run was started. In such a way, the sample was transferred from the injection loop to the trapping column; meanwhile, the effluent emerging from the trapping column was sent to waste. Next, the trapping and analytical columns were connected by rotating valve 2, and 0.5 min later valve 1 was turned to the load position.

For separation of the tested PAHs a mobile-phase of acetonitrile-water in gradient elution was used (see Table 1). The solvents were filtered through 0.45 μm nylon membranes (Teknokroma) before use. All the experiments were done in triplicate at ambient temperature.

IT-SPME-Capillary LC-DAD

A scheme of the system used for on-line enrichment by in-valve IT-SPME, separation and detection of PAHs is shown in Figure 1b. As observed, this system differs from that of Figure 1a in the injection device: the inner loop of the injection valve was replaced by a section of a capillary column such as those used in GC. This capillary column was a 70 cm x 0.32 mm i.d., 3-μm film thickness column coated with a polysiloxane polymer with 95 % methyl-5 % phenyl-substituted backbone (Teknokroma). The injection valve was connected directly to the analytical column (valve 2 was omitted). The other elements were those of the column switching-capillary LC-DAD system (Figure 1a). Conditions for separation and detection were those summarized in Table 1.The solvents were filtered through 0.45 μm nylon membranes (Teknokroma) before use. All the experiments were done in triplicate at ambient temperature.

IT-SPME-LC-FLD

The chromatographic system was the same as Figure 1b, but the analytical column was a Onyx Monolithic C_{18}, 100 mm × 3 mm i. d., (mesopores of ≈ 13 nm, macropores of ≈ 2 μm) column purchased from Phenomenex (Torrance, CA, USA); the signal was monitored with a programmable fluorimetric detector (Hewlett-Packard, 1046 Series). The detector was linked to a data system (Hewlett-Packard, HPLC ChemStation) for data acquisition and calculation.

The mobile-phase was a mixture of acetonitrile-water in gradient elution mode, and flow rate was 1.0 mL/min. Conditions for separation and detection are also summarized in Table 1. The solvents were filtered through 0.45 μm nylon membranes (Teknokroma) and degassed with helium before use. All the experiments were done in triplicate at ambient temperature.

Analysis of Real Water Samples

Different real water samples were analyzed by the proposed methods: river water, sea water, sewage water and harbour water. All samples were collected at different points of the coast of the Comunidad Valenciana (Spain). Samples were filtered through 0.45 μm nylon membranes (Teknokroma) after their arrival at the laboratory in order to remove any particulate matter. Each sample was analyzed in triplicate and all assays were carried out at ambient temperature.

RESULTS

SPE-Conventional LC-FLD Method

In this study SPE cartridges packed with a sorbent specially designed for PAHs as well as conventional LC with fluorimetric detection were used. In preliminary tests different elution and detection conditions were assayed for the separation of the selected PAHs. The elution program finally selected was that presented in Table 1. The FLD detector was programmed accordingly (see also Table 1).

Cartridges were previously activated with 2 mL of acetonitrile, and then washed with 5 mL of nanopure water. After sample loading the cartridges were dried with air, and then the analytes were desorbed and chromatographed. Acetonitrile was the first option tested to desorb the analytes from the SPE cartridges. Aliquots of 1 mL of acetonitrile were successively passed through the cartridges, and the extracts were collected separately. Finally, 20 μL of each extract were injected into the chromatographic system. The obtained chromatograms revealed that most PAHs were desorbed in the second and third 1 mL-fractions of acetonitrile. However, the most apolar PAHs benzo[*a*]pyrene, dibenzo[*a,h*]anthracene and benzo[*ghi*]perylene required up to 5 mL of solvent to be totally desorbed from the cartridges, and thus, the preconcentration achieved was considered unsuitable. In an attempt to avoid solvent evaporation THF instead of acetonitrile was tried as solvent for desorption. Nearly complete desorption of the analytes was observed with 1.5 mL of THF, which was then the option selected (Table 1). As an illustrative example, in Figure 2 are shown the chromatograms obtained for a blank and for an standard solution containing a mixture of the tested PAHs. As observed from this figure benzo[*b*]fluoranthene and benzo[*k*]fluoranthene partially overlapped. In an attempt to improve the resolution a variety of elution conditions were assayed but complete separation of these compounds could not be achieved in reasonable chromatographic times. Coelution between these PAHs seems to be a common problem in many of the procedures reported for PAHs using both LC and GC [6, 15,

17]. Nevertheless, the resolution achieved under conditions of Table 1 can be considered suitable for most purposes. The selectivity was also adequate as no matrix peaks were observed at the retention times of the analytes. This is illustrated in Figure 2c, which shows the chromatogram obtained for real sewage water in which none of the tested PAHs was found.

The optimized conditions were applied to standard solutions of PAHs containing 5.0 ng/mL of each compound. The percentages of PAHs recovered were established by comparing the peak areas obtained for each PAH with those obtained from the direct injection of 20 μL of standard solutions containing an equivalent amount of the analytes. As listed in Table 2 the analyte recoveries ranged from 87 to 112 %. The intra- and inter-day reproducibilities were also tested; the RSD values encountered are listed in Table 3. The limits of detection (LODs), established as the concentration required to generate a signal-to-noise ratio of 3, were $\leq$ 1 ng/mL (see Table 4).

CS-Capillary LC-DAD Method

The switching system used for on-line enrichment and chromatography of PAHs is shown in Figure 1a. Before each run the trapping and the analytical columns were equilibrated with the mobile phase. Then, the sample was placed in the injection loop by means of a syringe. At the beginning of the run valves 1 and 2 were rotated, so the sample was sent to the trapping column, and meanwhile, the effluent emerging from the trapping column was directly transferred to waste in order to protect the analytical column. After transfer the sample from the loop to the trapping column, both the extractive and the analytical columns were connected by changing the position of microvalve 2. At the same time or somewhat later the injection valve (valve 1) was also turned to its original position. In such a way, the mobile phase passed through the injection loop only to transfer the sample to the trapping column. In such a way, the time required for analyte desorption and separation could be significantly shortened, taking into account the the analytes were chromatographed under gradient elution.

Initially, different acetonitrile-water gradient elution programs, as well as different time intervals for connecting the extractive and the analytical columns were assayed. In this study standard solutions containing 10.0 ng/mL of each PAHs were processed, and the volume of the injection loop was 100 μL. The mobile-phase flow rate was 20.0 μL/min. Under such conditions the time required to transfer the sample plug from the injection loop to the trapping column was c.a. 5 min. In order to ensure the complete transfer of the sample to the trapping column and to minimize the time required for chromatography the injection valve was turned to its original position at 5.5 min. Maximum analyte responses were achieved by rotating valve 2 at 5.0 min. On the other hand, all the tested analytes could be resolved under a variety of conditions with the exception of benzo[*b*]fluoranthene and benzo[*k*]fluoranthene. As in the SPE-conventional LC-FLD approach, these compounds eluted at similar retention times under a variety of conditions. However, they could be easily distinguished by their UV spectra. Thus, the elution conditions were adjusted to obtain satisfactory resolution of the other PAHs in the minimum chromatographic time (see Table 1).

Table 1. Conditions used in the analysis of PAHs by the tested procedures

<table>
<tr><th>Conditions</th><th>SPE-conventional LC-FLD</th><th>CS-capillary LC-DAD</th><th>IT-SPME-capillary LC-DAD</th><th>IT-SPME-LC(monolithic)-FLD</th></tr>
<tr><td>Sample treatment</td><td>Sample volume: 25 mL

Procedure:

- conditioning of the cartridges with 2 mL of acetonitrile + 5 mL of water
- sample loading
- drying of the cartridges
- desorption of the analytes with 1.5 mL of THF
- injection onto the LC equipment of 20 μL of the extracts</td><td>Sample volume: 100 μL

Procedure:

- sample loading
- at 0 min - valve 1 rotation + valve 2 rotation
- at 5.0 min - valve 2 rotation
- at 5.5 min - valve 1 rotation</td><td colspan="2">Sample volume: 2 mL

Procedure:

- sample loading
- cleaning of the capillary with 100 μL of water
- valve rotation</td></tr>
<tr><td>Separation</td><td>Elution program (acetonitrile:water, v/v):

- at 0 min 60:40
- at 40 min 80:20
- at 42 min 100:0

Mobile-phase flow rate: 1.0 mL/min</td><td colspan="2">Elution program (acetonitrile:water, v/v):

- at 0-7 min 50:50
- at 20 min 70:30
- at 22 min 100:0

Mobile-phase flow rate: 20.0 μL/min</td><td>Elution program (acetonitrile:water, v/v):

- at 0 min 50:50
- at 12 min 80:20
- at 13 min 100:0

Mobile-phase flow rate: 1.0 mL/min</td></tr>
<tr><td>Detection</td><td>Time (min) / $\lambda_{excitation/emission}$ (nm)
0-7.5 224 330
7.5-10.0 251 395
10.0- 20.0 235 450
20.0-26.0 301 438
28.0-30.0 300 404
30.0 302 461</td><td colspan="2">λ (nm) Compound:
220 naphthalene
230 fluoranthene, benzo[k]fluoranthene
250 anthracene
260 benzo[b]fluoranthene
290 benzo[a]pyrene, dibenzo[a,h]anthracene, benzo[ghi]perylene</td><td>Time (min) / $\lambda_{excitation/emission}$ (nm)
0-3.9 224 330
3.9-4.9 251 395
4.9-6.2 235 450
6.2-13.0 301 438</td></tr>
</table>

Table 2. Absolute recoveries obtained by the SPE with disposable cartridges, CS and IT-SPME sample treatment procedures (values established at a concentration of 5.0 ng/mL in the SPE-LC-FLD and CS-capillary LC-DAD methods, and 1.0 ng/mL in the IT-SPME-FLD method; n = 3)

Compound	SPE	CS	IT-SPME
Naphthalene	104 ±11	89 ± 3	37 ± 4
Anthracene	98 ± 7	95 ± 6	45 ± 4
Fluoranthene	112 ± 10	101 ± 4	42 ± 2
Benzo[*a*]anthracene	-	-	32 ±2
Benzo[*b*]fluoranthene	99 ± 6	99 ± 3	26 ± 1
Benzo[*k*]fluorantene	101 ± 7	105 ± 2	28 ± 1
Benzo[*a*]pyrene	94 ± 7	93 ± 2	33 ± 1
Dibenzo[*a,h*]anthracene	91 ± 9	95 ± 2	22 ± 1
Benzo[*ghi*]perylene	87 ± 9	93 ± 2	26 ± 2

Table 3. RSD obtained with the SPE-LC-FLD, CS-capillary LC-DAD and IT-SPME-LC-FLD methods (values established at a concentration of 5.0 ng/mL in the SPE-LC-FLD and CS-capillary LC-DAD methods, and 0.3 ng/mL in the IT-SPME-FLD method)

Compound	SPE-LC-FLD		CS- capillary LC-DAD		IT-SPME-LC- FLD	
	Intra-day (n = 3)	Inter-day (n = 6)	Intra-day (n = 3)	Inter-day (n = 6)	Intra-day (n = 3)	Inter-day (n = 6)
Naphthalene	12	11	2	11	9	7
Anthracene	9	8	4	5	4	5
Fluoranthene	9	8	4	6	4	6
Benzo[*b*]fluoranthene	8	4	7	6	4	11
Benzo[*k*]fluorantene	2	9	3	7	3	5
Benzo[*a*]pyrene	9	11	6	5	1	7
Dibenzo[*a,h*]anthracene	6	11	2	11	5	9
Benzo[*ghi*]perylene	10	9	4	10	7	9

Table 4. LODs (ng/mL) obtained with the four optimized procedures

Compound	SPE-LC-FLD	CS-capillary LC-DAD	IT-SPME-capillary LC-DAD	IT-SPME-LC-FLD
Naphthalene	0.5	0.5	0.1	0.008
Anthracene	0.05	0.1	0.02	0.0015
Fluoranthene	0.5	0.5	0.05	0.008
Benzo[*a*]anthracene	-	-	0.1	0.006
Benzo[*b*]fluoranthene	0.5	1	0.1	0.006
Benzo[*k*]fluorantene	0.05	1	0.1	0.006
Benzo[*a*]pyrene	0.1	0.5	0.2	0.007
Dibenzo[*a,h*]anthracene	0.1	0.25	0.025	0.005
Benzo[*ghi*]perylene	1	1	0.025	0.0025

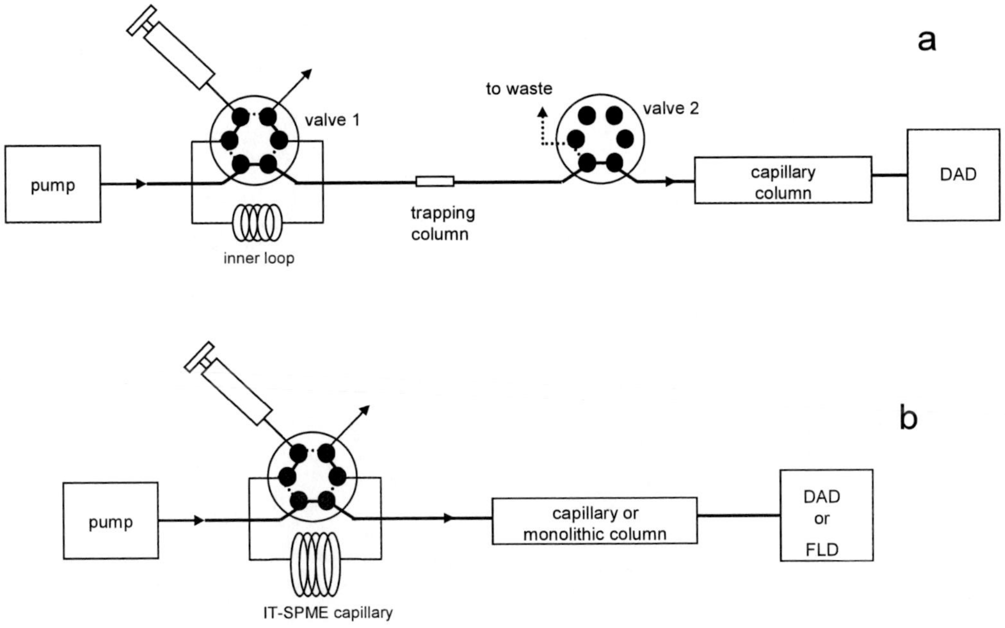

Figure 1. (a) Scheme of the column switching capillary chromatographic system used for PAHs. System conditioning and sample loading: valve 1 (—) and valve 2 (—); sample injection and on-line enrichment: valve 1 (---) and valve 2 (---); transfer of the analytes to the analytical column: valve 1 (---) and valve 2 (—); analyte separation and detection: valve 1 (—) and valve 2 (—). (b) Scheme of the IT-SPME device used for PAHs. For other experimental details, see text.

The effect of the sample volume was estimated by using external injection loops of 50, 100 and 200 μL. The times selected for rotating valve 2 were 2.5, 5.0 and 10.0 min, respectively, and in all instances the injection valve was returned to the load position 0.5 min later (see Figure 1a). The initial composition of the mobile-phase (50:50 acetonitrile-water, v/v) was kept constant in the 0-5, 0-7 and 0-12 min time intervals when injecting 50, 100 and 200 μL of the samples, respectively. Next, the gradient profile applied was equivalent to that shown in Table 1 for the method using 100 μL. Indeed, increasing the sample volume resulted in higher peak areas for the analytes, but also in longer retention times. When injecting 50 μL of the samples, the signals obtained for most PAHs were close to their respective LODs, and some of the analytes were not detected. In contrast, all the PAHs tested were detected when using the 100 μL and 200 μL loops. The peak areas obtained with 200 μL of samples were only 1.4-1.8 times greater than those obtained for a sample volume of 100 μL (data not shown). In other words, increasing the sample volume increased the sensitivity but the improvement was lower than that expected. This is because peak broadening became more marked as the time of residence of the analytes in the chromatographic system increased. Although not tested, a change in the elution conditions would most probably improve the LODs by shortening the times of retention. As a compromise between sensitivity and analysis time, a sample volume of 100 μL was the option selected. The entire CS-capillary LC-DAD method is summarized in Table 1. In Figure 3 are depicted the chromatograms obtained under the selected conditions for a blank (water) and a standard solutions of the analytes at different wavelengths.

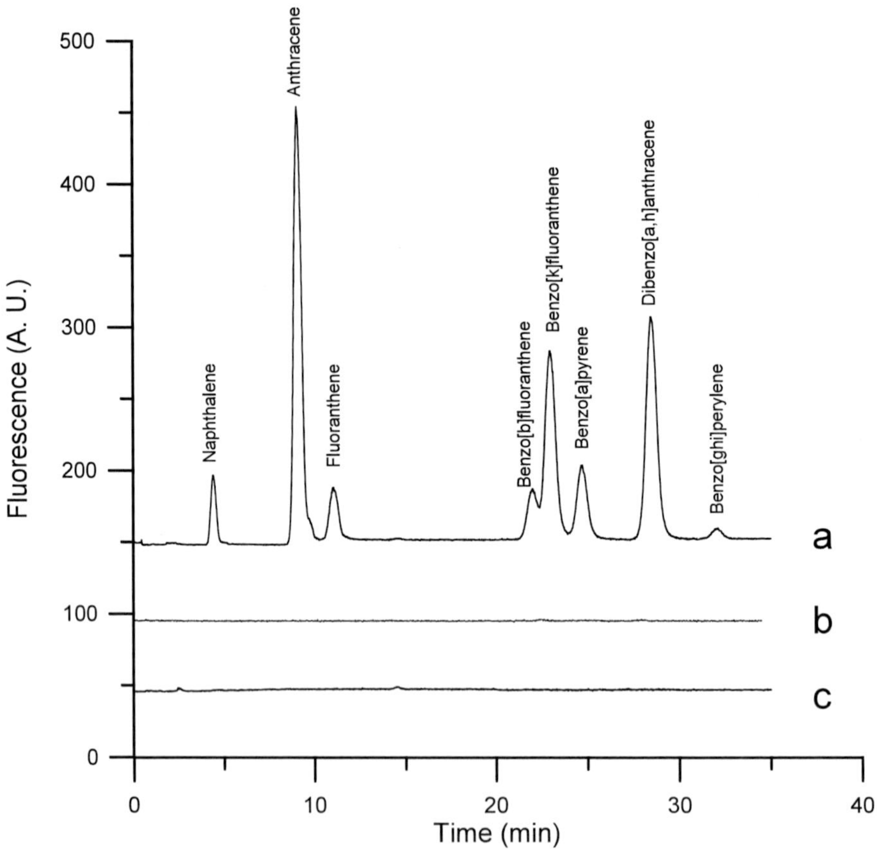

Figure 2. Chromatograms obtained by the optimized SPE-conventional LC-FLD method for: (a) a standard solution containing 20.0 ng/mL of each of the tested PAHs, (b) blank (nanopure water), and (c) a real waste water. For other details, see text.

The recoveries obtained under the proposed method were estimated by comparing the peak areas obtained for each PAH in the samples (5.0 ng/mL, each compound) with those obtained from the direct injection into the analytical column of 2 μL of standard solutions containing an equivalent amount of the analytes. The percentages obtained ranged from 89 to 105 % (Table 2). Data on the intra- and inter-day reproducibilities of the method are summarized in Table 3, whereas the LODs are listed in Table 4.

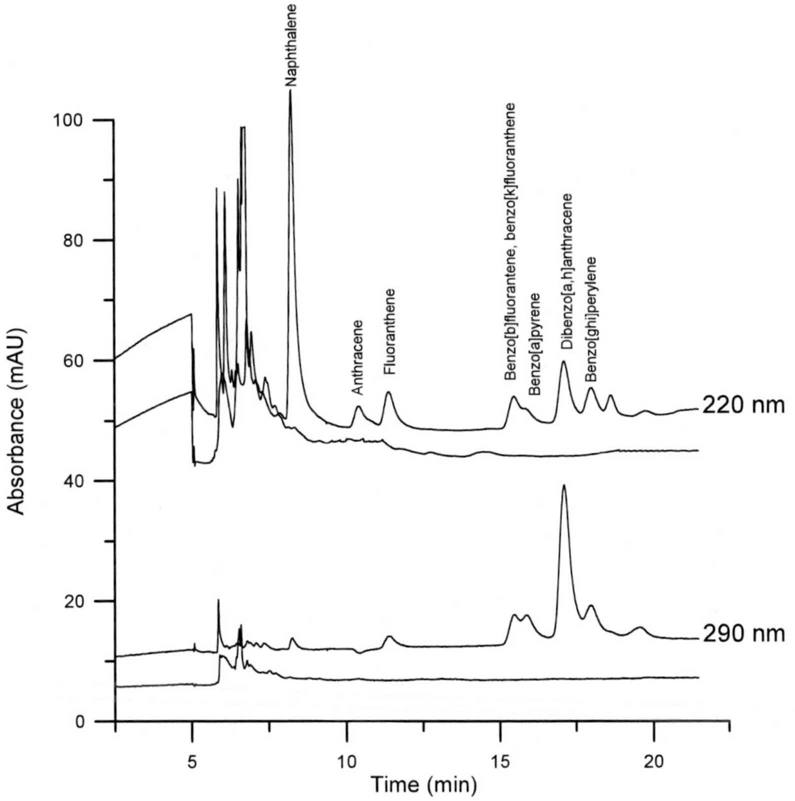

Figure 3. Chromatograms obtained by the optimized column switching capillary LC-UV(DAD) method for a standard solution containing 20.0 ng/mL (up) and for blank (nanopure water) (down) at 220 nm and 290 nm. For other details, see text.

IT-SPME Based Methods

IT-SPME is a well-established technique in which the analytes are extracted and concentrated into the stationary phase of a capillary column by aspirating and dispensing the samples from the capillary a number of times until equilibrium is reached or until sufficient analyte is extracted [30]. Alternatively, a volume as large as necessary of the sample is passed through the capillary until sufficient analyte is extracted to reach the desired sensitivity (one single extraction) [28]. For simplicity, the later approach was used in the present study.

Obviously, in this form of IT-SMPE and for a given concentration of analyte, the responses depend on the sample volume passed through the capillary. On the other hand, according to previously described results the presence of small percentages of an organic solvent (such as acetonitrile) in the sample modifies the retention properties of the capillary coating [28]. Consequently, a systematic study was carried out in order evaluate the effect of the volume of sample to be processed, as well as the usefulness of adding acetonitrile to the samples. The concentrations of the PAHs in the samples were selected to produce peak areas of about the same order for all compounds, and ranged from 0.03 to 0.5 ng/mL. In this study, the extractive capillary column was coupled to a monolithic column, and fluorescence detection was used (see Figure 1b). Volumes ranging from 1.0 to 4.0 mL of standard solutions of the tested PAHs, prepared in water or in different acetonitrile:water mixtures, were passed through the capillary. Then, the solution remaining into the capillary was flushed with 100 μL purified water. Finally, the extractive capillary column and the separative monolithic column were connected by rotating the injection valve.

As illustrative examples, in Figure 4 are depicted the results obtained for three PAHs of different polarity, fluoranthene (high-medium), benzo[*a*]pyrene (low-medium) and benzo[*ghi*]perilene (low). As expected, the increment of the sample volume had a positive effect on analyte responses, and nearly linear increment was observed for all the tested compounds within the tested interval. On the other hand, the addition of acetonitrile to the samples up to percentages of 10-20 % (v/v) increased the analytical signal for the most apolar compounds; this effect was not dependent on the sample volume. For example, a percentage of acetonitrile of 20 % (v/v) in the solutions loaded into the capillary column increased the peak areas of benzo[*ghi*]perilene by a factor of about 2.5. The percentage of acetonitrile that produced maximum analyte responses increased as the polarity of the analyte decreased. For example, maximum analyte responses for benzo[*ghi*]perilene were obtained when the solution loaded into the extractive capillary contained a percentage of acetonitrile of 20 % (v/v), whereas for benzo[a]pyrene a percentage 10 % produced the highest responses. For the most polar analytes such as fluoranthene, the presence of acetonitrile did not improve the signals. Percentages of acetonitrile of 30 % resulted in lower recoveries for all tested PAHs. On the other hand, good stability of the system was observed even when processing 4.0 mL of the samples, and the same capillary column was be repeatedly used without detecting losses in its extractive capabilities.

The results obtained thought out this study demonstrated that with the IT-SPME-LC-FLD method, a sample volume of 2.0 mL would be suitable for the screening analysis of PAHs in waters according to the maximum levels fixed by the European regulations [1]. If required, the sensitivity can be increased by processing larger sample volumes; for the most apolar PAHs, the sensitivity can be further improved by adding acetonitrile to the samples in percentages of about 10-20 % (v/v).

The percentages of PAHs recovered by the in-tube SPME procedure were established for a sample volume of 2.0 mL and without adding acetonitrile. The absolute recoveries were calculated by comparing the peak areas of the chromatograms obtained when injecting into the in-tube SPME device 2.0 mL of standard solutions of the PAHs (1 ng/mL, each compound) with those obtained by injecting directly 20 μL of a solution containing an equivalent amount of each PAH into the LC system. The results obtained are listed in Table 2. As observed, the absolute recoveries ranged from 22 to 45 % which are values typically encountered in this form of SPME [26, 28]. As an illustrative example, in Figure 5 are depicted the chromatograms obtained for a blank and for a standard solution of the PAHs under the optimized IT-SPME-LC-FLD approach. It is interesting to note that for the same mobile-phase flow rate, the employment of a monolithic column instead of a conventional column drastically reduced the time required for the separation. As it can be deduced by comparing Figures 2 and 5, the time required for the chromatographic run could be reduced from 35 min to 12 min, and with a minimum loss of resolution for the tested PAHs.

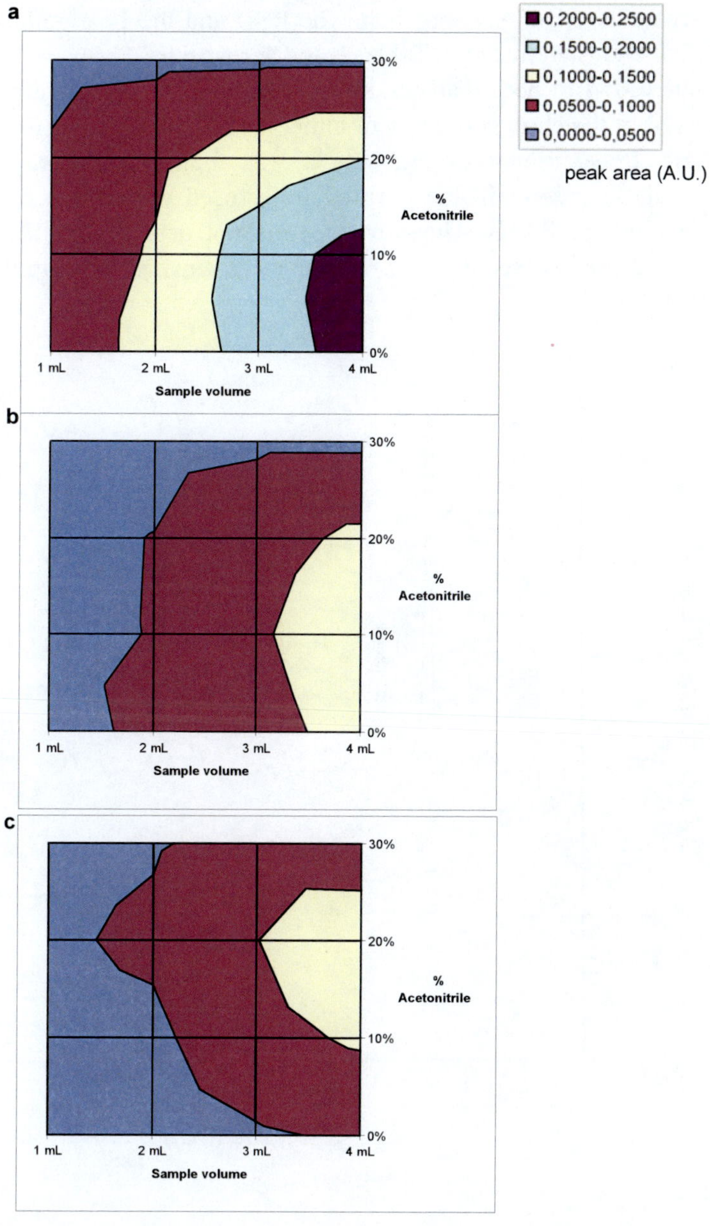

Figure 4. Effect of the sample volume and the percentage of acetonitrile on analyte responses obtained by IT-SPME of: (a) fluoranthene (0.18 ng/mL), (b) benzo[*a*]pyrene (0.074 ng/mL), and (c) benzo[*ghi*]perylene (0.44 ng/mL). For other details, see text.

The IT-SPME device was also coupled to the capillary LC-DAD system. Figure 6 shows the chromatograms obtained under optimized extraction and separation conditions for a blank and for a standard solution of the tested PAHs (1.0 ng/mL). As observed in Figure 6, the chromatograms showed a profile similar to those obtained when by the CS-capillary LC-DAD approach (see Figure 3). However, analyte responses were much higher when applying the IT-SPME-capillary LC-DAD approach. The RSD and the LODs obtained by the IT-SPME-based configurations are listed in Tables 3 and 4, respectively.

It is interesting to note that peaks others than those corresponding to the tested PAHs were observed in the chromatograms of either a blank and a standard solution containing the tested PAHs. In additional experiments it was found that those peaks corresponded to different phtalates present in the purified water used either as a blank and to prepare the working solutions of PAHs. Those peaks were not detected neither when using the more selective FLD detection (see Figure 5), nor the less sensitive CS-capillary LC-DAD approach (see Figure 3).

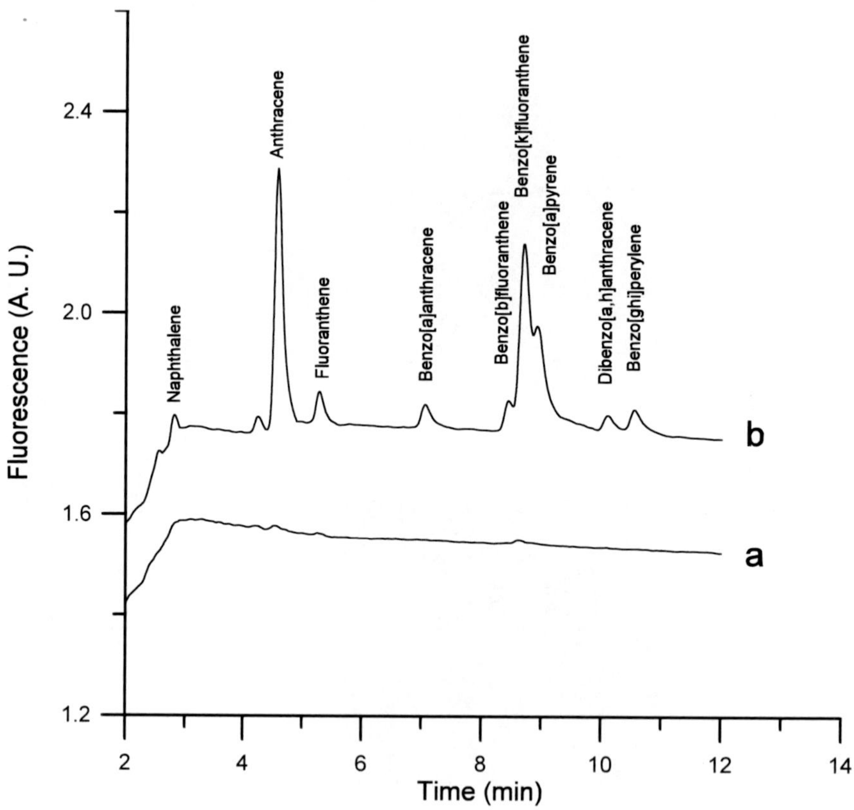

Figure 5. Chromatograms obtained by the optimized IT-SPME-LC-FLD method for: (a) standard solution containing 0.3 ng/mL of each PAHs, and (b) for blank (nanopure water). For other details, see text.

Analysis of Real Samples

Finally, the described methods were applied to the analysis of several samples of river water, sea water, sewage effluent and harbour water. All the analysed samples were collected at different points along the coast of the Comunidad Valenciana (Spain).

None of the PAHs included in this study was found when applying the SPE-conventional LC-FLD and the CS-capillary LC-DAD methods, except naphthalene. This compound was found in one of the harbour waters analyzed at a concentration close to its LOD, thus suggesting a high level of contamination due to fuel spills. In Figure 7a is shown the chromatogram obtained by the CS-capillary LC-DAD method for this sample.

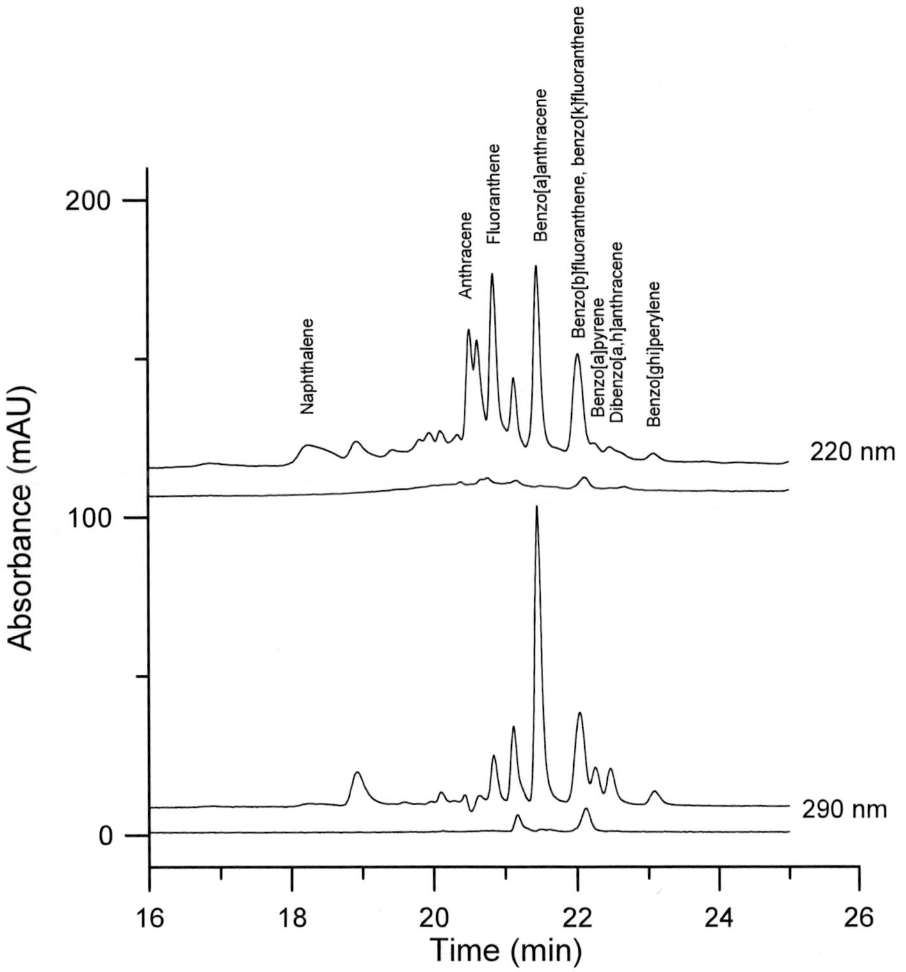

Figure 6. Chromatograms obtained by the optimized IT-SPME-capillary LC-DAD method for a standard solution containing 1.0 ng/mL (up) and for blank (nanopure water) (down) at 220 nm and 290 nm. For other details, see text.

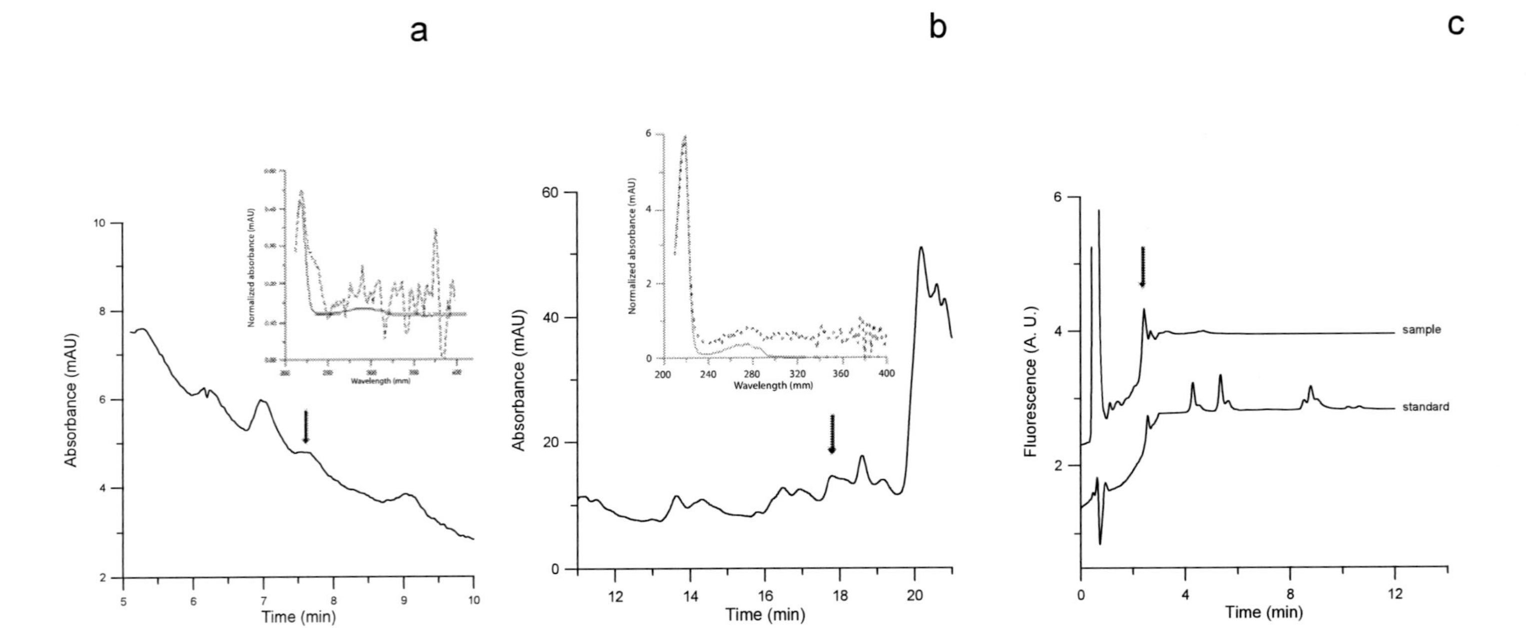

Figure 7. Chromatograms obtained for real water samples: (a) harbour water sample by the CS-capillary LC-UV(DAD) method; (b) sewage water by the IT-SPME-capillary LC-DAD, and (c) sewage water (up) and standard mixture of the tested PAHs (down) by the IT-SPME-LC-FLD methods. Inset: comparison of the UV spectra obtained for the sample (---) and for a standard solution of naphthalene (—). For other details, see text.

Naphtalene and in minor extension fluoranthene were found in some of the samples assayed when applying the IT-SPME-LC-FLD method, although the concentrations found were lower than the expected limits of quantification. In samples that contained the highest levels of naphthalene, this PAH was also detected when applying the IT-SPME-capillary LC-DAD method. This is illustrated for one of the sewage samples analysed in Figures 7b and 7c. The peak corresponding to naphthalene could be easily detected in the IT-SPME-LC-FLD chromatogram not only because of its great intensity, but also because the chromatogram show less peaks corresponding to matrix compounds. The identification of this PAH in the sample was confirmed by the IT-SPME-capillary LC-DAD method by matching the UV spectrum of the peak of suspected naphthalene.

It should be noted that not significant differences were found in the chromatogram profiles obtained for the different types of waters analysed.

Comparison of the Different Extraction/Separation/Detection Approaches Tested

The four methodologies used in the present study are valid alternatives for the analysis of PAHs in waters. All of them are comparable in terms of reproducibility (see Table 3). As observed in Table 2, the absolute recoveries obtained for the tested PAHs were close to 100 % when using either off-line SPE into cartridges and on-line SPE into a trapping column packed with a C_{18} phase. When using IT-SPME for analyte isolation and preconcentration the alsolute recoveries obtained were significantly lower due to the modality of IT-SPME used. However, the total amounts of PAHs retained were higher as this form of IT-SPME is compatible with the injection of much larger sample volumes. As a consequence, the sensitivity ca be considerably improved (Table 4).

The LODs obtained by using off-line SPE without solvent evaporation combined with conventional LC-FLD are comparable to those obtained by the CS-capillary LC-DAD method (see Table 4). In other words, although less sensitive detection is used the coupling of analyte enrichment by CS and capillary LC enables the detection of PAHs with sensitivity comparable to that obtained by the conventional method. The main advantage of the CS-capillary LC-DAD approach is that off-line preconcentration operations are not required. According to the literature both approaches can be considered adequate for the analysis of PAHs in waste water and polluted waters [4, 5, 16, 17].

As regards IT-SPME, with the extractive capillary column used in the present study at least up to 4.0 mL of the samples can be processed resulting in maximum analyte responses. Although not tested, the sensitivity would be most probably increased by injecting larger sample volumes. The sensitivity can be further improved for the most apolar PAHs by adding acetonitrile to the samples in final percentages of about 20 % (v/v). When coupling IT-SPME to LC-FLD the direct processing of only 2.0 mL of the samples provided LODs clearly lower than the maximum concentrations fixed by the European regulations for natural waters [1]. Although less sensitive, the IT-SPME-capillary LC-DAD method offers improved confirmation of the analytes by matching the UV spectra. The application of this method to control and screening analysis of PAHs in natural waters would required the utilization of larger sample volumes, probably combined with the addition of an organic solvent such as acetonitrile to the samples. On the other hand, the employment of a monolithic column for the chromatographic separation results in a very simple and fast method, particularly well-suited for routine screening analysis.

In Table 5 are also compared other relevant features of the four methodologies used in present study. When considering the off-line sample manipulation involved in the analysis, the three on-line methods proposed are clearly advantageous over the traditional approach, as samples only need to be filtered. Moreover, the employment of organic solvents for analyte enrichment and preconcentration is totally avoided. The consumption of extractive phases is also reduced to a minimum. For example, more than 500 real water samples have been processed in our lab by the IT-SMPE-LC-FLD method with the same extractive capillary column and without observing deterioration on its retention capabilities.

Indeed, a modification of the chromatographic system is required if the on-line methods are going to be applied; in the conventional off-line method the chromatographic system does not need to be changed, but additional instrumentation for sample treatment such as a vacuum manifold was used. The CS-capillary LC-DAD approach requires a capillary trapping column and an additional switching microvalve over an standard capillary LC equipment. The IT-SPME based methods only require the replacement of the inner loop of the injection valve by a capillary extractive column. Thus, the implementation of the IT-SPME-LC-FLD method only requires a minimum modification of conventional LC equipments. Thererefore, this is the most cost-effective approach.

Table 5. Comparison of the tested methods. The higher the number of asterisks, the better is the method

Features	SPE-LC-FLD	CS-capillary LC-DAD	IT-SPME-capillary LC-DAD	IT-SPME-LC*-FLD
Sensitivity	*	*	**	***
Time for sample preparation	*	**	***	***
Total time of analysis	*	**	**	***
Off-line manipulation	*	***	***	***
Consumption of solvents and extractive phases	*	**	***	***
Instrumental requirements	**	*	**	***
Cost of analysis	*	**	**	***

(*) LC with a monolithic column

Conclusion

This chapter shows the reliability of different alternatives for the LC analysis of common PAHs in water samples that have been developed taking into account novel tendencies of Analytical Chemistry either for sample preparation or for analyte separation: CS-capillary LC-DAD, IT-SPME-capillary LC-DAD and IT-SPME-LC with a monolithic column-FLD. Such methods are clearly advantageous over a classical procedure based on SPE with disposable cartridges for a number of reasons: *(i)* the minimum off-line sample manipulation, as the samples only needed to be filtered; *(ii)* the rapidity, as on-line analyte enrichment is accomplished in a few min, *(iii)* the minimum consumption of extractive phases, as the extractive columns used (a C_{18} packed microcolumn or a segment of a capillary GC column) can be reutilized for several hundred samples; and *(iv)* the total elimination of organic solvents in sample preparation.

The CS-capillary LC-DAD approach provides sensitivity adequate for the routine monitoring analysis of PAHs in sewage effluents, and identity confirmation by matching the UV spectra. Capillary LC-DAD is also compatible with IT-SPME, resulting in improved sensitivity. The best sensitivity is achieved when using FLD, even when combined with normal scale LC. The total analysis time can be further reduced by using a monolithic column for separation, thus resulting in a very rapid, sensitive, simple and cost- effective method.

It can be concluded that the on-line methods described in this chapter are reliable alternatives to the classical methodologies for controlling PAHs pollution in waters, and more attractive in the context of a Green Chemistry.

Acknowledgments

The authors are grateful to the Spanish Ministerio de Educación y Ciencia (Project CTQ2008-01329/BQU) and to the Conselleria de Territorio y Vivienda de la Generalitat Valenciana (Application of EU Water Framework Directive 2000/60/EC on endocrine disruptors and other priority substances in coastal areas in the Comunidad Valenciana) for the financial support received.

References

[1] European Community Directive 98/80/EC, Brussels, November 1998.

[2] Miège, C.; Bouzige, M.; Nicol, S.; Dugay, J.; Pichon, V.; Hennion, M. C. *J. Chromatogr. A* 1999, *859*, 29-39.

[3] Martínez, E.; Gros, M.; Lacorte, S.; Barceló, D. *J. Chromatogr. A* 2004, *1047*, 181-188.

[4] Filipowska, A.; Lubecki, L.; Kowalewska, G. *Anal. Chim. Acta* 2005, *547*, 243-254.

[5] Bussetti, F.; Heitz, A.; Cuomo, M.; Badoer, S.; Traverso P. *J. Chromatogr. A* 2006, *1102*, 104-115

[6] Zhou, Y.-Y.; Yan, X.-P.; Kim, K.-N.; Wang, S.-W.; Liu, M.-G. *J. Chromatogr. A* 2006, *1116*, 172-178.

[7] Gimeno, R. A.; Altelaar, A. F. M.; Marcé, R. M.; Borrull, F. *J. Chromatogr. A* 2002, *958*, 141-148.

[8] Oliferova, L.; Statkus, M.; Tsysin, G.; Shpigun, O.; Zolotov, Y. *Anal. Chim. Acta* 2005, *538*, 35-40.

[9] Wang, X.; Mitra, S. *J. Chromatogr. A* 2006, *1122*, 1-6 .

[10] Rezaee, M.; Assadi, Y.; Hosseini, M.-R. M.; Aghaee, E.; Ahmadi, F.; Berijani, S. *J. Chromatogr. A* 2006, *1116*, 1-9.

[11] Hsieh, Y.-N.; Huang, P.-C.; Sun, I.-W.; Whang, T.-J.; Hsu, C.-Y. ; Huang, H.-H.; Kuei, C.-H. *Anal. Chim. Acta* 2006, *557*, 321-328.

[12] El-Beqqali, A.; Kussak, A.; Abdel-Rehim, M. *J. Chromatogr. A* 2006, *1114*, 234-238.

[13] Negrão, M. R.; Alpendurada, M. F. *J. Chromatogr. A* 1998, *823*, 211-218.

[14] Djozan, D.; Assadi, Y. *Microchem. J.* 1999, *63*, 276-284.

[15] Doong, R.; Chang, S.; Sun, Y. *J. Chromatogr. A* 2000, 8*79*, 177-188.

[16] Popp, P.; Bauer, C.; Möder, M.; Paschke, A. *J. Chromatogr. A* 2000, *897*, 153-159.

[17] King, A. J.; Readman, J. W.; Zhou, J. L. *Anal. Chim. Acta* 2004, *523,* 259-267.

[18] Hu, Y.; Fu, Y.; Li, G.-K. *Anal. Chim. Acta* 2006, *567*, 211-217.

[19] León, V. M.; Álvarez, B.; Cobollo, M. A.; Muñoz, S.; Valor, I. *J. Chromatogr. A* 2003, *999*, 91-101.

[20] León, V. M.; Llorca-Pórcel, J.; Álvarez, B.; Cobollo, M. A.; Muñoz, S.; Valor, I. *Anal. Chim. Acta* 2006, *558*, 261-266.

[21] Titato, G. M.; Lancas, F. M. *J. Liq. Chromatogr. Relat. Techno.* 2005, *28*, 3045-3056.

[22] He, Y.; Lee, H. K. *J. Chromatogr. A* 1998, *808*, 79-86.

[23] Oliferova, L.; Ststkus, M.; Tsysin, G.; Zolotov, Y. *Talanta* 2007, *72*, 1386-1391.

[24] Lim, L. W.; Okouchi, Y.; Takeuchi, T. *Talanta* 2007, *7*, 1600-1608.

[25] Cháfer-Pericás, C.; Herráez-Hernández, R.; Campíns-Falcó, P. *J. Chromatogr. A* 2006, *1125*, 159-171.

[26] Cháfer-Pericás, C.; Herráez-Hernández, R.; Campíns-Falcó, P. *J. Chromatogr. A* 2007, *1141*, 10-21.

[27] Olejniczak, J.; Staniewski, J. *Anal. Chim. Acta* 2007, *588*, 64-72.

[28] Campíns-Falcó, P.; Verdú-Andrés, J.; Sevillano-Cabeza, A.; Molins-Legua, C.; Herráez-Hernández, R. *J. Chromatogr. A* 2008, *1211*, 13-21.

[29] Klodzińska, E.; Moravcova, D.; Jandera, P.; Buszewski, B. *J. Chromatogr. A* 2006, *1109*, 51-59.

[30] Kataoka, H. *Anal. Bioanal. Chem* 2002, *373*, 31-45.

In: Polycyclic Aromatic Hydrocardons: Pollution, Health... ISBN 978-1-60741-462-9
Editors: Pierre A. Haines and Milton D. Hendrickson

Chapter 8

PAHS IN SEDIMENTS ASSOCIATED WITH COAL AND COAL-DERIVED PARTICLES -- OCCURRENCE, MOBILITY AND RISK ASSESSMENT

***Yi Yang*[1] *and Thilo Hofmann*[2]**

[1]School of Resources and Environmental Sciences, East China Normal University
[2]Department of Environmental Geosciences, Vienna University

ABSTRACT

Polycyclic aromatic hydrocarbons (PAHs) are ubiquitous hydrophobic organic contaminants in the environment. They tend to be associated with particles and are widely transported by flooding and atmospheric pathways, resulting in elevated concentrations in sediments/soils. Coal and coal-derived particles in natural sediments/soils can act not only as strong sinks for the PAHs, but also as very important sources of PAHs in sediments/soils. The understanding of the mobility of these contaminants from the sediments/soils, especially the sequestration of PAHs by coal and coal-derived particles is very important, because they can control the transportation, bioavailability, degradation and hence the potential risk of these contaminants in the environment.

1. INTRODUCTION

Polycyclic aromatic hydrocarbons (PAHs) are of great concern, due to their persistence, toxicity and bioaccumulation. Once occurred in the aquatic system, these compounds tend to bound to particles because of their hydrophobic nature, and consequently accumulate in sediments. Thereby, sediments act as the important sink of these persistent organic pollutants (POPs), and pose the potential risk to the aquatic system. Coal and coal-derived particles with elevated PAHs concentrations produced by coal industry can be wild spread in the aquatic system through atmospheric transportation, runoff and flooding etc. Coal and coal-derived particles in natural sediments can act not only as strong sinks for the PAHs, but also as very important sources of PAHs in sediment. The occurence and mobility of these contaminants from the coal and coal-derived particles in sediments/soils is of significance, because sorption/desorption, especially sequestration of PAHs by these coal and coal-derived particles in sediments/soils can control their transportation, bioavailability, degradation and hence the potential risk in the environment.

2. Occurrence of Coal and Coal-Derived Particle Associated PAHs in Sediments

2.1. Identification of Coal and Coal-Derived Particles in Sediments

Hard coal product has increased from less than 1 billion ton to almost 5 million ton from 1900 to 2005 [1]. Due to coal mining, coal industry, atmospheric transport, runoff, and flooding etc, unburnt coal and coal-derived particles can be wild spread in the aquatic system, and consequently settled in sediments. For example, unburnt coal particles can be released by open pit mining; coal stored at industrial sites for the production of coke, gas or steam, is subjected to erosion; moreover, coal naturally eroded into aquatic systems for sedimentary rock outcroppings containing coal seams. Occurrence of coal in sediments was reported from harbors such as Hamilton Harbour [2] and Roberts Bank coal terminal, in Canada [3]. Coal particles present in sediments made up 10.5 to 11.9 % dry weight of the soil mass in the vicinity of coal-loading terminals and was reported as non-hydrolysable solids.

The classic method to identify coal and coal-derived particles in sediments is the coal petrography. Using optical microscopy enables the observation, identification, classification and quantification of coal and coal derived particles in natural sediments [4-8]. In a recent study, about 75 vol-% of the light fraction (i.e., the fraction with a density < 2 g/cm³, predominantly organic carbon) in a river floodplain soil were identified as hard coal particles [9]. Photomicrographs in Figure 1 show different coal and coal-derived particles identified by organic petrography techniques in sediment samples collected from Mosel River bank, where intensive coal mining activities happened in the former time.

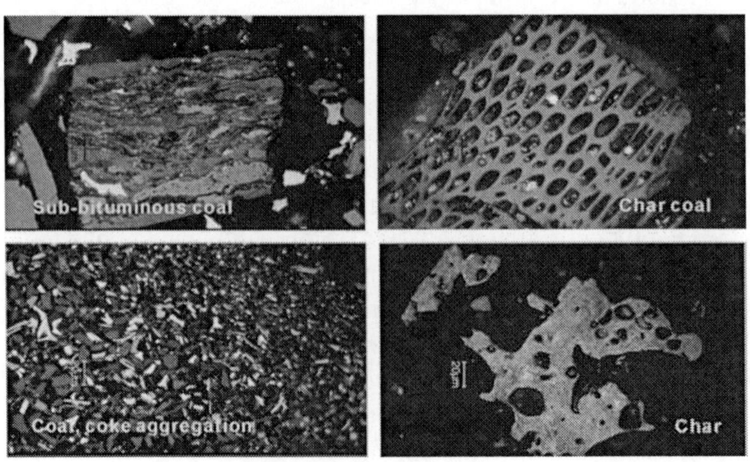

Figure 1. Coal and coal-derived particles identified by organic petrography techniques.

2.2. Elevated Concentration of Pahs in Sediments with Coal and Coal-Derived Particles

Coal is defined as a sedimentary rock containing more than 70% by volume of carbonaceous material with a three-dimensional macromolecular matrix 'network'. In addition to the network structure, a multitude of small molecules, i.e. a 'mobile phase', is present within the network and is of particular environmental interest. These molecules can be released from the coal network more rapidly because they are less bound to the macromolecular matrix compared to the cross-linked molecules within the network. The number of conjugated aromatic rings per structural unit within the matrix increases with increasing coalification (maturity), ultimately resulting in graphite [10].

PAHs were often determined in hard coals with the concentration up to hundreds, and even few thousands of mg/kg. The characteristics of coal, including its PAH content and distribution pattern depend on (1) original biological raw materials (coal type) and (2) the degree to which the coal-bearing strata have been altered by the affects of pressure and temperature associated with burial over geologic time (coal rank) [11-15].

Due to the low density, coal particles are preferentially transported and can accumulate in remote areas, resulting in elevated PAH concentrations in sediments [4, 15]. For example, up to 80 mg/kg EPA-PAHs were found in bank sediment samples from the Saar and Mosel River (Figure 2). The light fraction, dominated by coal and coal-derived particles was collected by density separation. Although contributing <5% of total sediment by weight, this light fraction concentrated 75% of total PAHs in the sediment, with the PAHs concentration of 460 mg/kg [9]. The results showed that predominantly former coal mining activities in the Saarland, Germany, upstream of the sampling sites, are responsible for the PAH contamination in the downstream river sediments and floodplain soils. These studies confirmed that coal mining had resulted in a serious problem of an extensive PAH contamination in river floodplain sediments.

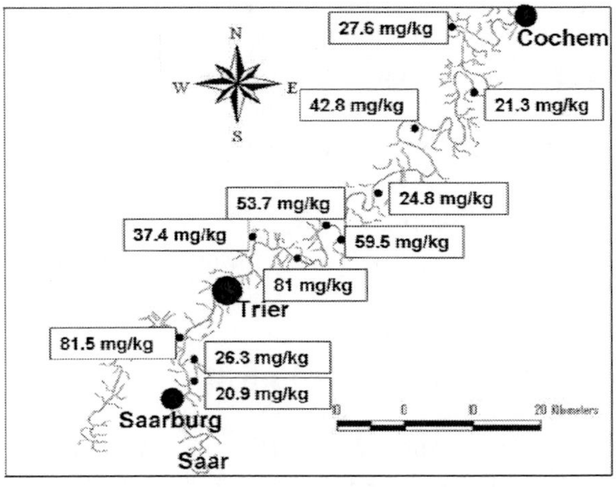

Figure 2 Elevated PAH concentrations (sum of EPA-PAHs, 1- and 2- methylated naphthalenes and perylene) in Saar and Mosel River floodplain soils, Germany [16]. The flow direction is from SW to NE and the Saar coal mining area is near Saarburg. Coal transportation by ships occurred from the mining area via the Saar River, towards NE into the Mosel River and further into the Rhine River.

The most obvious characteristics of native PAH from coals are the predominance of naphthalene, phenanthrene and their alkylated derivatives, however, these are also typical of PAH from other petrogenic sources such as oil. The PAH patterns of the coals investigated in the aftermath of the *Exxon Valdez* oil spill were also dominated by naphthalene, phenanthrene and alkylated derivatives [17, 18]. EPA-PAH concentrations are estimated at approx. 500 mg/kg in coals from the Kulthieth Formation [19]. In these coals, the presence of naphthalene, C1 and C2 naphthalenes tended to be greater than in seep oils, whereas the higher molecular weight PAHs were more abundant in seep oils. Low molecular weight PAH can easily volatilize during sample preparation, e. g. if the extract is dried, which was not the case in this study. Naphthalenes volatilize more easily from oil or tar compared to coal since coal is a very strong sorbent. Therefore, the decrease in naphthalene concentrations in soils and sediments, after being subject to gentle evaporation, could be regarded as an indicator for oil-bound PAH but not coalbound PAH. At the Mosel River, Germany, naphthalenes were also shown to be the key compounds for source apportionment of PAH. Native PAH in Saar coals (70 -165 mg/kg EPA-PAH) are the major reason for elevated PAH concentrations in Saar and Mosel River floodplain soils downstream from mining activity [16]. Although major mining activity occurred between 200 to 50 years ago, naphthalenes and alkylated derivatives still dominate the PAH patterns in the floodplain soils [20].

3. Mobility of PAHs in Sediments with Coal and Coal-Derived Particles

The classic partitioning theory implies a linear sorption of hydrophobic organic contaminants (HOCs) and is believed to be restricted to “normal” organic matter (NOM) of the soil [21-23]. These amorphous “soft” organic materials include gel-like humic and fulvic substances, as well as lignin which, in general, is regarded to be geologically young [24-26]. Recently, carbonaceous materials including coal and coal-derived particles occurred in sediments have increasingly gained attention because they play an important role in the sorption/desorption of HOCs, such as PAHs. Nonlinear sorption isotherms and sorption capacities above the partitioning limit are often reported in the literature with respect to carbonaceous materials [5, 27]. Furthermore, these materials are suspected of causing high sorption enthalpies, competitive sorption, slow desorption kinetics, sorption-desorption hysteresis, and hence low bioavailability of HOCs in soils and sediments [26, 28-30].

3.1. Pahs Sorption to Carbonaceous Sediments

A simple and widely adopted model to describe the partitioning of a substance between the aqueous and the OC phase makes use of the partitioning coefficient (K_p):

$$C_s = K_p C_w = f_{oc} K_{oc} C_w \tag{1}$$

where C_s and C_w are the equilibrium solid phase and aqueous phase solute concentrations, respectively and f_{oc} is the fraction of OC. If normalized to OC, this distribution coefficient

(K_{oc}) for water is often constant for a wide variety of geosorbents and under various aqueous phase solute concentration conditions. K_{oc} can be estimated from the solubility (S) and octanol-water partitioning coefficient (K_{ow}) of the target compound. For many compounds of low solubility, the K_{ow} seems more reliable or more frequently available than S.

The empirical Freundlich model is widely used to describe nonlinear sorption:

$$C_s = K_{Fr} C_w^n \tag{2}$$

where K_{Fr} [e. g. mg kg^{-1}/ (mg L^{-1})$^{1/n}$] and n [-] are the Freundlich sorption coefficients and the Freundlich exponent, respectively. K_{Fr} ($K_{oc}' = K_{Fr} / f_{oc}$).

Modified Freundlich sorption models were developed to describe sorption in heterogeneous geosorbents containing both NOM and carbonaceous materials. Composite or distributed reactivity models (DRM) were introduced [28] and applied in a manner which combines the linear partitioning (absorption) part with the nonlinear adsorption part [26, 27, 31, 32]. The overall sorption isotherm can be described in the form of:

$$C_s = X_L K_{p,L} C_w + \sum_{i=1}^{m} X_{NL}^i K_{Fr}^i C_w^{n_i} \tag{3}$$

where X_L and X_{NL} are the mass fractions of the solid phase exhibiting linear and nonlinear sorption behavior, respectively. m is the number of discrete reactive sorption domains. Accardi-Dey and Gschwend [27] successfully applied a BC-inclusive Freundlich sorption model in a sorption experiment of pyrene in the presence of environmental BC by the following term:

$$C_s = f_{OC} K_{OC} C_w + f_{BC} K_{BC} C_w^n \tag{4}$$

where f_{BC} and K_{BC} are the sediment mass fraction of BC, and the Freundlich BC-water distribution ratio, respectively. Furthermore, "soft carbon" and "hard carbon" were defined by Weber [28] in relation to their physical chemical characteristics. "Soft carbon", such as geologically younger humic type OM, can be hypothesized as amorphous and swollen SOM, analogous to a rubbery polymer showing low sorption enthalpies, linear sorption behavior, noncompetitive exchange, rapid kinetics, and no sorption-desorption hysteresis, whereas "hard carbon" characterized of high sorption enthalpies, nonlinear sorption behavior, competitive exchange, slow kinetics, and possible sorption-desorption hysteresis can be envisioned as condensed and relatively rigid organic matrix, analogous to a glassy polymer, such as kerogens and coals [26, 33, 34].

The Polanyi-Manes model is postulated to follow a pore-filling mechanism, which was first applied by Xia and Ball [35], and later applied by other research groups [34, 36] to describe sorption of several HOCs by selected natural soils and sediments. The Polanyi adsorption model originally was set up for the quantification of the adsorption of gas molecules to energetically heterogeneous solids, and was extended to a wide range of vapor and liquid phase systems by Manes and his co-workers. The Polanyi theory considers that, for a molecule located within the attractive force field of a micro-porous solid, there exists an adsorption potential (ε), which is defined as the energy level required removing the molecule from the location to a point outside the attractive force field of the solid surface. For adsorption of partially miscible solutes from aqueous solution, the effective adsorption potential (ε_{sw}, cal/mol) can be defined as:

$$\varepsilon_{sw} = RT\ln(S_w/C_{eq}) \tag{5}$$

where S_w is the water solubility and C_{eq} the aqueous solute equilibrium concentration. Dubinin suggested that a plot of adsorbate volume against adsorption potential density (adsorption potential divided by sorbate molar volume). Following, Crittenden et al., suggested the following empirically derived relation between adsorbed volume and adsorption potential (ε):

$$\log(Cs) = \log(V_0) + a(\varepsilon_{sw}/V_s)^b \tag{6}$$

where V_O is the adsorption volume capacity at saturation per unit mass of sorbents (cm^3/kg), V_S is the molar volume of the solute (cm^3/mol) and the exponent b is often set to an integer. For $b = 2$, the Dubinin Radushkevich equation is obtained, which corresponds to a log-normal distribution of the sorption energies [37]. Exponents of b larger than 2 are commonly found in some activated carbons showing a Weibull pore size distribution [38]. Hence, for heterogeneous geosorbents, the combined adsorption-partitioning model is described by a Polanyi-Manes type of modeling approach:

$$C_s = V_0 \times 10^{a(\varepsilon_{sw}/V_s)^b} \times \rho + f_{oc}K_pC_w \tag{7}$$

where a can be calculated as $-\frac{1}{2.3}(\frac{V_s}{E})^b$.

Kleineidam et al. [36] used this combined model in the form of:

$$C_s = V_0 \times \exp[\frac{-RT(-\ln(\frac{C_w}{S}))}{E}] \times \rho + f_{oc}K_pC_w \tag{8}$$

In a recent study, sorption experiments have been performed with coal impacted sediment samples. The light fraction (ρ <2 g/cm^3) showed high sorption capacity comparable to low rank coals [39] which was about a factor of 100 higher than that of the heavy, predominantly mineral fraction of the sediment. Sorption was strongly non-linear and the combined partitioning and pore filling model gave a better fit than the Freundlich sorption model (Figure 3). Adsorption (pore-filling) was responsible for 80 – 90 % of total sorption at C_w/S_{scl} < 0.001, and then decreased with higher concentrations towards aqueous solubility where partitioning is dominant. Moreover, the *log* K_{oc}' and n values yielded by Freundlich model in this study were around 4.3 and 0.75, in the range of those for carbonaceous materials reviewed by Cornelissen et al [30]. The highest K_{Fr}, in the light fraction with highest amount of coal and coal-derived particles and the similar K_{oc}' values in all fractions suggested that these carbonaceous coal and coal-derived particles presented in samples could be the dominant geosorbents for PAH sorption.

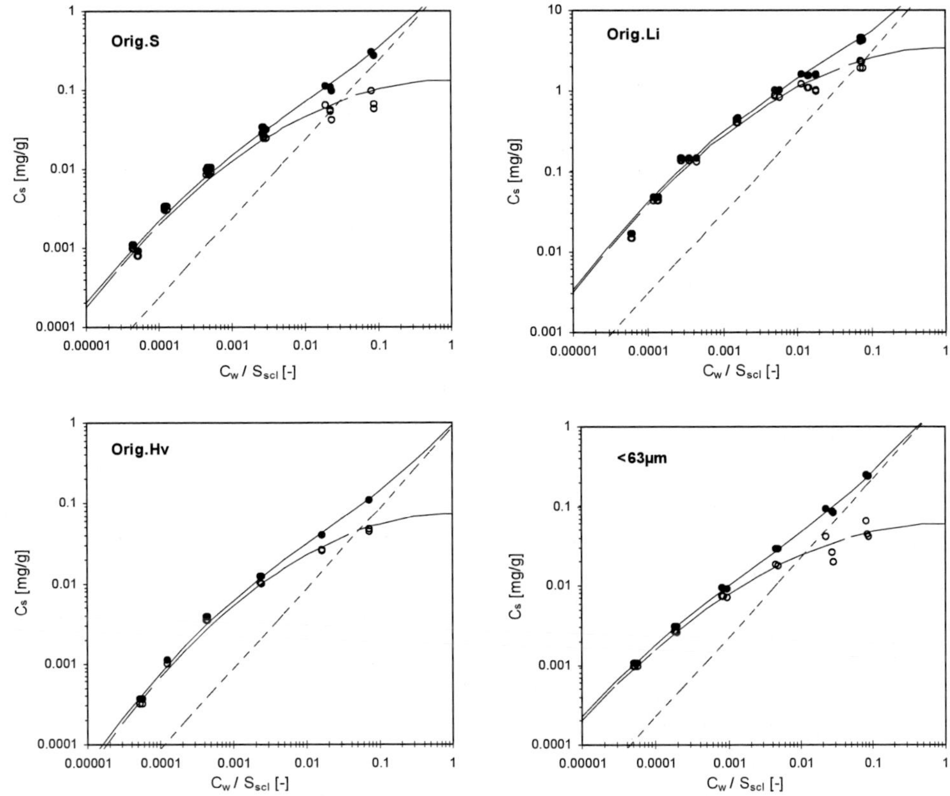

Figure 3 Combined pore-filling and partitioning sorption isotherm for phenanthrene (eq 6). Solid line: combined isotherm; dashed line: pore-filling part; dotted line: partitioning part. The partitioning part was predicted on the basis of the empirical relationship in eq 2.; open symbols indicate calculated data for the pore filling part. Orig.S: original soil sample; Orig.Li: light fraction (ρ <2 g/cm^3); Orig.Hv: heavy fraction (ρ <2 g/cm^3); <63µm: <63µm fraction.

3.2. Slow Desorption

The mobility and bioavailability of hydrophobic organic contaminants (HOCs) in aquatic systems, such as PAHs, depend on desorption of these compounds from the solid phase to water. In several studies a two-stage desorption kinetics were observed, in which rapid desorption was followed by slow release [40, 41]. Furthermore, a subsequent very slow desorption stage was defined by Cornelissen et al. [42], further supporting the hypothesis that bioavailability is limited by mass transfer kinetics [43]. The slow and very slow desorption of contaminants from natural sorbents has gained much attention, because this could be a severe problem in remediation that challenges concepts concerning cleanup standards and risks [29].

Extremely slow desorption of PAHs from carbonaceous particles were reported [44]. Moreover, the similar desorption behavior was found in natural sediments associated with coal and coal-derived particles [45]. In both studies, resulting desorption rate constants were in the order of 10^{-4} - 10^{-3} for k_1 and 10^{-6} - 10^{-5} for k_2, which is one order of magnitude lower than the slow and very slow desorption rate reported in other studies [42, 46]. These previous studies used Tenax beads and observed rate constants of 10^{-1} - 10^{0} for fast desorption, 10^{-3} - 10^{-2} for slow desorption and 10^{-5} - 10^{-4} for very slow desorption. Consequently no "fast" desorption of PAHs could be observed coal and coal-derived particles. Desorption rates of PAHs displayed "very slow" or even "extremely slow" desorption.

Coal and coal-derived particles can be responsible for the extremely slow desorption of PAHs at study area impacted by coal mining activities. Two physical interpretations explain very slow desorption: (1) thermodynamically controlled release from high energy adsorption sites and (2) retarded diffusion through the rigid sorbent matrix or microporous organic material [47-49]. These PAHs displayed little tendency to desorb with time and indicated effective sequestration or physical entrapment of PAHs in coal and coal-derived particles [20, 44]. As for coal particles, only PAH on or near the surface of the materials were available, while PAHs embedded inside cavities or pores of the cross-linked network of coal are probably not able to be reached by water. In addition, physical entrapment of PAHs could occur in ink-bottle shaped or sorbate-deformed pores in the other coal-derived materials [50]. These PAHs can be only extracted by organic solvents by swelling [44] or enhanced extraction techniques like ASE.

4. Risk Assessment Considering Coal and Coal-Derived Particles

Sorption/desorption concerned bioavailability is the major issue governing degradation and ecotoxicity of PAHs in soil/sediment system, moreover the slow desorption is involved in limiting the bioavailability and then the biodegradation of contaminated soils/sediments. Coal and coal-derived particles occurred in sediments show sequestration of PAHs with large sorption capacities, strong sorption affinities, and extremely slow desorption kinetics. Sequestration of PAHs by these particles indicates a lower (bio)availability and reduced environmental risk, although total (solvent) extractable concentrations are high. Therefore, for the establishment and development of sediment quality criteria (SQC) it is important to consider coal and coal-derived particles which occurred in natural sediments.

References

[1] Thielemann T., Schmidt S., Gerling C. *Int. J. Coal Geol.*, 2007, *72*,1-14.

[2] Curran K. J., Irvine K. N., Droppo I. G., et al. *Journal of Great Lakes Research*, 2000, *26*,18-30.

[3] Johnson R., Bustin R. M. *Int. J. Coal Geol.*, 2006, *68*,57-69.
[4] Ghosh U., Gillette S., Luthy R. G., et al. *Environ. Sci. Technol.*, 2000, *34*,1729-1736.
[5] Karapanagioti H. K., Childs J., Sabatini D. A. *Environ. Sci. Technol.*, 2001, *35*,4684-4690.
[6] Karapanagioti H. K., Sabatini D. A. *Environ. Sci. Technol.*, 2000, *34*,2453-2460.
[7] Kleineidam S., Rügner H., Ligouis B., et al. *Environ. Sci. Technol.*, 1999, *33*,1637-1644.
[8] Ligouis B., Kleineidam S., Karapanagioti H. K., et al. *in Environmental Chemistry*, 2005.
[9] Yang Y., Ligouis B., Pies C., et al. *Environmental Pollution*, 2008,doi:10.1016/j.envpol.2007.02.020.
[10] Stout S. A., Uhler A. D., McCarthy K. J., et al. *In: Introduction to Environmental Forensics*, 2002, *Murphy B. L., Morrison R. D., Ed.; Elsevier Academic Press: Amsterdam*,137-260.
[11] Radke M., Leythaeuser D., Teichmüller M. *Org. Geochem.*, 1984, *6*,423-430.
[12] Radke M., Welte D. H., Willsch H. *Org. Geochem.*, 1986, *10*,51-53.
[13] Radke M., Willsch H., Leythaeuser D. *Geochimica et Cosmochimica Acta*, 1982, *46*,1831-1848.
[14] Radke M., Willsch H., Teichmüller M. *Organic Geochemistry*, 1990, *15*,539-563.
[15] Stout S. A., Emsbo-Mattingly S., Uhler A. D., et al. *Soil Sediment and Water*, 2002,http://www.aehsmag.com/issues/2002/june/forensics.htm.
[16] Pies C., Yang Y., Hofmann T. *J. Soils & Sediments*, 2007, *7*,216-222.
[17] Boehm P. D., Page D. S., Burns W. A., et al. *Environ. Sci. Technol.*, 2001, *35*,471-479.
[18] Short J. W., Kvenvolden K. A., Carlson P. R., et al. *Environ. Sci. Technol.*, 1999, *33*,34-42.
[19] Van Kooten G. K., Short, J.W., Kolak, J.J. *Environ. Forensics*, 2002, *3*,227-241.
[20] Yang Y., Ligouis B., Pies C., et al. *Chemosphere*, 2008, *77*,2158-2167.
[21] Chiou C. T., Peters L. J., Freed V. H. A. *Science*, 1979, *206*,831.
[22] Karickhoff S. W. 1980, *2*,193.
[23] Schwarzenbach R. P., Gschwend P. M., Imboden D. M., *Environmental Geochemistry*. 1993, New York: John Wiley.
[24] Allen-King R. M., Grathwohl P., Ball W. P. *Advances in Water Resources*, 2002, *25*,985-1016.
[25] Huang W., Weber J., W.J. *Environ. Sci. Technol.*, 1997, *31*,2562-2569.
[26] Huang W., Young T. M., Schlautman M. A., et al. *Environ. Sci. Technol.*, 1997, *31*,1703-1710.
[27] Accardi-Dey A., Gschwend P. M. *Environ. Sci. Technol.*, 2003, *37*,99-106.
[28] Weber W. J., McGinley P. M., Katz L. E. *Environ. Sci. Technol.*, 1992, *26*,1955-1962.
[29] Luthy R. G., Aiken G. R., Brusseau M. L., et al. *Environ. Sci. Technol.*, 1997, *31*,3341-3347.
[30] Cornelissen G., Gustafsson Ö., Bucheli T. D., et al. *Environ. Sci. Technol.*, 2005, *39*,6881-6895.
[31] Accardi-Dey A., Gschwend P. M. *Environ. Sci. Technol.*, 2002, *36*,21-29.
[32] Cornelissen G., Gustafsson Ö. *Environ. Sci. Technol.*, 2004, *38*,148-155.

[33] Weber Jr. W. J., Huang W. *Environ. Sci. & Technol.*, 1996, *30*,881-888.
[34] Ran Y., Xing B., Rao P. S. C., et al. *Environ. Sci. Technol.*, 2004, *38*,4340-4348.
[35] Xia G., Ball W. P. *Environ. Sci. Technol.*, 1999, *33*,262-269.
[36] Kleineidam S., Schüth C., Grathwohl P. *Environ. Sci. Technol.*, 2002, *36*,4689-4697.
[37] Condon J. B. *Mircopor. Mesopor. Mater.*, 2000, *38*,359-376.
[38] Roque-Malherbe R. *Micropor. Mesopor. Mater.*, 2000, *41*,227-240.
[39] Yang Y., Hofmann T., Pies C., et al. 2008, *doi:10.1016/j.envpol.2008.02.024*.
[40] Carroll K. M., Harkness M. R., Bracco A. A., et al. *Environ. Sci. Technol.*, 1994, *28*,253-258.
[41] Cornelissen G., Van Noort P. C. M., Govers H. A. J. *Environmental Toxicology and Chemistry*, 1997, *16*,1351-1357.
[42] Cornelissen G., van Noort P. C. M., Parsons J. R., et al. *Environ. Sci. Technol.*, 1997, *31*,454-460.
[43] Cornelissen G., Van Noort P. C. M., Govers H. A. J. *Environ. Sci. Technol.*, 1998, *32*,3124-1313.
[44] Jonker M. T. O., Hawthorne S. B., Koelmans A. A. *Environ. Sci. Technol.*, 2005, *39*,7889-7895.
[45] Yang Y., Cajthaml T., Hofmann T. *Environmental Pollution*, 2008, *doi:10.1016/j.envpol.2008.06.007*.
[46] Cornelissen G., Rigterink H., Van Noort P. C. M., et al. *Environmental Toxicology and Chemistry*, 2000, *19*,1532-1539.
[47] Cornelissen G., Rigterink H., Ferdinandy M. M. A., et al. *Environ. Sci. Technol.*, 1998, *32*,966-970.
[48] Grathwohl P. 1997.
[49] Pignatello J. J., Xing B. *Environ. Sci. Technol.*, 1996, *30*,1-11.
[50] Yang C., Huang W., Xiao B., et al. *Environ. Sci. Technol.*, 2004, *38*,4396-4408.

In: Polycyclic Aromatic Hydrocardons: Pollution, Health… ISBN 978-1-60741-462-9
Editors: Pierre A. Haines and Milton D. Hendrickson

Chapter 9

Analysis of Polycyclic Aromatic Hydrocarbons Adsorbed on Soot Surface by Laser Desorption, Laser Ionization and Mass Spectrometry

***Michael Ziskind*[1]*, *Alessandro Faccinetto*[1,2], *Pascale Desgroux*[2], *Eric Therssen*[2], *Cristian Focsa*[1]**

[1]Laboratoire de Physique des Lasers, Atomes et Molécules, UMR 8523
[2] Laboratoire de Physico-Chimie des Processus de Combustion et de l'Atmosphère, UMR 8522
Centre d'Etudes et de Recherches Lasers et Applications, FR 2416
Université Lille 1 Sciences et Technologie, 59655 Villeneuve d'Ascq cedex, France

Abstract

Polycyclic aromatic hydrocarbons (PAHs) are notably formed and emitted by incomplete combustion of coal, oil, petrol, wood, tobacco… In addition, they participate to soot inception. The remaining gaseous PAHs are then adsorbed onto the porous soot matrix so formed, conditioning physicochemical properties of these particles and consequently their impact on human health and environment. This has motivated various studies. However, an extensive analysis of the adsorbed PAHs onto soot particles is still challenging. We are developing an experimental set-up, based on the combined use of 3 techniques - laser desorption, multi-photon ionization and time-of-flight mass spectrometry, dedicated to this study with particular focus on the dependence of the adsorbed species on the stage of combustion and the nature of the fuel. The complexity of the first obtained mass spectra led us to successively investigate the laser desorption of pure PAHs then "synthetic" soot samples- obtained by adsorbing PAHs on various carbonaous substrates- to characterize the processes involved in our method, and finally combustion soot collected from various diffusion flames. We present an overview of these studies together with their first practical implications, obtained on soot collected from Diesel and Diesel surrogate flames.

* E-mail: ziskind@phlam.univ-lille1.fr

A. INTRODUCTION

Soot is made of fine particles resulting of various incomplete combustions. Their formation in a flame is a complex (and not fully understood) process consisting of several physico-chemical reactions like nucleation, coagulation, agglomeration [1-4]. Polycyclic Aromatic Hydrocarbons (PAHs), present in the first stages of the combustion, are generally considered to be important precursors of these particles. Furthermore, the remaining PAHs present in combustion gas are adsorbed on the soot surface and then contribute to its potential impact on environment and human health [5]. Therefore, there is a scientific interest in characterizing and quantifying gaseous and adsorbed PAHs along the combustion process in order to provide key information about the mechanism of the soot formation and the future of these particles in atmosphere.

Characterization of PAHs in flame is usually achieved after probe sampling coupled with Gas Chromatography (GC) and possibly Mass Spectrometry (MS) analysis, performed on the gas phase. However, the limitations of the GC conventional method are well known: the sample preparation is long and tiresome, heavier PAHs (m/z>300 amu) do not go through standard columns due to their low volatility, a single measurement can take few hours, a significant quantity (at least 0.1 ng of a given PAH) is required into the column to allow its detection…

Laser mass spectrometry techniques (in various configurations, two-step or one-step desorption/ionization, Time-of-Flight (ToF) or Fourier Transform (FT) MS) overcome this limitation: no particular preparation is necessary, detection of heavy PAHs is possible, the measurement is fast – usually few seconds – and the detection limit is lower. This explains why these techniques have already been extensively used in the study of PAHs present in natural samples (or laboratory surrogates), like meteorites [6-7], interplanetary dust particles [8], ancient terrestrial rocks [9], sediments and soils [10-13], combustion soot and particulate matter [14-25], wood ash [26], aerosols [27-28], or water [29-31].

However Zare has pointed out in [32] that the nature of the recorded mass spectra (intensity and occurrence of the PAH signals, fragmentation…) highly depends on various parameters involved in this method (e.g. energy, beam geometry and wavelength of the desorption and – in case of two-steps configuration - ionization lasers, desorption/ionization delay, etc…).

We have developed an experimental set-up dedicated to the characterization of PAHs adsorbed on soot particles and based on the laser desorption/multi-photon ionization/ToF-MS method. The behaviour underlined by Zare has been evidenced in our studies, [33-36]. This indicates that a complete characterization of the technique should precede an unambiguous analysis of the laser desorption mass spectra of complex natural soot samples. This has been achieved in our laboratory through a three-step program:

1. the study of the laser desorption process on pure or mixed PAH samples
2. the use of 'surrogate' soot samples by adsorbing PAHs on carbon substrates (graphite, activated charcoal, black carbon…)
3. the analysis of soot samples following the stage and the nature of the combustion.

Moreover, this strategy allows to carry out investigation of complex natural soot samples in a controlled way. In the next paragraphs, we will present an overview of the recent advances on this project with a special focus on the influence of the desorption laser energy on the recorded mass spectra. The last paragraph will be dedicated to the first practical applications of our method, involving Diesel fuel and its surrogate.

B. Experimental Technique

The main features of the laser desorption mass spectrometer have been presented in detail elsewhere [33-36] and is schematized in Fig.1. Briefly, a solid soot sample is placed in a UHV chamber, separated from the mass-spectrometer chamber by a gate valve, on a liquid nitrogen cooled sample holder (to prevent sublimation of light PAHs due to their high vapour pressure at ambient temperature [37]). Once the pressure in the transfer chamber is low enough ($<10^{-7}$ Torr), the gate valve is opened and the sample is translated into the desorption zone of the mass spectrometer (residual pressure ~ 10^{-9} Torr).

The sample is then irradiated at normal incidence by a focused 10 ns frequency – doubled Nd:YAG laser (λ=532 nm, Φ ~ 1 mm on the surface target). Laser energies of ~ 0.1 to 10 mJ/pulse (corresponding to fluences of 0.01 to 1 J/cm²) are typically used and induce desorption from the sample of neutral species, forming a desorbed plume which propagates normal to the sample surface and between the mass spectrometer extraction plates.

Then, these neutrals are ionized by a 7 ns quadrupled Nd:YAG laser (λ = 266 nm, 4.7 eV, Continuum Powerlite) apertured using a 0.5 × 10 mm slit and relay imaged to the center of the extraction voltage plates using 50 and 30 cm plano-convex quartz lenses, forming a homogeneous fluence laser sheet with a cross section of ~ 0.3 ×6 mm². This geometry has been selected among various tested configurations since we could demonstrate in [36] that it greatly reduces the fragmentation related to the ionization step and can result in enhanced sensitivity. The fluence used is about 0.1 J/cm² (0.55 mJ/pulse) and, taking account of their ionization potential (typically 7-8 eV), two 266 nm-photons are required to ionize the PAHs. Timing of the desorption and ionization lasers is controlled using a digital four channels delay generator (SRS DG535) and set for a delay of 50 μs between the two lasers. The ions are mass-analyzed in a 1-m reflectron ToF-MS (RM Jordan). A mass resolution of $m/\Delta m \cong 1000$ is achieved throughout a typical spectrum. Ion detector signals are recorded using a 4 ns/point digital oscilloscope (LeCroy Waverunner) and analyzed using a homemade program under Labview environment.

C. Results

1. PAHs

Extensive studies on the desorption/ionization of pure or mixed PAH samples, prepared by pressing powder of different PAHs, have already been presented in various publications

[33-36]. We only remind the results necessary for a good comprehension of the rest of this chapter.

Typical laser desorption mass spectrum of pure pyrene sample is shown in Fig.2a. This fragment-free spectrum has sufficient resolution to distinguish the isotopomers of the PAHs. At higher desorption (and/or ionisation) laser pulse energy E_d, additional peaks become apparent while the PAH peak becomes weaker (cf. Fig 2b). In order to give a better insight into this process, we have performed a study by varying E_d (cf. Fig.3). The general evolution of the PAH signal intensity with E_d has been explained in terms of competition between desorption and fragmentation through an increase of the internal energy of the desorbed species: beyond a specific threshold, the laser energy is sufficient to eject the PAHs; the number of desorbed particles increases with E_d leading to the initial increase of the PAH signal. However, the "excess energy" can also be used to fragment the molecules or to increase the internal energy of the ejected particles, which further rises the fragmentation probability at the ionization stage. Finally, beyond a new threshold, the fragmentation process becomes predominant leading to the extinction of the PAH signal along with the progressive appearance of fragmentation products. Thus, this study allows to distinguish intact desorbed PAHs from fragments formed during the desorption process (and, consequently, to avoid fragmentation using appropriate laser energy).

Moreover, this parametric study evidences a characteristic response of each tested PAH, in agreement with Zare's remarks [32], defined by a desorption threshold, an optimum laser pulse energy for maximum yield and a working range. This provides the discrimination of PAHs of same mass (e.g. pyrene and fluoranthene). This specific behaviour is preserved in the case of mixtures of several PAHs, opening the possibility of selective analysis of samples of various PAHs.

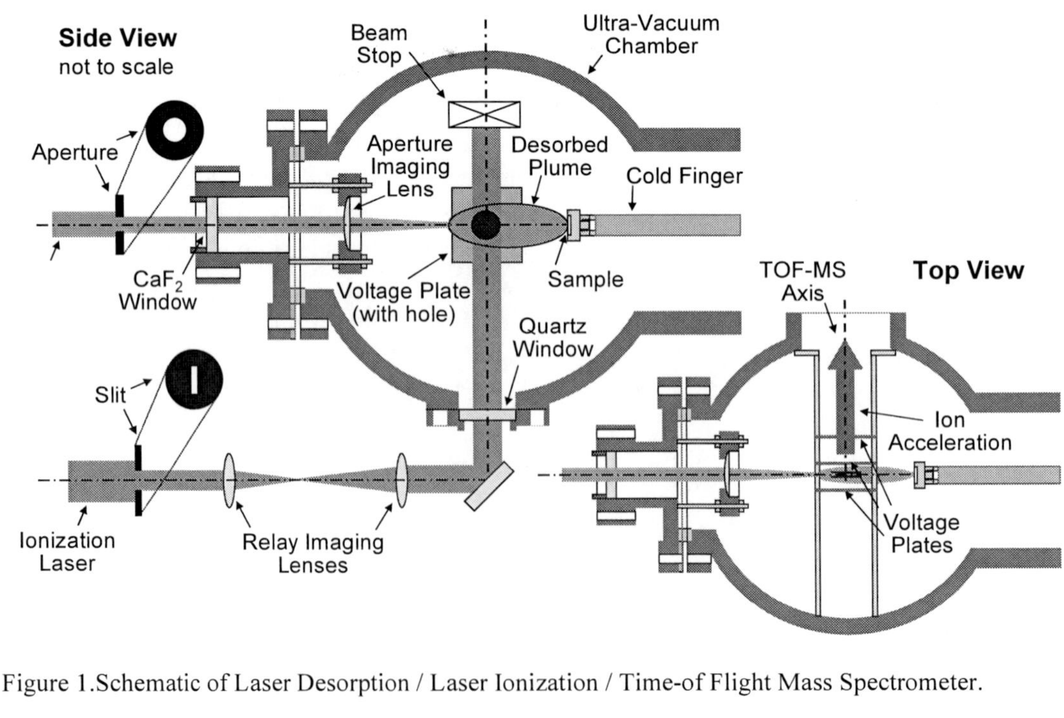

Figure 1.Schematic of Laser Desorption / Laser Ionization / Time-of Flight Mass Spectrometer.

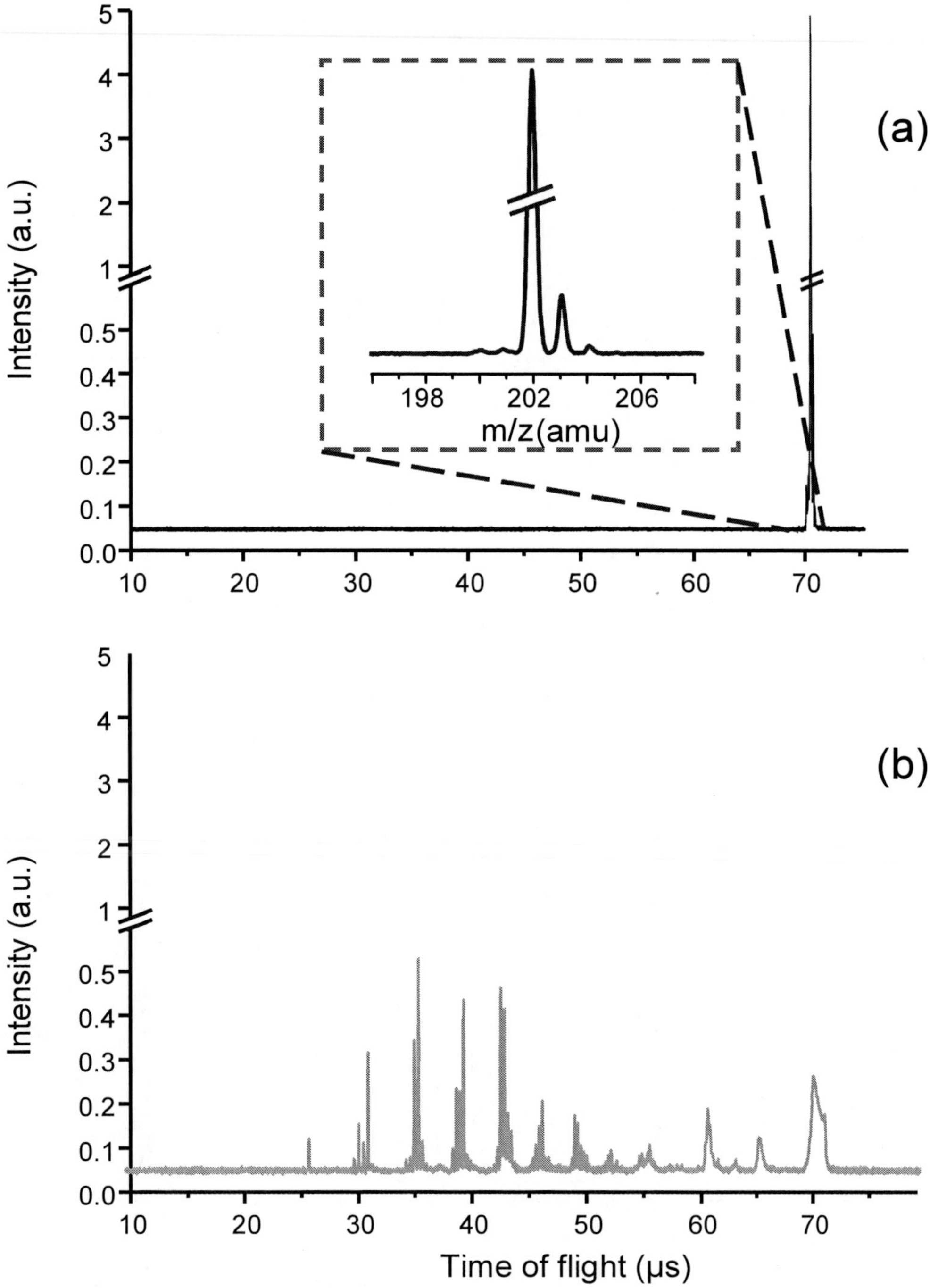

Figure 2. Laser Desorption Mass Spectra of a Pure Pyrene Sample, E_{ion} = 0.5 mJ/pulse, (a): E_d= 1 mJ/pulse; (b): E_d=2 mJ/pulse.

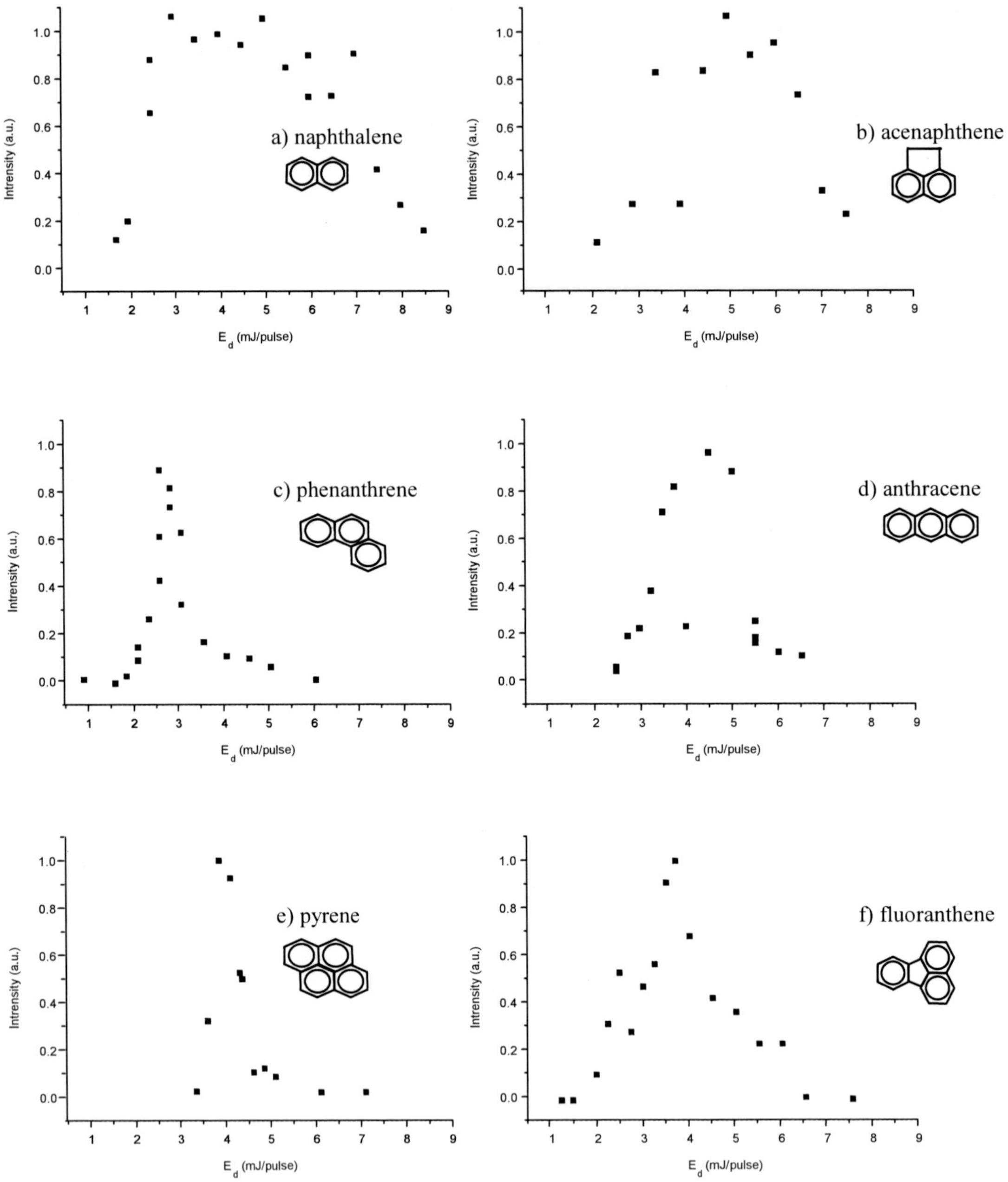

Figure 3. Normalized response of six PAH peak intensities to the desorption laser energy E_d. Reprinted from [35], with permission from IOP.

2.Synthetic soot

2.1. Preliminary Study on Graphite

The next step on the way to the analysis of combustion soot is the study of PAHs adsorbed on graphite surfaces.

Graphite is the most stable allotrope of carbon under standard conditions of temperature and pressure. At first sight, its planar multi-layer structure, with different graphene sheets being connected together by Van Der Waals forces, is different from the soot one which presents an onion-like organization made out of small graphitic crystallites [38]. However, graphite is often used as model for the study of reactions on soot surface or of adsorbed PAHs onto atmospheric particles [39] since soot particles can be roughly seen as fragments of graphite growing due to successive additions of acetylene to a radical species followed by cyclization [40].

2.1.1 Response of a Pure Graphite Sample to E_d

An important concern of our work is the determination of the origin of the detected molecules: are they representative of the adsorbed phase or artifacts coming from the destruction of the substrate? To give an insight on this issue, the response to E_d of a reference graphite sample (obtained by pressing few hundred mg of pure powder) has been studied (cf. Fig. 4).

The E_d distribution exhibits a desorption threshold for the carbon desorbed from the graphite matrix (in atomic form but also in C_2 and C_3 clusters) around 4.5 mJ/pulse. Therefore, in the E_d interval for the PAHs detection (cf. Fig. 3), the underneath carbon matrix should not be destroyed. Supposing a similar structure for a soot matrix, one can achieve a “smooth” desorption process, without appearance of species artificially produced by the ejection of soot particles fragment during desorption.

2.1.2. Behaviour of PAHs Adsorbed on Graphite with E_d

To validate this assumption, we have examined the behaviour of PAHs adsorbed onto graphite with respect to the desorption energy. For this purpose, 1g of pure graphite has been mixed to 5 ml of a 0.2 M solution of fluoranthene in dichloromethane (DCM). Considering the small value of the specific surface of the graphite (5m²/g- to be compared to ~1000 m²/g for the activated charcoal), this quantity corresponds to a coverage of about one hundred of monolayers on the surface of the substrate. This has been estimated by dividing the specific surface by the surface occupied by the PAHs (assimilated to parallelepipeds of dimensions given in [41]) present in the solution. The solution is filtered, and dried for 12 hours at room temperature and pressure until obtaining a powder. 500 mg of this powder is finally pressed using a hydraulic press to get a solid sample.

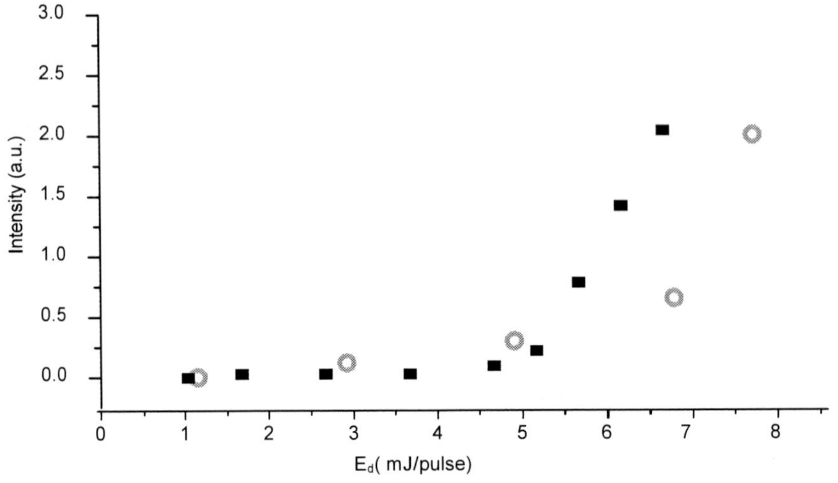

Figure 4. Evolution with E_d of the peak intensity of atomic carbon (m/z = 12 amu) desorbed from a pure graphite sample (square) and a black carbon sample (cercle).

This sample has been analyzed following the procedure described above. Results are in good agreement with the previous observations (see Fig. 5): at low energy, fluoranthene is desorbed with an optimum efficiency around 3 mJ/pulse, while signals corresponding to the carbon significantly rise around 4.5 mJ/pulse. Below this value, PAH signal disappears after about ten laser shots : the graphite surface has been cleaned and the irradiated point has to be changed. This confirms the possibility of a smooth desorption process, i.e. to address only the adsorbed phase.

From 4.5 mJ/pulse, the PAH signal increases again and does not disappear after thousand shots: the destruction of the carbon matrix releases PAHs present in the deeper graphene layers. Moreover, the desorption of fluoranthene is achieved in the same E_d interval as pure fluoranthene sample (see Fig.1-f and Fig.5). The origin of this common behaviour will be discussed in the next chapter, in the light of the results of a similar study realised on two other matrices: activated charcoal and black carbon.

2.2. Activated Charcoal/Black Carbon

Activated charcoal and black carbon are both raw forms of graphite formed through the incomplete combustion of fossil fuels and biofuels. They differentiate themselves by a additional physicochemical treatment of the activated charcoal, which make it extremely porous, resulting into a very large specific surface (500 to 1500 m²/g) available for adsorption or chemical reactions (100 m²/g for the black carbon). These substrates have structures closer to the soot one when compared to graphite, making analysis of its adsorbed phase an important step towards the study of natural samples.

In the following, no major distinction will be made between these two substrates (except the coverage) since their behaviours to the desorption processes are found to be similar.

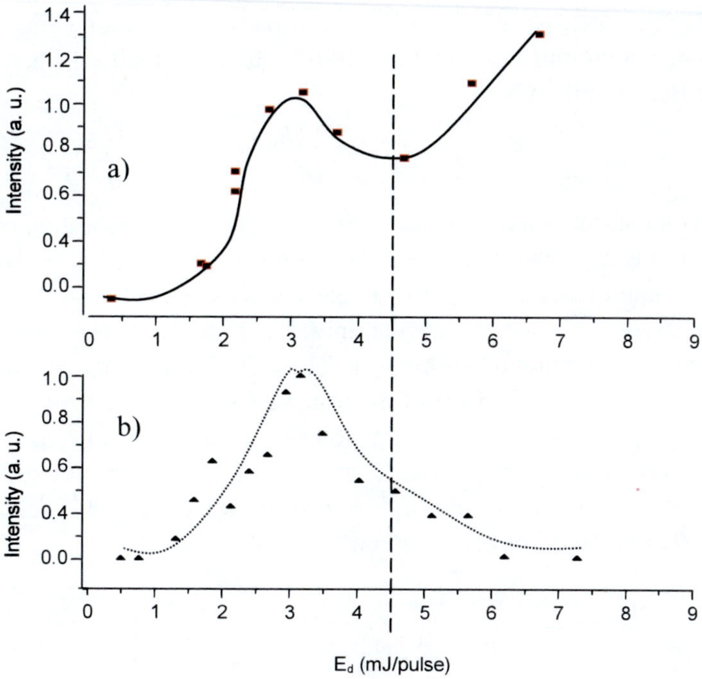

Figure 5. Evolution with E_d of the fluoranthene ($m/z = 202$ amu) signal for desorption of a) fluoranthene adsorbed on graphite and b) pure fluoranthene. The dashed vertical line corresponds to the graphite desorption threshold. Reprinted from [35], with permission from IOP.

2.2.1. Response of a Pure Black Carbon Sample to E_d

The response of black carbon (and activated charcoal) to desorption energy is comparable to the graphite one (cf. Fig. 4). Particularly, the desorption threshold is slightly higher – ~ 6 mJ/pulse vs 4.5 mJ/pulse – ensuring again the possibility to desorb PAHs without destruction of the carbon matrix in the low energy range.

2.2.2. Preparation of Synthetic Soot

Boosted by the experience we acquired on the graphite, the procedure for making a synthetic soot sample based on black carbon and activated charcoal has been improved on several points. First, a mother 10^{-1}mol.L^{-1} solution of PAHs in DCM is prepared. Solutions of different concentration (between 10^{-5} and 10^{-2} mol.L^{-1}) are then obtained by successive dilutions. 2g of substrate is added in 100 mL of solution. This volume is sufficient to completely submerge the charcoal, and small enough to avoid solvent waste. The small size of charcoal granules ensures the solution to stay in a colloidal form, even without using a magnetic agitator. After 15 minutes, the obtained solution is filtered under vacuum to guarantee a quick separation of the charcoal present in solution, preventing evaporation of the solvent and thus the crystallization of PAHs possibly remained in solution. About 150 mg of the residue is finally pressed to get an homogeneous solid sample,.

In addition, this procedure allows the determination of the quantity of the adsorbed PAHs by Vis-NIR absorption spectroscopy. Actually, the initial concentration being established, one can measure the concentration in the solution after the treatment with the active charcoal and the filtering using Beer-Lambert law:

$$A = \varepsilon(\lambda) bc,$$

where A represents the absorbance of the solution, $\varepsilon(\lambda)$ the molar absorption coefficient, b the optical path length, and c the molar concentration of the active species. Finally the determination of the matter amount adsorbed in carbon is calculated from the difference in the intensities of the absorption spectra of the initial and the exhaust solution (cf. Fig.6). Calibration curves have been plotted in this way. These studies have demonstrated that, for an initial concentration of 10^{-3}mol.L^{-1}, 97% of the tested PAHs (anthracene and pyrene) remains on the charcoal. This percentage increases for lower concentrations, up to about ~100% for 10^{-5}mol.L^{-1}. The coverage has been evaluated in the same way as for the graphite from 10^{-2} monolayer for a concentration of 10^{-5}mol.L^{-1}, to one monolayer for 10^{-3}mol.L^{-1} for activated charcoal, ~1 to 100 monolayer for black carbon.

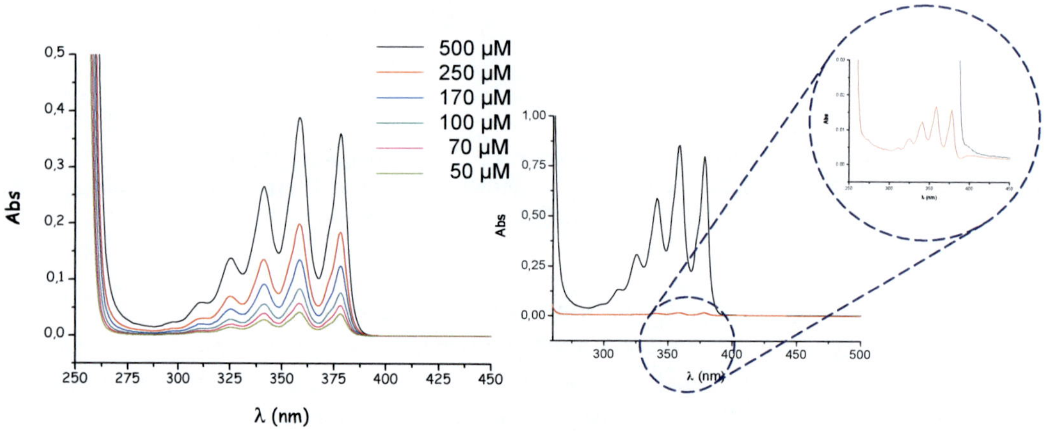

Figure 6. UV-Vis absorption spectra of anthracene solutions in DCM: a) initial solutions of different concentrations ; b) 1 mM initial solution (black curve) and remaining solution (red curve).

2.2.3. Behaviour of PAHs Adsorbed on Black Carbon with E_d

Fig. 7 shows the response of different PAHs adsorbed onto black carbon to E_d. In broad outline the adsorbed species has a similar behaviour to that previously observed: by progressively increasing the energy, one can notice a raise of the PAH signal, then a decrease caused by the fragmentation and finally a new increase beyond the matrix desorption threshold. One finds again the possibility to achieve a smooth desorption (cf.§-2.2).

Two main differences are evidenced at low energy:

1. the PAH peak disappears after few laser shots (even only one - depending on the concentration). Let us note that the good homogeneity of the sample ensures a good reproducibility of the observed signal when the irradiated point is changed.
2. the E_d distribution is obviously shifted compared to graphite

The first difference has to be naturally linked to the high coverage of the charcoal or black carbon. The second one is relatively unexpected since a *priori* the response to desorption energy oughts to notably depend on the strength of the Van der Waals bond between the desorbed molecules and the substrate (PAH-PAH in §-1, PAH-graphite in §-2, PAH-Charcoal/black carbon in the present section). At microscopic level, graphite and charcoal have a similar chemical nature. One can thus expect the same strength of the bond, resulting in a similar behaviour under laser irradiation, which does not correspond to our experimental observations. Actually, the evidenced shift could again be related to the different value of the coverage. In the case of the charcoal, the small number of monolayers (≤ 1) ensures the desorption of PAHs directly linked to he carbon substrate. Conversely, the majority of the molecules adsorbed on graphite should not be ejected directly from the top of the carbon matrix but from another PAH layer, recalling the desorption of a pure PAH sample. This assumption (which couldn't have been tested in the case of graphite because the sample became very inhomogeneous at lower concentration, making systematic studies complex) explains at the same time the differences in the E_d distribution of the PAHs desorbed from charcoal and graphite and similarities between this last distribution and the pure PAH ones.

Moreover, while all the studied PAHs have their own distribution, Fig. 7 indicates they are all identical in the case of PAHs adsorbed on carbon. Origin of this common behaviour has to be linked to identical PAH-carbon bonding. From a practical point of view, the selectivity depending on the desorption energy is lost. This will simplify the analysis of the natural soot samples since, by contrast with the mixtures of several PAHs, the energy which optimizes a given PAH peak in a mass spectrum is the same whatever the PAH under consideration.

2.2.4. Determination of the Potentialities of the Set-Up

The procedure for making a synthetic soot sample provides a good reproducibility of the sample: we could have checked the homogeneity of a given sample and the signal reproducibility in the desorption of these surrogate, using samples of various mother solutions and concentration to a relative value of ~ 30% enough to perform systematic (and qualitative) measurements.

The knowledge of the quantity of the adsorbed PAHs also offers the possibility to determine the sensitivity of our set-up. Taking account of the specific surface of the charcoal, the minimal concentration which allows the achievement of significant signal (about 10^{-5}mol/L^{-1}) corresponds to a coverage of about 10^{15} molecules/m² that is largely inferior to a monolayer. All the molecules being desorbed after a single laser pulse, focused on about 1 mm-diameter surface, one can detected less than 10^9 molecules (one femtomole) on average per laser shot, well over conventional techniques such like CG. As direct consequence, the use of synthetic soot of known concentration could lead to quantitative measurements through acquisition of calibration curves. From this perspective, a systematic study on the 16 PAHs classified by the US EPA [42] as priority pollutants is underway together with a comparative thermogravimetric analysis, using a thermobalance available in our laboratory.

Finally, this gives the possibility of shot-by-shot studies. Possible applications will be presented at the end of the next paragraph.

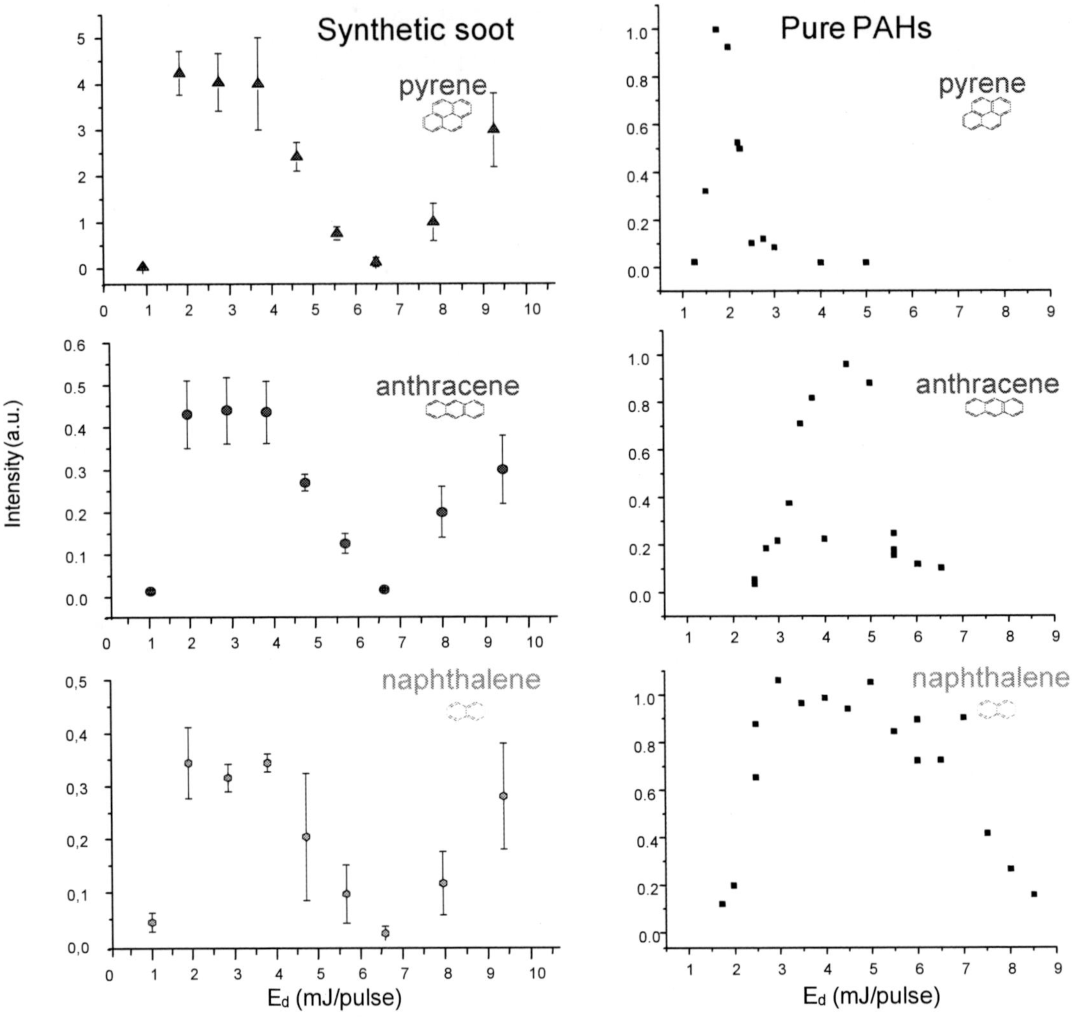

Figure 7. Response of 3 PAH peak intensities to E_d during the desorption of pure PAH samples (left) and corresponding synthetic soot (right).

3. Natural Soot

The last part of this chapter describes the very first practical applications of this method on the desorption of soot particles formed during the combustion of a Diesel turbulent flame. These studies have naturally benefited from works presented in previous paragraph but from spatially-resolved measurements of PAHs and soot volume fraction using a technique developed in Lille as well. This technique will be briefly depicted after the description of the used burner, flame, collecting probe and fuel.

3.1. Description of the Burner and Properties of the Investigated Flames

The *Holthuis* burner we used is a modified commercial water-cooled McKenna burner, a conceptually simple device originally used for studying gaseous laminar or turbulent jet flames [43-45].

Since LDI TOF-MS is an ex-situ technique, soot must be sampled from flames. For that purpose, we use a sampling probe perpendicular to the main flame axis, which allows the direct deposition of soot on the surface of a porous borosilicate filter, suitable for the following mass analysis (no signal has been evidenced during the desorption of a blank filter).

Soot is collected through a double-wall quartz probe to achieve local sampling and fast dilution with gaseous nitrogen, in order to minimize the gas condensation on soot and to hinder most chemical reactions.

The low-sulfur commercial Diesel studied in this paragraph has been analysed by gas-chromatography/mass-spectrometry. Its main composition includes 23% of n-paraffin, 25% of branched paraffin, 7% of cycloparaffin, 17% of aromatics, 8% of esters and 20% of unknowns (in Volume %). More than 400 other compounds have been detected.

3.2. Analysis of the Species Adsorbed on Soot Surface

3.2.1. Choice of the Analysis Zone: Laser Induced Irradiation (LII) & Laser Induced Fluorescence (LIF) mapping

The sampling has been performed in the most significant regions of the turbulent flame [46], determined by spatially-resolved measurements of PAHs and soot volume fraction using a technique developed in our laboratory. Briefly, this method (namely the two colour LII/LIF) is based on the assumption that the radiation emission from 1064 nm heated soot is only assignable to soot (LII signal) while the emission from 532 nm irradiation results from both LII signal and fluorescent high-number-ring PAHs (LIF signal). Consequently, by subtracting the two signals, one can isolate characteristic signal of PAHs from soot [44,46].

This leads to 2D images of the soot and PAH distributions using imaging techniques (cf. Fig 8a & b). Let's note that the LIF technique only concerns the heavy PAHs (containing at least 4 aromatic rings). Nevertheless, our measurements confirm their roles of soot precursors and guide the interpretations about the soot mass spectra.

Four significant zones have been identified in the flame in this way (cf. Fig 8c) :

1. The gaseous PAHs region, before the formation of the first soot particles.
2. The young soot region, corresponding to the maximum of LIF signal.
3. The soot peak region corresponding to the maximum LII signal and then to the maximum soot volume fraction Fv.
4. The beginning of the oxidation region.

In the following, we will only compare the gaseous PAHs and soot peak regions. As the first zone only includes gaseous PAHs, the sampled species have been deposed on a black carbon substrate whereas the soot (and the PAHs adsorbed on its surface) are collected on the porous borosilicate substrate described above.

3.2.2. Comparaison Gaseous/Adsorbed PAHs

Fig. 8d shows two typical mass spectra recorded during the desorption of soot or trapped gaseous PAHs. The chosen parameters (E_d=0.3 mJ/pulse, E_{ion}=0.5 mJ/pulse) simultaneously ensure smooth desorption (the signals disappear after few laser shots) and fragment-free spectra (no signal has been evidenced at low mass). These spectra are at least qualitively reproducible. Generally, a very regular mass sequence has been detected (each peak is spaced out by 12 or 14 amu from its neighbors). Peaks are detected up to 450 amu. Finally, it must be emphasized that thanks to the cooling procedure of the sample, the set-up allows the detection of the more volatile compounds (like e.g. naphthalene).

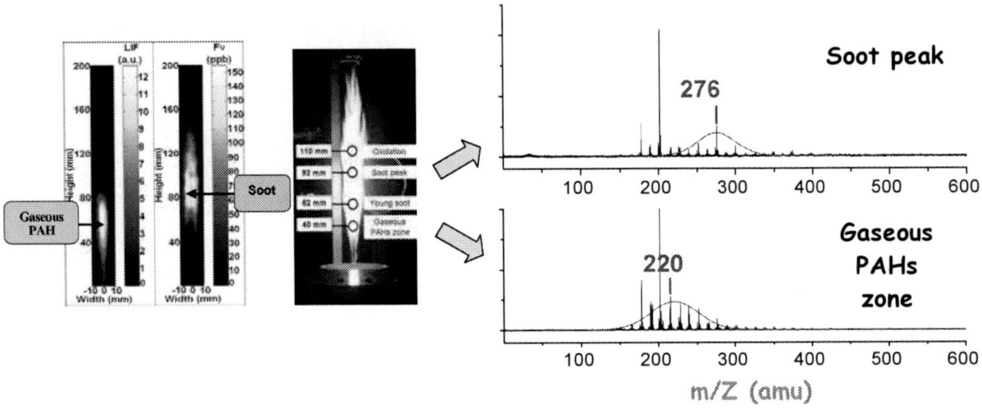

Figure 8. LIF (a) and LII (b) mapping of a Diesel flame. Reprinted from [46], with permission from Elsevier. (c) sampling heights (d) corresponding desorption mass spectra for the soot peak and gaseous PAHs regions.

By contrast to the mass spectrum issued from gaseous PAHs, soot content reveals higher-mass PAHs. This tendency was found along the entire sooting zone and has to be linked to the transit time of the soot particles in the flame. The quality of the mass spectra demonstrates the potential of our set-up for the analysis of soot particles and for further improvement of the soot growth process understanding. The difference between light and large PAHs will be discussed in the forthcoming paragraph in the context of a comparative study between Diesel and one of its substitutes, namely the Surrogate 2.

3.2.3. Comparison between Diesel and Surrogate 2

The substantial number of compounds present in Diesel makes its analysis very complex. Convenient two-compound fuel, the Surrogate 2 (containing 80% of n-decane and 20% of α-methylnaphthalene in volume), has been proposed to replace Diesel in numerical simulations. It is derived from Surrogate1 [47] initially formulated within the framework of the Integrated Diesel European Action program (IDEA), from which it differs by a closer Threshold Soot Index (TSI) to Diesel [47-48].

An extensive analysis of this Surrogate has been performed, evidencing its excellent agreement with Diesel in terms of physical and optical properties [46]. For instance, one could demonstrate the good match for the two LII and LIF mappings in shape and intensity. We have then analyzed the species adsorbed on soot particles formed during the combustion. The comparison with Diesel was carried out with the parameters given in §-2.1, at the same sampling height, corresponding to the soot peak region in both cases. The obtained spectrum is undoubtedly less rich for the Surrogate since only light PAHs could be observed (cf. Fig. 9). This confirms our previous observation on the difference between light and heavy PAHs.

An attractive hypothesis to explain this discrepancy is proposed in [49]. Following Dobbins one could distinguish the so called "pyrogenic" PAHs which result from reaction kinetics in the flame from “petrogenic” PAHs originating from Diesel composition which survive to the combustion process. However, one cannot exclude the possibility of condensation of residual gaseous PAHs during the sampling to explain the presence of the same light PAHs in all the obtained mass spectra.

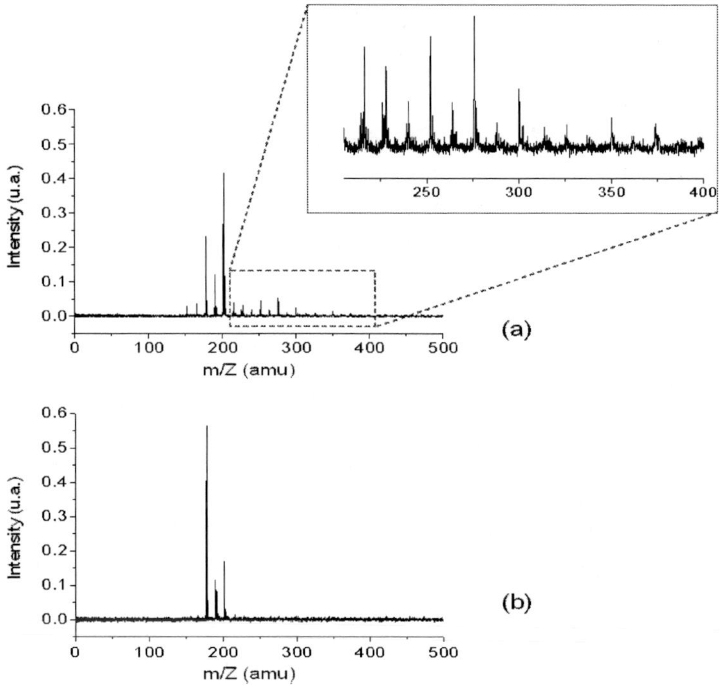

Figure 9. Five-shot averaged mass spectra of PAHs desorbed from Diesel soot (a) and surrogate soot (b). In both case Soot was collected at 92 mm above the burner. Reprinted from [46], with permission from Elsevier

Shot-by-shot study could give some insight on this issue. Actually, according to §-2.3 and by supposing a difference in the nature of the bond of the condensed PAHs and the adsorbed ones, the responses to E_d could be different in these two cases. Consequently, the E_d distribution for the very first laser shots (corresponding to possible condensed species) and for the next ones (corresponding to adsorbed species) could be shifted relative to each others. Such behavior has previously been observed during the desorption of soot collected but no definitive conclusion has been given at this time since this turbulent flame did not allow accurate shot-by-shot measurements. A better reproducibility is expected in current investigations performed in low pressure-premixed sooting flames stabilized on a porous burner.

CONCLUSION

Characterization of PAHs adsorbed on soot samples has been performed by laser desorption combined with multi-photon ionization and mass spectrometry. The complexity of the processes involved in this method led us to successively investigate the laser desorption of pure PAH samples, "synthetic" soot samples and combustion soot collected from a Holthuis burner. We have presented an overview of these studies with particular focus on the influence of the laser desorption energy on the recorded spectra. Some conclusions on the fragmentation mechanism have been reached in this frame. The complete optimization of the technique offers the ability to obtain with one single shot fragment-free PAH mass spectra with an excellent sensitivity (better than 1 fmol/laser shot). The possibility to desorb molecules from the adsorbed phase only without destruction of the soot matrix has also been demonstrated. First practical implications, obtained on soot collected from Diesel and Diesel surrogate flame, indicate the influence of the fuel composition on soot content (despite similar optical and morphological characteristics) and evidence the possibility to distinguish petrogenic PAHs from pyrogenic ones.

REFERENCES

[1] H. Bockhorn, Berlin: *Springer Series in Chemical Physics* (1994) 59

[2] J.A. Miller, J.V. Volponi and J.F. Pauwels, *Combust. Flame* 105 (1996) 454

[3] H. Richter, J.B. Howard, *Prog. Energy Combust.Sci.* 26 (2000) 565

[4] C.S. McEnally, L.D. Pfefferle, B. Atakan and K. Kohse-Höinghaus, *Prog. Energy Combust.Sci*, 32 (2006) 247

[5] P. Gerde, B.A. Muggenbourg, M. Lundborg and A.R. Dahl, *Carcinogenesis* 22 (2001) 741

[6] S. Messenger, S. Amari, X. Gao, R.M. Walker, S.J. Clemett, X.D.F.Chillier, R.N. Zare and R.S. Lewis, *Astrophys. J.* 502 (1998) 284

[7] S.J. Clemett, X.D.F. Chillier, S. Gillette, R.N. Zare, M. Maurette, C. Engrand and G. Kurat, *Orig. Life Evol. Biosph.* 28 (1998) 425

[8] S.J. Clemett, C.R. Maechling, R.N. Zare, P.D. Swan and R.M. Walker, *Science* 262 (1993) 721

[9] T.B. Mahajan, F.L. Plows, J.S. Gillette, R.N. Zare and G.A.Logan, *J. Am. Soc. Mass Spectrom.* 12 (2001) 975

[10] J.S. Gillette, R.G. Luthy, S.J. Clemett and R.N. Zare, *Environ. Sci. Technol.* 33 (1999) 1185

[11] A. Specht and M.W. Blades, *J. Am. Soc. Mass Spectrom.* 14 (2003) 562

[12] M.L. Alexander, P.H. Hemberger, M.E. Cisper and N.S. Nogar, *Anal. Chem.* 65 (1993) 1609

[13] M.J. Dale, A.C. Jones, S.J.T. Pollard, P.R.R. Langridge-Smith and A.G. Rowley, *Environ. Sci. Technol.* 27 (1993) 1693

[14] R.A. Dobbins, R.A. Fletcher and W. Lu, *Combust. Flame* 100 (1995) 301

[15] R.A. Dobbins, R.A. Fletcher and H.C. Chang, *Combust. Flame* 115 (1998) 285

[16] S.M. Hankin and P.John, *Anal. Chem.* 71 (1999) 1100

[17] L.C. Marr, T.W. Kirchstetter, R.A. Harley, S.V. Hering and S.K. Hammond, *Environ. Sci. Technol.* 33, (1999) 3091
[18] R. Zimmermann, L. Van Vaek, M. Davidovic, M. Beckmann and F. Adams, *Environ. Sci. Technol.*, 34 (2000) 4780
[19] O.P. Haefliger, T.D. Bucheli and R. Zenobi, *Environ. Sci. Technol.* 34 (2000) 2178
[20] R. Zimmermann, T. Ferge, M. Gälli and R. Karlsson, *Rapid Commun. Mass. Spectrom.* 17 (2003) 851
[21] V. Carré, L. Vernex-Loset, G. Krier, P. Manuelli and J.F. Muller, *Anal. Chem.* 76 (2004) 3979
[22] B. Oktem, M.P. Tolocka, B. Zhao, H. Wang and M.V. Johnston, *Combust. Flame* 142 (2005) 364
[23] Y. Bouvier, C. Mihesan, M. Ziskind, E. Therssen, C. Focsa, J.F. Pauwels and P. Desgroux, *Proc. Combust. Inst.* 31 (2007) 841
[24] B. Apicella, A. Carpentieri, M. Alfè, R. Barbella, A. Tregrossi, P. Pucci and A. Ciajolo, *Proc. Combust. Inst.* 31 (2007) 547
[25] O. Mathieu, G. Frache, N. Djebaïli-Chaumeix, C.-E. Paillard, G. Krier, J.-F.Muller, F. Douce and P. Manuelli, *Proc. Combust. Inst.* 31 (2007) 511
[26] M. Bente, T. Adam, T. Ferge, S. Gallavardin, M. Sklorz, T. Streibel and R. Zimmermann, *Int. J. Mass Spectrom*. 258 (2006) 86
[27] B.D. Morrical, D.P. Fergenson and K.A. Prather, *J Am Soc Mass Spectrom* 9 (1998) 1068
[28] B. Oktem, M.P. Tolocka and M.V. Johnston, *Anal. Chem.* 76 (2004) 253
[29] T.D. Bucheli, O.P. Haefliger, R. Dietiker Jr. and R. Zenobi, *Anal. Chem.* 72 (2000) 3671
[30] C. Weikhardt, K. Tönnies and D. Globig, *Anal. Chem.*74 (2002) 4861
[31] C. Emmenegger, M. Kalberer, B. Morrical and R. Zenobi, *Anal. Chem.* 75 (2003) 4508
[32] J. E. Elsila, N. P. de Leon and R. N. Zare, *Anal. Chem.*76 (2004) 2430
[33] C. Mihesan, M. Ziskind, E. Therssen, P. Desgroux and C. Focsa, *Chem. Phys. Lett.* 423 (2006) 407
[34] K. Thomson, M. Ziskind, C. Mihesan, E. Therssen, P. Desgroux and C. Focsa, *Appl. Surf. Sci.* 253 (2007) 6435
[35] C. Mihesan, M. Ziskind, E. Therssen, P. Desgroux and C. Focsa, *J. Phys.: Condens. Matter* 20 (2008) 025221
[36] A. Faccinetto, K. Thomson, M. Ziskind and C. Focsa, *Appl. Phys. A* 92 (2008) 969
[37] K. Nass, D. Lenoir and A. Kettrup, *Angew.Chem. Int. Edn Engl.* 34 (1995) 1735
[38] O.B. Popovicheva, N.M. Persiantseva, B.V. Kuznetsov, T.A. Rakhmanova, N.K. Shonija, J. Suzanne and D. Ferry *J. Phys. Chem.* A 107 (2003) 10046
[39] W. Esteve, H. Budzinskia and E. Villenave, *Atmos. Environ.* 38 (2004) 6063
[40] G.M. Florio, T.L. Werblowsky, T. Muller, B.J. Berne and G.W. Flynn, *J. Phys. Chem.* B 109 (2005) 4520
[41] C. Lane, S.A. Sander and A. Wise, NIST Special Publication (1997) 922
[42] *Evaluation and Estimation of Potential Carcinogenic Risks of Polynuclear Aromatic Hydrocarbons* (Washington, DC: Carcinogen Assessment Group, Office of Health and

Environmental Assessment, Office of Research and Development, US Environmental Agency) (1985)

[43] Y. Bouvier, C. Mihesan, M. Ziskind, E. Therssen, C. Focsa, J.F. Pauwels and P. Desgroux., *Proc. Combust. Inst.* 31 (2007) 841

[44] E. Therssen, Y. Bouvier, C. Schoemaecker-Moreau, X. Mercier, P. Desgroux, M. Ziskind and C. Focsa., *Appl. Phys. B* 89 (2007) 417

[45] R. Lemaire, E. Therssen, J.F. Pauwels and P. Desgroux, *Experimental comparison of soot formation in turbulent flames of kerosene and surrogate model fuels*, in: A. Sarofim, A. D'Anna, H. Wang (Eds.), *Combustion-Generated Fine Carbon Particles*, in press.

[46] R. Lemaire; A. Faccinetto, E. Therssen; M. Ziskind, C. Focsa and P. Desgroux, *Proc. Combust. Inst.* 32 (2009) 737

[47] H. Barths, H. Pitsch and N. Peters, *Oil Gas Sci. Technol.* 54 (1999) 233

[48] D.B. Olson, J.C. Pickens and R.J. Gill, *Combust. Flame* 62 (1985) 43

[49] R.A. Dobbins, R.A. Flechter, B.A. Benner Jr. and S. Hoeft, *Combust. Flame* 144, (2006) 773

In: Polycyclic Aromatic Hydrocardons: Pollution, Health… ISBN 978-1-60741-462-9
Editors: Pierre A. Haines and Milton D. Hendrickson

Chapter 10

TOXICITY AND TOXICOLOGY OF BENZO[A]PYRENE

M.S. Yang and W.S. To
Department of Biology, Hong Kong Baptist University, Kowloon, Hong Kong

ABSTRACT

Benzo[a]pyrene (BaP) is one of the most well studied PAHs. It is a product of incomplete combustion of fossil fuel. BaP can be deposited into food, soil and water. The major route of human exposure to BaP is via food consumption. BaP can interact with DNA to form bulky adducts, or biotransformed to produce free radicals. Both processes may lead to mutation and cancer. Despite the clarity of its actions, it is still difficult to establish a clear dose-response relation between BaP and cancer in human. Currently, BaP is considered class 2A carcinogen by the International Agency for Research in Cancer (IARC), a chemical with well documented evidence as an animal carcinogen, but limited evidence as human carcinogen. It is believed that only by understanding the basic difference between animal and human biochemistry will it be possible to address the toxicity of BaP in the various biological systems. In the present review, experiments pertaining to studying the response to BaP in human and animal cells are discussed.

INTRODUCTION

Polycyclic aromatic hydrocarbons (PAHs) are widespread environmental contaminants originated from incomplete combustion of fossil fuels. These chemicals can be found in food, air, water and soil and are shown to be carcinogenic. PAHs include a wide spectrum of organic toxicants ranging from a molecule composed of two aromatic rings (naphthalene) to large multiple aromatic rings molecules such as benzo[a]pyrene and benzo[e]pyrene. Of all the PAHs studied, benzo[a]pyrene (Figure 1) is the prototype of this family of toxicants.

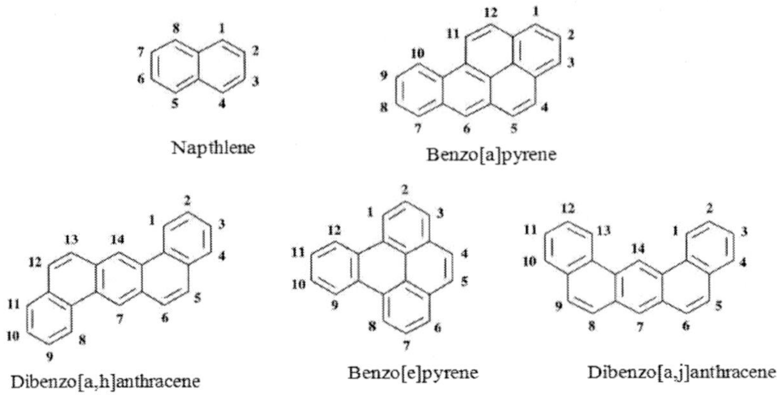

Figure 1. Chemical structures of some PAHs.

Benzo[a]pyrene (BaP) was first isolated from coal tar by Hieger in 1933. Because several synthetic autofluorescence pentacyclic hydrocarbons, such as dibenz[a,j]anthracene and dibenzo[a,h]anthracene (Figure 1), were found to cause cancer in mouse skin test, the isolation and identification of purified autofluorescent BaP from coal tar was considered the PAH congener responsible for inducing cancer. BaP has a molecular weight of 252.3 g/mol. Unlike the low molecular weight PAH, naphthalene (Figure 1), BaP is not volatile [88]. Under atmospheric pressure, purified BaP has a melting point of 176 °C. Compared to its isomer benzo[e]pyrene, a minor component with melting point of 187 ^{o}C, BaP was proven to possess higher potency in causing cancer in a variety experimental models [20,23,37,48].

ENVIRONMENTAL EXPOSURE TO BAP

Extensive monitoring studies have shown that the average level of BaP is high in urban areas with abundant human and industrial activities. In U.S.A., air level could be as high as 100 ng/m^3 in heavily polluted sites [34]. In developing countries such as China, BaP level is high at sites near industrial areas when compared to sites far from the city center and industrial areas [24,52]. In sites with heavy traffic in a city like Beijing, BaP level in air could reach over 8 folds higher than the national standard of 10 ng/m^3 and more than 5 folds in heavily industrialized area such as Tianjin. BaP attaches to dust and particulates, which can deposit onto soil and water [52,58,100]. In Beijing suburb, soil BaP level can reach 270 μg/kg [58], which is higher than that in industrial areas (9.5 μg/kg) in Hong Kong [18]. In water, sediment is the major sink for BaP. Correspondingly, BaP level is high in sediment of industrialized regions [53,57,78].

Besides through the air, water and soil, human can be exposed to BaP via food ingestion. In fact, the food chain has been shown to be the dominant exposure pathway, accounting for about 97% of the total daily intake of BaP [34]. Aquatic animals living in heavily polluted sites can take up organic toxicants from the sediment [60] and, in turn, transfers to human. BaP in air can also be a source of food contaminant. Li et al., [51] demonstrated that while rice plants grown under control green house condition were unable to take up BaP from water, the BaP content of rice grain, husk, and stem with leaf sampled from outdoor field near heavily industrialized area was significantly elevated. BaP can also be detected in foodstuff after the process of smoking or barbecuing. The level of BaP could reach 100 μg/kg in smoked food [82].

Inhalation of cigarette smoke is another major pathway of human exposure to BaP [87]. There is strong evidence linking cigarette smoking and lung cancer although the relative contribution of the different toxicants are still under debate [48]. It has been estimated that each cigarette contains approximately 25 ng of BaP [56]. Unfiltered and filtered cigarette contributes to 0.7 μg/day or 0.4 μg/day of BaP intake in an average smoker. A survey of 20 brands of cigarette marketed in China prior to 2004 indicated that BaP level in cigarettes made in China was 2.5 folds higher than the European brands [1]. Tar and nicotine levels have been made to conform to EU standards as of 1 July 2004. As tar level was found to correlate with carcinogenic PAH ($r = 0.53$, $P < 0.01$), BaP and PAHs in cigarettes may have also reduced [1].

Biotransformation of BaP

BaP is not a direct mutagen. In *in vitro* test for mutagenicity, BaP becomes mutagenic in the presence of liver homogenate [10,38]. Thus, the conversion of BaP to active metabolites by liver enzymes accounts for its mutagenicity [4,59,61]. Metabolism of BaP depends on the microsomal mixed function oxygenase system involving the cytochrome P450 enzymes [3]. BaP can be sequentially converted to epoxidated and hydroxylated products and with further conversion to quinone (BPQ) (Figure 2) [28]. The preferred sites of conversion are the 7-8, 4-5 and 9-10 positions where steric hindrance is lowest. Hydroxylation facilitates the excretion through urine [29,91,99]. The major biotransformed products of BaP is the BaP-7,8-diol-9,10 epoxide (BPDE).

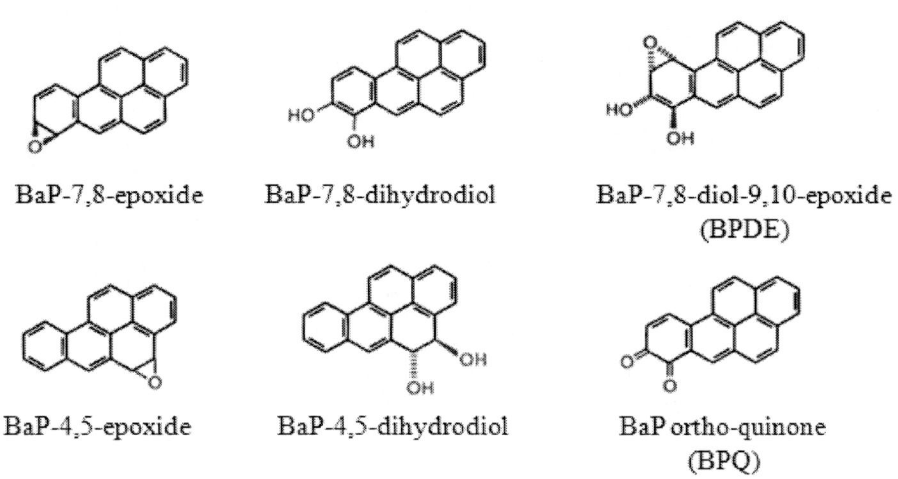

Figure 2. Some important metabolites of BaP.

In addition to the phase I biotransformation by cytochrome P450s, BaP can also undergo phase II enzymatic conjugation by glutathione S-transferase or uridine diphosphate [UDP]-glucuronosyltransferase (GST) [7]. The conjugated products could also be excretion through the bile. The phase I and phase II enzymes are inducible and serve as an adaptive response to exposure to organic toxicants [7]. Two possible nuclear receptors were identified for enzyme induction: one involves the xenobiotic-response element (XRE) and the other one involves the antioxidant response element (ARE), also known as the electrophilic response element (EpRE). The cytochrome P450 gene, CYP1A1, are largely regulated by the XRE, whereas other such as GST are largely regulated by ARE. BaP could also induce DT-diaphorase [79] via the XRE, whereas electrophilic and /or redox active metabolites of BaP are responsible for inducing glutathione S-transferase as well as DT-diaphorase via ARE.

Induction of cytochrome P450 can be reflected by an increased transcription of the CYP1A1 gene. Both the AhR and the AhR nuclear translocator (Arnt), mediates induction [98] (Figure 3). Pretreatment with alpha-naphthoflavone, the inhibitor of the aryl hydrocarbon receptor (AhR) prevented the induction of cytochorme P450 [107]. BaP treatment induced expression of the CYP1A1 gene in the skin and liver of AhR-positive mice bearing +/+ and +/- genotypes and no induction was detected in AhR-null (-/-) mouse liver. The presence of AhR correlates with carcinogenicity in these animals. Tumor at the site of injection was evidenced in the AhR +/+ and +/- mice but not in the AhR(-/-) strain [85].

The importance of AhR and Arnt in mediating BaP toxicity was conducted using two human lung adenocarcinoma cell lines, A427 and CL3 [13]. In the A427 cells, AhR and/or Arnt function is impaired while those in the CL3 cells were normal. Protein levels of AhR and Arnt were similar in A427 and CL3 cells before BaP treatment. Both enzyme activity and mRNA levels of CYP1A1 were highly induced following treatment with 1 or 10 μM BaP for 24 hr in CL3 cells but not in A427 cells. DNA adducts were induced by BaP treatment in CL3 cells, however, A427 cells were much less responsive to BaP treatment. Co-treatment with alpha-naphthoflavone, a CYP1A1 inhibitor, abolished DNA adduct formation by BaP in CL3 cells. Nevertheless, CYP1A1 inducer beta-naphthoflavone, enhanced DNA adduct formation by BaP in both A427 and CL3 cells.

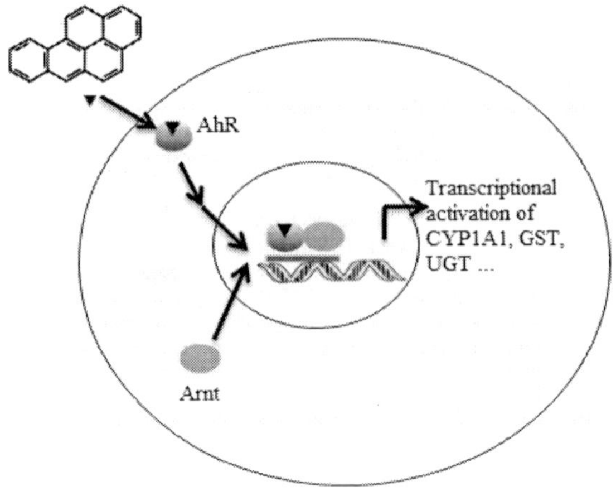

Figure 3. A simplified diagram of AhR mediated CYP450 gene induction.

TOXICITY AND TOXICOLOGY OF BAP

Carcinogenicity

Carcinogenicity and cytotoxicity are the major toxic resposnes of most PAHs. BaP has been shown to cause cancer in mouse skin and lung [22,36,84,86] at a dose-dependent manner [30,35,77]. Studies have also shown that exposure of isolated rat hepatocytes to [^{3}H] benzo[a]pyrene-7,8-dione (BPQ) resulted in a sequestration of radioactivity in cell pellet. Approximate 30 adducts were identified per 10^6 base pairs. There were significant DNA fragmentation and the effect was attenuated by indomethacin, a dihydro-dehydrogenase inhibitor. Treated with either 7,8-dihydroxy benzo[a]pyrene [28] or benzo[a]pyrene-7,8-dione (BPQ) can produce large quantities of superoxide anion radical ($O_2^{\cdot-}$) [27]. The amount of $O_2^{\cdot-}$ generated by BPQ can also be blocked by indomethacin suggesting that reactive oxygen species (ROS) were generated and it was the free radicals which caused DNA fragmentation. Furthermore, BPQ could cause a concentration-dependent (0.05-10 mM) DNA strand scission in the supercoiled phi X174 DNA. Strand scission was enhanced under conditions which promoted redox-cycling (e.g. presence of NADPH and $CuCl_2$) and prevented by catalase and hydroxyl radical scavengers but not by superoxide dismutase [27]. Anitoxidants such as polyphenol [70,95] could suppress BaP-induced skin cancer further support the role of free radical in mediating BaP-induced DNA damage.

Immunosuppressant

Benzo[a]pyrene is a potential immunosuppressant. The action is caused by death, leading to a reduction of the immune cells number. BaP has been shown to cause cell death in lymphocytes [33], macrophages [15,92], monocytes [94] and bone marrow cells [104]. Macrophages are the cell type that can metabolize BaP to BPDE [49]. Thus, production of BPDE caused macrophage cell death and reduce phagocytic activity.

Atherosclerosis

BaP is also toxic to smooth muscle cells [41,42,63,65,68,69,103] leading to atherosclerotic lesion. Again, the origin of atherosclerosis is associated with CYP450-mediated conversion of the parent compound to toxic metabolite. Smooth muscle cells respond to BaP via binding to the AhR. Mice that are AhR-sensitive are more susceptible to atherosclerosis than the Ah-resistant mice [64]. There are two possible hypotheses for BaP-mediated atherosclerosis [64]. Benditt and Benditt [6] suggested that the atherosclerotic plaque may be a neoplastic growth of transformed smooth muscle cells. Another hypothesis is that BaP could produce ROS, oxidize proteins that are potentially involved in the ARE/EpRE protein complex. The oxidation of protein activation induces binding to transcriptional element and causing transcriptional activation of proto-oncogene such as *c-Ha-ras*, resulting in elevated growth and proliferation of vascular smooth muscle cells involved in plaque formation [63].

Reproduction

BaP has been considered an endocrine disruptor as it exerts it action by mimicking hormones. Acute exposure to BaP could lower plasma progesterone, estrogen, and prolactin concentrations and compromised fetal survival in F-344 rats [2]. In adult female mice exposed to BaP, free radical production is a major factors that affects embryo development [97]. In addition, deficiency in p53-phenotype is also a risk factor [97]. Transgenic mice with +/- or -/- in the p53 tumor suppressor gene are more susceptible to BaP teratogenicity. Incidentally, DNA adduct from interacting with BPDE could be transferred from heavy smoking father to the child via sperms [106].

BaP-induced Cytotoxicity

While the carcinogenic potential was observed at nanomolar concentration, the ability of BaP to induce cell death was mostly at micromolar concentration. BaP can cause death in the lung cells [12,16,45,101,102,108], liver cells [31,32,43,55], breast tumor cells [5,9,62,76,81,89,90], skin cells [25,66,71,75] and other types of cell [21,72,73,80,83]. In majority of the cases, BaP induces cell death through apoptosis. The binding of BaP to AhR leading to translocation and binding to Arnt in the promoter region of the CYP1A genes (Figure 3). Induction of cytochrome P450 subsequently leads to an increase in BaP biotransformation and production of oxidative metabolties (BPQ), which in turn, cause cell death via oxidative injuries.

Assessment of BaP Toxicity in Human vs. Animal Cell Models

Although it is well established that BaP can cause cancer in animal models, there is a lack of clear experimental evidence to demonstrate a dose-dependent increase in cancer in human subjects. Epidemiological data, although numerous, is often difficult to interpret as it is difficult to isolate incidence in which BaP is the single organic toxicant involved. Another way for assessing BaP exposure is to quantify BaP metabolites in human urine. Studies were conducted in workers engaged in high exposure industries or in smokers to correlate BaP exposure to urinary BPDE level. In many cases, urinary BPDE were not detected or detected at very low level (fmole/μg creatinine) [14]. Correlation between BaP level in the environment and the urinary metabolites level cannot be confirmed despite continuous improvement in the detection methods [26,93]. However, by understanding the mode of carcinogenic action of BaP, it is relatively safe to imply that BaP is also mutagenic to human and thus, must be under continuous monitoring and regulation. Using mathematical modeling methods, human respiratory tract tumor was assessed upon exposure to BaP [19,67]. The question remains: is it justified to extrapolate human toxicity using animal data? There are still controversies on the does-dependent changes in cells from different animal species to

BaP exposure. The abundancy in AhR receptor, the level of enzyme induction, the level of proteins along the signaling pathways, etc. have all been considered essential to distinguish the susceptibility of cells to BaP toxicity.

Diversity in Enzyme Induction

The liver is the major site of BaP biotransformation. In the liver, BaP can be rapidly transformed into hydrophilic metabolites and excreted. Genetic features that regulate cytochrome P450 induction may underlies the differential susceptibility of animals to BaP toxicity. Animals with higher enzyme inducing potential may be more susceptible to BaP-induced toxicity [96]. In cell culture studies, Zeiger et al., [105] showed that the cytochrome P450 induction potency in cells from human is lower than those from rats. The rat hepatocytes and rat hepatoma cells were more sensitive to various PCB congeners including TCDD in cytochrome P450 induction when compare to the human hepatoma HepG2 cells. The difference ranges from 5 to 1200 folds depending on the congener tested. Studies in both the mouse hepatoma cells Hepa1c1c7 and the human HepG2 cells showed that both cell lines resposne to BaP by activation of the AhR receptor and cytochrome P450 induction. In the Hepa1c1c7 cell line, BaP causes apoptotic cell death [17,46,47] while in the HepG2 cells, BaP caused cellular necrosis [55]. Despite the different mode of cell death, the signaling pathway involves the MAPK and p53 phosphorylation [54]. Although there is still uncertainties on the difference in the response between the two cell lines, it was suggested that the response of the Hepa1c1c7 may be due to the abundance in AhR which potentiates BaP-induced apoptosis [17,55].

Diversity in Biochemical and Molecular Biological Responses

Besides the difference in response to BaP between animal species, response of different cell types from the same species were also observed. Comparison was made between the human liver HepG2 and MCF-7 breast tumor cells exposed to BaP using DNA microarray analysis [39]. MCF-7 and HepG2 cells are both human cell lines competent in metabolically activating BaP and express wild-type p53. However, similar treatment in HepG2 and MCF-7 resulted in different patterns of gene expression. Of over 200 genes that reported changes in their expression level, only 27 were altered in both cells. They are predominantly involved in xenobiotic metabolism, cell cycle regulation, chromatin assembly and oxidative response. Detail examination of the overall response in the HepG2 cells and the MCF-7 breast tumor cells showed that both AhR and p53 are important in regulating expression of these genes [40]. In another study, Castorena-Torres et al. [11] showed that there were 293 commonly expressed genes amongst the over 1952 gene in HepG2 and 784 genes in the human lung A549 cells following exposure to BaP. Similar to the study by Hockley et al. [39], genes for xenobiotic metabolism, cell proliferation, DNA repair were detected. In addition, genes for oxidative stress, protein degradation, ion transport, steroid synthesis were also up regulated.

Another function affected following BaP exposure was a repression of the histone genes for DNA repair. BaP caused a transient repression of the histone mRNA in the HepG2 cells while in the MCF-7 cells, the suppression did not recover even at 24 hr following exposure [39]. Other studies have also demonstrated a deficiency in DNA repair being important features of BaP-induced injury. Braithwaite et al. [8] reported that the nucleotide excision repair (NER) is an efficient method for removing DNA adducts formed from various BaP metabolites. Mice deficient in the nucleotide excision repair gene XPA have elevated sensitivity to BaP-induced lung tumors [44]. Coke oven workers with reduced NER genotype showed higher level of DNA adduct levels in mononuclear white blood cells [74]. Similarly, a reduced glutathione S-transferase μ1 (phase II conjugation enzyme) phenotype in these workers were also more susceptible to such injury [74]. Lin and Yang [55] showed that DNA damage as determined by the presence of 8-oxoguanine in the HepG2 cells occurred following the induction of biotransformation enzymes cytochrome P4501A and glutathione S-transferase. The DNA repairing enzyme poly-(ADP-ribose) polymerase (PARP-1) was also activated. Both NAD^+, the substrate of PARP-1, and pyruvate were able to reverse the BaP-induced cell death. Glutamine and glutamate were unable to protect BaP-induced cell death. These results were taken to indicate that cellular metabolism may also play a role in BaP-induced cell death, thus, contributing to yet another dimension to be considered in studying BaP toxicity. Finally, Largequvist et al. [50] demonstrated that a reduced in the fidelity of DNA replication by enhancing the error prone DNA replication may also increase the number of DNA adduct following BaP exposure in hamster cells lines.

CONCLUSION

The present review provides a current research status on the toxicity and toxicology of BaP. Ample evidences have demonstrated that there are differences in biochemical and molecular biological features amongst the different animal species as well as amongst individuals of the same species. These differences may underlie their differential response to BaP exposure. The use of DNA microarray analysis has also revealed some novel features (e.g. DNA repair) which may be involved in pathways that are still to be investigated. Are these pathways universal or are they cell types or species specific? The continuous use of microarray studies, the time course analysis on the progression of different pathways in different cell lines from different animal species may offer final data for assessing the toxicity of BaP in human. Another question in assessing BaP toxicity is that under normal condition, BaP coexists with other PAHs. In mouse experiment, application of sublethal concentration of BaP becomes more toxic if it was co-applied with other PAHs. Although BaP has been considered the major cancer causing congeners of PAHs, its action may be affected by other congeners present in the environment. Thus, the interrelation between BaP and other PAHs must be addressed. While it may not be necessary to consider BaP an independent environmental toxicant in assessing human health risk, the chemical represents a unique model for studying the response of cell to organic chemicals and contributes to the understanding of cellular biochemical and molecular biological response to chemically induced stress.

REFERENCES

[1] Akpan V, Huang S, Lodovici M, Dolara P: High levels of carcinogenic polycyclic aromatic hydrocarbons (PAH) in 20 brands of Chinese cigarettes. *J Appl Toxicol* 2006, 26: 480-483.

[2] Archibong AE, Inyang F, Ramesh A, Greenwood M, Nayyar T, Kopsombut P *et al.*: Alteration of pregnancy related hormones and fetal survival in F-344 rats exposed by inhalation to benzo(a)pyrene. *Reprod Toxicol* 2002, 16: 801-808.

[3] Ayrton AD, McFarlane M, Walker R, Neville S, Coombs MM, Ioannides C: Induction of the P-450 I family of proteins by polycyclic aromatic hydrocarbons: possible relationship to their carcinogenicity. *Toxicology* 1990, 60: 173-186.

[4] Balk L, Depierre JW, Sundvall A, Rannug U: Formation of mutagenic metabolites from benzo(a)pyrene and 2-aminoanthracene by the s-9 fraction from the liver of the Northern pike (Esox lucius): inducibility with 3-methylcholanthrene and correlation with benzo[a]pyrene monooxygenase activity. *Chem Biol Interact* 1982, 41: 1-13.

[5] Barbisan LF, Mello ML, Russo J, Vidal BC: Apoptosis and catastrophic cell death in benzo[a]pyrene-transformed human breast epithelial cells. *Mutat Res* 1999, 431: 133-139.

[6] Benditt EP, Benditt JM: Evidence for a monoclonal origin of human atherosclerotic plaques. *Proc Natl Acad Sci U S A* 1973, 70: 1753-1756.

[7] Bock KW, Lipp HP, Bock-Hennig BS: Induction of drug-metabolizing enzymes by xenobiotics. *Xenobiotica* 1990, 20: 1101-1111.

[8] Braithwaite E, Wu X, Wang Z: Repair of DNA lesions: mechanisms and relative repair efficiencies. *Mutat Res* 1999, 424: 207-219.

[9] Calaf GM, Alvarado ME, Hei TK: Beta catenin is associated with breast cancer progression in vitro. *Int J Oncol* 2005, 26: 913-921.

[10] Calle LM, Sullivan PD: Screening of antioxidants and other compounds for antimutagenic properties towards benzo[a]pyrene-induced mutagenicity in strain TA98 of Salmonella typhimurium. *Mutat Res* 1982, 101: 99-114.

[11] Castorena-Torres F, Bermudez dL, Cisneros B, Zapata-Perez O, Salinas JE, Albores A: Changes in gene expression induced by polycyclic aromatic hydrocarbons in the human cell lines HepG2 and A549. *Toxicol In Vitro* 2008, 22: 411-421.

[12] Catallo WJ, Kennedy CH, Henk W, Barker SA, Grace SC, Penn A: Combustion products of 1,3-butadiene are cytotoxic and genotoxic to human bronchial epithelial cells. *Environ Health Perspect* 2001, 109: 965-971.

[13] Chang KW, Lee H, Wang HJ, Chen SY, Lin P: Differential response to benzo[a]pyrene in human lung adenocarcinoma cell lines: the absence of aryl hydrocarbon receptor activation. *Life Sci* 1999, 65: 1339-1349.

[14] Chen YL, Wang CJ, Wu KY: Analysis of N7-(benzo[a]pyrene-6-yl)guanine in urine using two-step solid-phase extraction and isotope dilution with liquid chromatography/tandem mass spectrometry. *Rapid Commun Mass Spectrom* 2005, 19: 893-898.

[15] Chin BY, Choi ME, Burdick MD, Strieter RM, Risby TH, Choi AM: Induction of apoptosis by particulate matter: role of TNF-alpha and MAPK. *Am J Physiol* 1998, 275: L942-L949.

[16] Chu CY, Chang JP, Wang CJ: Modulatory effect of piperine on benzo[a]pyrene cytotoxicity and DNA adduct formation in V-79 lung fibroblast cells. *Food Chem Toxicol* 1994, 32: 373-377.

[17] Chung JY, Kim JY, Kim WR, Lee SG, Kim YJ, Park JE, Hong YP, Chun, YJ, Park YC, Oh S, Yoo YH, Kim JM.: Abundance of aryl hydrocarbon receptor potentiates benzo[a]pyrene-induced apoptosis in Hepa1c1c7 cells via CYP1A1 activation. *Toxicology* 2007, 235: 62-72.

[18] Chung MK, Hu R, Cheung KC, Wong MH: Pollutants in Hong Kong soils: polycyclic aromatic hydrocarbons. *Chemosphere* 2007, 67: 464-473.

[19] Cohen SM, Ellwein LB: Proliferative and genotoxic cellular effects in 2-acetylaminofluorene bladder and liver carcinogenesis: biological modeling of the ED01 study. *Toxicol Appl Pharmacol* 1990, 104: 79-93.

[20] Daynes RA, Harris CC, Connor RJ, Eichwald EJ: Skin cancer development in mice exposed chronically to immunosuppressive agents. *J Natl Cancer Inst* 1979, 62: 1075-1081.

[21] Dhalluin S, Gate L, Vasseur P, Tapiero H, Nguyen-Ba G: Dysregulation of ornithine decarboxylase activity, apoptosis and Bcl-2 oncoprotein in Syrian hamster embryo cells stage-exposed to di(2-ethylhexyl)phthalate and tetradecanoylphorbol acetate. *Carcinogenesis* 1997, 18: 2217-2223.

[22] Evans CD, LaDow K, Schumann BL, Savage RE, Jr., Caruso J, Vonderheide A *et al.*: Effect of arsenic on benzo[a]pyrene DNA adduct levels in mouse skin and lung. *Carcinogenesis* 2004, 25: 493-497.

[23] Evans CD, LaDow K, Schumann BL, Savage RE, Jr., Caruso J, Vonderheide A, Succop P, Talaska G: Effect of arsenic on benzo[a]pyrene DNA adduct levels in mouse skin and lung. *Carcinogenesis* 2004, 25: 493-497.

[24] Feng Y, Shi G, Wu J, Wang Y, Zhu T, Dai S, Pei Y: Source analysis of particulate-phase polycyclic aromatic hydrocarbons in an urban atmosphere of a northern city in China. *J Air Waste Manag Assoc* 2007, 57: 164-171.

[25] Feo F, Pirisi L, Pascale R, Daino L, Frassetto S, Garcea R, Gaspal L: Modulatory effect of glucose-6-phosphate dehydrogenase deficiency on benzo(a)pyrene toxicity and transforming activity for in vitro-cultured human skin fibroblasts. *Cancer Res* 1984, 44: 3419-3425.

[26] Ferreira Junior MF, Tas S, dell'Omo M, Goormans G, Buchet JP, Lauwerys R: Determinants of benzo(a)pyrenediol epoxide adducts to haemoglobin in workers exposed to polycyclic aromatic hydrocarbons. *Occup Environ Med* 1994, 51: 451-455.

[27] Flowers L, Bleczinski WF, Burczynski ME, Harvey RG, Penning TM: Disposition and biological activity of benzo[a]pyrene-7,8-dione. A genotoxic metabolite generated by dihydrodiol dehydrogenase. *Biochemistry* 1996, 35: 13664-13672.

[28] Flowers-Geary L, Harvey RG, Penning TM: Examination of diols and diol epoxides of polycyclic aromatic hydrocarbons as substrates for rat liver dihydrodiol dehydrogenase. *Chem Res Toxicol* 1992, 5: 576-583.

[29] Forster K, Preuss R, Rossbach B, Bruning T, Angerer J, Simon P: 3-Hydroxybenzo[a]pyrene in the urine of workers with occupational exposure to

polycyclic aromatic hydrocarbons in different industries. *Occup Environ Med* 2008, 65: 224-229.

[30] Gallagher JE, Jackson MA, George MH, Lewtas J: Dose-related differences in DNA adduct levels in rodent tissues following skin application of complex mixtures from air pollution sources. *Carcinogenesis* 1990, 11: 63-68.

[31] Gayda DP, Pariza MW: Effects of carcinogens on hormonal regulation of gene expression in primary cultures of adult rat hepatocytes. *Carcinogenesis* 1983, 4: 1127-1131.

[32] Gorria M, Tekpli X, Rissel M, Sergent O, Huc L, Landvik N, Fardel O, Dimanche-Roitrel MT, Holme JA, Lagadic-Gossmann D: A new lactoferrin- and iron-dependent lysosomal death pathway is induced by benzo[a]pyrene in hepatic epithelial cells. *Toxicol Appl Pharmacol* 2008, 228: 212-224.

[33] Hardin JA, Hinoshita F, Sherr DH: Mechanisms by which benzo[a]pyrene, an environmental carcinogen, suppresses B cell lymphopoiesis. *Toxicol Appl Pharmacol* 1992, 117: 155-164.

[34] Hattemer-Frey HA, Travis CC: Benzo-a-pyrene: environmental partitioning and human exposure. *Toxicol Ind Health* 1991, 7: 141-157.

[35] Hecht SS, Isaacs S, Trushin N: Lung tumor induction in A/J mice by the tobacco smoke carcinogens 4-(methylnitrosamino)-1-(3-pyridyl)-1-butanone and benzo[a]pyrene: a potentially useful model for evaluation of chemopreventive agents. *Carcinogenesis* 1994, 15: 2721-2725.

[36] Hecht SS, Kenney PM, Wang M, Upadhyaya P: Benzyl isothiocyanate: an effective inhibitor of polycyclic aromatic hydrocarbon tumorigenesis in A/J mouse lung. *Cancer Lett* 2002, 187: 87-94.

[37] Hecht SS, Kenney PM, Wang M, Upadhyaya P: Benzyl isothiocyanate: an effective inhibitor of polycyclic aromatic hydrocarbon tumorigenesis in A/J mouse lung. *Cancer Lett* 2002, 187: 87-94.

[38] Hermann M, Chaude O, Weill N, Bedouelle H, Hofnung M: Adaptation of the salmonella/mammalian microsome test to the determination of the mutagenic properties of mineral oils. *Mutat Res* 1980, 77: 327-339.

[39] Hockley SL, Arlt VM, Brewer D, Giddings I, Phillips DH: Time- and concentration-dependent changes in gene expression induced by benzo(a)pyrene in two human cell lines, MCF-7 and HepG2. *Bmc Genomics* 2006, 7.

[40] Hockley SL, Arlt VM, Brewer D, Te PR, Workman P, Giddings I, Giddings I, Phillips DH: AHR- and DNA-damage-mediated gene expression responses induced by benzo(a)pyrene in human cell lines. *Chem Res Toxicol* 2007, 20: 1797-1810.

[41] Holderman MT, Miller KP, Dangott LJ, Ramos KS: Identification of albumin precursor protein, Phi AP3, and alpha-smooth muscle actin as novel components of redox sensing machinery in vascular smooth muscle cells. *Mol Pharmacol* 2002, 61: 1174-1183.

[42] Holderman MT, Miller KP, Ramos KS: Activation of nuclear protein binding to the antioxidant/electrophile response element in vascular smooth muscle cells by benzo(a)pyrene. *Biochem Biophys Res Commun* 2000, 267: 12-16.

[43] Holme JA, Gorria M, Arlt VM, Ovrebo S, Solhaug A, Tekpli X, landvik NE, Huc L, Fardel O, Lagadic-Gossmann D: Different mechanisms involved in apoptosis following

exposure to benzo[a]pyrene in F258 and Hepa1c1c7 cells. *Chem Biol Interact* 2007, 167: 41-55.

[44] Ide F, Iida N, Nakatsuru Y, Oda H, Tanaka K, Ishikawa T: Mice deficient in the nucleotide excision repair gene XPA have elevated sensitivity to benzo[a]pyrene induction of lung tumors. *Carcinogenesis* 2000, 21: 1263-1265.

[45] Jacobsen NR, Saber AT, White P, Moller P, Pojana G, Vogel U, Loft S, Gingerich J, Soper, L, Douglas GR, Wallin H: Increased mutant frequency by carbon black, but not quartz, in the lacZ and cII transgenes of muta mouse lung epithelial cells. *Environ Mol Mutagen* 2007, 48: 451-461.

[46] Kim JY, Chung JY, Park JE, Lee SG, Kim YJ, Cha MS, Han MS, Lee HJ, Yoo YH, Kim JM: Benzo[a]pyrene induces apoptosis in RL95-2 human endometrial cancer cells by cytochrome P450 1A1 activation. *Endocrinology* 2007, 148: 5112-5122.

[47] Ko CB, Kim SJ, Park C, Kim BR, Shin CH, Choi S, Chung SY, Noh JH, Jeun JH, Kim NS, Park R: Benzo(a)pyrene-induced apoptotic death of mouse hepatoma Hepa1c1c7 cells via activation of intrinsic caspase cascade and mitochondrial dysfunction. *Toxicology* 2004, 199: 35-46.

[48] Kouri RE, Rude TH, Curren RD, Brandt KR, Sosnowski RG, Schechtman LM *et al.*: Biological activity of tobacco smoke and tobacco smoke-related chemicals. *Environ Health Perspect* 1979, 29: 63-69.

[49] Ladics GS, Kawabata TT, Munson AE, White KL, Jr.: Generation of 7,8-dihydroxy-9,10-epoxy-7,8,9,10-tetrahydrobenzo[a]pyrene by murine splenic macrophages. *Toxicol Appl Pharmacol* 1992, 115: 72-79.

[50] Lagerqvist A, Hakansson D, Prochazka G, Lundin C, Dreij K, Segerback D, Jernstrom B, Tomqvist M, Seidel A, Erixon K, Jenssen D: Both replication bypass fidelity and repair efficiency influence the yield of mutations per target dose in intact mammalian cells induced by benzo[a]pyrene-diol-epoxide and dibenzo[a,l]pyrene-diol-epoxide. *DNA Repair (Amst)* 2008, 7: 1202-1212.

[51] Li P, Li X, Stagnitti F, Zhang H, Lin X, Zang S, Zhou XD, Liu JM: Studies on the sources of benzo[a]pyrene in grain and aboveground tissues of rice plants. *J Hazard Mater* 2008.

[52] Li XH, Ma LL, Liu XF, Fu S, Cheng HX, Xu XB: Polycyclic aromatic hydrocarbon in urban soil from Beijing, China. *J Environ Sci (China)* 2006, 18: 944-950.

[53] Li YT, Li FB, Chen JJ, Yang GY, Wan HF, Zhang TB, Zheng XD, Liu JM: The concentrations, distribution and sources of PAHs in agricultural soils and vegetables from Shunde, Guangdong, China. *Environ Monit Assess* 2008, 139: 61-76.

[54] Lin T, Mak NK, Yang MS: MAPK regulate p53-dependent cell death induced by benzo[a]pyrene: involvement of p53 phosphorylation and acetylation. *Toxicology* 2008, 247: 145-153.

[55] Lin T, Yang MS: Benzo[a]pyrene-induced necrosis in the HepG(2) cells via PARP-1 activation and NAD(+) depletion. *Toxicology* 2008, 245: 147-153.

[56] Liu YY, Schmeltz I, Hoffmann D: Chemical studies on tobacco smoke. Quantitative analysis of hydrazine in tobacco and cigarette smoke. *Anal Chem* 1974, 46: 885-889.

[57] Luo XJ, Chen SJ, Mai BX, Yang QS, Sheng GY, Fu JM: Polycyclic aromatic hydrocarbons in suspended particulate matter and sediments from the Pearl River Estuary and adjacent coastal areas, China. *Environ Pollut* 2006, 139: 9-20.

[58] Ma LL, Chu SG, Wang XT, Cheng HX, Liu XF, Xu XB: Polycyclic aromatic hydrocarbons in the surface soils from outskirts of Beijing, China. *Chemosphere* 2005, 58: 1355-1363.

[59] Maack CA, Silva MH, Petrakis NL, Lee RE, Lyon M: Procarcinogen activation by rat and human mammary extracts. *Carcinogenesis* 1986, 7: 899-905.

[60] Meador JP, Stein JE, Reichert WL, Varanasi U: Bioaccumulation of polycyclic aromatic hydrocarbons by marine organisms. *Rev Environ Contam Toxicol* 1995, 143: 79-165.

[61] Melius P: Comparative benzo[a]pyrene metabolite patterns in fish and rodents. *Natl Cancer Inst Monogr* 1984, 65: 387-390.

[62] Mello ML, Barbisan LF, Lareef MH, Russo J, Vidal BC: Cell death evaluation in benzo[a]pyrene-transformed human breast epithelial cells after microcell-mediated transfer of chromosomes 11 and 17. *Mutat Res* 2004, 546: 39-43.

[63] Miller KP, Chen YH, Hastings VL, Bral CM, Ramos KS: Profiles of antioxidant/electrophile response element (ARE/EpRE) nuclear protein binding and c-Ha-ras transactivation in vascular smooth muscle cells treated with oxidative metabolites of benzo[a]pyrene. *Biochem Pharmacol* 2000, 60: 1285-1296.

[64] Miller KP, Ramos KS: Impact of cellular metabolism on the biological effects of benzo[a]pyrene and related hydrocarbons. *Drug Metab Rev* 2001, 33: 1-35.

[65] Miller KP, Ramos KS: DNA sequence determinants of nuclear protein binding to the c-Ha-ras antioxidant/electrophile response element in vascular smooth muscle cells: identification of Nrf2 and heat shock protein 90 beta as heterocomplex components. *Cell Stress Chaperones* 2005, 10: 114-125.

[66] Miller ML, Andringa A, Cody T, Dixon K, Albert RE: Cell proliferation and nuclear abnormalities are increased and apoptosis is decreased in the epidermis of the p53 null mouse after topical application of benzo[a]pyrene. *Cell Prolif* 1996, 29: 561-576.

[67] Moolgavkar SH, Luebeck G: Two-event model for carcinogenesis: biological, mathematical, and statistical considerations. *Risk Anal* 1990, 10: 323-341.

[68] Moorthy B, Miller KP, Jiang W, Ramos KS: The atherogen 3-methylcholanthrene induces multiple DNA adducts in mouse aortic smooth muscle cells: role of cytochrome P4501B1. *Cardiovasc Res* 2002, 53: 1002-1009.

[69] Moorthy B, Miller KP, Jiang W, Williams ES, Kondraganti SR, Ramos KS: Role of cytochrome P4501B1 in benzo[a]pyrene bioactivation to DNA-binding metabolites in mouse vascular smooth muscle cells: evidence from 32P-postlabeling for formation of 3-hydroxybenzo[a]pyrene and benzo[a]pyrene-3,6-quinone as major proximate genotoxic intermediates. *J Pharmacol Exp Ther* 2003, 305: 394-401.

[70] Mukhtar H, Wang ZY, Katiyar SK, Agarwal R: Tea components: antimutagenic and anticarcinogenic effects. *Prev Med* 1992, 21: 351-360.

[71] Nakayama J, Yuspa SH, Poirier MC: Benzo(a)pyrene-DNA adduct formation and removal in mouse epidermis in vivo and in vitro: relationship of DNA binding to initiation of skin carcinogenesis. *Cancer Res* 1984, 44: 4087-4095.

[72] Nwagbara O, Darling-Reed SF, Tucker A, Harris C, Abazinge M, Thomas RD, Gragg RD: Induction of cell death, DNA strand breaks, and cell cycle arrest in DU145 human

prostate carcinoma cell line by benzo[a]pyrene and benzo[a]pyrene-7,8-diol-9,10-epoxide. *Int J Environ Res Public Health* 2007, 4: 10-14.

[73] Patton WP, Routledge MN, Jones GD, Lewis SE, Archer DB, Davies RJ, Chakravarthy U: Retinal pigment epithelial cell DNA is damaged by exposure to benzo[a]pyrene, a constituent of cigarette smoke. *Exp Eye Res* 2002, 74: 513-522.

[74] Pavanello S, Pulliero A, Siwinska E, Mielzynska D, Clonfero E: Reduced nucleotide excision repair and GSTM1-null genotypes influence anti-B[a]PDE-DNA adduct levels in mononuclear white blood cells of highly PAH-exposed coke oven workers. *Carcinogenesis* 2005, 26: 169-175.

[75] Pirisi L, Pascale R, Daino L, Frassetto S, La S, V, Zanetti S, Gaspa L, Ledda GM, Garcea R, Feo F: Effect of glucose-6-phosphate dehydrogenase deficiency on the benz(a)pyrene toxicity for in vitro cultured human skin fibroblasts. *Res Commun Chem Pathol Pharmacol* 1982, 38: 301-311.

[76] Pliskova M, Vondracek J, Vojtesek B, Kozubik A, Machala M: Deregulation of cell proliferation by polycyclic aromatic hydrocarbons in human breast carcinoma MCF-7 cells reflects both genotoxic and nongenotoxic events. *Toxicol Sci* 2005, 83: 246-256.

[77] Pott F, Stober W: Carcinogenicity of airborne combustion products observed in subcutaneous tissue and lungs of laboratory rodents. *Environ Health Perspect* 1983, 47: 293-303.

[78] Qiao M, Wang C, Huang S, Wang D, Wang Z: Composition, sources, and potential toxicological significance of PAHs in the surface sediments of the Meiliang Bay, Taihu Lake, China. *Environ Int* 2006, 32: 28-33.

[79] Radjendirane V, Jaiswal AK: Coordinated induction of the c-jun gene with genes encoding quinone oxidoreductases in response to xenobiotics and antioxidants. *Biochem Pharmacol* 1999, 58: 597-603.

[80] Rorke EA, Sizemore N, Mukhtar H, Couch LH, Howard PC: Polycyclic aromatic hydrocarbons enhance terminal cell death of human ectocervical cells. *Int J Oncol* 1998, 13: 557-563.

[81] Sadikovic B, Rodenhiser DI: Benzopyrene exposure disrupts DNA methylation and growth dynamics in breast cancer cells. *Toxicol Appl Pharmacol* 2006, 216: 458-468.

[82] Santodonato J, Howard P, Basu D: Health and ecological assessment of polynuclear aromatic hydrocarbons. *J Environ Pathol Toxicol* 1981, 5: 1-364.

[83] Sharma A, Neekhra A, Gramajo AL, Patil J, Chwa M, Kuppermann B, Kenney MC: Benzo(e)Pyrene Induces Apoptosis in Human Retinal Pigment Epithelial Cells in vitro. *Invest Ophthalmol Vis Sci* 2008.

[84] Sharma S, Gao P, Steele VE: The chemopreventive efficacy of inhaled oltipraz particulates in the B[a]P-induced A/J mouse lung adenoma model. *Carcinogenesis* 2006, 27: 1721-1727.

[85] Shimizu Y, Nakatsuru Y, Ichinose M, Takahashi Y, Kume H, Mimura J, Fujii-Kuriyama Y, Ishikawa T: Benzo[a]pyrene carcinogenicity is lost in mice lacking the aryl hydrocarbon receptor. *Proc Natl Acad Sci U S A* 2000, 97: 779-782.

[86] Slaga TJ: Overview of tumor promotion in animals. *Environ Health Perspect* 1983, 50: 3-14.

[87] Smith CJ, Perfetti TA, Rumple MA, Rodgman A, Doolittle DJ: "IARC group 2A Carcinogens" reported in cigarette mainstream smoke. *Food Chem Toxicol* 2000, 38: 371-383.

[88] Southworth GR: The role of volatilization in removing polycyclic aromatic hydrocarbons from aquatic environments. *Bull Environ Contam Toxicol* 1979, 21: 507-514.

[89] Tampio M, Loikkanen J, Myllynen P, Mertanen A, Vahakangas KH: Benzo(a)pyrene increases phosphorylation of p53 at serine 392 in relation to p53 induction and cell death in MCF-7 cells. *Toxicol Lett* 2008, 178: 152-159.

[90] Tsuruda L, Hou Y, Penning TM: Stable expression of rat dihydrodiol dehydrogenase (AKR1C9) in human breast MCF-7 cells results in the formation of PAH-o-quinones and enzyme mediated cell death. *Chem Res Toxicol* 2001, 14: 856-862.

[91] Tzekova A, Thuot R, Viau C: Correlation between biomarkers of polycyclic aromatic hydrocarbon exposure and electrophilic tissue burden in a rat model. *Arch Toxicol* 2004, 78: 351-361.

[92] Urso P, Zhang W, Cobb JR: Immunological consequences from exposure to benzo(a)pyrene during pregnancy. *Scand J Immunol Suppl* 1992, 11: 203-206.

[93] Varga C, Horvath G, Pocsai Z, Timbrell V: On the mechanism of cogenotoxic action between ingested amphibole asbestos fibres and benzo[a]pyrene: I. Urinary and serum mutagenicity studies with rats. *Cancer Lett* 1998, 128: 165-169.

[94] Vayssier-Taussat M, Camilli T, Aron Y, Meplan C, Hainaut P, Polla BS, Weksler B: Effects of tobacco smoke and benzo[a]pyrene on human endothelial cell and monocyte stress responses. *Am J Physiol Heart Circ Physiol* 2001, 280: H1293-H1300.

[95] Wang ZY, Khan WA, Bickers DR, Mukhtar H: Protection against polycyclic aromatic hydrocarbon-induced skin tumor initiation in mice by green tea polyphenols. *Carcinogenesis* 1989, 10: 411-415.

[96] Watanabe M, Watanabe K, Konno K, Sato H: Genetic differences in the induction of aryl hydrocarbon hydroxylase and benzo(a)pyrene carcinogenesis in C3H/He and DBA/2 strains of mice. *Gann* 1975, 66: 217-226.

[97] Wells PG, Kim PM, Laposa RR, Nicol CJ, Parman T, Winn LM: Oxidative damage in chemical teratogenesis. *Mutat Res* 1997, 396: 65-78.

[98] Whitlock JP, Jr.: Induction of cytochrome P4501A1. *Annu Rev Pharmacol Toxicol* 1999, 39: 103-125.

[99] Wu MT, Simpson CD, Christiani DC, Hecht SS: Relationship of exposure to coke-oven emissions and urinary metabolites of benzo(a)pyrene and pyrene in coke-oven workers. *Cancer Epidemiol Biomarkers Prev* 2002, 11: 311-314.

[100] Wu SP, Tao S, Xu FL, Dawson R, Lan T, Li BG, Cao J: Polycyclic aromatic hydrocarbons in dustfall in Tianjin, China. *Sci Total Environ* 2005, 345: 115-126.

[101] Wu X, Yu H, Amos CI, Hong WK, Spitz MR: Joint effect of insulin-like growth factors and mutagen sensitivity in lung cancer risk. *J Natl Cancer Inst* 2000, 92: 737-743.

[102] Xiao H, Rawal M, Hahm ER, Singh SV: Benzo[a]pyrene-7,8-diol-9,10-epoxide causes caspase-mediated apoptosis in H460 human lung cancer cell line. *Cell Cycle* 2007, 6: 2826-2834.

[103] Xu CB, Stavenow L, Pessah-Rasmussen H: Interactions between cultured bovine arterial endothelial and smooth muscle cells; further studies on the effects of injury and modification of the consequences of injury. *Artery* 1993, 20: 163-179.

[104] Yamaguchi K, Matulka RA, Shneider AM, Toselli P, Trombino AF, Yang S, Hafer LJ, Mann KK, Tao XJ, Tilly JL, Near RI, Sherr DH: Induction of PreB cell apoptosis by 7,12-dimethylbenz[a]anthracene in long-term primary murine bone marrow cultures. *Toxicol Appl Pharmacol* 1997, 147: 190-203.

[105] Zeiger M, Haag R, Hockel J, Schrenk D, Schmitz HJ: Inducing effects of dioxin-like polychlorinated biphenyls on CYP1A in the human hepatoblastoma cell line HepG2, the rat hepatoma cell line H4IIE, and rat primary hepatocytes: comparison of relative potencies. *Toxicol Sci* 2001, 63: 65-73.

[106] Zenzes MT, Puy LA, Bielecki R, Reed TE: Detection of benzo[a]pyrene diol epoxide-DNA adducts in embryos from smoking couples: evidence for transmission by spermatozoa. *Mol Hum Reprod* 1999, 5: 125-131.

[107] Zhao W, Ramos KS: Inhibition of DNA synthesis in primary cultures of adult rat hepatocytes by benzo[a]pyrene and related aromatic hydrocarbons: role of Ah receptor-dependent events. *Toxicology* 1995, 99: 179-189.

[108] Zhu H, Gooderham N: Neoplastic transformation of human lung fibroblast MRC-5 SV2 cells induced by benzo[a]pyrene and confluence culture. *Cancer Res* 2002, 62: 4605-4609.

In: Polycyclic Aromatic Hydrocardons: Pollution, Health... ISBN 978-1-60741-462-9
Editors: Pierre A. Haines and Milton D. Hendrickson

Chapter 11

CHROMATOGRAPHIC ANALYSIS OF POLYCYCLIC AROMATIC HYDROCARBONS BY Π-ELECTRON CONTAINING POLYMERIC ORGANIC PHASE-GRAFTED ON SILICA

M. Mizanur Rahman [1†,*], Makoto Takafuji [1], Ashequl A. Rana [1], Abul K. Mallik [1], Usha G. Gautam [1], Hirotaka Ihara [1†]

[1]Department of Applied Chemistry and Biochemistry,
Kumamoto University, Kumamoto, Japan

ABSTRACT

Polycyclic aromatic hydrocarbons (PAHs) are well-known organic compounds derived from incomplete combustion of organic matters. PAHs have been widely studied in a broad field such as material science, industrial technologies and pharmaceutical sciences, as well known environmental pollutants, and an increasing particular concern has been paid to their adverse harmful effects to human due to the carcinogenic and mutagenic properties of many PAH species. Clearly, the concerns over adverse health effects of PAHs in human and wildlife provide the impetus for research on their sources, environmental fate and food chain transfer. High performance liquid chromatography (HPLC) is useful and conventional method to monitor PAHs. More than 60% of all HPLC separations are carried out under reversed-phase (RP) conditions and alkylated-silicas were successfully employed for the separation of mixtures of PAHs. In RP-HPLC chromatography, not only long alkyl chains but also aromatic compounds such as benzene, pyrene, and hetero aromatic selector such as vinylpyridine have been investigated for organic phase of stationary phase so far. However, development of new stationary phases has been required for separation of similar compounds such as structural isomers, positional isomers and optical isomers. In this review article, we focus on the chromatographic analysis of PAHs by π-electron containing ordered and

* Present Address. Department of Applied Chemistry and Chemical Engineering, Faculty of Engineering and Technology, University of Dhaka, Dhaka-1000, Bangladesh

† Tel: +81-96-342-3661, fax: +81-96-342-3662, E-mail: ihara@kumamoto-u.ac.jp, mizan607@gmail.com

disordered polymeric organic phase-grafted silica. Polymeric phase-grafted silica stationary phases exhibits unexpected enhancement of separation for PAHs by a multiple-interaction mechanism, even if the polymer does not form any specific secondary structure. Poly(octadecylacrylate) and poly(4-vinylpyridine)-grafted mesoporous silica particles (Sil-ODA$_n$ and Sil-VP$_n$, respectively) will be described as typical examples of such stationary phases. Effects of grafting method and grafting-density of polymeric phases on the separation of PAHs in HPLC will be also discussed. Extensive chromatographic studies showed that selectivity for PAHs can be enhanced by multiple π-π interactions between the π-electron containing polymeric phases and the guest PAHs molecules. The presence of aligned carbonyl groups in a crystalline state of poly(octadecylacrylate) on silica significantly enhanced the selectivity for PAHs by π-π interaction mechanism. However, Sil-VP$_n$ recognizes molecular shape (planar and disk-like) of PAHs through π-π interactions between pyridyl groups of grafted-polymer and PAHs, along with electrostatic interactions involving a inductive effects and /or a quadrupolar effect in the normal phase HPLC. Comparison of the HPLC separations for PAHs with conventional octadecylsilylated silica (ODS) demonstrates the efficiencies and effectiveness of these π-electron containing polymeric organic phases grafted silica particles.

1. Sources of Polycyclic Aromatic Hydrocarbons (PAHs)

Polycyclic aromatic hydrocarbons (PAHs) are a large group of organic compounds with two or more fused aromatic rings. They have a relatively low solubility in water, but are highly lipophilic. Most of the PAHs with low vapor pressure in the air are adsorbed on particles. When dissolved in water or adsorbed on particulate matter, PAHs can undergo photodecomposition under ultraviolet light from solar radiation. In the atmosphere, PAHs can react with pollutants such as ozone, nitrogen oxides and sulfur dioxide, yielding diones, nitro- and dinitro-PAHs, and sulfonic acids, respectively. PAHs may also be degraded by some microorganisms in the soil [1]. PAHs are formed mainly as a result of pyrolytic processes, especially the incomplete combustion of organic materials during industrial and other human activities, such as processing of coal and crude oil, combustion of natural gas, including for heating, combustion of refuse, vehicle traffic, cooking and tobacco smoking, as well as in natural processes such as carbonization. There are several hundred PAHs; the best known is benzo[*a*]pyrene (BaP). The contributions from the different important sources, such as residential heating (coal, wood, and oil), vehicle exhausts, industrial power generation, incinerators, the production of coal tar, coke and asphalt, and petroleum catalytic cracking, are very difficult to estimate. Eight carcinogenic PAHs in the environment are: benzo[*a*]anthracene, chrysene, benzo[*b*]fluoranthene, benzo[*k*]fluoranthene, BaP, indeno[1,2,3-*cd*]pyrene, dibenzo[*ah*]anthracene and benzo[*ghi*]perylene. The exposures calculated were based on US statistics and analysis, and thus related to the situation in North America. Estimates were made for a "reference man" aged between 19 and 50 years and presented on a total body basis. In nonsmokers a mean total intake of 3.12 μg/day was estimated, of which food contributed 96.2%, air 1.6%, water 0.2% and soil 0.4%. Smokers consuming one pack of nonfiltered cigarettes per day had an estimated additional intake of 1-5 μg/day [2].

2. HEALTH AND MUTAGENIC EFFECT OF PAHS

Human exposure to PAHs occurs principally by direct inhalation, ingestion or dermal contact, as a result of the widespread presence and persistence of PAHs in the urban environment. PAHs are important environmental and occupational carcinogens and figure prominently in risk assessments for hazardous waste sites; coal-fired, municipal and hazardous waste incinerators and industrial property transfers. For example, exposure in iron and steel foundries entails exposure to PAHs and other potentially carcinogenic substances such as nickel, chromium, silica, soot, asbestos and benzene. Consequently, estimating the risk associated with exposure to specific PAHs requires controlled experimental data; otherwise, other substances may be responsible for the health outcome or may be interacting with the PAH of interest. Substances besides PAHs may account for a more significant portion of the carcinogenicity of some mixtures, such as cigarette smoke, diesel emissions and urban aerosol [3]. Several PAHs, including benz[*a*]anthracene, benzo(a)pyrene, benzo[*b*]fluoranthene, chrysene, dibenz[*ah*]anthracene and indeno[1,2,3-*cd*]pyrene have been shown to cause tumors in laboratory animals. Benzo[*a*]pyrene and pyrene are the most important carcinogenic PAHs and are components of combustion processes, coke oven and foundry emissions, cigarette smoke and charcoal-grilled meats [4]. The highest PAH levels are probably found in coke ovens, although the levels to which workers are exposed are not described in most epidemiological studies. Numerous studies of coke oven workers have provided evidence of a dose-response relationship between numerous PAHs and health endpoints, including lung cancer and depressed immune function. However, uncertainty surrounds the presence of a dose response relationship between PAHs and lung cancer in aluminium plant workers. Other health outcomes associated with PAHs exposure in aluminium plant workers includes urinary bladder cancer, asthma-like symptoms, lung function abnormalities and chronic bronchitis. In foundry workers, research indicates that PAHs exposure is associated with a greater risk of lung cancer, silicosis and other chronic respiratory abnormalities. One study found 83% of the total carcinogenic activity of side stream smoke was due to PAHs [5-7]. Indoor benzo[*a*]pyrene levels, from the burning of smoky coal, approached levels typically found in coke oven factories (14.7 μgm^{-3}). Given the long term evidence suggesting PAHs elevate the risk of various cancers, immunotoxic and respiratory problems, the IPCS made a number of recommendations for the protection of human health [3]. These recommendations included eliminating or minimizing emissions in occupational settings, improved monitoring of urban air pollution and public education. The complexity of investigating the health effects associated with PAHs exposure in the ambient air is reflected in the absence of research addressing this issue to date. Grimmer et al. [5-7] studied extracts of particulate matter from several air pollutant emission sources using carcinogenic effects in rats as an end point, and found unsubstituted PAHs with three rings to account for the total carcinogenic activity.

Bioassay-directed chemical analysis is a technique used to isolate important chemical mutagens in environmental samples, which contain a complex mixture of thousands of chemical species. Bacterial assays have been used as biological end points for most bioassay-directed chemical analysis for mutagens. For example, for emission sources, Nishioka et al. [8] found that nitro-polycyclic aromatic compounds (nitro-PACs) accounted for 20-25% of the bacterial mutagenic activity observed without further enzymatic activation of the assay (ie, no "-S9 test") of diesel exhaust extracts. Salmeen et al. [9] found that mono- and dinitro-PACs account for 30-40% of the bacterial mutagenicity (-S9) of diesel exhaust emissions. Recently Hannigan et al. [10] examined the human cell mutagenicity of Los Angeles airborne fine particulate matter, and found sub fractions containing unsubstituted polycyclic aromatic compounds (PACs) responsible for a considerable portion of the mutagenic potency of the whole atmospheric sample. Six unsubstituted PAHs (cyclopenta[*cd*]pyrene, benzo[*a*]pyrene, benzo[*ghi*]perylene, benzo[*b*]fluoranthene, indeno[1,2,3-*cd*]pyrene and benzo[*k*]fluoranthene) accounted for most of the mutagenic potencies that could be assigned to specific compounds within the atmospheric samples.

3. Analytical Methods for PAHs

The literature shows that analysis by high-performance liquid chromatography (HPLC), using fluorescence or UV-visible detection, or gas chromatography (GC), with mass spectroscopy (MS) or flame ionization detection (FID), are employed for identifying and quantifying PAH extracts from air samples. Capillary electrophoresis (CE) methods have emerged as important separation tools in chemical analysis and have the potential to rival traditional separation techniques such as GC and HPLC. These techniques involve the use of electro-osmotic flow (EOF), an electrically induced force that moves a mobile phase (electrophoretic buffer) through a packed bed column, to effect separation of different electrical species on the column. The most commonly employed method for PAHs investigation is the HPLC analysis. Chromatographic methods for PAH analyses in environmental media have been developed and evaluated extensively over the past few decades, as reviewed in the literature [11-14].

4. Brief History of Analysis of PAHs by HPLC

HPLC is the most used analytic technique for the examination of environmentally relevant matrices. For the chromatographic separation of PAHs conventional reversed-phases with alkyl chains that are chemically bonded to silica gel are usually used (as shown in Scheme 1). These reversed-phases differ in the length of their alkyl chains, the functionality of the used alkyl silanes, the coverage density of the bonded phases or the physical characteristics of the underlying silica gels. All of these factors affect the properties of stationary phases [15-17]. Pfleiderer et al. [18] were able to demonstrate, by means of solid-state NMR spectroscopy, that the mobility of chemically bonded alkyl chains depends on their length. Maximum mobility is reached for short chains containing six to eight carbon atoms. A further increase of the chain length causes a decline in the mobility due to a decrease in the order of the alkyl chains. The investigation by Pursch et al. [19] showed that, in the case of long alkyl chains of up to 30 carbon atoms, two domains of different orders exist simultaneously, differing in the chemical shifts of the main chain signals in the ^{13}C NMR spectra. These two domains have been assigned on the one hand to chains with all-*trans* conformations and on the other hand to chains with gauche conformations. The ratio between all-*trans* and *gauche* conformations is strongly influenced by the coverage density of the ligands and the temperature and has been found to control the elution order of vitamin A acetate isomers in HPLC. Thus, it is not surprising that a lot of conventional reversed-phase octadecylsilylated silica (ODS) or C_{18} phases are commercially available, showing totally different elution orders, e.g., in the case of the separation of PAHs [20]. These polymeric C_{18} phases seem to be more suitable for this separation task than monomeric phases. Sander [21] and Sander and Wise [22] developed a test mixture of three PAHs in order to rank the stationary phases into the categories monomeric, polymeric or intermediate, depending on the selectivity, $\alpha_{BaP/TBN}$, of the pair benzo[*a*]pyrene and tetrabenzonaphthalene. The problems arising from different properties of the reversed-phase (RP) materials also exist here. Besides, some pairs of PAHs cannot be resolved completely on such stationary phases [23, 24] which need rigorous chromatographic conditions. Since the adjustment of the mobile phase did not solve all problems arising in the chromatographic separation of the PAHs, it was necessary to adapt the stationary phase. The development of a stationary phase that could benefit from the possibility of entering into a further interaction with the solutes was an obvious step. In this review we will demonstrate highly-ordered self-assembled and dis-ordered non-self assembled polymer grafted stationary phases for PAHs investigation.

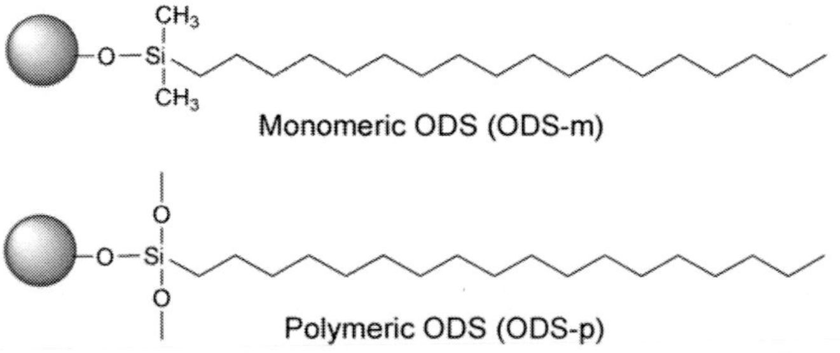

Scheme 1. Structure of commercial ODS column materials.

5. Polymer Grafting for HPLC Analysis

Polymer grafting on silica represents a well-established and useful procedure for the preparation of certain polymer architectures [25-27]. It may be accomplished by many polymerization techniques, e.g., metathesis polymerization [28], atom-transfer radical polymerization [29], Ziegler-Natta [30] or cationic ring-opening polymerization [31] and carried out either homogeneously, heterogeneously or from the gas phase [32]. In terms of surface grafting, mainly styrene- and acrylate-based resins have been used so far [25,33]. Despite all progress and advantages, grafted supports often suffer from insufficient pH stability. One approach that circumvents these problems is the use of soluble prepolymers for coating. This technique has been reported as early as in 1987 by Kolla et al. who prepared poly(butadiene-*block*-maleic acid)-coated silica columns for ion chromatography [34]. Later, Laubli and Kampus reported on poly-maleic acid-based columns with improved durability [35, 36]. Optimum deposition conditions [37] have also been developed for pure hydrophobic coatings of silica [38].

However, grafting of polymer onto silica ("grafting to" method) can be usually carried out by activation of silica surface and subsequent radical polymerization with a vinyl monomer. Nevertheless, the recommendations from the authors are to prepare a polymer (telomer) with a reactive terminal group at one side, and then immobilizing it onto the silica surface. Some examples for "grafting to" for the preparation of HPLC packing materials are illustrated in Scheme 2. This method provides many advantages which can be mentioned as:

a) A polymer with a terminal reactive group can be obtained by one-step telomerization. By choosing monomer, a function of the resultant polymer can be tunable.
b) Usual spectroscopy is applicable for determination of the chemical structure before immobilization onto silica. The stereo regularity is also estimated by NMR spectroscopy
c) Telomerization usually leads narrow poly dispersity of degree of polymerization. This promises homogeneity as an organic phase and
d) Polymerization degree can be controlled by the initial molar ratio of a vinyl monomer to a telogen. This is an essentially important feature because large polymers cannot penetrate into the pores of silica and, thus, this causes heterogeneity in surface modification.

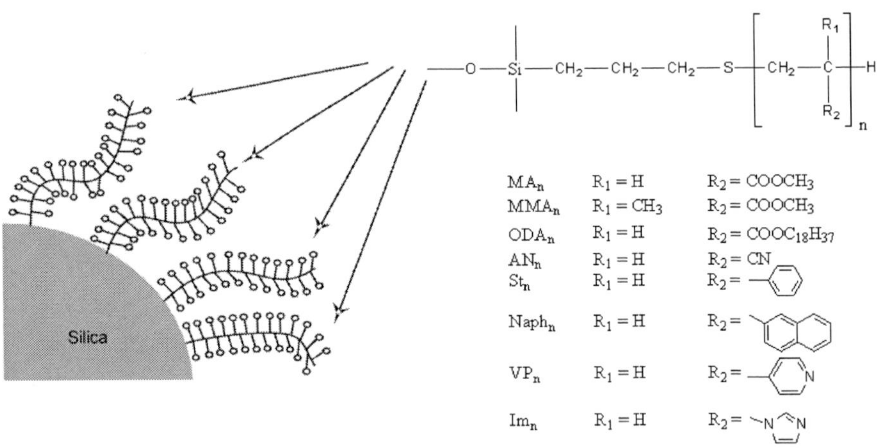

Scheme 2. General structures of polymer grafted silica stationary phases for HPLC.

As shown in Scheme 2 a variety of polymers were grafted on silica. Out of those the major successful result was observed from highly-ordered lipid membrane analogue; poly(octadecylacrylate) having π-electrons that enhanced separation by multiple π-π interactions with the guest PAHs was grafted onto silica for reversed-phase mode and reported in several papers [39]. Grafting of π-electrons containing polymers derived from octadecylacrylate were prepared by both "grafting to" and "grafting from" method. However, disordered or less ordered polymer derived from vinylpyridine also contain π-electrons from its hetero cycle moiety was also grafted on silica and exceptionally enhanced selectivity for PAHs isomers were obtained. The preparations, characteristics of these two phenomena having highly-ordered polymer and disordered polymer grafted stationary phases will be discussed in the following sections of this review. The resultant polymer-grafted silica shows, not only extremely high separation of PAHs, but also specific temperature dependency on the selectivity, which is induced by an ordered-to-disordered transition of the grafted polymer for self-assembled poly(octadecylacrylate), however no such phenomena was observed for the disordered polymers.

6. Ordered Stationary Phase for HPLC

Self-assembled systems such as lipid membrane aggregates can provide a highly ordered microenvironment leading to unique host-guest chemistry exceeding the functions of the original lipid. Further specificity has been desirable in RP-HPLC with the development of biotechnology and combinatorial chemistry. There have been many reports dealing with the development of new ligands immobilized onto silicas [40-45]. Typical examples are seen in fullerene- and porphyrin immobilized silicas, for which unique selectivity has been realized. However, it is certain that their separation modes are not only from simple reversed-phase mode, but also exceed the functions of the original ligands. On the other hand, self-assembling systems are very attractive, as they yield supramolecular functions beyond those of unit segment molecules in a solution state. The most successful examples can be seen in bio-organic systems, such as lipid membranes, proteins, and nucleic acids. In this article, Ihara et al. introduced self-assembled organic systems composed of simple hydrocarbons, which provide unique functions as organic phases that exceeded selectivity in reversed-phase mode than conventional polymeric and monomeric ODS phases. We have designed and efficiently synthesized the comb-shaped polymer, ODA_n [39] as a lipid membrane analogous organic phase. This polymer can not form bilayer structures in water but forms ''molecular gels'', which undergo temperature-responsive phase transition between ordered and disordered structures like aqueous lipid membrane systems. Another advantage of this polymer is due to the presence of a reactive terminal group at one side of the polymer main chain. Therefore, even after immobilization of the polymer with the terminal group, the bonded organic phase can maintain flexibility without being constrained by silica.

6.1. Grafting of Ordered Polymer on Porous Silica Surface

6.1.1. "Grafting To" Method

Comb-shaped poly(octadecylacrylate) grafted HPLC packing materials (Sil-to-ODA_n) demonstrated extremely large selectivity for PAHs. Sil-ODA_n can be synthesized by telomerization ("grafting to") method as follows[63]: the polymer is synthesized by one-step telomerization of octadecylacrylate (ODA) with silane coupling agent, 3-mercaptopropyltrimethoxysilane (MPS); thus the polymer can be immobilized through a terminal trimethoxy groups at one side of polymer main chain (as shown in Scheme 3). The average degree of polymerization (n) can be readily controlled by the initial molar ratio of ODA to MPS because of the good chain transfer constant of the mercapto group. The value of n can be estimated by the proton ratio, based on the terminal methoxy group using ^{1}H-NMR spectroscopy, and the value of M_w=M_n was determined by SEC to be approximately 1.4-1.6.

Immobilization of ODA_n usually carried out by mixing with silica at reflux temperature for 72 hrs. Toluene and tetrachloromethane were found suitable for the grafting process for ODA_n. Silica gels with different pore size and diameter can be employed for the grafting process. For example if grafting is done by silica with diameter 5 μm, pore size 120Å, and specific surface area 300 m^2g^{-1} (YMC gel), amount of immobilization can be reached up to values of 22 wt % .

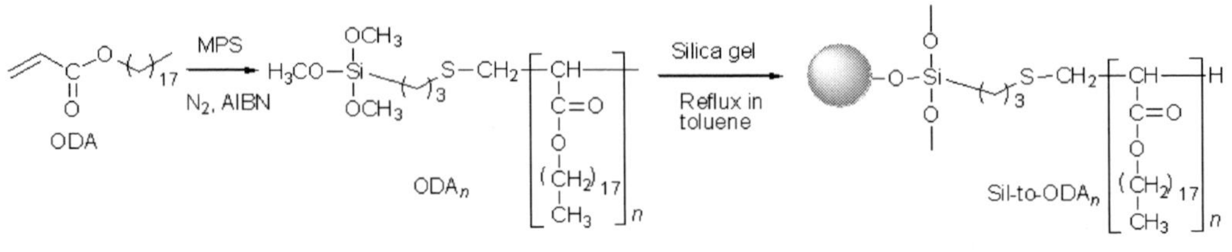

Scheme 3. Telomerization of ODA with MPS and subsequent grafting process on to porous silica gel.

6.1.2. Phase Transition Behavior of Sil-to-Oda$_n$

When differential scanning calorimetric (DSC) measurements were carried out with the Sil-ODA_n prepared by "grafting to" approach, the results indicated that crystalline-to-isotropic phase transitions occurred as temperature (T) was varied. The first transition was observed at a temperature, T_{C1}, of 30 °C using 80% methanol/water (v/v) as the phase underwent a transition from the crystalline-to-liquid-crystalline state, and the second transition was observed at a temperature, T_{C2}, of 40 °C for the transition of the phase from the crystalline-to-isotropic state. The phase transition temperatures, T_{C1} and T_{C2}, gradually shifted to higher values with higher percentages of water in the mobile phase. Thus, the corresponding T_{C2} values at 60% (v/v) and 40% (v/v) methanol/water were 42 and 44.5 °C, respectively (Figure 1). In an 80% (v/v) methanol/water system, the crystalline state was detected at T values below 30 °C, whereas the crystalline-to-liquid-crystalline state occurred between 30 and 40 °C. Liquid crystals represent a specific state between a crystalline solid (most ordered) and an isotropic liquid (least ordered). Liquid crystals are systems that display liquid like translational order but solid like orientation order, with the liquid-crystalline state formed after the melting of the solid. The resulting liquid preserves properties intermediate between the solid and the liquid over a certain T range. A transition takes place at a specific T value, called the clearing temperature, to the isotropic liquid. The T difference between the melting point of the solid and the isotropic liquid determines the meso-phase range (i.e., for the Sil-to-ODA_n phase, this was between 30 and 40 °C). As apparent from the DSC experiments, the ratio of the liquid-crystalline to crystalline phase for the Sil-to-ODA_n increased within the T range of 30-40 °C, whereas above 40 °C, the ratio of the isotropic to liquid-crystalline phase increased. The decreases in phase organization followed the order of crystalline>liquid-crystalline>isotropic. Furthermore, the phase transition T gradually shifted to a higher value with a higher percentage(s) of water. These results indicated that the polarity of the mobile phase had a significant influence on the orientation order of the immobilized, silica supported, highly-ordered comb-shaped stationary phase [46].

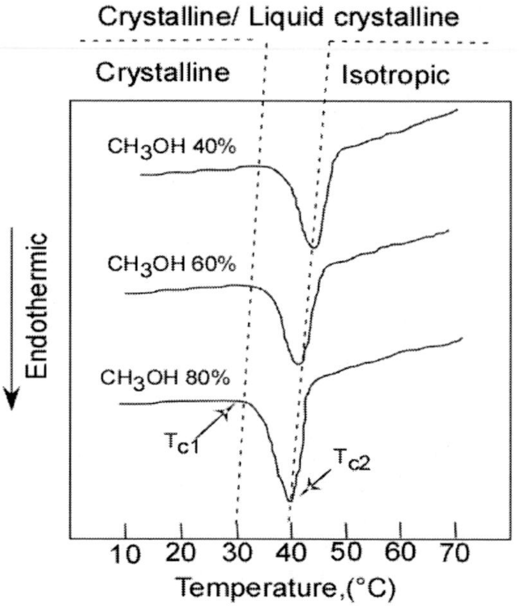

Figure 1. Differential scanning microcalorimetric thermograms of the Sil-to-ODA$_n$ sorbent as a methanol-water dispersion at 40, 60, and 80% (v/v) methanol (MeOH) in water, illustrating the shift in the phase transition temperature to higher values with increasing percentages of water.

Furthermore, the mobility of the alkyl moieties in Sil-to-ODA$_n$ can be realized by suspended-state ^{1}H NMR spectroscopic measurements at variable temperature. Neither half-height width (line width) of methylene groups nor spin-spin relaxation time (T_2 = 100 ms) showed any significant change with temperature. However, intensity of NMR peaks representing terminal methyl and methylene groups of octadecyl moieties increases significantly for Sil-to-ODA$_n$ with increasing measuring temperature as mentioned in a recent report [47]. Nevertheless, for monomeric ODS, intensity of NMR peaks representing terminal methyl and methylene groups increased slightly, but for polymeric ODS, only the peak of methylene groups was detectable when a very high vertical scale was used for graphical presentation. As shown in Figure 2, intensity of NMR signals from methylene groups of Sil-to-ODA$_n$ increase distinctly around an endothermic peak of DSC. Peak-top temperature was measured to be 38.4 °C for Sil-to-ODA$_n$ in presence of methanol. Intensities, to be comparable between different samples and at different temperatures, were normalized according to the equation mentioned in our previous report [47, 48]. Normalized intensity of ^{1}H NMR peak of methylene groups of Sil-to-ODA$_n$ begins to increase distinctly around 38 °C but no similar style of increase can be seen in case of monomeric ODS and polymeric ODS. The detailed characterization by suspended-state, solid CP/MAS NMR and PAHs separation by Sil-to-ODA$_n$ phase that prepared with "grafting to" method will be discussed in the following section.

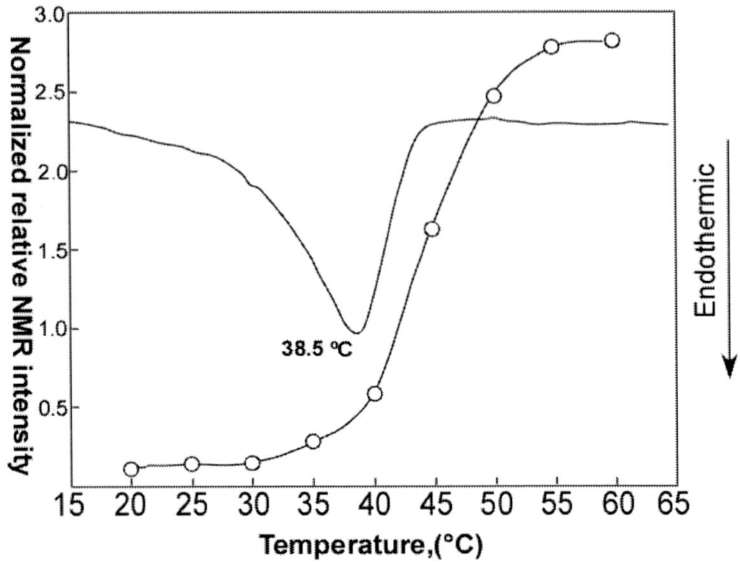

Figure 2. Normalized relative ^{1}H NMR intensity of methylene groups of Sil-to-ODA$_n$ increase distinctly around 38 °C, which is just around the peak-top temperature of its DSC thermogram.

6.1.3. "Grafting From" Strategies for Oda$_n$ Grafted Silica

The most notably disadvantage for telomerization followed by grafting or "grafting to" method is low grafting densities because of steric crowding of reactive sites [49]. Because of the steric hindrance imposed by the grafted polymer chains, it is difficult to prepare high-density polymer-grafted particles using "grafting-to" methods [50, 51]. Thus the problem of low grafting density can be circumvented by "grafting from" approach which usually possess high grafting of organic polymer on inorganic particle surface. Out of several "grafting from" techniques, the most commonly employed is "living"/controlled radical polymerization (CRP) methods, such as nitroxide-mediated polymerization (NMP), atom transfer radical polymerization (ATRP), reversible addition-fragmentation chain transfer (RAFT) polymerization, living anionic polymerization (LAP), and living cationic polymerization (LCP), provide an ability to produce particular polymer architectures with controlled molecular weight and molecular weight distributions [52]. The major difference between the conventional radical polymerization techniques and CRP is the lifetime of the propagating radicals. While radicals derived from the thermal decomposition of conventional radical polymerization initiators (azobisisobutyronitrile type) may undergo termination reactions within a few seconds, the lifetime of the propagating radicals can be extended to several hours in controlled processes. Among the CRP processes, increasing attention has been paid to ATRP since its discovery by Matyjaszewski and Sawamoto, [53,54] because this method does not require precise experimental conditions like LAP or LCP.

Table 1. Elemental analysis data of Sil-from-ODA$_n$ and Sil-to-ODA$_n$.

	C (%)	H (%)	N (%)
Sil-fromODA$_n$	21.64	3.26	0
Sil-to-ODA$_n$	15.70	3.28	0

To overcome the limitation of "grafting to" method we applied the surface-initiated atom transfer radical polymerization (ATRP) method ("grafting from" technique) to prepare Sil-from-ODA$_n$. The ATRP initiator was synthesized from 2-bromo-2-methyl propionic acid undecyl ester by hydrosilation with trichlorosilane and then immobilized onto porous silica gel. The reaction between a trichlorosilane anchoring group and a surface OH-group of silica resulted in the formation of self-assembled monolayer (SAM) of the initiator. The surface-initiated polymerization of ODA was carried out from the initiator-grafted silica using copper(I)bromide and *N,N,N',N',N''*-pentamethyldiethylenetriamine (PMDETA) as catalyst precursors (as shown in Scheme 4). The polymer-grafted stationary phase was characterized by DRIFT, elemental analysis, thermogravimetry, suspended-state ^{1}H NMR, solid-state ^{13}C and ^{29}Si CP/MAS NMR and DSC measurements. The surface coverage of ATRP polymerized octadecylacrylate; Sil-from-ODA$_n$ was calculated on the basis of elemental analysis results according to our previously reported method [47]. Table 1 shows the elemental analysis data of Sil-to-ODA$_n$ and Sil-from-ODA$_n$. The grafting density of ODA moiety on Sil-from-ODA$_n$ is 3.75 μmolm^{-2} which is significantly higher than that obtained from Sil-to-ODA$_n$ prepared by the "grafting to" method (2.63 μmolm^{-2}).

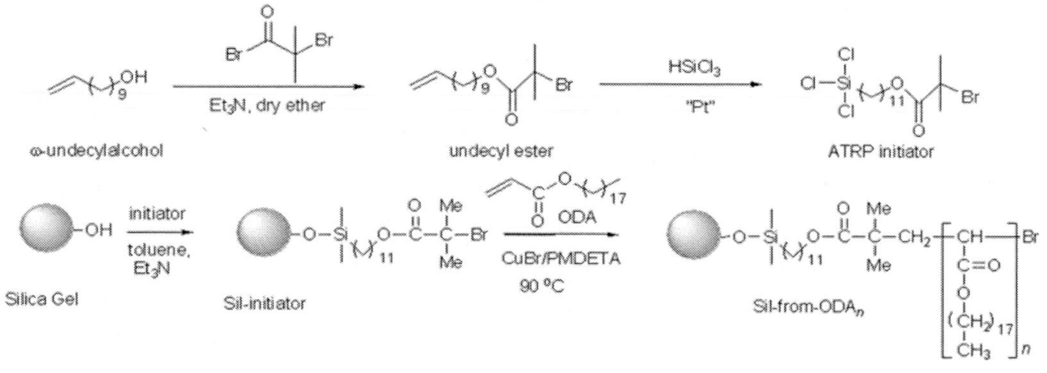

Scheme 4. Preparation poly(octadecyacrylate) grafted silica by atom transfer radical polymerization method.

Differential scanning calorimetry for polymeric ODA (ODA$_n$) provided an endothermic peak at 47 °C (T_{c2}, peak-top temperature) with a shoulder around 42 °C (T_{c1}) in the heating process (Fig. 3). By polarization microscopic analysis T_{c1} and T_{c2} can be assigned as crystalline-to-liquid crystalline and liquid crystalline-to-isotropic phase transitions respectively as mentioned above. We have mentioned in earlier section that for Sil-to-ODA$_n$ peak-top temperature (phase transition temperature) decreased from 47 to 38.42 °C in methanol [47], however, Sil-from-ODA$_n$ yielded less reduction of peak top temperature occurred from 47 to 43 °C in the methanol-water mixture. These results clearly indicated that the polymer chains in Sil-from-ODA$_n$ can retain ordered structure to a higher temperature than Sil-to-ODA$_n$.

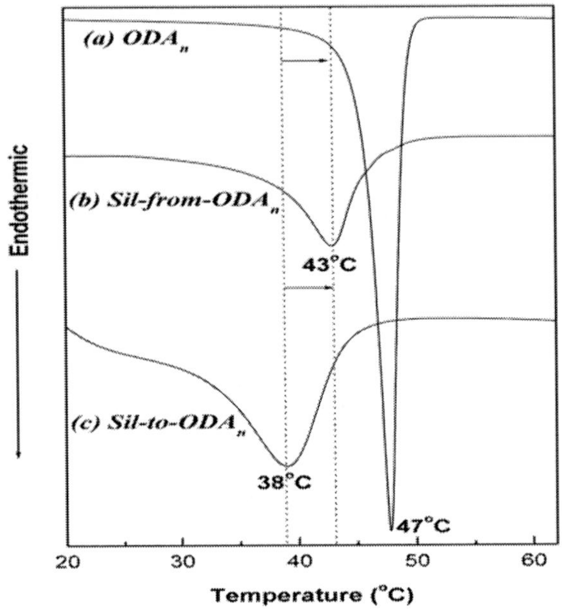

Figure 3. DSC thermograms of poly (octadecylacrylate) grafted silica prepared by both “grafting to” and“grafting from”methods.

One of the most interesting phenomena observed for octadecyl moieties as observed in ^{13}C CP/MAS NMR measurements for both Sil-to-ODA$_n$ and/or Sil-from-ODA$_n$ that demonstrated transformation from ordered to disordered state (*trans* to *gauche*) as the temperature is increased as shown in Fig. 4. The ^{13}C signal for $(CH_2)_n$ groups is observed at two resonances, one is at 32.6 ppm attributed to *trans* conformation, indicating crystalline and rigid state, and the other at 30.0 ppm attributed to *gauche* conformation, indicating disordered and mobile state [55,56]. The NMR signal for the main methylene group occurs at 33.3 ppm corresponding to *trans* conformation and at 30.6 ppm corresponding to *gauche* conformation. Comparing between two approaches the “grafting from” method is more dominated by *trans* conformation than Sil-to-ODA$_n$ as shown in Fig. 5.

The observation from DSC, suspended-state ^{1}H NMR and ^{13}C CP MAS NMR confirms the existence of higher order of alkyl chain in Sil-from-ODA$_n$ than Sil-to-ODA$_n$. Phase characterization method, in conjunction with conformation determination drawn from NMR spectroscopy, is important for a better understanding of the structure, dynamics, and separation behavior of organic layers grafted onto the silica surface. Interestingly, this method clarified that most of the octadecyl chains in the case of monomeric ODS, despite having a *gauche* conformation, were still in a solid phase and only 12.3%-18.5% of them have enough mobility to be considered as being in a liquid phase [47].

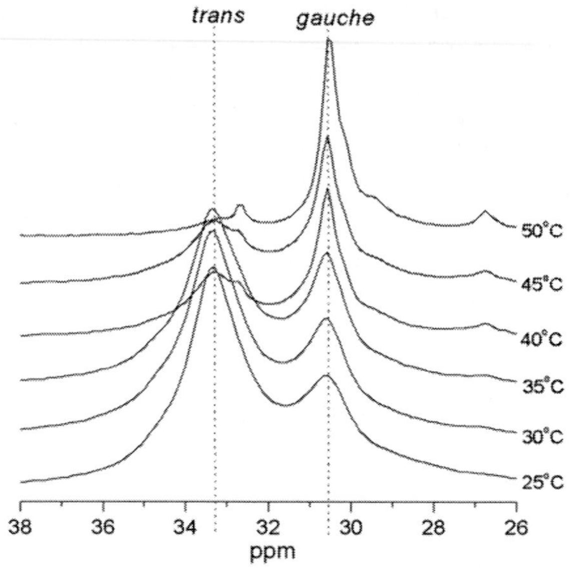

Figure 4. Variation of ordered to disordered conformation of alkyl chain with temperature for poly(octadecylacrylate) grafted silica.

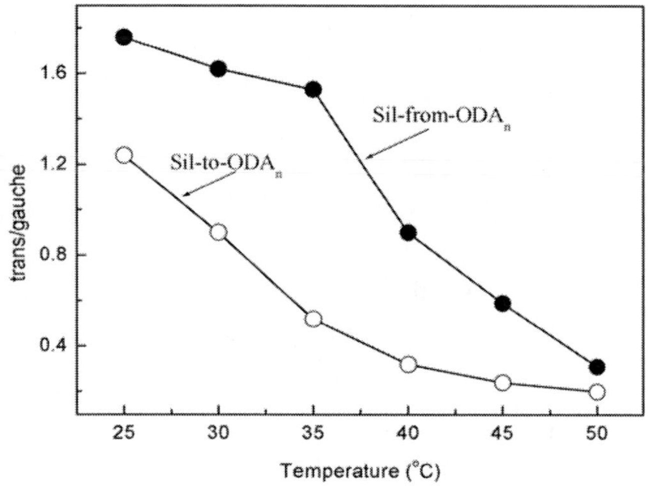

Figure 5. Variation of *trans-gauche* ratio for ordered stationary phase prepared by "grafting to" and "grafting from" method.

6.1.4. Retention Mode and Temperature Dependence on Selectivity for Ordered Phases

It is known that conventional ODS or alkyl phases can recognize the hydrophobicity of solutes in HPLC and this hydrophobicity is measured by the methylene group selectivity of the stationary phases. This reflects the possibility of the phase being able to separate two molecules that differ only in methylene groups, e.g., amylbenzene and butylbenzene or ethylbenzene or toluene. The retention mode as well as the extent of hydrophobic interaction among the solutes and the packing materials in HPLC can be determined by retention studies of alkylbenzenes as solutes [57, 58]. Fig. 6 shows the correlation between log *k* and log *P* for Sil-from-ODA$_n$, Sil-to-ODA$_n$ and ODS-m phase. It is noteworthy that the retention mode of Sil-from-ODA$_n$ and Sil-to-ODA$_n$ showed a reversed-phase mode to that of ODS-m phase. Furthermore, log *P* and log *k* plots for alkylbenzenes almost intersect with each other however; significant deviation was observed for PAHs and alkylbenzenes on Sil-to-ODA$_n$ and Sil-from-ODA$_n$ demonstrating the existence of other possible interaction sites besides molecular hydrophobicity. It also observed that Sil-from-ODA$_n$ showed higher retention for PAHs compared to its values for alkylbenzenes with similar log *P*. For instance, the log *P* of naphthacene (5.71) is smaller than that of octylbenzene (6.30), while log *k* value of naphthacene (1.53) is higher than that of octylbenzene (0.93).

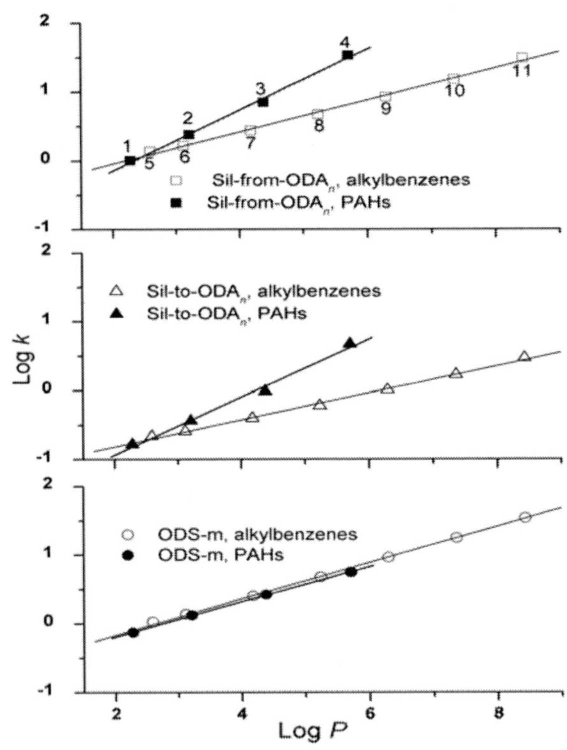

Figure 6. Log *P* and log *k* plot of alkylbenzenes and linear PAHs on a Sil-to-ODA$_n$, Sil-from-ODA$_n$, ODS-m columns for mobile phase methanol/water (90:10v/v), flow rate 1.00 mL min^{-1}, and column temperature 30 °C (1 benzene, 2 naphthalene, 3 anthracene, 4 naphthacene, 5 toluene, 6 ethylbenzene, 7 butylbenzene, 8 hexylbenzene, 9 octylbenzene, 10 decylbenzene, 11 dodecylbenzene).

In addition, highly-ordered poly(octadecylacrylate) grafted silica particles prepared in both methods performed complete separation for mixtures of various isomeric PAHs polycyclic aromatic hydrocarbons in a methanol-water as a mobile phase as mentioned in many of our previous investigation [59]. Their elution orders are the same as those that can be observed with ODS. It is also showed that the retention factor (*k*) increases with an increase in mobile phase polarity. The performance of Sil-to-ODA$_n$ and Sil-from-ODA$_n$ is unique because of its attractive temperature dependencies behavior. As shown in Fig. 7, both Sil-to-ODA$_n$ and Sil-from-ODA$_n$ shows distinct bend in the plots of temperature vs. retention factor (k) for triphenylene. No such remarkable temperature dependence is observed with ODS or any other phases. As shown by the in Fig. 7, Tc$_2$ of Sil-to-ODA$_n$ and Sil-from-ODA$_n$ and that observed in DSC (in methanol dispersion) almost completely acceded with the temperature of the bending point. This strongly suggests that the typical temperature dependencies a property by polyoctadecylacrylate grafted silica occurs at the crystalline-to-isotropic phase transition temperature of the bonded phase.

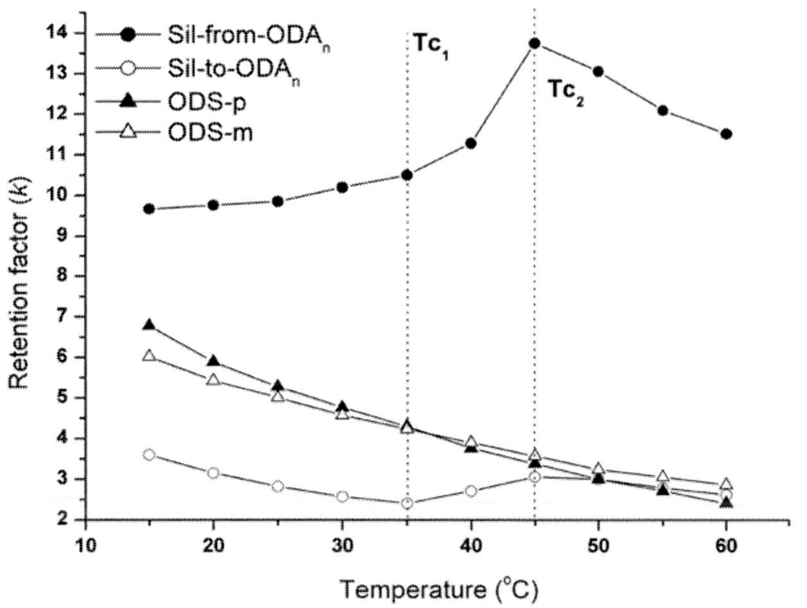

Figure 7. Temperature dependencies on retention factors for triphenylene on various HPLC columns.

6.1.5. Selectivity for Pahs on Ordered Polymeric Phase

The shape selectivity of a given stationary phase can be measured from the selectivity results of pairs of PAHs test solutes having the same carbon number but different shape, such as triphenylene/*o*-terphenyl, chrysene/pyrene and *trans*-stilbene/ *cis*-stilbene. On a typical monomeric ODS phase, the separation factor triphenylene/*o*-terpheny ranges from 1.0 to 1.7, while the value on a polymeric ODS phase lies between 2.0 and 2.7. Jinno et al. [60, 61] suggested that a separation factor ˛triphenylene/*o*-terpheny ≥2.0 is an indication of significant solute planarity recognition. triphenylene and *o*-terphenyl have the same carbon number per molecule and, thus, their molecular hydrophobicities are similar but different shape. The chromatogram for *o*-terphenyl and triphenylene on different columns are shown Figure 8. The selectivity at 30 °C are 2.28, 3.6 and 4.6 for ODS-p, Sil-to-ODA$_n$ and Sil-from-ODA$_n$ clearly indicating higher selectivity by ODA$_n$ prepared from "grafting from" method, however "grafting to" approach also yielded higher selectivity than ODS-p. Table 2 shows selectivity results for a series of PAHs on ordered stationary phase prepared from different grafting approaches.

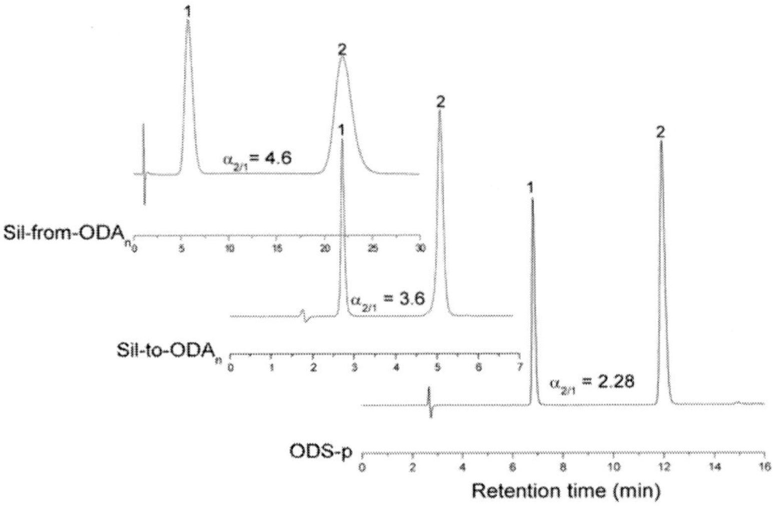

Figure 8. Chromatogram for *o*-terphenyl (*1*) and triphenylene (*2*) mixture on various columns at 30C. Mobile phase methanol-water=90/10, flowrate 1.00 mLmin^{-1}.

Table 2. Selectivity results for PAHs on Sil-from-ODAn, Sil-to-ODAn and ODS-p stationary phases.

Solute	Sil-from-ODA$_n$		Sil-to-ODA$_n$		ODS-p	
	k	*A*	*k*	*α*	*k*	*α*
Benzene	0.668		0.225		0.56	
		2.12		**1.95**		**2.02**
Naphthalene	1.418		0.440		1.13	
		6.85		**5.78**		**5.35**
Anthracene	4.573		1.300		3.00	
Triphenylene	9.761		3.153		5.89	
		1.41		**1.35**		**1.14**
Benz[a]anthracene	13.751		4.240		6.72	
		1.60		**1.47**		**1.17**
Chrysene	15.631		4.625		6.87	
		3.32		**2.73**		**1.60**
Naphthacene	32.365		8.624		9.36	
o-Terphenyl	1.187		0.710		2.29	
		3.78		**1.86**		**1.68**
m-Terphenyl	4.485		1.320		3.85	
		6.87		**3.52**		**2.10**
p-Terphenyl	8.815		2.500		4.82	

Mobile phase: methanol:water (90:10). Column temperature: 20 °C. Flow rate: 1.00 ml min^{-1}

Remarkable temperature dependence on separation factors are also observed for this sample set. As shown in Fig. 9, the selectivity is significantly higher at the crystalline temperature of Sil-to-ODA$_n$ than at the isotropic temperature. On the other hand, ODS shows no significant temperature dependence. Therefore, the enhanced selectivity at the crystalline temperature cannot be explained by only hydrophobicity recognition. It is known that planar compounds, such as triphenylene, are more strongly incorporated into an oriented medium than nonplanar compounds, such as *o*-terphenyl. It was reported by us that that α-helical poly(γ-methyl-L-glutamate) produces highly oriented bundles and shows surprisingly enhanced retention capacity for planar PAHs which can explained from molecular slot model [62]. To verify the separation mechanism of poly(octadecylacrylate) grafted silica, we investigated the selectivity with geometrical isomers from various substituted azobenzene compounds [63]. As a result, it was found that the separation factor between the *trans*- and *cis*-isomers was remarkably dependent on the electron-donating property of the substituted group. This indicates that bonded ODA$_n$ works as an electron acceptor. Therefore, we focused on the role of carbonyl groups in ODA$_n$. As, expected ODA$_n$ derived stationary phases showed higher selectivity at its crystalline temperature than ODS.

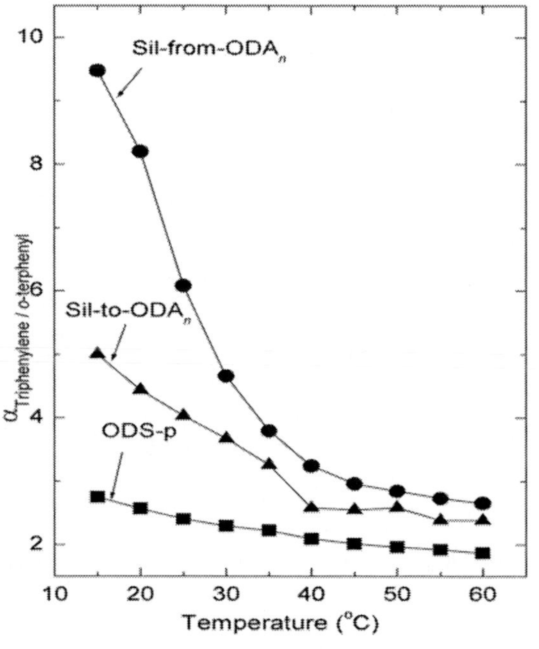

Figure 9. Temperature dependencies on selectivity for different columns. Mobile phase, methanol-water=90/10, flow rate 1.00 mLmin^{-1}.

However, further significant information is observed at its isotropic temperature: The selectivity is still higher than in ODS. This must be due, in part, to effects of carbonyl groups because Sil-to-MA$_n$, which prepared from methyl acrylate without long-chain alkyl groups, shows selectivity similar to that of Sil-to-ODA$_n$/Sil-from-ODA$_n$ that always kept isotropic state [64]. Additional supporting data can be obtained by calculation [65-67]. A carbonyl π-benzene π complex was evaluated against a formaldehyde-benzene complex model by ab initio MO=MP2 calculations performed with the Gaussian 94 package. The binding energy was calculated as a function of distance, R, between the carbon atom of formaldehyde and benzene plane, in which formaldehyde was, moved perpendicular to the benzene plane (plane-to-plane interaction) with the orientation fixed to that of the optimized geometry. According to our calculated results, a formaldehyde-benzene interaction (1.83 kcal mol^{-1}) is more effective than CH_4-benzene [65], (0.57 kcal mol^{-1}) and benzene-benzene (0.49 and 1.78 kcal mol^{-1})in the parallel and perpendicular interactions) complexes. Therefore, we conclude that the higher selectivity of ODA_n grafted silica is brought about by a carbonyl π-benzene π-interaction, especially on a highly oriented state. Based on these observations, we propose the modified mechanism schematically illustrated in Fig. 9: A carbonyl π-benzene π-interaction as an essential driving force for bonded phase-solute interaction. This estimation is supported both experimentally and theoretically. For multiple carbonyl π-solute π-interactions, as illustrated above, it is clear that if the bonded organic phase is in crystalline state that can interact more effectively with linear and planar compounds (as shown in Fig 10). It is also important to mention that Sil-from-ODA_n prepared by "grafting from "approach showed higher selectivity up to higher temperature since it has tendency to keep crystalline temperature higher than Sil-to-ODA_n prepared by "grafting to" method. Moreover, the uniqueness of selectivity observed for ODA polymerized silica can not be visualized for commercial ODS phase.

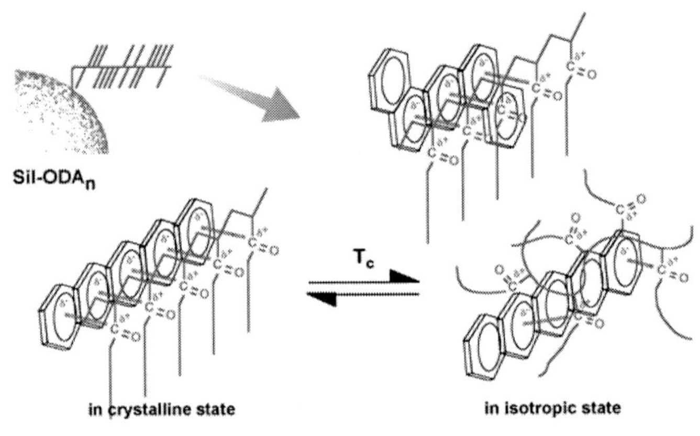

Figure 10. Schematic illustration of the separation mechanism for PAHs on poly(octadecylacrylate) grafted silica.

7. Π-Electron Containing Disordered Polymer Grafting for HPLC

Separation of PAHs is posing a challenge using conventional columns as the structural PAH isomers have similar hydrophobicity. Besides hydrophobic effect, direct π-π interaction between organic stationary phase and solutes would increase selectivity for PAHs. On this view point we have developed many π-electron containing stationary phases to enhance the selectivity for this class of compounds [68, 69]. Furthermore analysis of basic compounds with HPLC is very important because many pharmaceuticals and bio-related materials often include amino groups. However serious problem have been noted during separation specifically peak-tailing and poor reproducibility through residual silanol groups on silica surfaces. To eliminate these undesirable properties, many end-capping reagents and methods have been developed [70-71] We have introduced poly(4-vinylpyridine) as a novel π-electron-containing organic phase which do not show characteristic highly-ordered behavior likewise poly(octadecylacrylate) grafted silica. However, the polymer shows disorderness on silica surface. The chromatographic separation by poly(4-vinylpyridine) is extremely unique and will be discussed in the following sections. Moreover undesirable silanol effect against basic compounds in HPLC may be reduced by the fact that a masking effect is brought about by poly(4-vinylpyridine) as a weak basic polymer which provides exclusion effect for basic compounds [72]

7.1. Preparation of Sil-To-Vp$_n$ by "Grafting To" Method

A typical procedure for VP$_n$ preparation followed by grafting on to silica is: 10 mL of 4-vinylpyridine and 0.91 mL of MPS were mixed and N_2 gas was bubbled for 20 min at room temperature. After addition of 100 mg of AIBN, the mixture was stirred with bubbling N_2 gas at 65˚C for 6 h. The obtained solid was collected and dissolved in chloroform and precipitated from n-hexane. The precipitation was repeated several times using chloroform and n-hexane to remove homopolymer and unreacted monomer from the telomer. The polymer was then grafted onto the silica surface by mixing in chloroform at the reflux temperature. The polymer-grafted silica (Sil-to-VP$_n$) was packed into stainless-steel column (as shown in Scheme 5).

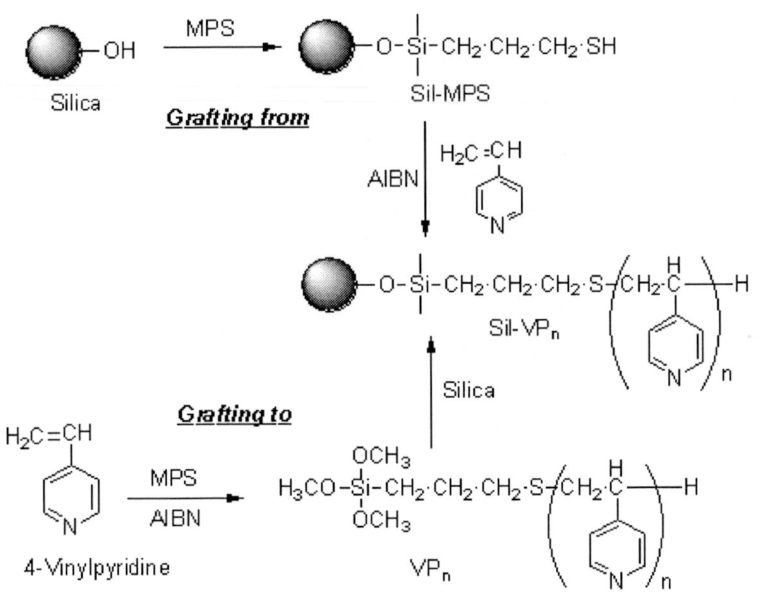

Scheme 5. Preparation of poly(4-vinylpyridine) grafted silica for PAHs analysis by both "grafting to" and "grafting from" methods.

7.1.1. Analysis of PAHs by NP-HPLC

As PAHs have very poor solubility in polar solvents so normal-phase high-performance liquid chromatographic separation (NP-HPLC) could be ideal one since it uses nonpolar solvents as mobile phase. Moreover, Sil-to-VP_n showed less sensitivity for molecular hydrophobicity and higher sensitivity for molecular aromaticity, in NP-HPLC higher retentivity along with higher selectivity for the PAHs were observed as compared to the others stationary phase as given in Table 3 and Table 4 and was reported in our recent investigations [73,74].

Table 3. Retention data for linear PAHs on disorder polymeric phase and reference columns at 35 °C.

PAHs	No. of double bonds	Sil-to-VP_n		Sil-to-PS_n		Silica		ODS-m	
		k	A	k	α	k	α	k	α
Benzene	3	0.17		0.12		0.42		0.15	
			2.06		1.75		1.69		1.27
Naphthalene	5	0.35		0.21		0.71		0.19	
			2.17		1.62		1.62		1.32
Anthracene	7	0.76		0.34		1.15		0.25	
			2.17		1.53		1.61		1.44
Naphthacene	9	1.65		0.52		1.85		0.36	

Mobile phase: n-hexane/2-propanol (90:10) in Sil-to-VP_n and Sil-to-PS_n; n-hexane in silica and ODS-m

Table 4. Retention (*k*) and separation factors (□) for PAHs isomers on different columns at 35 °C.

PAHs		Sil-to-VP_n		Sil-to-PS_n		Silica		ODS-m	
		k	α	k	α	k	α	k	α
Naphthacene	$C_{18}H_{12}$	1.65		0.52		0.12		0.36	
Benz[a]anthracene	$C_{18}H_{12}$	1.95		0.53		0.12		0.33	
			1.09		1.04		1.00		1.06
Chrysene	$C_{18}H_{12}$	2.13		0.55		0.12		0.35	
			1.20		1.07		1.08		1.00
Triphenylene	$C_{18}H_{12}$	2.56		0.59		0.13		0.35	
Benzo[a]pyrene	$C_{20}H_{12}$	3.14		0.74		0.13		0.45	
			1.17		1.07		1.00		1.00
Benzo[e]pyrene	$C_{20}H_{12}$	3.68		0.79		0.13		0.45	
Dibenzo[a,h]anthracene	$C_{22}H_{14}$	4.14		0.84		0.13		0.45	
			1.22		1.07		1.00		1.04
Dibenzo[a,c]anthracene	$C_{22}H_{14}$	5.06		0.90		0.13		0.47	

Mobile phase : *n*-hexane/2-propanol (9:1) in Sil-to-VP_n, Sil-to-PS_n and silica; *n*-hexane in ODS-m

Furthermore, when chromatograms were recorded for alkylbenzenes and PAHs solutes good resolution was obtained for PAHs but not for alkylbenzenes as shown in Fig.11.

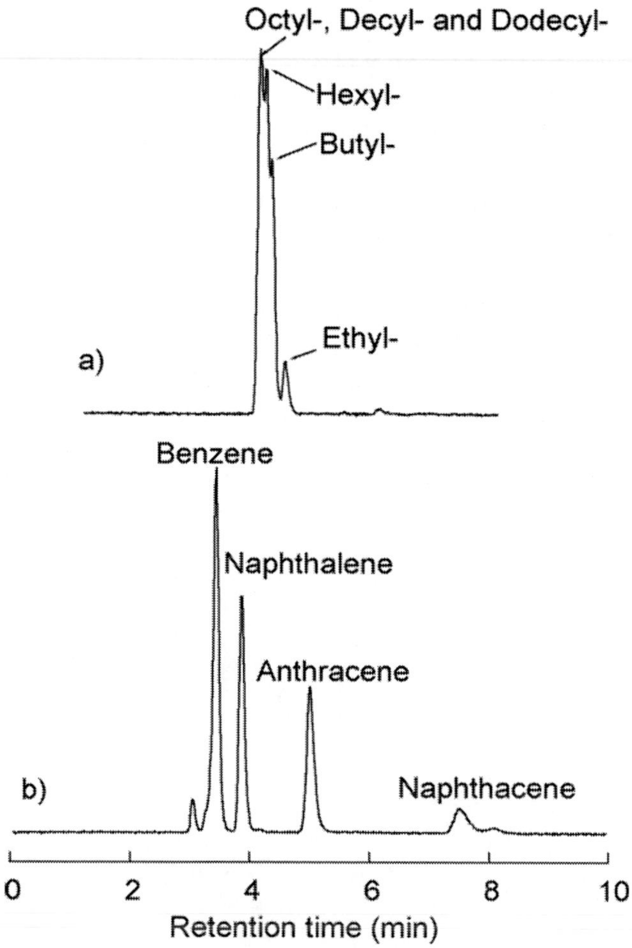

Figure 11. Typical chromatograms for a mixture of alkylbenzenes (a) and polycyclic aromatic hydrocarbons (b) with Sil-to-VP_n, mobile phase: n-hexane: 2-propanol (9:1), flow rate: 1.0 mL min^{-1}, temp: 35°C.

The special preference towards aromatic solutes can also be observed in the graph plotted between the retention factors and number of carbon atoms for polycyclic aromatic hydrocarbons and alkylbenzenes (Fig. 12). The retention factor for alkylbenzenes decreased with an increase in the number of carbon atoms due to hydrophobic effect while the retention factor for PAHs increased with the number of carbon atoms. Interestingly Sil-to-VP_n showed much higher retention factors than other reference columns which indicates that Sil-to-VP_n provides a specific interacting site for PAHs.

7.2. "Grafting From" Strategy of Sil-Vp_n

In order to circumvent low grafting density of "grafting to" approach, "grafting from" method can be applied. A typical synthetic procedure for the synthesis was as follows: 5 g of porous silica was first suspended in 50 mL of toluene then 0.91 mL of 3-mercaptopropyltrimethoxysilane (MPS) was added in the suspension. The mixture was mechanically stirred under solvent reflux in an inert nitrogen atmosphere for 72 h. MPS modified silica was washed several times with toluene and acetone and dried in vacuum. In the next stage, 4.5 g of the modified silica was suspended in 40 ml 4-vinylpyridine monomer and bubbled with nitrogen gas at room temperature for 10 min. After adding 100 mg of AIBN (dissolved in 5 ml of 4-vinylpyridine monomer) the suspension was mechanically stirred at 70 °C for 6 h. The obtained solid was washed several times with hot chloroform, methanol, and acetic acid and lastly with water then dried in vacuum. The polymer-grafted silica (Sil-from-VP_n) was then packed into two stain-less steel columns. The grafting amount obtained from elemental analysis and TGA measurements shows that about 22% can be grafted by "grafting from" method, however "grafting to" approach only yield 16% grafting density. A typical chromatogram for Sil-to-VP_n and Sil-from-VP_n for NP-HPLC are shown.

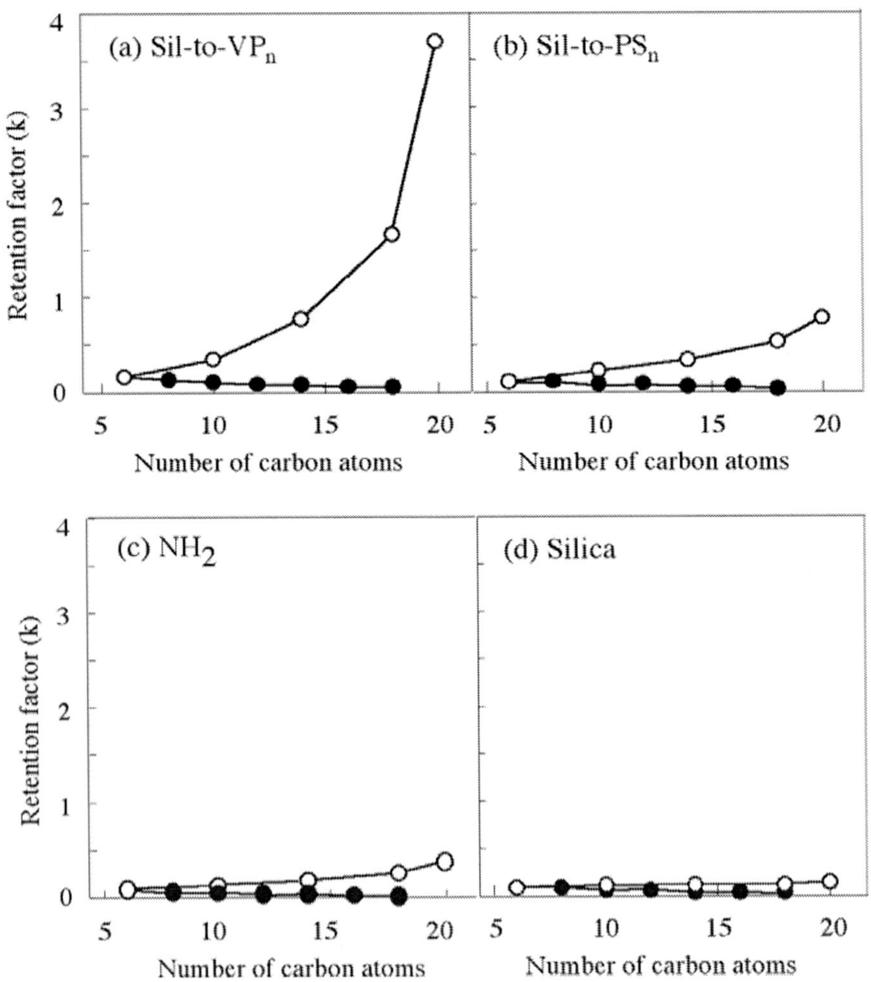

Figure 12. Relationships between the retention factor (k) and number of carbon atoms with Sil-to-VP_n (a), Sil-to-PS_n (b), NH_2 (c), and silica (d). Elutes: benzene, naphthalene, anthracene, naphthacene and benzo[e]pyrene (open circles); ethyl-, butyl-, hexyl-, octyl-, decyl-, and dodecyl-benzenes (solid circles).

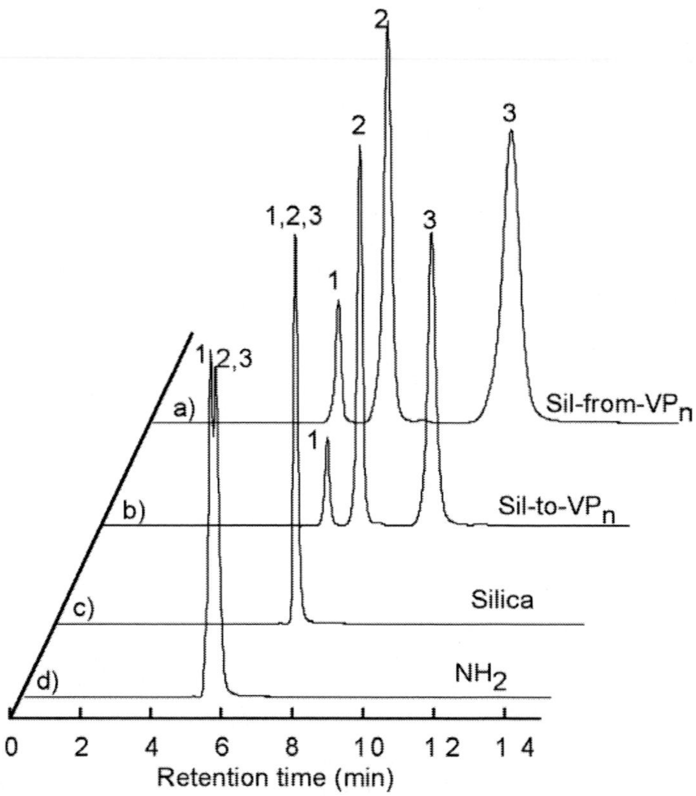

Figure 13. Typical chromatograms for a mixture of polycyclic aromatic hydrocarbons with (a) Sil-from-VP_n, (b) Sil-to-VP_n, (c) Silica and (d) NH_2, mobile phase: n-hexane:2-propanol (7:3), flow rate: 0.5 $mLmin^{-1}$, temp: ambient 1: Benzene, 2: Naphthalene, 3: Anthracene.

8. Selectivity for PAHs by Disordered VP_N Phase

The results from the chromatographic retention behaviors of PAHs with VP_n showed its higher sensitivity towards molecular aromaticity and lower sensitivity towards molecular hydrophobicity, despite the fact that the main polymeric chain is hydrophobic in nature. VP_n phases prefer disc-like solutes (having less L/B ratio) compared to the linear one (having higher L/B ratio). Moreover VP_n phase showed very high selectivity towards molecular planarity. When the retention behaviors of Sil-to-VP_n and Sil-from-VP_n were compared, the elution order was similar in both NP-HPLC as well as in RP-HPLC. Higher retentivity and selectivity for the PAHs were observed with Sil-from-VP_n having higher phase load compared to Sil-to-VP_n. On comparison with ODS phases, distinctly difference in chromatographic preference for the spatial shape of the solutes had been observed between the VP_n and ODS phases. VP_n phases showed greater preference towards disc like molecules (having lower L/B ratio), whereas ODS phases showed the greater preference for the linear ones (having higher L/B ratio). In addition to the preference towards the disc like molecules, elution order of PAHs with VP_n phases was in the reverse order of hydrophobicity, which is quite opposite to ODS phases. The retention results for PAHs in RP mode for VP_n prepared in different mode are given in Table 5.

Table 5. Selectivity results for PAHs isomers in RP-HPLC for VP_n phases prepared by "grafting to" and "grafting-from" method.

PAHs		L/B	Sil-from-VP_n		Sil-to-VP_n		ODS-m		ODS-p	
			K	α	*k*	α	*k*	α	*k*	α
Anthracene	$C_{14}H_{10}$	1.57	2.13		0.77		2.21		2.04	
				1.06		1.06		0.95		0.92
Phenanthrene	$C_{14}H_{10}$	1.46	2.25		0.82		2.10		1.88	
Naphthacene	$C_{18}H_{12}$	1.89	5.47		1.85		4.54		5.23	
Benz[a]anthracene	$C_{18}H_{12}$	1.58	5.99		1.99		4.01		4.15	
				1.08		1.08		0.99		1.01
Chrysene	$C_{18}H_{12}$	1.72	6.47		2.14		3.96		4.21	
				1.24		1.23		0.96		0.90
Triphenylene	$C_{18}H_{12}$	1.12	8.01		2.63		3.79		3.79	
Benzo[a]pyrene	$C_{20}H_{12}$	1.49	13.38		4.23		6.38		7.46	
				1.18		1.17		0.92		0.87
Benzo[e]pyrene	$C_{20}H_{12}$	1.12	15.74		4.96		5.84		6.48	
Dibenzo[a,h]anthracene	$C_{22}H_{14}$	1.82	16.43		5.00		7.42		8.98	
				1.32		1.30		0.98		0.90
Dibenzo[a,c]anthracene	$C_{22}H_{14}$	1.24	21.61		6.51		7.24		8.11	

Mobile phase: Methanol/water (9:1), flow rate: 0.5 ml min^{-1} in Sil-from-VP_n and 1.0 ml min^{-1} in Sil-to-VP_n, ODS-m and ODS-p, column size: (250mm×4.6mm).

Separation of PAHs by VP_n phases can be summarized as follows:

a) VP_n phases showed very low retention factor for alkylbenzenes but very high retention factor was observed for PAHs, which can be explained by π-π interactions between the solute and stationary phases.
b) VP_n phases showed higher selectiviy for the molecular planarity, such behaviors of stationary phases are often described by charge-transfer interaction between the solute and stationary phases. Thus, charge-transfer interaction between the VP_n phase and stationary phase can also be expected. However, almost no retention for PAHs solutes has also been reported with electron donor phase (pyrene) in NP-HPLC. In contrast to this report, very good retention factor for PAHs solutes has been obtained with VP_n phases in NP-HPLC as well.
c) Electrostatic interaction between the VP_n phase and PAHs solutes can also be predicted. Because of non-polar nature of PAHs solute, electrostatic interaction between VP_n phases and PAHs solutes can be explained in terms of inductive effects and/or quadrupolar effects. Inductive effect is related with the average molecular polarizability of the solute. When average molecular polarizability and quadrupole moment of solutes was correlated with the retention factor of PAHs solutes in NP-HPLC, good correlation coefficient was observed. Furhtermore, VP_n phases showed

the selective retention for the *ortho* isomers of dicyano- and dinitrobenzenes and this *ortho* isomer selectivity is brought about by the electrostatic interactions.

d) Due to the presence of point negative charge on the nitrogen atom of pyridine moiety, nitrogen atom can interact with residual silanol groups of silica and this phenomenon brings the masking effect for the residual silanol groups resulting in a better asymmetric factor for the basic compounds.

The elution order of PAHs solute with VP_n phases is in the reversed order of hydrophobicity and is quite opposite to ODS phases. Moreover, in contrast to ODS phases, VP_n phases prefer disc-like molecule compared to the linear-one. Such a property of VP_n phases can be explained by the different alignment of polymeric chains on silica surface. VP_n phases do not have any ordered structure like that of crystalline octadecylated acrylate, however pyridine moiety of polymer can interact with the residual silanol groups of silica through hydrogen bonding and reduce the mobility of the polymers. We envisioned that the alignment of VP_n polymers on the silica surface is oblique and slightly flexible. Such a model brings effective interaction for disc-like molecules than the linear molecules as shown in Fig. 14.

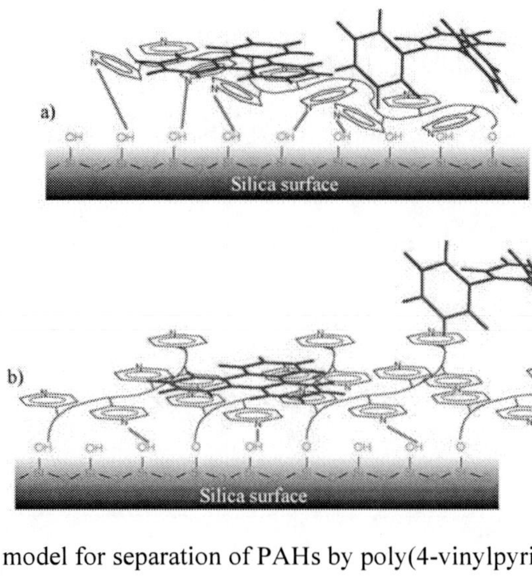

Figure 14. Schematic model for separation of PAHs by poly(4-vinylpyridine) grafted silica phases in HPLC.

9. Comparison between Π Electron Containing Ordered and Disordered Stationary Phases

We have mentioned both ordered and disordered stationary phase (Sil-to-ODA_{20} and Sil-to-VP_{22}) showed enhanced selectivity for PAHs and its positional isomers compare to the commercial columns regardless of their retention order. However, between these two stationary phases remarkable selectivity differences were observed as shown in Table 6. For example, for the analogues of four-ring compounds (Table 6) higher selectivity was observed on Sil-to-ODA_{20} than Sil-to-VP_{22} although they have differences in elution order which cannot be explained only by π-π interaction with a solute, but should be rather discussed with conformational effect of the polymer chains [76]. It is noteworthy to mention that, through highly-oriented polymers as bonded phase in Sil-to-ODA_{20}, multiple interaction is more effective for linear solutes with the large aspect ratio such as naphthacene and pentacene (Fig. 10) than disk-like solutes with the small aspect ratio such as pyrene and coronene when π-π interaction sources such as carbonyl groups are one-dimentionally ordered on silica [76]. On the other hand, we assume that the Sil-to-VP_{22}, as a bonded phase, be disordered on silica because of no driving force for ordering but tightly interact with the residual silanol group. Any direct observation is not yet obtained for this assumption, but it is assumed that the disordered structure might have preferences to contact with disk-like solutes having small aspect ratio such as pyrene and triphenylene (Fig. 14), rather than linear (naphthacene) and nonplanar (*o*-terphenyl) solutes. Moreover, it is interesting to mention that for di-substituted benzenes (dinitrobenzenes), Sil-to-VP_{22} showed amazingly higher *o*-selectivity than Sil-to-ODA_{20} (Table 6). These results suggets that the selectivity due to VP_{22} cannot only be explanined by π-π interaction and charge transfer interaction. In addition, the selectivity for dinitro- and dicyano-substituents can be characterized as strong electron-withdrawing groups which is not observed in dimethyl and dimethoxy-substituents those are electron donating groups [72]. Therefore, higher *o*-selectivity on Sil-to-VP_{22} can be brought by dipole-dipole interaction on the otherhand Sil-to-ODA_{20} is independent of the difference of dipole.

Table 6. Comparison of selectivity of four-ring PAHs, di-substituted nitrobenzene and planar/nonplanar compounds on Sil-to-ODA_{20} and Sil-to-VP_{22} stationary phases.

	PAHs	Sil-to-ODA_{20}		Sil-to-VP_{22}	
		k	α	*k*	α
4-ring PAHs	Triphenylene	3.45		6.29	
			1.34		0.79
	Chrysene	[a]4.62		[b]4.98	
			1.26		0.74
	Benz[a]anthracene	4.34		4.64	
			2.25		0.68
	Naphthacene	7.77		4.25	
Di-substituted nitrobenzene	*p*-dinitrobenzene	1.57		0.67	
			1.48		1.61
	m-dinitrobenzene	[c]2.33		[c]1.08	
			1.50		3.76
	o-dinitrobenzene	2.35		2.53	
Planar/non planar compounds	Triphenylene	4.33		7.73	
			5.06		12.9
	o-terphenyl	0.86		0.60	
	Phenanthrene	[d]1.30		[e]1.62	
			2.32		4.50
	cis-stilbene	0.56		0.36	
	Fluorene	1.01		0.85	
			2.24		3.04
	Diphenylmethane	0.45		0.28	

Mobile phases: [a]methanol-water = 9:1 at 20 °C, [b]methanol-water = 9:1 at 25 °C, [c]methanol-water = 6:4 at 25 °C, [d]methanol-water = 9:1 at 10 °C and [e]methanol-water = 8:2 at 10°C

CONCLUSION

Polycyclic aromatic hydrocarbons are ubiquitous in the biosphere and because of their impact on health and environment; much attention is needed for these compounds. Although HPLC with reversed-phase mode is quite popular, separation of PAHs having similar polarity is still a big challenge for the scientist dealing with analysis of these compounds. Herein we present two different types of π-electron containing organic phases and their selectivity towards PAHs. One is lipid bilayer membrane systems, having gel (solvated crystalline state)-to-liquid crystalline phase transitions are attractive as specific organic media for separation chemistry. Poly(octadecylacrylate) grafted on silica surface by different grafting approaches have been introduced. The ODA_n phase does not make lipid membrane structures in an aqueous system, but it possesses similar functions, such as side-chain ordering and phase transition behavior between ordered and disordered states. However, the resultant selectivity in HPLC often excelled, better than expected. This is due to a sort of a side effect and, then, it may be called a polymeric effect, which promotes multiple interactions to increase the selectivity. Molecular ordering of functional groups particularly enhances it. Moreover, the selectivity can be further pushed ahead by grafting higher amount of polymer and in this case we applied surface initiated ATRP which comprises higher order of the organic phase and enhanced selectivity than that prepared by conventional telomerization process. Furthermore, we have discussed another type of π-electron containing hetroatomic monomer (4-vinylpyridine) that do not displayed such ordered structure like ODA_n stationary phases and in which hydrophobicity of the stationary phase can be diminished and separation was brought by π-π interactions and inductive effect and /or quadraupolar effects. In addition to RP chromatography, Sil-to-VP_n, also showed remarkable separations for PAHs in NP-HPLC. As the solubility of PAHs is very limited to polar solvents so NP retention mode would be another choice. Both of the polymeric π-electron containing ordered and disordered stationary phases showed significant efficiency for separation of PAHs.

REFERENCES

[1] Polynuclear aromatic hydrocarbons (PAH). *In*: *Air quality guidelines for Europe*. Copenhagen,World Health Organization Regional Office for Europe, 1987, pp. 105–117. Agency for Toxic Substances and Disease Registry. *Toxicological profile for polycyclic aromatic hydrocarbons (PAHs): update.* Atlanta, GA, US Department of Health and Human Services, Public Health Services, 1994.

[2] Menzie, C. A.; Potocki, B. B.; Santodonato, J. *Environ. Sci. Technol.* 1992, *26,* 1278-1284.

[3] "Selected Non-heterocyclic Polycyclic Aromatic Hydrocarbons" Environmental Health Criteria 202, International Program on Chemical Safety, WHO, Geneva.

[4] Dell'Omo, M.; Lauwerys, R. R. *Crit. Rev. Toxicol.* 1993, *23,* 111-126.

[5] Grimmer, G.; Brune, H.; Deutsch-Wenzel, R.; Naujack, K. W.; Dettbarn, G.; Misfeld, J.; Abel, U.; Trimm, J. *Cancer Lett.* 1984, *23,* 167-176.

[6] Grimmer, G.; Brune, H.; Deutsch-Wenzel, R.; Naujack, K.W.; Misfeld, J.; Trimm, J. *Cancer Lett*. 1983, *21,* 105-113.

[7] Grimmer, G.; Dettbarn, G.; Brune, H.; Deutsch-Wenzel, R.; Misfeld, J. *Int. Arch. Occup. Environ. Health*, 1982, *50,* 95-100.

[8] Nishioka, M.G., Petersoen, B.A. and Lewtas, J. 1982. "PAC: Physical and Biological Chemistry", Batelle Press, Columbus, OH, pp. 603-614.

[9] Salmeen, I. T.; Pero, A. M.; Zator, R.; Schuetzle, D.; Riley, T. L. *Environ. Sci. Technol.* 1984, *18,* 375-382.

[10] Hannigan, M. P.; Cass, G. R.; Penman, B. W.; Crespi, C. L.; Lafleur, L. A.; Busby Jr., W. F.; Thilly, W. G.; Simoneit, B. R. T. *Environ. Sci. Technol.* 1998, *32,* 3502-3514.

[11] De Boer, J.; Law, R. J. *J. Chromatogr. A* 2003, *1000 (1-2),* 223–251.

[12] Fetzer, J. C. (1989) In Chemical analysis of polycyclic aromatic compounds. Chemical analysis, Vo-Dinh T. Ed. Wiley, New York, pp 59-109

[13] Poster, D.L.; Sander, L. C.; Wise, S. A. (1998) In The handbook of environmental chemistry, Hutzinger O. Ed. vol 3 Part I, PAHs and related compounds. Springer, Berlin Heidelberg New York, pp 77-135

[14] Lee H.K. (2001) In Environmental analysis. Handbook of analytical separations, Kleiböhmer W. Ed. Elsevier Science, Amsterdam, The Netherlands, pp 39-74

[15] Buszewski, B.; Schmid, J.; Albert, K.; Bayer, E. *J. Chromatogr.* 1991, *552 (1-2),* 415-427.

[16] Sindorf, D. W.; Maciel, G. E. *J. Am. Chem. Soc.* 1983, *105 (12),* 3767-3776.

[17] Dorsey, J. G.; Dill, K. A. *Chem. Rev.* 1989, *89 (2),* 331-346.

[18] Pfleiderer, B.; Albert, K.; Lork, K. D.; Unger, K. K.; Briickner, H.; Bayer, E. *Angew. Chem.* 1989, *101,* 336-340.

[19] Pursch, M.; Strohschein, S.; Handel, H.; Albert, K. *Anal. Chem.* 1996, *68 (2),* 386-393.

[20] Sander, L. C.; Wise, S. A. In Retention and Selectivity in Liquid Chromatography; Smith, R. M.; Ed.; Journal of Chromatography Library; Elsevier: Amsterdam, 1995; Vol. 57, p. 337.

[21] Sander, L. C. *J. Chromatogr. Sci.* 1988, *26 (8),* 380-387.

[22] Sander, L. C.; Wise, S. A. *LC-GC* 1990, *8*, 378- 390.

[23] Bouvier, E.S.P.; Oehrle, S.A. *LC.GC Int.*, 1990, *8,* 338.

[24] Bouvier, Edouard S. P.; Oehrle, Stuart A.. Bouvier, Edouard S. P.; Oehrle, Stuart A. *LC-GC* 1995, *13(2)*, 120-130.

[25] R.C. Schulz, H. Waniczek, in: R.C. Schulz, H. Waniczek-(Eds.), 4th ed., Preparation of Macromolecular Materials by Reactions at Macromolecules (General Chemical Reactions, Graft Polymerization), Vol. E20, Georg Thieme Verlag, Stuttgart, 1987.

[26] Edgecombe, B. D.; Stein, J. A.; Fre´chet, J. M. J.; Xu, Z.; Kramer, E. J. *Macromolecules* 1998, *31,* 1292-1304.

[27] Kiyohara, S.; Kim, M.; Toida, Y.; Sugita, K.; Sugo, T. *J. Chromatogr. A* 1997, *758,* 209-215.

[28] Buchmeiser, M. R.; Sinner, F.; Mupa, M.; Wurst, K. *Macromolecules* 2000, *33,* 32-39.

[29] Beers, K. L.; Gaynor, S. G.; Matyjaszewski, K.; Sheiko, S.; Moller, M. *Macromolecules* 1998, *31,* 9413-9415.

[30] Stehling, U. M.; Malmstrom, E. E.; Waymouth, R. M.; Hawker, C. J. *Macromolecules* 1998, *31,* 4396-4398.

[31] Nuyken, O.; Rueda-Sanchez, J.; Voit, B. *Polym. Bull.* 1997, *38,* 657-664.

[32] Ogiwara, Y.; Torikoshi, K.; Kubota, H. *J. Polym. Sci.* 1982, *20,* 17-21.
[33] Dhal, P. K.; Vidyasankar, S.; Arnold, F. A. *Chem. Mater.* 1995, *7,* 154-162.
[34] Kolla, P.; Kohler, J.; Schomburg, G. *Chromatographia* 1987, *23,* 465-472.
[35] Laubli, M. W.; Kampus, B. *J. Chromatogr. A* 1995, *706,* 99-102.
[36] Laubli, M. W.; Kampus, B. *J. Chromatogr. A* 1995, *706,* 103-107.
[37] Reeder, D. H.; Li, J.; Carr, P. W.; Flickinger, M. C.; McCormick, V. *J. Chromatogr. A* 1997, *760,* 71-79.
[38] Hanson, M.; Unger, K. K.; Schomburg, G. *J. Chromatogr.* 1990, *517,* 269-284.
[39] Hirayama, C.; Ihara, H.; Mukai, T. *Macromolecules* 1992, *25*, 6375-6376.
[40] Sander, L. C.; Wise, S. A. *Anal. Chem.* 1984, *56,* 504-510.
[41] Sander, L.C.; Wise, S.A. *Anal. Chem.* 1987, *59,* 2309-2313.
[42] Sentell, K. B.; Dorsey, J. G. *Anal. Chem.* 1989, *61,* 930-934.
[43] Kimata, K.; Hirose, T.; Morichi, K.; Hosoya, K.; Araki, T.; Tanaka, N. *Anal. Chem.* 1995, *67,* 2556-2561.
[44] Jinno, K.; Okumura, C.; Harada, M.; Saito, Y. *J. Liq. Chromatogr.* 1996, *19,* 2883-2899.
[45] Chen, S.; Meyerhoff, *Anal. Chem.* 1998, *70,* 2523-2529.
[46] Chowdhury, M. A. J.; Boysen, R. I.; Ihara, H.; Hearn, M. T. W. *J. Phys. Chem. B* 2002, *106,* 11936-11950.
[47] Ansarian, H. R.; Derakhsan, M.; Rahman, M. M.; Sakurai, T.; Takafuji, M.; Taniguchi, I.; Ihara, H. *Anal. Chim. Acta.* 2005, *547,*179-187.
[48] Ansarian, H. R.; Derakhsan, M.; Rahman, M. M.; Sakurai, T.; Takafuji, M.; Ihara, H. *Canadian J. Chem.* 2005,*77,*1792-1798.
[49] Lyatskaya, Y.; Balazs, A. C. *Macromolecules* 1998, *31(19)*:6676-6680.
[50] Zajac, R. ; Chakrabarti, A. *Phys. Rev.* 1995, *52(6-B)*, 6536-49.
[51] Husseman, M.; Malmström, E.E.; McNamara, M.; Mate, M.; Mecerreyes, D.; Benoit, D.G.; Hedrick, J.L; Mansky, P.; Huang, E.; Russel, T.P.; Hawker, C.J. *Macromolecules* 1999, *32(5)* 1424-1431.
[52] Braunecker, W. A.; Matyjaszewski, K. *Prog. Polym. Sci.* 2007, *32(1)*, 93-146.
[53] Wang, J . S.; Matyjaszewski, K. *J. Am. Chem. Soc.* 1995, *117*, 5614-5615.
[54] Kato, M.; Kamigato, M.; Sawamoto, M.; Higashimura, T. *Macromolecules* 1995, *28*, 1721-1723.
[55] Pursch, M.; Strohschein, S.; Handel, H.; Albert, K. *Anal. Chem.* 1996, *68(2)*, 386-93.
[56] Tonelli, A.E.; Schiling, F.C.; Bovey, F.A. *J. Am. Chem. Soc.*1984, *106*, 1157.
[57] Shundo, A.; Sakurai, T.; Takafuji, M.; Nagaoka, S.; Ihara, H. *J. Chromatogr. A*, 2005, *1073* 169-174.
[58] Claessens, A.; Van Straten, M.A.; Cramers, C.A.; Jezierska, M.; Buszewski, B. *J. Chromatogr. A,* 1998, *826(2)*, 135-156.
[59] Chowdhury, M. A. J.; Ihara, H.; Sagawa. T.; Hirayama, C. *J. Chromatogr. A.*2000, *877,* 71-85.
[60] Jinno, K.; Yamamoto, K.; Nagashima, H.; Ueda, T.; Itoh, K. *J. Chromatogr.*1990, *517*, 193-207.
[61] Jinno,K.; Ibuki, T.; Tanaka, N.; Okamoto, M.; Fretzer, J.C.; Biggs, W.R.; Griffiths, P.R.; Olinger, J.M. *J. Chromatogr.*1989, *461*, 209-27.

[62] Hirayama, C.; Ihara, H.; Nagaoka, S.; Syono, T. *Chem. Lett.* 1992, *6*, 971-974.

[63] Ihara, H.; Sagawa, T.; Goto, Y.; Nagaoka, S. *Polymer* 1999, *40*, 2555-2560.

[64] Ihara, H.; Uemura, S.; Okazaki, S.; Hirayama, C. *Polym. J.* 1998, *30*, 394-397.

[65] Sakaki, S.; Kato, K.; Miyazaki, T.; Musahi, Y.; Ohkubo, K.; Ihara, H.; Hirayama, C. J. *Chem. Soc. Faraday Trans.* 1993, *9*, 659-664.

[66] Ihara, H.; Sagawa, T.; Nakashima, K.; Mitsuishi, K.; Goto, Y.; Chowdhury, J. *Chem. Lett.* 2000, *2,* 128-129.

[67] Goto, Y.; Nakashima, K.;Mitsuishi, K.; Takafuji, M.; Sakaki, S.; Ihara, H. *Chromatographia* 2002, *56*, 19-23.

[68] Ihara, H.; Uemura, S.; Okazaki, S.; Hirayama, C. *Polym. J.* 1998, *30(5)*, 394-97

[69] Ihara, H.; Okazaki, S.; Ohmori, K.; Uemura, S.; Hirayama, C.; Nagaoka, S. Anal. Sci. 1998, *14(2)*, 349-354. .

[70] Tanaka, N.; Kinoshita, H.; Araki, M.; Tsuda, T. *J. Chromatogr.* 1985, *332,* 57-69

[71] Belyakova, L. A.; Varvarin, A. M. *Colloids Surf.* 1999, *154,*285-294

[72] Ihara, H.; Fukui, M.; Mimaki, T.; Shundo, A.; Dong, W.; Derakhshan, M.; Sakurai, T.; Takafuji, M.; Nagaoka, S. *Anal. Chim. Acta* 2005, *548, 51*-57.

[73] Gautam, U. G.; Shundo, A.; Gautam, M. P.; Takafuji, M.; Ihara, H. *J. Chromatogr. A.* 2008, *1189*, 77-82.

[74] Gautam, U. G.; Gautam, M. P.; Sawada, T.; Takafuji, M.; Ihara, H. *Anal. Sci.*. 2008, *24*, 615-621.

[75] Takafuji, M.; Dong, W.; Sagawa, T.; Sakurai, T.; Nagaoka, S.; Ihara, H. *Polym. J.* 2002, *34,* 437-442.

[76] Ihara, H.; Goto, Y.; Sakurai, T.; Takafuji, M.; Sagawa, T.; Nagaoka, S. *Chem. Lett.* 2001, 1252-1253.

In: Polycyclic Aromatic Hydrocardons: Pollution, Health… ISBN 978-1-60741-462-9
Editors: Pierre A. Haines and Milton D. Hendrickson

Chapter 12

DETERMINATION OF POLYCYCLIC AROMATIC HYDROCARBONS IN SOLID-MATRIX FOODS USING A NON-DESTRUCTIVE METHOD

Diana Martin and Jorge Ruiz
Tecnología de los Alimentos, Facultad de Veterinaria, Universidad de Extremadura, Caceres, Spain

ABSTRACT

The use of a direct extraction device (DED) for coupling to solid phase microextraction (SPME) is described. Such a device enables the introduction of the SPME fibre in the core of solid matrixes, allowing the analysis of volatile compounds, such as polycyclic aromatic hydrocarbons (PAHs), with no or little deterioration of the solid product. The feasibility of using SPME-DED for extraction of PAHs from solid matrixes mimicking solid foodstuffs (gelatine alone, gelatine containing fat and salt, or cooked chicken patties) and from commercial smoked meat products (smoked tenderloin, smoked jowl, smoked dry cured sausage and smoked potato sausage) was tested. Moreover, the diffusion of PAHs in solid matrixes by SPME-DED was also evaluated. SPME-DED appeared as a rapid, non-destructive technique for primary screening of low molecular weight (MW) PAHs in all solid matrixes assayed, as well as for the study of the diffusion process of PAHs within solid systems, without sampling and any laboratorial manipulation.

INTRODUCTION

Most methodologies for analysis of polycyclic aromatic hydrocarbons (PAHs) in food systems are laborious, time consuming and need organic solvents, specially for complex matrixes in which PAHs are strongly trapped due to their interaction with different components of the sample (Moret and Conte, 2000). Moreover, the sampling step may deteriorate the quality of several solid foods. Solid-phase microextraction (SPME) is an alternative technique that can overcome some of these disadvantages. SPME integrates sampling, extraction, concentration and sample introduction in a single process, and uses no solvent during extraction (Arthur and Pawliszyn, 1990). Previous studies have demonstrated the feasibility of SPME for analysing PAHs in different types of samples, such as environmental (water, oil and soil) (Eriksson, Fäldt, Dalhammar and Borg-Karlson, 2001; Langenfeld, Hawthorne and Miller, 1996; Llompart and Fingas, 1998), biological (urine, blood serum) (Poon, Lam and Lam, 1999; Waidyanatha, Zheng and Rappaport, 2003) and foodstuffs (smoked fish and smoked cheese) (Guillen and Errecalde, 2002; Guillen, Ibargoitia, Sopelana, Palencia and Fresno, 2004).

Despite the advantages of SPME, this technique implies, as traditional methods, a sampling step that could deteriorate the integrity of solid foods, and consequently their quality. We have described the use of a direct extraction device (DED) (Andres, Cava and Ruiz, 2002; Ruiz, Ventanas and Cava, 2001), which enables the introduction of the SPME fibre in the core of solid matrixes, allowing a rapid screening of volatile compounds from solid foods with no or little deterioration of the product. Such device (Figure 1) has an internal chamber in which the SPME fibre is exposed. This small chamber has several holes that allow volatile compounds to equilibrate between the solid matrix and the internal headspace of the DED. The SPME-DED method has been successfully applied on the extraction of volatile compounds from dry cured ham and pâté in one step, without sampling and thus, with only minor or none physical damage of the foodstuffs (Andres et al., 2002; Ruiz et al., 2001). Moreover, Ventanas and Ruiz (2006) effectively extracted volatile nitrosamines by SPME-DED from solid model systems.

The feasibility of using SPME-DED for extraction of PAHs from solid matrixes mimicking solid foodstuffs (gelatine alone, gelatine containing fat and salt, or cooked chicken patties) and from commercial smoked meat products (smoked tenderloin, smoked jowl, smoked dry cured sausage and smoked potato sausage) has been carried out in the present study. Moreover, the diffusion of PAHs in solid matrixes has also been evaluated by SPME-DED without sampling and any laboratorial manipulation.

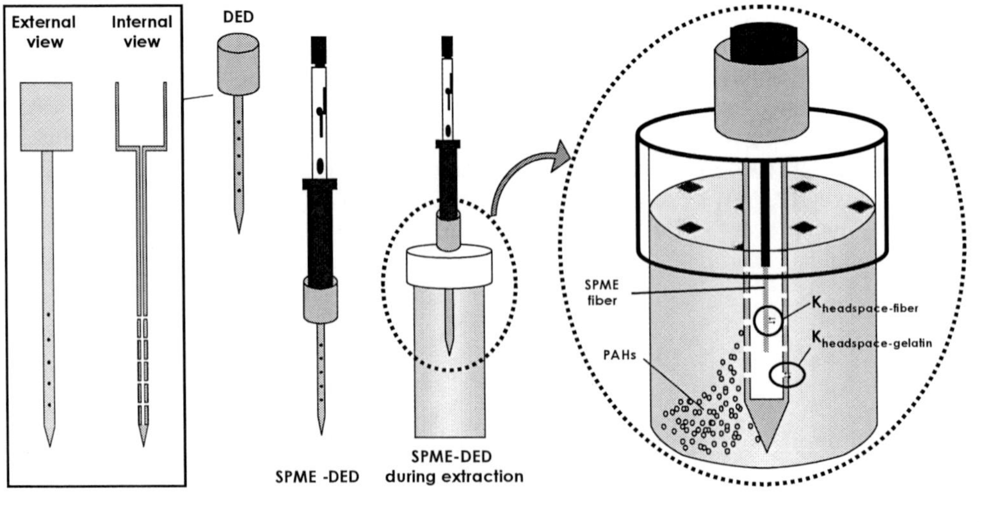

Figure 1. Scheme of the use of SPME-DED in model systems of gelatine. An enlargement of a section of the gelatine shows the diffusion process of some PAHs from the matrix to the headspace of the DED. The equilibriums that are implied in the process are also showed.

MATERIAL AND METHODS

Reagents and Materials

PAHs standard was supplied by Sigma-Aldrich (QTM PAHs Mix, Sigma Aldrich, St. Louis, USA). This solution contained 16 analytes: acenaphthene (AC), acenaphthylene (ACL), anthracene (AN), benz[*a*]anthracene (BaA), benzo[*a*]pyrene (BaP), benzo[*b*]fluoranthene (BbFA), benzo[*ghi*]perylene (BghiP), 2-bromonaphthalene (2-BrNA), chrysene (CHR), dibenz[*ah*]anthracene (DBahA), fluoranthene (FA), fluorene (FL), indeno[*1,2,3-cd*]pyrene (IP), naphthalene (NA), phenanthrene (PHE) and pyrene (PY).

Methanol (HPLC grade) was obtained from Scharlau Chemie (Barcelona, Spain). Gelatine (300 bloom, A type) was supplied by Sigma Aldrich (St Louis, USA). Sunflower oil and chicken meat were obtained from a local supermarket and sodium chloride was from Panreac (Panreac Quimica S.A., Barcelona). 100 μm PDMS SPME stationary phase was provided by Supelco.

Smoked meats products were obtained in a small village from an area of Extremadura (Spain) in which home processing of this type of products is common. These products were: smoked tenderloin, smoked jowl, smoked dry cured sausage and smoked potato sausage. All of them had been directly exposed to the smoke for 30 days.

Preparation of Solid Gelatines Containing PAHs

PAHs stock solution at 8 μg mL^{-1} was prepared by diluting 4 μL of the PAHs standard (2000 μg mL^{-1}) in methanol.

Gelatine solutions were prepared by addition of the appropriate amount of gelatine to distilled water and microwave heating. Subsequently, different amounts of salt (20% in distilled water) and sunflower oil were added. All systems showed a final concentration of 20% (w/v) of gelatine. The selected values of fat and salt were: 0, 3 and 6% of sunflower oil combined with 0, 1 and 2 % of sodium chloride. These values were chosen since they are common proportions of fat and salt in smoked fish and meat products. Gelatines with each combination of salt and oil were prepared by triplicate. All components were mixed using a homogenizer (Omni 5000 model International) during 1 min at 3000 rpm. The PAHs standard stock solution was added to the gelatine systems to obtain a final concentration of 10 ng/mL. Working gelatines were immediately sealed, vortexed for 30s and stored at 2-4°C until analysis. Analysis was performed within next 3 days after preparation.

The evaluation of the diffusion process of PAHs in solid gelatines was carried out by adding the PAHs standard (50 μL) on the surface of solid gelatines. Extraction was carried out at 0, 1, 4 and 8 days after PAHs addition and at depths of 1.5, 3 and 4.5 cm measured from the surface of the gelatine for each of the times evaluated. Each combination of time × depth was made by triplicate.

Preparation of Cooked Chicken Patties Containing PAHs

Cooked chicken patties were made up of 57% of chicken meat, 40% of distilled water and 3% of salt. These ingredients were minced until achieving an homogeneous paste and subsequently PAHs standard stock solution was added to obtain a final concentration of 5 ng/ml. Polypropylene tubes with polyethylene caps were filled with 32 ml of this mix. Gelling of meat proteins was achieved by cooking the tubes at 70°C during 15 minutes. Meat gels were stored at 2-4°C until analysis.

SPME-DED Extraction of PAHs

To carry out the analysis of PAHs by SPME-DED, the DED was introduced into the core of solid gelatines and thereafter, the needle of the SPME holder was placed into the DED. Once inside the solid material, the fibre was exposed to the small space inside the DED, which has small holes that allow the volatiles from the sample to enter into the space in which the fibre is kept. Figure 1 shows the DED, the way in which it is coupled to SPME, and how SPME-DED is used to extract PAHs from gelatines. The used SPME fibre was that made up of polydimethylsiloxane (PDMS 100 μm). PDMS is a liquid and non-polar polymer that is commonly used in the extraction of PAHs due to the non-polar nature of these compounds. However, since there are a variety of stationary phases for SPME fibres, we previously carried out the selection of the most appropriate SPME fibre for PAHs analysis (Martin and Ruiz, 2007). In such study, different SPME fibres were evaluated (PDMS 100 μm; divinylbenzene/polydimethylsiloxane 65 μm; and polyacrilate 85μm). The three SPME fibres showed good and similar efficacy, reproducibility, linearity and limit of detection for extraction of PAHs by SPME-DED, but PDMS stand out from the others because its better results for most studied PAHs and other desirable characteristics like its higher resistance or lower conditioning time. Extractions were carried out at 25°C for 60 min.

PAHs from cooked chicken patties were extracted following the same procedure for gelatines. PAHs from smoked meat products were extracted by directly inserting the DED into the product and subsequently inserting the SPME fibre.

Gas Chromatography-Mass Spectrometry Conditions

After extraction, analytes were desorbed onto the gas chromatography column by inserting the fibre in the injector at 250°C for PDMS. Analyses were performed using an Agilent 6890 series gas chromatograph (Agilent, Avondale, USA) coupled to a mass selective detector (Agilent 5973, Agilent, Avondale, USA). Analytes were separated using a 5% Phenyl-Methyl Silicone (HP-5) bonded phase fused silica capillary column (Hewlett-Packard, 50m x 0.32mm id., film thickness 1.05μm), operating at 60.2 kPa of column head pressure, resulting in a flow of 1.2 mL min^{-1} at 40°C. The SPME fibre was desorbed during 30 min. The injection port was in splitless mode. The temperature program was 130°C during 0.5 min, raised to 290°C at a rate of 5°C min^{-1} and maintained at this temperature for 42 min.

The transfer line to the mass spectrometer (MS) was maintained at 290 °C. The mass spectra were obtained by electronic impact at 70eV, a multiplier voltage of 1824V and collecting data at a rate of 2.83 scan s^{-1}. Detection of compounds was carried out in SIM mode. Selected ions chosen for each compound and retention times are shown in Table 1. In order to establish the retention time for each analyte, stock solutions were previously injected under the same chromatographic conditions as those for detection of PAHs in SIM mode but using the mass selective detector in scan mode over the m/z range of 30 to 550.

RESULTS AND DISCUSSION

PAHs from Solid Matrixes Extracted by SPME-DED

The feasibility of using SPME-DED for extraction of PAHs from solid matrixes mimicking solid foodstuffs was carried out in the present study. Given that it would be very difficult to achieve an homogeneous concentration of PAHs in those real food samples for which SPME-DED was initially developed (mainly whole piece meat and fish products), we have tested the feasibility of using SPME-DED for analysing PAHs in solid model systems made with gelatine. This allows a correct dilution of PAHs while the gelatine is still liquid, while showing similar characteristics to muscle foods when it is solid, since it is quite elastic and the matrix is constituted by proteins.

The obtained chromatogram for extracted PAHs from model systems of gelatine (0% salt and 0% oil) is shown in Figure 2. The SPME-DED only extracted 9 of the 16 PAHs present in the standard, more specifically those with lower molecular weights (MW) (NA, ACL, 2-BrNA, AC, FL, PHE, AN, FA and PY) (Table 1).

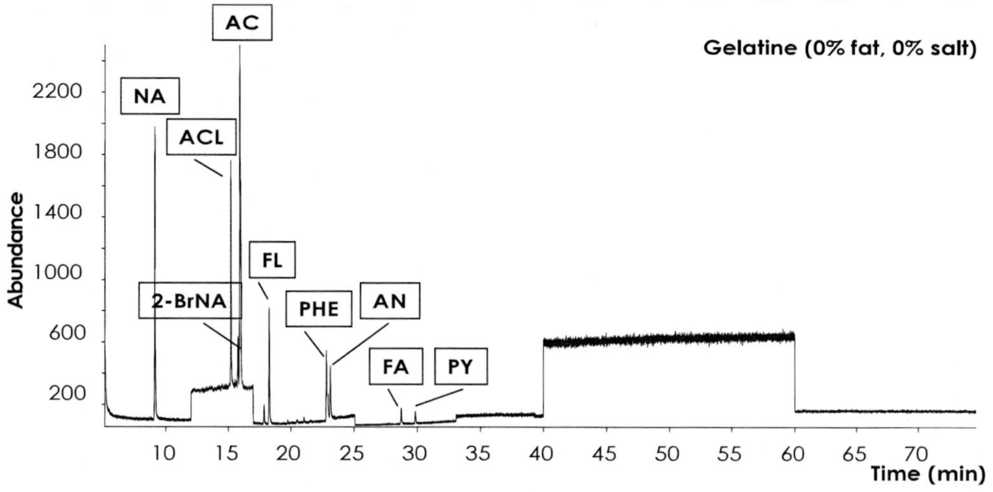

Figure 2. Chromatograms corresponding to the extracted PAHs from gelatine model systems (10 ng/ml of PAHs) by SPME-DED.

Table 1. PAHs included in the studied standard, showing the selected ion for detection in MS (SIM mode) and obtained retention times.

PAH	MW and m/z	Retention time (min)
Naphthalene (NA)	128	9.12
Acenaphthylene (ACL)	152	15.24
2-Bromonaphthalene (2BrNA)	206	15.80
Acenaphthene (AC)	154	16.03
Fluorene (FL)	166	18.32
Phenanthrene (PHE)	178	22.88
Anthracene (AN)	178	23.10
Fluoranthene (FA)	202	28.72
Pyrene (PY)	202	29.84
Benz[a]anthracene (BaA)	228	36.19
Chrysene (CHR)	228	36.47
Benzo[b]fluoranthene (BbFA)	252	45.32
Benzo[a]pyrene (BaP)	252	49.00
Indeno[1,2,3-cd]pyrene (IP)	276	66.51
Dibenzo[ah]anthracene (DBahA)	278	67.04
Benzo[ghi]perilene (BghiP)	276	71.62

These obtained results were most likely caused by the physicochemical properties of the studied PAHs, such as low volatility, high molecular weight and low polarity, as well as the features of the matrix, such as chemical composition and microstructure. PAHs show a wide range of volatility, so the release of those with lower volatility to the headspace inside the DED during extraction is limited. This could explain the extraction of only 9 PAHs but not the whole 16 PAHs in the standard and the decrease in chromatographic area of the 9 PAHs extracted due to decreasing volatility of the analytes (Figure 3). This fact was also observed by other authors in the extraction of PAHs by headspace-SPME. Doong, Chang and Sun (2000) satisfactorily extracted only 8 of 16 PAHs present in water samples with 100 μm PDMS phase and PY was the heaviest PAH found. Similarly, Eriksson et al., (2001) studying the presence of PAHs in soils by SPME, were not able to extract analytes with heavier MW than that of PY.

High MW PAHs are those mainly recognized in carcinogenic and mutagenic processes (Agency for Toxic Substances and Disease Registry, ATSDR). However, the non carcinogenicity of most low MW PAHs has not been proved and they are still under research. On the other hand, Hansen et al. (1991) pointed out that NA and PY might function as markers for carcinogenic PAHs. Both NA and PY were satisfactorily detected in solid matrixes by SPME-DED (Figure 3). Thus, the SPME-DED technique would be a rapid and interesting method for a primary screening in the determination of the presence of low MW PAHs, both in solid foodstuffs or other kind of solid matrixes in which the presence of these specific toxic compounds would be of interest, without sampling and any laboratorial manipulation.

After evaluating the feasibility of SPME-DED for the extraction of PAHs from simple models mimicking solid foodstuffs, we evaluated the use of SPME-DED for the extraction of

PAHs from more complex matrixes, more similar to solid foods, containing different components (meat proteins, nucleotides, peptides, fat, salt…). Thus, chicken patties similar to frankfurters were manufactured and a volume of the PAHs stock solution was added during elaboration. The chromatographic areas obtained from the extraction of PAHs from chicken patties are shown in Figure 3. All the 16 PAHs present in the stock solution were detected. Moreover, the chromatographic areas were rather similar for all the PAHs and in this case, there was not a trend to a lower extraction due to increasing MW of analytes, as it was observed in model systems of gelatine. The reason that could explain these results is the nature of the matrix, since there are differences between the meat and the gelatine gels, both in structure and composition. Chicken patties showed low strength and thus evident exudates, unlike the gelatine systems, which had a great consistency and no exudates. The great exudates of meat systems are most likely the main reason that explains the detection of all the 16 PAHs and the similar areas obtained for all of them. Liquid released from the chicken patties could get into the DED through the holes during extraction, being in contact with the stationary phase of the SPME and thus turning the method into direct SPME instead of headspace SPME. However, it should be considered that direct contact between the exudates and the stationary phase does not invalidate the application of SPME-DED for extraction of PAHs from solid matrixes, since the main objective of this technique is not the quantification but the qualitative analysis of the presence of these compounds, without a sampling step and without deterioration of foodstuffs. This aim is not affected by the way of extraction by SPME (direct or headspace) that takes place inside the DED.

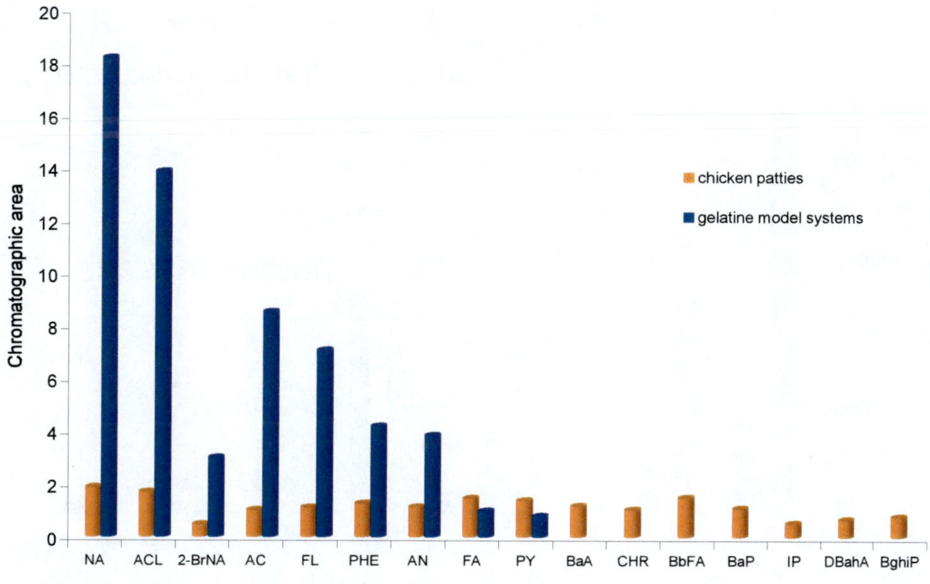

Figure 3. Chromatographic areas ($x10^4$) obtained for PAHs extracted by SPME-DED from chicken patties at 5 ng/ml compared with those obtained from model systems of gelatine (0% of fat and 0% of salt) at 10 ng/ml.

After testing the feasibility of the SPME-DED method on the analysis of PAHs from simple matrixes of gelatine models and complex matrixes of meat model systems, we finally applied the SPME-DED technique on the analysis of PAHs from real foodstuffs likely contaminated with these toxic compounds, such as traditional smoked meat products (Figure 4). Similarly to gelatine gels, PAHs with MW above PY were not detected. Main reasons explaining this fact are the same as those previously discussed for gelatines. Curiously, despite all products were smoked together at the same conditions, different PAH profiles were obtained for each one, both qualitatively and quantitatively. Thus, NA, ACL, FL, PHE and AN were detected in all products, but AC was only extracted from smoked tenderloin and FA and PY were not found in smoked dry cured sausage. NA showed the highest chromatographic area in the case of smoked dry cured sausage and smoked tenderloin, whereas the highest value corresponded to FA for smoked jowl and to PHE for smoked potato sausage. These different results might be related to the different nature, structure or composition of the matrixes assayed. Moreover, their different content in lipids or salt could lead to a different release of PAHs into the headspace. In addition, the presence of natural casings in some of them (smoked dry cured sausage or smoked potato sausage) could have limited, but not totally prevented, diffusion of PAHs from the surface to inside the product (Maga, 1988). Furthermore, it should be considered that during ripening and smoking of these products, a great variety of characteristic aroma compounds are generated (Maga, 1988). Such volatile compounds might interfere in the extraction process (Roberts, Pollien and Milo, 2000), since the concentration of volatile compounds derived from the ripening and smoked processes is higher than PAHs levels.

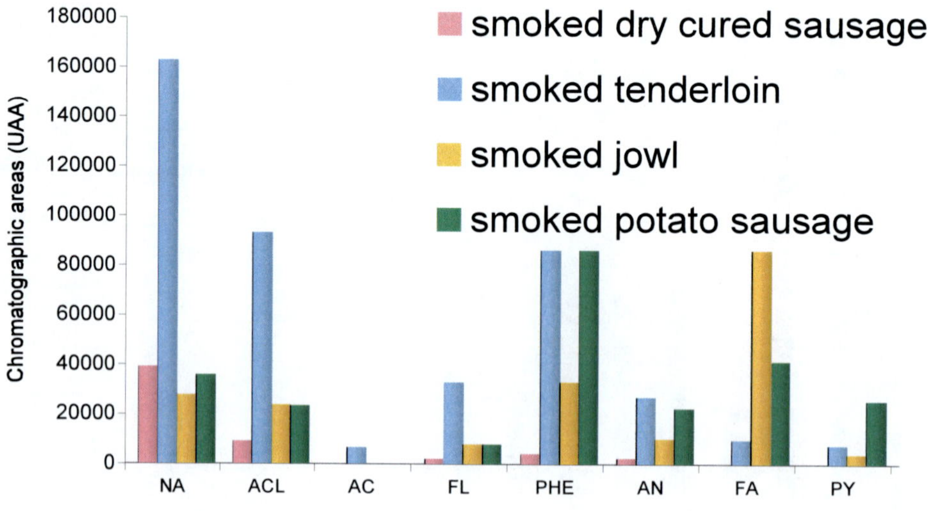

Figure 4. PAHs found in smoked meats products by SPME-DED.

In summary, SPME-DED satisfactorily extracts low MW PAHs from smoked foodstuffs but it seems that the application of SPME-DED on solid matrixes and foodstuffs is limited by the volatility of the targeted compounds. However, this circumstance does not invalidate the potential application of SPME-DED on PAHs analysis from solid matrixes, since SPME-DED would be a useful screening method for those studies in which the aim is the evaluation of the presence of low MW PAHs, without a sampling step and without deterioration of solid matrixes. This is especially interesting in solid foodstuffs, in which the depreciation of the product due to sampling could be an important drawback.

Effect of Fat and Salt on the Extraction of PAHs from Solid Matrixes by SPME-DED

As suggested above, the presence of salt and fat on the solid matrixes may influence the extraction of PAHs from solid matrixes. We applied the SPME-DED method for analysing the effect of the presence of salt and fat in the solid matrix on the analysis of PAHs. It is known that the presence of lipids in the matrixes hinders the release of some compounds during the extraction by SPME (Keszler and Heberger, 1999) or by other traditional methods such as saponification or liquid-liquid partition (Phillips, 1999; Moret and Conte, 2000) since lipophilic compounds, as PAHs, interact with the non-polar groups of lipids.

The results of the study of the effect of oil on the extraction of PAHs from model systems of gelatine by SPME-DED are shown in Table 2. There was a decrease in the efficacy of extraction of the 9 PAHs detected when the proportion of sunflower oil increased. This effect was more marked for low MW PAHs and when the increasing of fat was from 0 to 3% than when it was from 3% to 6%.

Table 2. Chromatographic areas (x10^{4}) obtained for each PAH extracted by SPME-DED from model systems of gelatine with different levels of oil (sunflower oil) and salt (sodium chloride).

	0% oil			3% oil			6% oil		
	0% salt	1% salt	2% salt	0% salt	1% salt	2% salt	0% salt	1% salt	2% salt
NA	18.2	13.7	11.0	2.2	2.3	1.9	1.7	1.4	1.9
ACL	13.9	9.8	8.6	1.5	1.3	1.0	0.7	0.8	1.1
2-BrNA	3.0	1.9	1.8	0.4	0.3	0.3	0.2	0.2	0.2
AC	8.6	5.2	4.7	0.9	0.8	0.7	0.5	0.6	0.6
FL	7.1	4.6	4.3	0.7	0.5	0.6	0.3	0.3	0.4
PHE	4.2	2.7	2.8	0.8	0.6	0.6	0.5	0.4	0.5
AN	3.9	2.6	2.4	0.6	0.5	0.4	0.2	0.3	0.3
FA	1.0	0.7	0.7	0.4	0.3	0.2	0.2	0.2	0.2
PY	0.8	0.7	0.5	0.4	0.3	0.3	0.2	0.3	0.2
BaA	-	-	-	-	-	-	-	-	-
CHR	-	-	-	-	-	-	-	-	-
BbFA	-	-	-	-	-	-	-	-	-
BaP	-	-	-	-	-	-	-	-	-
IP	-	-	-	-	-	-	-	-	-
DbahA	-	-	-	-	-	-	-	-	-
BghiP	-	-	-	-	-	-	-	-	-

On the other hand, the addition of different salts to the matrix during SPME extraction is usually considered a useful practice, since it may enhance the release of volatile compounds to the headspace (Pawliszyn, 1997). However, in the present study, a decrease in the extraction of PAHs was observed when salt concentration increased (Table 2). Some authors have observed a better extraction of PAHs from water (Djozan and Assadi, 1999) or urine (Waidyanatha et al., 2003) by SPME when the levels of salt were increased. However, the opposite effect of the presence of salt in the extraction of volatile compounds by headspace SPME has been evidenced at high salt concentrations. In the present study, the tested levels of salt were lower, similar to those present in meat and fish products.

The nature of the matrix should also be considered for explaining the extraction results. In the case of gelatine matrixes, the ionic strength can change the equilibrium of the intermolecular bonds between the proteins forming the gel with a consequent effect on the microstructure (Smyth, O'Neill and Smith, 1999; Totosaus, Montejano, Salazar and Guerrero, 2002). Accordingly, Yang et al. (1997) found a higher strength in gelatine gels with higher salt concentration. This could explain the worse extraction of PAHs due to increasing salt content in the present study, since gels with higher strength have shown lower release of volatile compounds (Boland, Buhr, Giannouli and van Ruth, 2004; Guinard and Marty, 1995).

Regarding the combination of different levels of salt and sunflower oil, a 3% of fat in the matrix caused a lower extraction of most PAHs as salt concentration increased (Table 2). The opposite trend was observed for most PAHs when there was a 6% of fat in the matrix and increasing salt concentrations were considered, showing a recovery of the extraction of PAHs. It seems that salt and fat slightly tend to enhance the PAHs extraction from solid matrixes of gelatine only when both ingredients were present at high levels. The combined effect of salt and fat on the release of PAHs from solid matrixes is not simple and most likely depends upon the physicochemical nature of the matrix.

Study of the Diffusion Process of PAHs in Solid Matrixes by SPME-DED

PAHs can be found in unprocessed foods such as vegetables, fruits or seafood, deposited from contaminated air, water or soil. In processed foods, PAHs can be formed during the thermal treatment at high temperatures, in technological processes such as smoking, drying or grilling (Guillen and Sopelana, 2003). Regardless the mechanism of contamination, PAHs can be found both at the surface of foodstuffs, but also within them, due to a diffusion process. However, the scientific information about the extent of the diffusion of PAHs in different foods is scarce. A deeper knowledge about diffusion of PAHs in foodstuffs would be of interest in order to have a better understanding of the presence of these toxic compounds in foods, with information concerning about the existence of penetration of PAHs from the surface of the product, the diffusion rate or the depth at which these compounds could be found. In the present work, the diffusion process of PAHs in model systems of solid gelatine was studied using SPME-DED, without sampling and without deterioration of the solid matrixes.

The chromatographic areas obtained for each extracted PAH at different depths and times evaluated are shown in Figure 5. Only those PAHs showing a MW below 206 could be detected. As previously explained, factors such as the physicochemical characteristics of these compounds (high MW, low volatility, low polarity) and interactions with the network of the solid matrix might be the main reasons explaining this fact.

As it can be observed in Figure 5, at the beginning of the study (day 0), NA was the only compound found, both at 1.5 and 3 cm of depth. However, one day after the preparation of the model systems (day 1), most of the PAHs were found at the first depth measured (1.5 cm). At 3 cm, 5 PAHs were detected on day 1 (NA, PHE, AN, FA and PY), with lower chromatographic areas than those found at 1.5 cm depth. No compound was found at 4.5 cm on day 1.

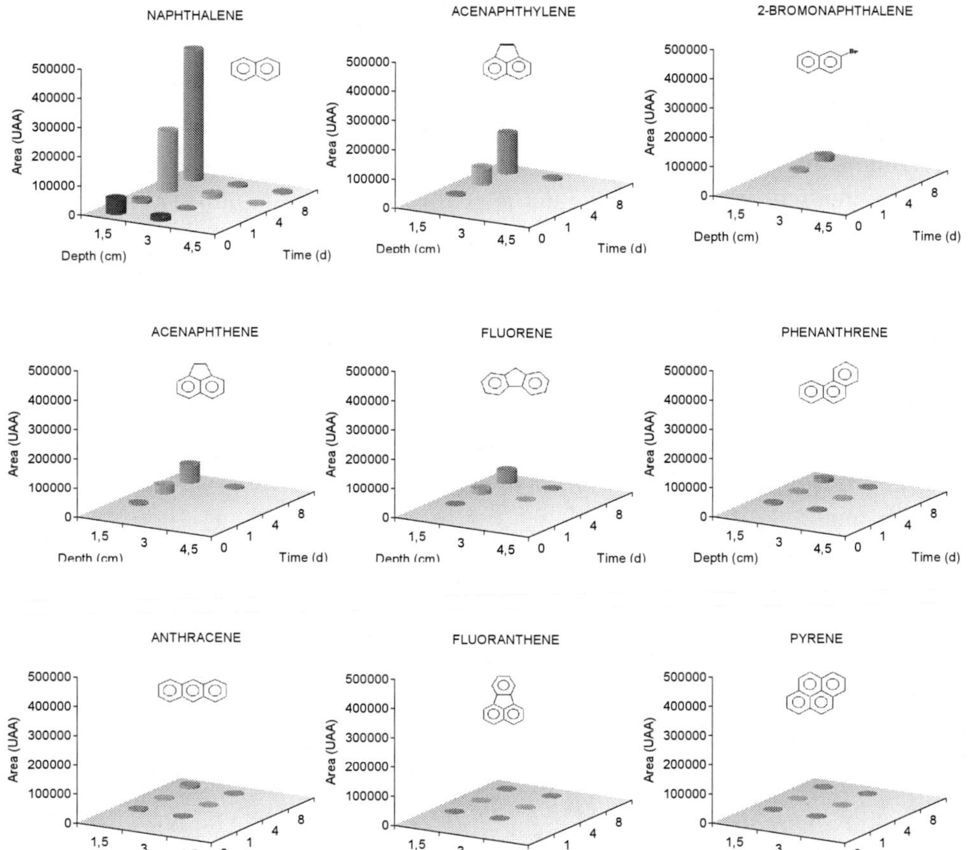

Figure 5. Chromatographic areas obtained for each PAH extracted by SPME-DED from model systems of solid gelatine at different times and depths.

4 days after addition of the PAHs on the surface of the solid gelatines, the 9 detected PAHs were found at 1.5 cm. Those analytes with lower MW (NA, ACL, AC and FL) showed the highest chromatographic areas values at this depth. At 3 cm from the surface of the gelatine, only 6 PAHs were extracted at day 4 (NA, FL, PHE, AN, FA and PY) and at 4.5 cm NA was the only analyte detected at day 4 of storage.

The results obtained after 8 days of the addition of the PAHs were similar to those observed after 4 days, the main difference being the higher chromatographic areas registered at the 8 days. Moreover, most of the detected PAHs diffused up to 3 cm of depth after 8 days.

These obtained results indicate that an initial contamination with PAHs in the surface of foodstuffs implies their subsequent penetration within the product over the storage time. In addition, a lower diffusion of such compounds due to increasing MWs of the analytes was observed. The physical and chemical interactions of analytes with the protein network of the microstructure of the gelatine imply a complex diffusion process (Baines and Morris, 1987; Boland et al. 2004). This might explain why the PAH with the lowest MW (NA) was the only one detected at the deepest depth evaluated (4.5 cm) after 8 days of addition of the analytes on the surface of the solid gelatine and also the general trend to decreasing chromatographic areas and number of found PAHs with increasing MWs. Simko and Simon (1999) also observed a lower diffusion of higher MW PAHs through polyethylene.

The number of aromatic rings that constitute the chemical structure of each PAH is directly connected with its MW and due to that, it might be related to the diffusion capacity of these compounds. This fact was shown up in the results obtained in this work (Figure 5), since those PAHs with the same number of rings in their chemical structure seemed to present a similar behaviour in the diffusion process on the solid matrix.

The solubility of PAHs in water is another factor that should be considered for explaining the diffusion process of PAHs, since water is the major component of the solid gel (80%). Diffusion of PAHs through the gel implies not only diffusion through the protein network, but the previous solubilization of analytes in the water fraction. The low water solubility of PAHs (which is even lower for high MW PAHs) surely contributes to the low diffusion rates found for almost all analytes.

In summary, solid foodstuffs exposed to an external source of PAHs (such as smoked or barbecued foods) are potential sources of these carcinogenic compounds even if the surface (or the casing) is trimmed. SPME-DED appears as a useful technique for the study of the diffusion of volatile compounds within solid foodstuffs, without sampling and any laboratorial manipulation.

REFERENCES

Agency for Toxic Substances and Disease Registry, available at: http://www.atsdr.cdc.gov/toxprofiles/tp69.html.

Andres A I, Cava R, Ruiz J (2002) *J Chromatogr* A, 963, 83.

Arthur CL, Pawliszy J (1990) *Anal Chem* 62, 2145.

Baines Z, Morris E. (1987). *Food Hydrocolloids* 1, 197-205.

Boland A, Buhr K, Giannouli P, van Ruth S (2004) *Food Chem* 86, 401-411

Djozan D, Assadi Y (1999) *Microchemical J* 63, 276-284

Doong R, Chang S, Sun Y (2000) *J Chromatogr* A 879, 177

Eriksson M, Fäldt J, Dalhammar G, Borg-Karlson A K (2001) Chemosphere 44 1641.

Guillen M D, Errecalde M C (2002) *J Sci Food Agric* 82, 945.

Guillen M D, Ibargoitia M L, Sopelana P, Palencia G, Fresno M (2004) *J Dairy Sci* 87, 284-299

Guillen MD, Sopelana P. (2003) Polycyclic aromatic hydrocarbons in diverse foods. In: D'Mello JPF, editor. *Food Safety*. Cambridge:MA. p 175.
Guinard JX, Marty C (1995) *J Food Sci* 60, 727-730
Hansen A M, Poulsen O M, Christensen J M (1991) *Int Arch Occup Environ Health* 63, 247
Keszler A, Heberger K (1999) *J Chromatogr* A 845, 337-347
Langenfeld J J, Hawthorne S B, Miller D J (1996) Anal Chem 68, 144.
Llompart M, Li K, Fingas M (1998) *J Chromatog* A 824, 53.
Maga J A (1988) *Smoke in Food Processing*, CRC Press, Boca Raton, FL, 1988.
Martin D, Ruiz J (2007) Talanta 71, 751-757.
Moret S, Conte L S (2000) *J Chromatogr* A 882, 245-253
Pawliszyn, J. (1997) Solid Phase Microextraction: Theory and Practice. New York: Wiley-VCH.
Phillips D H (1999) *Mutation Res* 443, 139-147
Poon K, Lam P K S, Lam M H W (1999) *Anal Chim Acta* 396, 303.
Roberts D D, Pollien P, Milo C (2000) *J Agric Food Chem* 48, 2430.
Ruiz J, Ventanas J, Cava R (2001) *J Agric Food Chem* 49, 5115.
Simko P, Simon P, Khunová V. (1999) *Food Chem* 64, 157-161.
Smyth AB, O'Neill E, Smith DM (1999) Functional properties of muscle proteins in processed poultry products. In: R.I. Richardson (Ed.) *Poultry Meat Science*. (pp. 377-396) Wallingford (UK): CABI Publishing.
Totosaus A, Montejano J, Salazar J, Guerrero I (2002) *Int J Food Sci Tech* 37, 589-601
Ventanas S, Ruiz J (2006) Talanta 70, 1017-1023
Waidyanatha S, Zheng Y, Rappaport S M (2003) *Chem Biol Internat* 145, 165.
Yang X, Zheng P, Cui Z, Zhao N, Wang Y, Yao K (1997) *Polymers Int* 44, 448-452.

In: Polycyclic Aromatic Hydrocardons: Pollution, Health... ISBN 978-1-60741-462-9
Editors: Pierre A. Haines and Milton D. Hendrickson

Chapter 13

PAH REMOVAL BY TWO NATIVE TROPICAL PLANTS CULTURED ON MODEL CONTAMINATED SOIL

***E. Escalante-Espinosa*[1], *L. Rodríguez-García*[2] *and M. Gutiérrez-Rojas*[2*]**

[1]División Académica de Ciencias Biológicas, Universidad Juárez Autónoma de Tabasco, Villahermosa, Tabasco, México
[2]Departamento de Biotecnología. Universidad Autónoma Metropolitana Unidad Iztapalapa, Vicentina, México

ABSTRACT

Cyperus laxus and *Cyperus ligularis* are native plants growing at swamp highly contaminated with hydrocarbons. Anthracene and pyrene phytoremediation from a hydrocarbon mixture (hexadecane 50%; pristane 14%; anthracene 18% and pyrene 18%) by both species were evaluated. The aim of this work was to demonstrate the ability of two native plants to remove PAH from contaminated model soils under greenhouse controlled conditions. Experiments were carried out in a model system using perlite to simulate soil conditions. Initial hydrocarbon mixture concentration was 4 000 mg kg^{-1} of dry perlite for *C. laxus* and 1 000 mg kg^{-1} of dry perlite for *C. ligularis*. Plants were inoculated with hydrocarbon-degrader microorganisms. Root dry biomass, total microbial viable counts, hydrocarbon removal and hydrocarbon sorption by roots were determined during 150 days of culture. Root dry biomass from inoculated plants was 1.3 times higher than those from non inoculated plants. At the end of culture, microbial viable counts for all treatments were around 10^9 CFU g^{-1} of dry perlite for bacteria and 10^7 CFU g^{-1} of dry perlite for fungi. PAH and aliphatics were completely removed by *C. ligularis* during first 60 days of culture. *C. laxus* removed pyrene totally until 150 days of culture while anthracene only 75%. Hydrocarbons from *C. ligularis* roots were negligible. PAH sorption by roots was detected for *C. laxus* since 75 days of culture, 125 and 215 mg kg^{-1} of dry roots for pyrene and anthracene, respectively. At the end of culture, pyrene sorption was reduced (60 mg kg^{-1} of dry roots) possibly due to plant metabolism. For

*Author for correspondence: e-mail: mgr@xanum.uam.mx.

species, hydrocarbon removal and accumulation in roots, at 150 days of culture, was independent on the microbial inoculum added. *C. laxus* and *C. ligularis* were capable to remove efficiently anthracene and pyrene, these plants can be useful for phytoremediation of PAH contaminated soils.

Keywords: PAH, phytoremediation, model contaminated soil, *Cyperus laxus*, *Cyperus ligularis*.

1. Introduction

The huge expansion of the chemical and petroleum industries in the 20th century has resulted in the production of a vast array of chemical compounds and materials that have transformed our lives (Ward and Singh, 2004). Specifically petroleum and derived fuels, as a result of their widespread worldwide use, are probably the most ubiquitous pollutants found in soil (Huesemann, 2004). Particularly, polycyclic aromatic hydrocarbons (PAH) are a group of more than 100 different organic compounds comprised of fused aromatic rings that are resistant to biodegradation in the environment. The increasing presence of PAH in all environmental compartments (air, water, soil and sediments) represents a serious health and ecological problem that motivated the interest of public health authorities and nature conservationists. PAH are in fact toxic and hazardous chemicals (Aina et al., 2006).

Physical, chemical and biological methods can be used for the remediation of hydrocarbon contaminated sites, but phytoremediation has long been recognized as a cost-effective method for removal of PAH from soil, and appears to have greater potential for the treatment of soils contaminated with residual concentrations of PAH (Xu et al., 2005). Phytoremediation is the term used to describe those methodologies that employ living higher organisms, which include green vegetation, plants, aquatic plants, trees and grasses, to remove toxic compounds. This technology has the advantage of *in situ* treatment of contaminated soils, sediments, groundwater, surface water and external atmosphere (Singh and Jain, 2003). Plants can be used to accumulate inorganic and organic contaminants, metabolize organic contaminants, and encourage microbial degradation of organic contaminants in the root zone. In a phytoremediation setting, degradation can happen in the rhizosphere (soil surrounding plant roots), as well as within the plant itself. The latter, phytodegradation, occurs when a plant has taken up the contaminant into its tissues, and enzymes within the plant complete the task transforming the contaminant, often into molecules that can be more readily broken down or released in root exudates. Rhizodegradation, or transformation of the contaminant in the rhizosphere, can occur in soil microorganisms such as fungi or bacteria, or via enzymes released from microorganisms or plants (Arthur et al, 2005). Recently, several studies have been focused on phytoremediation of PAH contaminated soils (Smith et al., 2008; Olson et al., 2007; Álvarez-Bernal et al., 2007; Maila et al., 2006; Xu et al., 2005; Corgié et al., 2003). Some general aspects can be summarized as follows: (i) rhizodegradation is the main mechanism for PAH removal, (ii) presence of plants enhanced removal rate during first stage of culture compared to unplanted treatments; however, in some cases, removal extent can be similar, (iii) hydrocarbons removal rates differ significantly respect to plant specie used in single or multispecies cultivation, (iv) PAH removal decreased with increasing number of aromatic rings, and (v) bioavailability also defines the rate and extent of removal. Various types of vegetation have been assayed, but few studies have been carried out with tropical species. *Cyperus laxus* and *Cyperus ligularis* are native plants growing in a highly hydrocarbon-contaminated tropical swamp in Veracruz, México. The aim of this work was to demonstrate the ability of two native plants to remove PAH from contaminated model soils under greenhouse controlled conditions. A key feature of this study was the use of perlite as a model soil to simulate soil and prevent bioavailability restrictions.

2. MATERIALS AND METHODS

2.1. Plants

Cyperus laxus and *Cyperus ligularis* were collected with the surrounding soil (1 m of diameter approximately), placed into pots and maintained at greenhouse conditions (16-28 °C, 75-90% relative humidity). Seeds were air-dried and stored within dark paper bags at room conditions.

2.2. Model Soil

Perlite (Dicalite, México) was used to simulate soil. The perlite was sieved through 4.76 and 1.19 mm mesh, the remaining fraction was washed with hot tap water and air-dried. Perlite was artificially spiked with a hydrocarbon mixture (HM) containing: hexadecane (HXD), 50%; pristane (PRIS), 14%; anthracene (ANT), 18%; and pyrene (PYR) 18%. Hydrocarbons were purchased from Sigma- Aldrich Chemie GmbH, Steinheim, Germany. Initial HM concentration was 4 000 mg kg^{-1} of dry perlite for *C. laxus* and 1 000 mg kg^{-1} of dry perlite for *C. ligularis*. HM was added to perlite previously solubilized in dichloromethane. Spiked perlite was air dried at room temperature to evaporate dichloromethane excess.

2.3. Microorganisms

A microbial consortium composed of 10 bacterial strains (*Bacillus cereus, Pseudomonas sp, Gordonia rubripertincta, Kocuria rosea, Arthrobacter oxydans, Bacillus subtilis* A, *Bacillus subtilis* B, *Micrococcus luteus* and two unidentified strains) and three fungal strains (*Penicillium janthinellum, Aspergillus carneus* and *Aspergillus terreus*) was used as inoculum. Microbial strains were previously isolated from *C. laxus* rhizosphere and selected based on their individual hydrocarbon degradation ability (Díaz-Ramírez et al., 2003). In order to standardize the inoculum, the consortium was re-constituted according to Escalante-Espinosa et al. (2005). Briefly, single bacterial and fungal strains were cultured separately (nutrient broth and propagation medium respectively), cell suspensions were washed three times with isotonic solution and equal volumes of each strain were blended until a homogeneous inoculum was observed; then added to spiked perlite to obtain an initial microbial total count of 10^6 CFU g^{-1} of dry perlite for bacteria and 10^5 CFU g^{-1} of dry perlite for fungi.

2.4. Experimental Procedure

Experiments were carried out in dark cylindrical glass pots containing 35 g (dry weight) of hydrocarbon-contaminated perlite. *C. laxus* and *C. ligularis* seedlings (60-day-old), previously grown in peat moss, were transplanted to each perlite pot. Two experimental sets for each species (12 pots per experiment) were assayed simultaneously: plants inoculated with the microbial consortium and plants not inoculated. Long-Ahston modified solution was added as nutrient solution once a week (in mg L^{-1}): KNO_3, 808; $Ca(NO_3)_2 \cdot 4H_2O$, 944; $NaH_2PO_4 \cdot H_2O$, 184; $MgSO_4 \cdot 7H_2O$, 368; and 1 mL of oligoelements solution (in mg L^{-1}): $MnSO_4 \cdot 4H_2O$, 2.23; $CuSO_4 \cdot 5H_2O$, 0.25; $ZnSO_4 \cdot 7H_2O$, 0.29; H_3BO_3, 3.10; NaCl, 5.90; $(NH_4)_6Mo_7O_{24} \cdot 4H_2O$, 0.088; and $FeSO_4 \cdot 7H_2O$, 0.02. Pots were maintained under greenhouse conditions (16-28 °C, 75-90% relative humidity).

2.5. Plants Measurements

Plants from non-inoculated pots were harvested at 135 days of culture and inoculated plants at 150 days. Roots were washed to eliminate perlite. Shoot and root dry weights were determined up to constant weight, after drying at 60 °C.

2.6. Microbial Counts

Microorganisms were enumerated by the spread plate technique (Lorch et al., 1995) using tripticasein soy agar (TSA; Becton Dickinson, México) for bacteria, and potato dextrose agar (PDA; Becton Dickinson, México) with chloramphenicol (Sigma), 100 mg L^{-1}, for fungi.

2.7. Hydrocarbons from Plants and Perlite

Residual hydrocarbons were quantified from roots (all roots from experimental units previously dried) and perlite (4 g of dry perlite) for each treatment. Hydrocarbons were extracted using the conventional Soxhlet method (acetone:hexane 1:1, 160 mL, 8 h). The extracts were vacuum evaporated and re-dissolved into 10 mL hexane. Chromatographic analyses were carried out using a Varian 3900 gas chromatograph with a Flame Ionization Detector. The PAH and alkanes were separated on a DB-5 fused silica capillary column (J & W Scientific, Folsom, CA, USA). All analyses were done on a split-splitless injector with helium as carrier. The initial oven temperature (150°C) was maintained for 2.5 minutes, then heated at 3°C per minute, up to 200°C and maintained for 71 min.

2.8. Statistical Analysis

Hydrocarbon concentration in spiked perlite and roots were subjected to one-way analysis of variance using SPSS (Windows version 10.0) to test significant differences between treatments ($P <0 .05$).

3. Results and Discussion

Cyperus laxus and *Cyperus ligularis* are both native species growing on weathered oil-contaminated sites in tropical Mexican swamps. In such sites, a number of species of the Cyperaceae and Graminae families are able to develop entire life cycles (germination, growth, flowering and seedling) on a wide total petroleum hydrocarbon (TPH) concentration from 10 000 to 434 000 mg of TPH kg^{-1} of dry soil (Gallegos-Martínez et al., 2000). Hydrocarbons phytoremediation capability of *C. laxus* has been assayed in previous studies (Escalante-Espinosa et al., 2005; López-Martínez et al., 2008). In order to evaluate PAH removal from a model mixture we carried out phytoremediation experiments in a model contaminated soil (perlite as support) to determine the ability of species to be used in bioremediation strategies. Previous bioassays were performed to establish hydrocarbons concentration in which no phytotoxic effects were observed. Initial HM concentrations were 1 000 mg kg^{-1} of dry perlite for *C. ligularis* and 4 000 mg kg^{-1} of dry perlite for *C. laxus* (unpublished results). At such concentration values our two plant species were able to growth healthy and developed vigorous roots in spiked perlite pots.

3.1. Plant Biomass and Microbial Counts

Several authors have mentioned that features of root system are a principal point for select species for successful phytoremediation claims (Aprill and Sims, 1990; Günther et al., 1996, Kirkpatrick et al., 2006). For example, a positive correlation between root biomass production of tropical grasses and legumes and oil degradation was found by Merkl et al. (2005). Regarding PAH phytotoxicity, recent studies showed that benzo(a)pyrene and naphthalene were both genotoxic for white clover, specie considered to be sensitive to pollutants, inducing significant changes in root and shoot DNA sequence. Damage was more severe in the roots than in the shoots suggesting that the translocation of these compounds and their genotoxic metabolites was limited (Aina et al, 2006). Our native species showed no signs of stress during the experiment. *C. ligularis* growing into contaminated perlite produced greater biomass of shoots values than *C. laxus* (Figure 1a); however, *C. ligularis* was less tolerant to hydrocarbons (initial HM was 1 000 mg of HM kg^{-1} of dry perlite, i.e 4-fold lower than initial concentration for *C. laxus*). In both species, microbial consortium added to plants enhanced root biomass 1.3 times higher than those obtained with non-inoculated plants (Figure 1b).

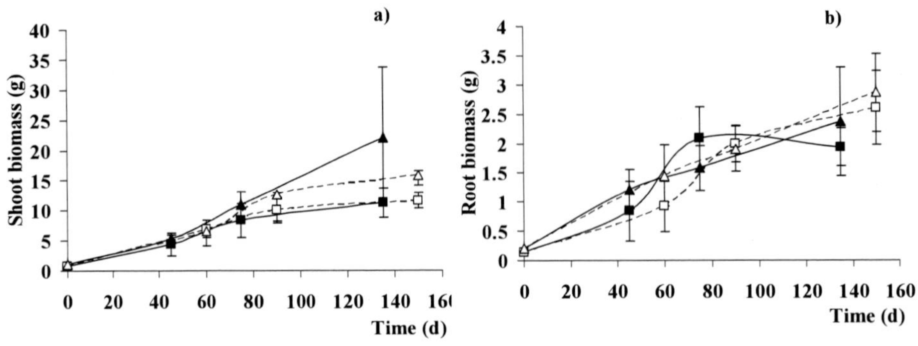

Figure 1. *C. laxus* and C. *ligularis* dry weight of shoots (a) and roots (b) during cultivation (■: *C. laxus* non inoculated; □: *C. laxus* inoculated; ▲: *C. ligularis* non inoculated; △: *C. ligularis* inoculated). Error bars represent the standard deviation of three plants.

Secondary plant effects on organic contaminants result from plant–microbe–contaminant interactions; wherein plant supplied substrates and/or physicochemical alterations caused by plant growth and metabolism, stimulates microbial activities enhancing biodegradation in the rhizosphere (Dzantor, 2007). In our experiments, microbial hydrocarbon-degraders were added to *C. laxus* and *C. ligularis*. At the end of culture, microbial viable counts for all treatments (inoculated and non-inoculated plants) were around 10^9 CFU g^{-1} of dry perlite for bacteria and 10^7 CFU g^{-1} of dry perlite for fungi. Fang et al (2001) reported similar numbers of cultivable heterotrophic microorganisms in Sudan grass, crested wheat grass and switch grass. Inoculation has been recognized as a tool to enhance plant growth. Kuiper et al. (2001) selected a pair plant-microorganism (*Lolium multiflorum* and *Pseudomonas putida* PCL1444) to reach efficient naphthalene biodegradation, enhancing hydrocarbon removal due to protection from hydrocarbon toxicity provided by bacteria. Joner and Leyval (2003) evaluated the effect of *Glomus mosseae* added to ryegrass and clover on PAH removal. Biomass of inoculated plants was up to 2.7 times higher than non-inoculated plants after 180 days of culture; at the end of culture, PAH removal was similar for both treatments. Huang et al. (2004) added plant growth promoting rhizobacteria (PGPR) including three strains of *Pseudomonas putida* UW3, *Azospirillum brasilense* Cd and *Enterobacter cloacae* CAL2 to *Festuca arundinacea*, *Poa pratensis* and *Elymus canadensis* grown in soil contaminated with creosote. They concluded that PGPR promoted root growth, providing a greater sink for the contaminants. Plants with PGPR were able to survive better and grow rapidly, thereby accumulating sufficient biomass for remediation. PGPR are free-living soil bacteria that positively influence plant growth and development (Glick, 2003). Additionally, microbial inocula can have the ability to relieve environmental stress in plants. In our study, probably some included strains in added microbial consortium to native plants, could be considered as potential PGPR; then, mutual benefits between plants and hydrocarbon-degrading microorganisms were improved in the presence of hydrocarbons.

3.2. Hydrocarbon Removal

Phytoremediation is a viable choice for hydrocarbons remediation if sufficient time is allowed for plant establishment and contaminant degradation. In the process, plants could be used to extract, detoxify, and/or sequester toxic pollutants from soil (Xu et al., 2005). Initial HM for *C. ligularis* was 1 000 mg kg^{-1} perlite. PAH and aliphatics were completely removed for inoculated and non-inoculated plants during first 60 days of culture (results not shown). This pattern probably can be due to low hydrocarbon concentrations resulting in a high hydrocarbon rate (4.75 mg HM kg^{-1} perlite d^{-1}) and extent (100%). Similar results were obtained by Álvarez-Bernal et al. (2007) using *Mimosa monancistra* to remove PAH in soil contaminated with 200 mg phenanthrene kg^{-1} soil, 100 mg anthracene kg^{-1} soil, and 50 mg benzo(a)pyrene kg^{-1} soil. Concentration of PHE dropped sharply in the first 14 days, after 56 days no PHE was detectable in soil. Anthracene showed the same pattern but the decrease was slower. In contrast, *C. laxus* exhibited different hydrocarbon removal pattern as depicted in Figure 2. Initial hydrocarbon concentrations were higher than those used for *C. ligularis.* Aliphatics (HXD and PRIS) were removed totally until 150 days of culture; however, the main removal occurred during the first 60 days, similar to *C. ligularis* reaching removal values close to 95% for inoculated and non-inoculated plants. PAH removal was not considerable during first stage of phytoremediation experiment for inoculated plants, after that, removal rate was two and four times higher for ANT and PYR respectively, compared to non-inoculated plants (14 mg ANT kg^{-1} perlite d^{-1} and 25 mg PYR kg^{-1} perlite d^{-1}) at 90 days of culture. Later, PAH removal rate diminished near to zero for non-inoculated and inoculated plants reaching final degradation extent values of 70% for ANT and 88% for PYR. At the end of experiment, removal extent was similar for both treatments. Similar pattern has been reported previously. For example, Olson et al. (2007) compared PAH removal from contaminated soil by 18 hydrocarbon species, they found that planting significantly enhanced the dissipations rates of hydrocarbons within the first seven months, but this effect was not significant after 14 months; they also concluded that some plant families are more effective than others. Maila et al. (2005) investigated the potential of multispecies rhizoremediation and monoculture of PAH (naphthalene, fluorene and pyrene) contaminated soil. There was no significant difference between PAH removal (naphthalene, 96%; fluorene, 81% and pyrene, 49%) in the monoplanted treatments with grass species (*Brachiaria serrata* and *Eleusine coronana*). Xu et al. (2005) carried out studies of phenanthrene and pyrene dissipation in presence of maize, annual ryegrass and white clover in spiked soils. Results demonstrated that extents did not differ significantly among three tested species, but the rates of degradation were different at 60 days of culture. The maize treatment had the highest extent of contaminant removal after two months (92% for phenanthrene and 88% for pyrene), followed by white clover and annual ryegrass.

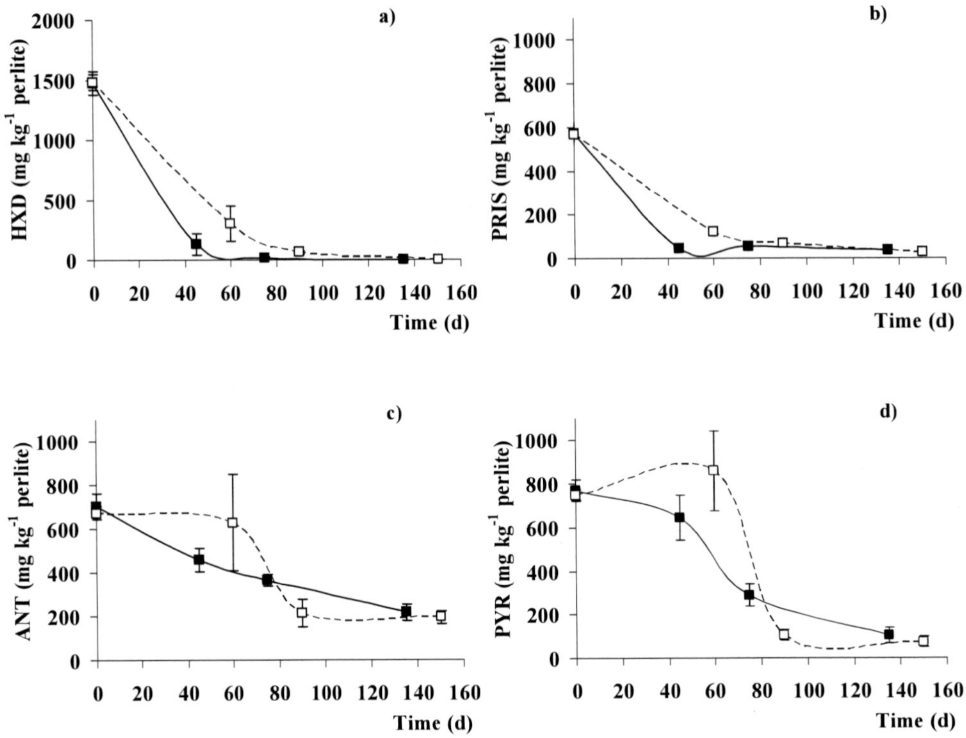

Figure 2. Residual hydrocarbons: hexadecane (a), pristane (b), anthracene (c) and pyrene (d) during phytoremediation with *C. laxus* inoculated plants (■) and non-inoculated plants (□). Error bars represent the standard deviation of three sampled pots.

3.3. PAH Sorption by Roots

When considering the whole-plant metabolism of organics, the organic chemical must first enter the plant tissue. Entry can occur by translocation of soil water with the transpiration stream, by transfer for soil gas, or from atmospheric gases or water to the foliar tissues. After being transported into the plant cells, an organic chemical can be metabolized if serves as a substrate for the plant enzymes (Burken, 2003). The main property controlling uptake of organics by plants roots is lipophilicity, this property is directly related to the 1-octanol-water partition coefficient, k_{ow} (Briggs et al., 1982). Additionally, sorption by roots may be affected by stage of plant growth, because the chemical and physical properties of roots change significantly over the life of the plant. Several studies have shown the plant compositional effects on plant uptake of lipophilic contaminants from soils, and results are suggestive of plants lipids as the major for the observed differences in plant uptake of compounds as aldrin, dieldrin, heptachlor and heptachlorepoxide (Chiou et al., 2001). Hydrocarbons from *C. ligularis* roots were detected as negligible for all sampling times. PAH sorption by roots was detected for non-inoculated *C. laxus* ever since 75 days of culture, 125 and 215 mg kg-1 of dry roots for PYR and ANT, respectively (Figure 3a and 3b). For inoculated plants, there are not significant differences for PAH sorption at 90 days of culture. At the end of experiment, concentration of ANT remained around 215 mg kg^{-1} of dry roots while PYR sorption decreased close to one half for inoculated and non-inoculated plants, probably due to plant metabolism.

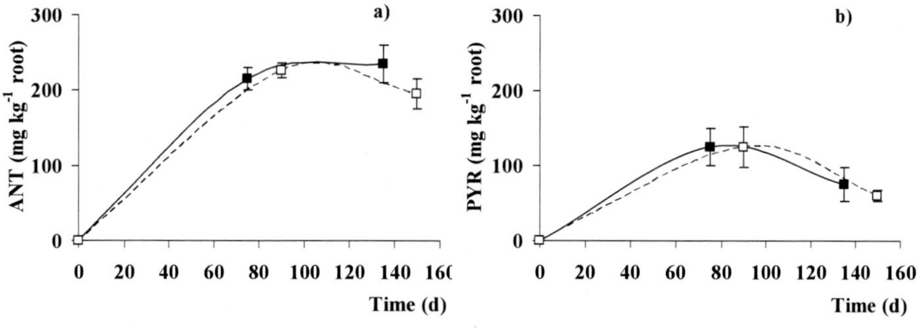

Figure 3. PAH sorption by *C. laxus* roots during 150 days of culture (■, inoculated plants; □, non-inoculated plants). Error bars represent the standard deviation of three sampled plants.

In addition, root concentration factor (RCF) was calculated and shown in Table 1. RCF is defined as the ratio of the PAH concentration in roots and in soils, on a dry weight basis (Briggs et al., 1982). For ANT, RCF increased during the culture for non-inoculated plants, hydrocarbon concentration in perlite diminished at the end of experiment although hydrocarbon concentration in roots remains constant. In contrast, for inoculated plants, RCF during the culture is 1.0, i.e. hydrocarbon concentrations in both perlite and roots are similar at 90 and 150 days. For PYR, RCF for non-inoculated plants follows same pattern during the culture as compared to ANT; whereas for inoculated plants, values of RCF were higher (around 1.0) with respect to non-inoculated and decreased lightly at 150 days of culture. Gao and Zhu (2004) reported accumulation of PHE and PYR for several plant species grown in contaminated soil; the RCF values of PAH, with initial concentrations of 133 mg kg^{-1} for PHE and 172 mg kg^{-1} for PYR, were 0.05-0.67 and 0.23-4.44, respectively. Significantly positive correlations were shown between RCF and root lipid contents of several plant species. For example, Watts et al. (2006) measured PAH in *Spartina alterniflora* grown in pots of contaminated sediment and plants grown in native sediment at a contaminated marshland. RCF ranged from 0.009 to 0.97 in the potted plants for PAH compounds, and do not show any correlation with chemical properties (*i.e.* K_{ow}). In our study, PAH sorption by roots was independent of the hydrocarbons degraders microorganisms added.

Table 1. Root concentration factor for anthracene and pyrene during hydrocarbon removal by *Cyperus laxus*.

	Root Concentration Factor			
	Anthracene		Pyrene	
Time	Non-inoculated	Inoculated	Non-inoculated	Inoculated
T1	0.6 ± 0.03	1.1 ± 0.04	0.4 ± 0.01	1.2 ± 0.03
T2	1.1 ± 0.01	1.0 ± 0.01	0.7 ± 0.02	0.8 ± 0.01

T1: 75 d, non-inoculated plants; 90 d, inoculated plants.
T2: 135 d, non-inoculated plants; 150 d, inoculated plants.

4. Conclusion

Phytoremediation is a viable choice for hydrocarbon contaminated soils if plants can establish and survive to stress conditions. The present study demonstrated that *Cyperus laxus* and *Cyperus ligularis*, native plants from hydrocarbon contaminated site, were capable to remove efficiently hydrocarbons including anthracene and pyrene. Hydrocarbon removal extent was similar for non-inoculated and inoculated plants; however, hydrocarbon removal rate was faster during first stage of cultivation for inoculated plants. PAH sorption in roots was detected only for *C. laxus* and was independent on the microbial inoculum added. This native species can be used for phytoremediation of hydrocarbons contaminated swamps. Further studies including plants and the inoculum proposed here should be addressed to phytoremediation of contaminated soils.

Acknowledgments

This research was partially supported by PEMEX-Refinación, México.

References

Aina, R., Palin, L. & Citterio, S. 2006. Molecular evidence for benzo[a]pyrene and naphthalene genotoxicity in *Trifolium repens* L. *Chemosphere*, *65*, 666-673.

Álvarez-Bernal, D., Contreras-Ramos, S., Marsch, R. & Dendooven, L. 2007. Influence of catclaw *Mimosa monancistra* on the dissipation of soil PAHs, *Int J Phytorem*, *9*, 79–90.

Aprill, W. & Sims, R. C. 1990. Evaluation of the use of prairie grasses for stimulating polycyclic aromatic hydrocarbons treatment in soil. *Chemosphere*, *20*, 253-265.

Arthur, E., Rice P., Rice P., Anderson, T., Baladi, S., Henderson, K. & Coats, J. 2005. Phytoremediation - An Overview. *Critical Reviews in Plant Sciences*, *24*, 109–122.

Briggs, G., Bromilow, R. & Evans, A. 1982. Relationships between lipophilicity and root uptake and translocation of non-ionized chemicals by barley. *Pesticide Science*, *13*, 405-504.

Burken, J. 2003. Uptake and metabolism of organic compounds: green-liver model. In Phytoremediation transformation and control (McCutcheon and Schnoor Eds.) Wiley-Interscience, U.S.A., pp.59-84.

Chiou, C., Sheng, G. & Manes, M., 2001. A partition-limited model for the plant uptake of organic contaminants from soil and water. *Environ Sci Technol*, ***35***, 1437-1444.

Corgié, S., Joner, E. & Leyval, C., 2003. Rhizospheric degradation of phenanthrene is a function of proximity to roots. *Plant and soil*, 257, 143-150.

Díaz-Ramírez, I., Ramírez-Saad, H., Gutiérrez-Rojas, M. & Favela-Torres, E. 2003. Biodegradation of Maya crude oil fractions by bacterial strains and a defined mixed culture isolated from *Cyperus laxus* rhizosphere soil in a contaminated site. *Can J Microbiol,* 49, 755-761.

Dzantor, E. K. (2007). Phytoremediation: the state of rhizosphere ‘engineering’ for accelerated rhizodegradation of xenobiotic contaminants. *J Chem Technol Biotechnol,* 82, 228–232.

Fang, C., Radosevich, M. & Fuhrmann, J., 2001. Atrazine and phenanthrene degradation in grass rhizosphere soil. *Soil Biol Biochem,* 33, 671-678.

Gallegos Martínez, M., Gómez Santos, A., González Cruz, L., Montes de Oca García, M.A., Yáñez Trujillo, L., Zermeño Eguía Lis, J.A. & Gutiérrez-Rojas, M. 2000. Diagnostic and resulting approaches to restore petroleum-contaminated soil in a Mexican tropical swamp. *Water Sci Technol,* 42, (5,6), 377-384.

Gao, & Zhu, L., 2004. Plant uptake, accumulation and translocation of phenantrene and pyrene in soils. *Chemosphere*, 55, 1169-1178.

Glick, B., 2003. Phytoremediation: synergistic use of plants and bacteria to clean up the environment. *Biotechnol Adv*, 21, 383-393.

Günther, T., Dornberger, U., & Fritsche, W. 1996. Effects of ryegrass on biodegration of hydrocarbons in soil, *Chemosphere*, 33, (2), 203-215.

Huang, X., El-Alawi, Y., Penrose, D.M., Glick, B. R. & Greenberg, B.M., 2004. Responses of three grass species to creosote during phytoremediation. *Environmental Pollution,* 130, 453-463.

Huesemann, M., (2004). Biodegradation and bioremediation of petroleum pollutants in soil. In Applied Bioremediation and Phytoremediation (O. Ward and A. Singh Eds., pp. 13-15). Germany, Springer – Verlag.

Kirkpatrick, W., White, P., Wolf, D., Thoma, G. & Reynolds, C. 2006. Selecting plants and nitrogen rates to vegetate crude-oil-contaminated soil. *Int J Phytorem*, 8, 285-297.

Kuiper, I., Bloemberg, G.V. & Lugtenberg, B.J.J., 2001. Selection of a plant-bacterium pair as a novel tool for rhizostimulation of polycyclic aromatic hydrocarbon-degrading bacteria. *Molecular Plant-Microbial Interactions*, *14*, 1197-1205.

López-Martínez, S., Gallegos-Martínez, M. E., Pérez-Flores, L. J. & Gutiérrez-Rojas, M. 2008. Phytoremediation of hydrocarbon contaminated soils by *Cyperus laxus* Lam.: induced EROD-activity in roots. *Int J Phytorem,* 10, 289-301.

Lorch, H. J., Benckieser, G., & Ottow, J.C.G. (1995). Basic methods for counting microorganisms in soil and water. In Methods in Applied Soil Microbiology and Biochemistry (Alef and Nannipieri Eds., pp. 146-161). Great Britain, Academic Press.

Maila, M., Randima, P. & Cloete, T., 2005. Multispecies and monoculture rhizoremediation of polycyclic aromatic hydrocarbons (PAHs) from the soil. *Int J Phytorem*, *7*, 87-98.

Merkl, N., Schultze-Kraft, R. & Infante, C., 2005. Assessment of tropical grasses and legumes for phytoremediation of petroleum-contaminated soils. *Water, Air and Soil Pollution*, *165*, 195–209.

Olson, P., Castro, A., Joern, M., DuTeau, N., Pilon-Smits, E. & Reardon, K., 2007. Comparison of plant families in a greenhouse phytoremediation study on an aged polycyclic aromatic hydrocarbon-contaminated soil. *J Environ Qual*, ***36***, 1461-1469.

Singh, O. & Jain, R., 2003. Phytoremediation of toxic aromatic pollutants from soil. *Applied Microbiol Biotechnol*, 63,128–135.

Smith, K., Schwab, A. & Banks, M., 2008. Dissipation of PAHs in saturated, dredged sediments: a field trial. *Chemosphere*, 72, 1614-1619.

Ward, O. & Singh, A. (2004). Soil Bioremediation and Phytoremediation – An overiew. In Applied Bioremediation and Phytoremediation (O. Ward & A. Singh Eds., pp. 1-4). Germany, Springer – Verlag.

Watts, A., Ballestero, T. & Gardner, K. (2006). Uptake of polycyclic aromatic hydrocarbons (PAHs) in salt marsh plants *Spartina alterniflora* grown in contaminated sediments. *Chemosphere*, *62*, 1253–1260.

Xu, S.Y., Chen, Y.X., Wu, W.X., Wang, K.X., Lin, Q. & Liang, X.Q. (2005). Enhanced dissipation of phenanthrene and pyrene in spiked soils by combined plants cultivation. *Sci Total Environ*, *363*, 206-215.

In: Polycyclic Aromatic Hydrocardons: Pollution, Health... ISBN 978-1-60741-462-9
Editors: Pierre A. Haines and Milton D. Hendrickson

Chapter 14

IN VITRO PHYTOREMEDIATION MECHANISMS OF PAH REMOVAL BY TWO PLANT SPECIES

L. Reynoso-Cuevas, F. Cruz-Sosa, and M. Gutiérrez-Rojas*
Departamento de Biotecnología. Universidad Autónoma Metropolitana-Iztapalapa, Col Vicentina, Iztapalapa, México

ABSTRACT

Polycyclic aromatic hydrocarbons (PAH) are widespread environmental pollutants of natural and anthropogenic origins. Despite their poor water solubility, they can be taken up and bioaccumulated by plants. The effect of PAH on selected plant (*Festuca arundinacea* and *Bouteloua curtipendula*) at early grown stage was evaluated under *in vitro* conditions. Plant species were cultured on Murashige and Skoog (MS) medium with three different PAH: phenanthrene (PHE), pyrene (PYR), and benzo[a]pyrene (BaP). PAH plant accumulation and distribution were evaluated with GC and HPLC. To date, number phytoremediation mechanisms of pollutants removal are known, including: (i) degradation (phyto- and rhizodegradation), (ii) removal (phytoextraction, phytovolatilization) and (iii) immobilization (phytoestabilization). The objective of this study was to identify the phytoremediation mechanisms of selected plants use to control and/or remove PAH from MS medium. To accomplish the task, plant samples, after 90 d culture, were collected and analyzed periodically to determine the residual PAH content in MS medium, roots and plants stems. *F. arundinacea* was able to transform 40.40% of initial PHE, accumulated 6.99% in stems and almost three times in roots (20.66%). With *B. curtipendula*, total removal was 66.14% and PHE was found in roots (14.38%) and stems (6.45%). Both species were able to remove and transform PYR, higher concentrations were found in roots than in stems. BaP outlined in a different pattern: both species were able to remove BaP from MS medium, *B. curtipendula* was more efficient than *F. arundinacea* 57.41% and 38.93%, respectively; bioaccumulation of BaP was the higher while transformation the lower. The dominant mechanisms observed for PHE and PYR in both plant species were phytodegradation and phytoextraction. *B. curtipendula* employed phytoextraction for BaP; it was able to bioaccumulate 45.05% in roots, 1.8 times more than *F. arundinacea* (24.99%). Several studies investigated the effect of

* mgr@xanum.uam.mx

plant-microbe interactions on the degradation of organic contaminants. In our study, plants were able to remove and transform PAH under sterile *in vitro* conditions, *i.e.* without microorganisms.

Keywords: PAH, phytoremediation, in vitro, Festuca arundinacea, Bouteloua curtipendula.

INTRODUCTION

We have to face the situation that hundreds of thousands of sites across the world are characterized as polluted with organic chemicals because of industrial processes, spills, and accidents, or resulting from improper use of chemicals. Among them are chemicals of remarkable stability and recalcitrance in soils and water bodies (Schröder, 2007). Many aromatic hydrocarbons, nitroaromatic compounds, polycyclic aromatic hydrocarbons (PAH), polychlorinated biphenyls, diauxins and their derivatives are considered as highly toxic, mutagenic and/or carcinogenic to natural microflora as well as to higher living systems including humans (Singh and Jain, 2003).

Phytoremediation is emerging as an efficient technology that uses plants to bioremediate pollutants from soil environments. A choice of modern tools and analytical devices has provided insight into the selection and optimization of remediation processes by various plant species (Macek et al., 2000). Plants use photosynthetic energy to scavenge organic compounds from the soil and to concentrate them in their biomass according to nutritional requirements. The stimulation of removal of pollutant contaminants from soil could be brought about through several mechanisms of plant/soil interaction: (i) increase in soil microbial activity, (ii) increase in microbial association with the root and toxic compounds, and (iii) changes in the physical and chemical properties of the contaminated soil (Meagher, 2000). Phytoremediation has been employed to remediate environments contaminated with metals, radionuclides and organics, as well (Alkorta and Garbisu, 2001). However, lack of information regarding the particular mechanisms involved does not allow go further in practical phytoremediation claims.

Contaminant attenuation mechanisms involved in phytoremediation are complex and not limited only to the direct metabolism of contaminants by plants. Certain indirect attenuation mechanisms are implicated in phytoremediation, such as the metabolism of contaminants by plant-associated microorganisms, and plant-induced changes in the contaminated environment. In terrestrial species, transport of contaminants to the plant is dominated by uptaking water by roots, and distribution within the plant relies on xylem or phloem transport (Macek et al., 2000). Various terms, reflecting each specific attenuation mechanism, have been extensively used to better describe specific applications of phytoremediation. These include phytoextraction, phytodegradation, phytotransformation, phytovolatilization and rhizodegradation (Burken and Ma, 2006).

Plants must not only be resistant to the pollutant, but also be able to remove it from the environment and to transform it into non-toxic metabolites or end-products (Schwitzguébel et al., 2006). The present study was conducted to investigate the phytoremediation mechanisms used by two plants species to promote PAH removal from MS *in vitro* culture medium.

Materials and Methods

2.1. Plants

Seeds of "tall fescue" (*Festuca arundinacea*) were obtained from the wholesale food market of Mexico City and kept dry at room temperature for subsequent desinfestation and culture.

Seeds of "side oats grama" (*Bouteloua curtipendula* cvs. Tenius Gould & Kapadia) were collected from different wild plants located at 4.5 km of highway Querétaro-Huimilpan, Mexico. The specie was identified at the herbarium of Biology Faculty of Universidad Autonoma de Queretaro. The seeds were separated and kept dry at room temperature for subsequent desinfestation and culture.

2.2. Culture Preparation

2.2.1. Seeds Desinfestation

Prior to desinfestation damaged and small seeds were hand-removed. Selected seeds were placed into a filter paper envelope, then submerged into commercial powder detergent solution (60 g detergent L^{-1} water) for 20 min with constant agitation, and rinsed with tap water for 10 min. Under laminar flow hood the envelope with seeds was submerged into 70% (v/v) ethanol for 30 s, follow by immersion in 0.6% (v/v) sodium hypochlorite with 0.1 mL of Tween-20 for 20 min with constant agitation. The envelope was washed five-fold with 150 mL deionized sterile water. The envelope was opened under sterile conditions and stored into sterile empty Petri dishes.

2.2.2. Culture Tubes

Murashige and Skoog (MS) culture medium was prepared from micro and macro nutrient stock solutions (Murashige and Skoog, 1962), pH was adjusted with 0.1 M HCl to 5.8 and Phytagel (2 g L^{-1}) (agar substitute, Sigma Chemical Co., St. Louis, MO, USA) was added and then boiled. MS medium (10 mL) was added to each culture tube (diam. 2.5 cm, height 15 cm), prior to sterilization for 15 min at 1.2 kg cm^{-2}).

2.3. Hydrocarbons

PHE (purity >96%), PYR (purity >98%) and BaP (purity >97%) were purchased from Sigma-Aldrich Chemie GmbH, Steinheim, Germany. The substances were dissolved in dichloromethane (DCM) (Mallinckrodt Baker SA de CV EDOMEX, Mexico).

Test substances (PAH) were dissolved in DCM. PAH diffusion in MS medium was previously determined (data not shown), therefore, test concentrations to evaluate removal of PAH varied as follows: 200 mg (kg of MS medium)$^{-1}$ for PHE and PYR; 2 000 mg (kg of MS medium)$^{-1}$ for BaP. In all cases MS medium was "as is" weight, *i. e,* not dry weight.

2.4. Boiassays

In vitro germination, root and stem length of plant species were evaluated with and without hydrocarbons (Reynoso-Cuevas et al., 2008). Pure DCM was used for the control samples; controls without solvent were also included. PAH solution (0.5 ml) was added to culture tubes just before hardening of the medium, shaken and maintained at room temperature for three days to evaporate the solvent; one seed was aseptically placed into each culture tube containing semi-solid MS medium supplemented with PAH test concentrations, the culture tube was closed with a plastic cap and sealed with parafilm. Sealed culture tubes were kept at 25 °C, with 16 h photoperiod for 90 days for PAH removal studies.

2.5. Analytical Methods

2.5.1. Plants

The two plant species were tested simultaneously, to achieve the best possible basis for comparing their individual outcome after 90 d culture. The plants were observed and monitored every 5 d to take note of changes in appearance and growth. The plants were sampled destructively five replicates per PAH concentration were analyzed and separated into roots and stems.

2.5.2. Hydrocarbons

The liquid-liquid extraction of hydrocarbons of culture medium was made with DCM; three washings with 10 ml each, every 5 days during the 90 days of study. The reproducibility and recovery of the extraction method for the samples were satisfactory, with recovery averaged 96.7%±3.82 (standard deviation) for PHE; 98.2%±2.57 for PYR; 96.8%±2.74 for BaP. PAH identification and PHE and PYR quantification was carried out with a gas chromatograph, a Varian 3900 provided with FID detector and DB-PETRO (100 m, I.D. 0.25 mm) column was used; helium was the carrier gas (30 kPa). The initial oven temperature was 100°C (holding time 1 min), increased to 180°C at 20°C min^{-1} and to 230°C at 15°C min^{-1} (holding time 5 min); then increased at 4°C min^{-1} until 290°C (holding time 23 min) and complete with 300°C at 3°C min^{-1} with a final holding time of 5 min. Injector and transfer lines were heated at 290°C. Injection (1 μL) was performed with Varian CP8410 auto-injector.

HPLC analysis for BaP was carried out on a Thermo separation products system consisting of a model ConstaMetric 3200 pump, an auto-sampler (model 545), and a UV detector (model 4100), λ 254 nm. BaP were separated on a Spherisorb ODS2 C18 column (15 cm x 2.1 mm, i.d.; 5 μm; film thickness), operated at 25 °C, at a flow rate of 2 mL min^{-1} with acetonitrile:water (70:30) as the mobile phase. The sample injection volume was 20 μL. For calibration, standard solutions of BaP in hexane were used. The detection limit was 3 μg L^{-1}.

2.5.3. Phytoremediation Factors

The root concentration factor (RCF) describes the potential of a given xenobiotic to accumulate in the plant root, and makes no discrimination between surfaces accumulation and uptake into the root tissue.

$$RCF = \frac{\text{Root final concentration (mg kg}^{-1})}{MS \text{ medium concentration (mg kg}^{-1})}$$

Stem concentration factor (SCF) indicates the potential of a given xenobiotic to accumulate in the plant stem.

$$SCF = \frac{\text{Stem final concentration (mg kg}^{-1})}{MS \text{ medium concentration (mg kg}^{-1})}$$

Translocation factor (TF) points to the total of PAH that can be transported to the aerial parts of the plant.

$$Translocation\,(\%) = \frac{\text{PAH content in stem (mg kg}^{-1})}{\text{PAH total content in whole plant (mg kg}^{-1})} \times 100$$

2.6. Data Analyses

All values presented for the chemical and biological analyses of plants are the means of five replicates. The mean and standard error were calculated for all data. PAH distributions in plants were one-way ANOVA tested. Statistical differences among individual treatments were assessed with Tukey-Kramer multiple comparison test at a significance level of $\alpha \leq 0.05$. All statistical analyses were performed with NCSS-2000, version 2001 (Copyright 2001 Jerry Hintze).

RESULTS AND DISCUSSION

3.1. PAH Removal and Distribution

The concept of using plants to remediate soils contaminated with organic pollutants is based on observations that disappearance of organic chemicals is accelerated in vegetated soils compared to surrounding non-vegetated bulk soils (Alkorta and Garbisu, 2001). The specific interaction of a pollutant with soil and plants will vary depending on the chemical properties of the contaminant, the physiological properties of the introduced plant species and the contaminated medium (Singh and Jain, 2003). Organic compounds can be translocated to other plant tissues as shows in Figure 1; they may be transformed to probably less phytotoxic compounds and bound in plant tissues.

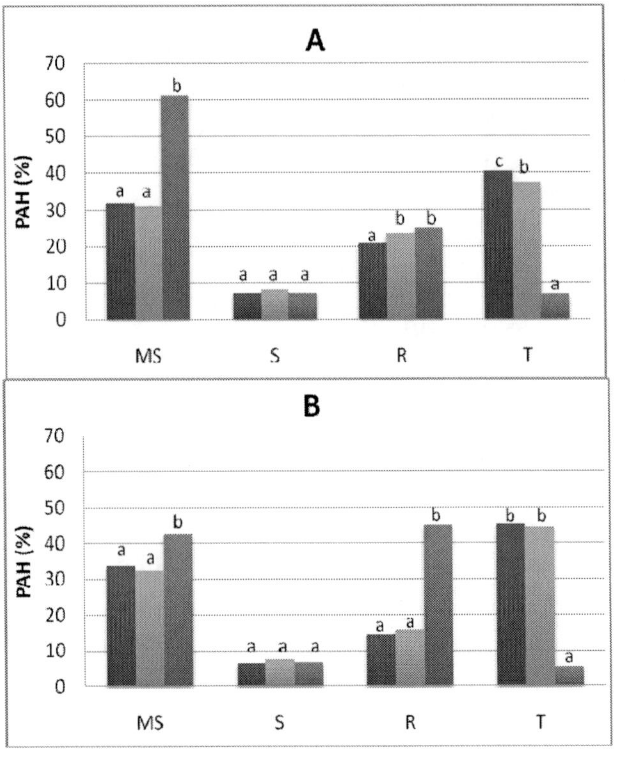

Figure 1. A) *F. arundinacea* (F.a) and B) *B. curtipendula* (B.c) PAH distribution. Transformation (%T), stem (%S), root (%R) and medium MS (%MS). Dark gray, PHE; soft gray, PYR; gray, BaP. Lowercase letters indicate significant differences among accumulation place (MS medium, stem, root and transformation).

PAH presence in plant stem clearly indicated the uptake occurs for PHE, PYR and BaP with both studied plants (Figure 1). Plants can remediate environments contaminated with PAH directly via root uptake, detoxification by phytotransformation and conjugation with sugars or glutathione, and subsequent storage of nonphytotoxic metabolites in plant tissues; and/or indirectly by the release of exudates or enzymes which can enhance degradation by rhizosphere microorganisms, when they are present (Schwitzguébel et al., 2006). *F. arundinacea* and *B. curtipendula* could transform PHE (40.40±1.19% and 45.31±1.41%, respectively) and PYR (37.50±0.94 and 44.39±1.30) respectively, according to Schwitzguébel et al. (2006); *in vitro* conditions allowed, in our work, to evaluate just plant action.

Accumulation for *F. arundinacea* roots were as follows: BaP same as PYR but PYR greater than PHE (Figure 1A); the concentration of PHE and PYR in *B. curtipendula* roots were lower and significantly different than *F. arundinacea*, but BaP concentration in *B. curtipendula* roots were the highest (Figure 1B). There is no significant difference for stem accumulations among PAH and between the two of the studied plants.

Kipopoulou et al., (1999) reported that PAH would be adsorbed through the epidermis of roots in contact with soil particles, but not drawn along the inner root; in contrast, Fismes et al., (2002) report a direct relationship between PAH concentrations in soil and plants and suggested a soil-to-root transfer according to Figure 1, where PAH were found in roots and a slighter amount in the stems of both plants. On the other hand, Oleszczuk and Baran (2005) suggested that the most probable pathway for plant pollution by PAH was the uptake of compounds that are volatilized from contaminated soil, particularly in the lower aerial parts of the plant; nevertheless, volatilization could be a possible source of plant contamination by PAH with relatively low molecular weight, not for "heavy" PAH as BaP.

The transformation pattern for *F. arundinacea* was: PHE greater than PYR, and PYR greater than BaP and they all were significantly different. *B. curtipendula* could transform statistically similar amounts of PHE and PYR and different of BaP; according with this, White et al. (2006) observed that the degradation pattern in their study was 2-ring greater than 3-ring and 3-ring greater than 4-ring and decreased with increased alkylation of larger ringed structures.

3.2. Phytoremediation Factors

The process of phytoremediation begins with contaminant transport to the plant; roots can take up PAH, up taken PAH may be transported, stored, converted, and accumulated in the different cells and tissues of the plant (Marmiroli et al., 2006). To evaluate uptake potential of *F. arundinacea* and *B. curtipendula* two concentration factors (root and stem) and translocation factor of three PAH were calculated as shown in Table 1. RCF generally proved to be a good indicator of whether a plant retains a contaminant in the root.

Table 1. Concentration factors. RCF: root concentration factor; SCF: stem concentration factor; TF: translocation factor. *F. arundinacea* (F.a) and *B. curtipendula* (B.c). PHE –phenanthrene; PYR –pyrene; BaP –benzo[a]pyrene.

PAH	Specie	RCF	SCF	TF (%)
PHE	F. a	0.65 bB	0.22 bA	25.32 bA
	B. c	0.42 aA	0.19 aA	30.96 bB
PYR	F. a	0.76 cB	0.26 cA	25.46 bA
	B. c	0.49 bA	0.23 bA	32.18 bB
BaP	F. a	0.41 aA	0.11 aA	21.87 aB
	B. c	1.06 cB	0.16 aB	13.16 aA

Lowercase letters indicate significant differences among treatments (PHE; PYR; BaP) for each plant species; uppercase letters indicate significant differences between species (F.a and B.c) for each treatment; within columns.

When considering vegetative uptake in remediation, uptake and entry into plants is an universal step to all plant mediated removal or metabolic degradation. Burken y Ma (2006) evaluate the phytoremediation of volatile organic compounds; their experiments showed that diffusive flux was clearly related to height, with the upper trap accumulating less TCE (trichloroethylene) than the lower trap in all sampling events for all plants. This trend is exhibited in Table 1 where SCF were lower than RCF in all cases.

Watts et al. (2006), reported values of RCF about 0.1 for some PAH compounds and they did not find any correlation with chemical properties of PAH, suggesting that PAH detected in roots are primarily present in soil adhered to the outer surface of the root, rather than transferred to the root tissue. In our work, all root samples of *F. arundinacea* and *B.*

curtipendula were carefully washed with DCM before PAH extraction and quantification; in our study RCF values are related to PAH root absorption.

It is generally assumed that the solubility of organic compounds in water is decisive for their uptake by plants. Pollutants dissolved in water can be transported in the soil and then, through the roots, translocate to other plant tissues (Oleszczuk and Baran, 2005). The partitioning of organics between roots and aboveground tissues varies considerably depending on the chemical in question (Alkorta and Garbisu, 2001). However, translocation factors for both plants with PHE and PYR were not statistically different, and decreased with BaP; it may possibly be because BaP was bound to insoluble cellular root structures such as lignin.

3.3. PAH Accumulation

The uptake of PAH is influenced by the properties of both organic chemicals and plant species (Li et al., 2002). Once the organic xenobiotic enters the plant system, it is partitioned to different plant parts through translocation; then, any number of reactions within the following series may occur: oxidations, reductions or hydrolysis. However, PAH can be conjugated to other products, transported and segregated into vacuoles or bound to the cell wall material (Kamath et al., 2004); as Figure 2 shows, PHE was absorbed into the roots and transported to the stem of both assayed plants.

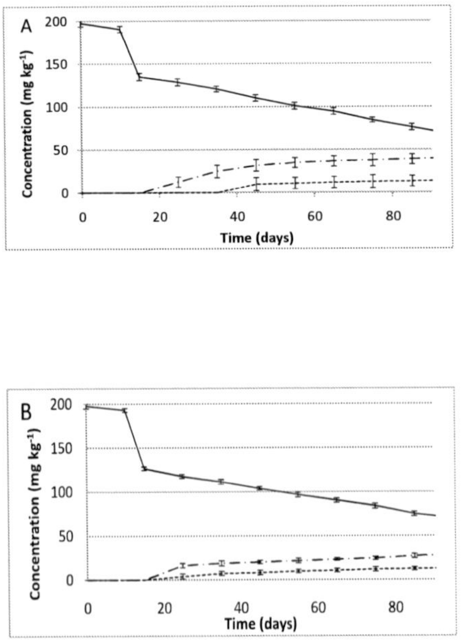

Figure 2. A) *F. arundinacea* and B) *B. curtipendula* PHE removal (continuous line), root accumulation (discontinuous line) and stem accumulation (dotted line).

The PAH are characteristically very persistent in soils; enhance the biodegradation of those compounds is thus a major challenge (Liste and Alexander, 2000). PHE concentration declined with time in the MS medium with both plants (Figure 2). PHE root accumulation with *F. arundinacea* was significantly higher than with *B. curtipendula*. There is no significant difference between stem accumulations between plant species.

The bioaccumulation of PAH from oil-polluted sediments in the shoots of *Salicornia fragilis* was early reported (Meudec et al., 2006). Since high concentrations of PAH were detected in plants, a soil-root-shoot transfer was suggested. Total PAH in roots of plants were found to be about ten times the level in shoots. In our work, PHE level in *F. arundinacea* roots were nearly three times higher than those in stems; for *B. curtipendula*, PHE in roots were almost two times higher than the observed level in stems.

The two plants assayed show similar removal and accumulation tendency for PHE and PYR. In rice plants, Tao et al. (2006) evaluated PAH concentration and found that the plant roots PAH concentration were much higher than those in aerial tissues; according with Tao et al. (2006) there was similarity in PYR accumulation profile (Figure 3) with *F. arundinacea* and *B. curtipendula*.

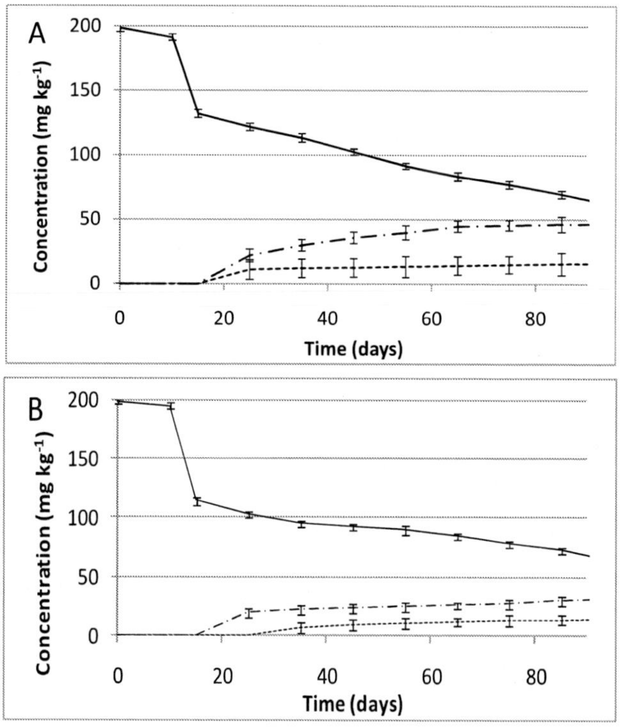

Figure 3. A) *F. arundinacea* and B) *B. curtipendula* PYR removal (continuous line), root accumulation (discontinuous line) and stem accumulation (dotted line).

PYR phytoextraction should not be favored, since they are hydrophobic chemicals, theoretically bound to plant roots, thus preventing plant uptake. However, *F. arundinacea* could remove PYR from MS medium (where the pollutant is highly bioavailable) and accumulate into their roots (Figure 3A). Lalande et al. (2003) studied the effect of annual ryegrass (*Lolium multiflorum*) on the dissipation of PYR in a 10 month experiment under field conditions; PYR level was found to decrease below its detection limit after 301 days. Under *in vitro* conditions, *F. arundinacea* in 90 days could remove 69.03%±0.57 of PYR from MS medium and *B. curtipendula* 67.83%±0.63. Suggesting that both plants could remove PYR and accumulate mainly into roots, and then into stems. Due to the lipophilic character of PYR, root concentration could be explained by adsorption. However, PYR concentration in stems was derived from root absorption and resulting in a root-to-stem transfer.

High abundance of benzo[a]pyrene was found in foliar extracts (Fismes et al. 2002), after translocation from roots to aerial parts; contrary to that, both studied plants species could uptake BaP from MS medium to roots; once in the roots, the chemical may be tend to bind with lipid membranes or oil possibly present in plant roots, because translocation of BaP to stem via xylem is limited as shows Figure 4.

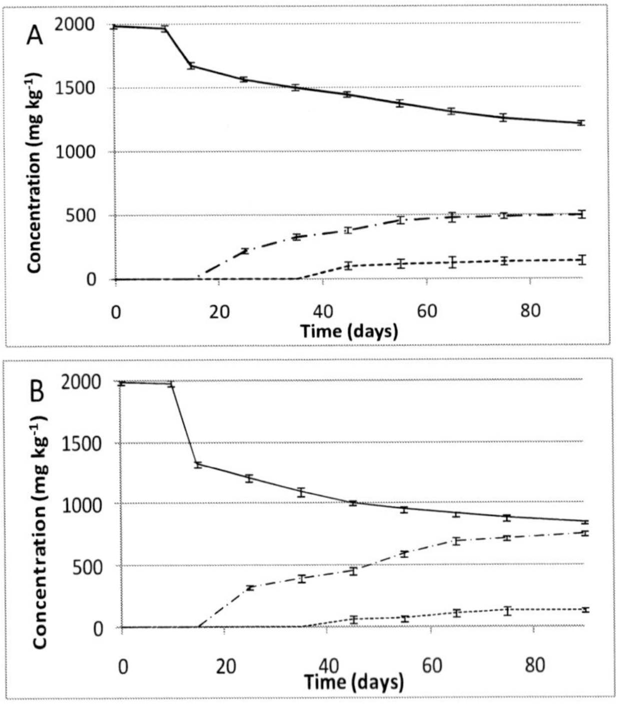

Figure 4. A) *F. arundinacea* and B) *B. curtipendula* BaP removal (continuous line), root accumulation (discontinuous line) and stem accumulation (dotted line).

The translocation of BaP from roots to shoots should be low, due to BaP hydrophobicity (Figure 4), unless special molecules are produced by plants, able to increase the apparent aqueous solubility of hydrophobic pollutants, such as 2,3,7,8-TCDD-binding proteins, shown by Campanella and Paul (2000) for dioxin absorption by zucchini (*Cucurbita pepo* L.); may be *B. curtipendula* root exudates can enhance the acquisition of BaP, then accumulate it into their roots; however, it could not translocate great amounts of BaP into their stems (Figure 4A).

It has been shown that the concentration of lindane in ryegrass, cultivated in hydroponics, slowly increases with uptake time to reach a plateau after a few days indicating that plant metabolism and formation of bound residues are minimal in such a system (Li et al., 2002); *F. arundinacea* performs alike with BaP (Figure 4A) where plant roots uptake of BaP increase slowly and after 20 days culture the pollutant become detectable into their stems and slowly increases too.

CONCLUSION

Plants have certain limitations with respect to their individual abilities to remove and transform organic pollutants like PAH. However, data shown that *F. arundinacea* and *B. curtipendula* could remove and transform PHE and PYR in MS medium where the pollutants are more availables. In all cases, the concentration of PAH declined with time, first appear into their roots and then into the stems.

The main PAH removal mechanisms was noted to be phytoextraction for both studied plants, then phytodegradation mainly for PHE and PYR. Differences in the ability of *F. arundinacea* and *B. curtipendula* to accumulate and transform a particular PAH have been shown, indicating that natural biodiversity should be better explored in order to choose the most appropriate plant species in any phytoremediation process; both plants were able to phytodegradate PHE and PYR; however, BaP is a more complex molecule and plants principally phytoaccumulate it.

Plants could use different mechanisms for the remediation of a wide variety of contaminants. Phytoremediation could therefore be used for the remediation of numerous contaminated sites. However, not much is well-known about contaminant fate and transformation pathways, including the identity of metabolites. Therefore, further research is required to reply these questions.

ACKNOWLEDGMENTS

This research was partially supported by PEMEX-Refinación and Consejo Nacional de Ciencia y Tecnología (CONACyT), Mexico: Project No. P50086536 and L. Reynoso-Cuevas received a fellowship from CONACyT.

REFERENCES

Alkorta, I. & Garbisu, C. (2001). Phytoremediation of organic contaminants in soils. *Bioresource Techn*, *79*, 273-276.

Burken, J. G. & Ma, X. (2006). Phytoremediation of volatile organic compounds. In: Mackova M., Dowling D. & Macek T. (Eds.), *Phytoremediation Rhizoremediation. Theoretical background* (First Edition 199-216) New York: Springer.

Campanella, B. & Paul, R. (2000). Presence, in the rhizosphere and leaf extracts of zucchini (*Cucurbita pepo* L.) and melon (*Cucumis melo* L.), of molecules capable of increasing the apparent aqueous solubility of hydrophobic pollutants. *Int J Phytoremed*, *2*, 145-158.

Fismes, J.; Perrin-Ganier, C.; Empereur-Bissonnet, P. & Morel, J.L. (2002). Soil-to-root transfer and translocation of polycyclic aromatic hydrocarbons by vegetables grown on industrial contaminated soils. *J Environ Qual*, *31*, 1649-1656.

Kamath, R; Rentz, J.A.; Schnoor, J.L. & Alvarez, P.J.J. (2004) Phytoremediation of hydrocarbon-contaminated soils: principles and applications. In: Vazquez-Duhalt R, Quintero-Ramirez R (Eds.), *Studies in Surface Science and Catalysis 151*, 447-478.

Kipopoulou, A.M.; Manoli, E.; Samara, C. (1999). Bioconcentration of polycyclic aromatic hydrocarbons in vegetables grown in an industrial area. *Environ Pollut*, *106*, 369-380.

Lalande, T. L.; Skipper, H. D.; Wolf, D. C.; Reynolds, C. M.; Freedman, D. L.; Pinkerton, B. W.; Hartel, P .G. & Grimes, L. W. (2003). Phytoremediation of pyrene in a Cecil soil under field conditions. *Int J Phytoremed*, *5*, 1-12.

Li, H.; Sheng, G.; Sheng, W. & Xu, O. (2002). Uptake of trifluralin and lindane from water by ryegrass. *Chemosphere*, *48*, 335-341.

Liste, H. & Alexander, M. (2000). Plant-promoted pyrene degradation in soil. *Chemosphere*, *40*, 7-10.

Macek, T.; Macková, M. & Kás, J. (2000). Exploitation of plants for the removal of organics in environmental remediation. *Biotech Adv*, *18*, 23-34.

Marmiroli, N; Marmiroli, M & Maestri, E. (2006). Phytoremediation and phytotechnologies: a review for the present and the future. In: Twardowska I, Allen H. E. & Hägbloom M. (Eds.), Soil and Water Pollution Monitoring, Protection and Remediation. (First Edition 403-416), Paris, Lavoisier Series.

Meagher, R. B. (2000). Phytoremediation of toxic elemental and organic pollutants. *Curr Opin Plant Biol*, *3*, 153-162.

Meudec, A.; Dussauze, J.; Deslandes, E. & Poupart, N. (2006). Evidence for bioaccumulation of PAHs within internal shoot tissues by a halophytic plant artificially exposed to petroleum-polluted sediments. *Chemosphere*, *65*, 474-481.

Murashige, T. & Skoog, F. (1962). A revised medium for rapid growth and bioassays with tobacco tissue cultures. *Plant Physiol*, *15*, 473-497.

Oleszczuk, P. & Baran, S. (2005). Polycyclic aromatic hydrocarbons content in shoots and leaves of willow (*Salix viminalis*) cultivated on the sewage sludge-amended soil. *Water Air Soil Pollut*, *168*, 91-111.

Reynoso-Cuevas, L.; Gallegos-Martínez, M.E.; Cruz-Sosa, F. & Gutiérrez-Rojas, M. (2008). *In vitro* evaluation of germination and growth of five plant species on medium supplemented with hydrocarbons associated with contaminated soils. *Bioresource Technol*, *99*, 6379-6385.

Schröder, P. (2007). Exploiting plant metabolism for the phytoremediation of organic xenobiotics. In: Willey N, editor. Methods in Biotechnology Vol. 23: Phytoremediation: Methods and reviews, (First edition 251-263), New Jersey, Humana Press Inc.

Schwitzguébel, J.P.; Meyer, J. & Kidd, P. (2006). Pesticides removal using plants: phytodegradation versus phytostimulation. In: Mackova M., Dowling D. & Macek T. (Eds.), *Phytoremediation Rhizoremediation. Theoretical background* (First Edition, 179-198) New York: Springer.

Singh, O.V.; Jain, R.K. 2003. Phytoremediation of toxic aromatic pollutants from soil. Appl Microbiol Biotechnol, 63, 128-135.

Tao, S.; Jiao, X. C.; Chen, S. H.; Liu, W. X.; Coveney Jr, R. M.; Zhu, L. Z. & Lou, Y. M. (2006). Accumulation and distribution of polycyclic aromatic hydrocarbons in rice (*Oryza sativa*). *Environ Pollut*, *140*, 406-415.

Watts, A. W.; Ballestero, T. P. & Gardner, K. H. (2006). Uptake of polycyclic aromatic hydrocarbons (PAHs) in salt marsh plants *Spartina alterniflora* grown in contaminated sediments. *Chemosphere*, *62*, 1253-1260.

White Jr., P. M.; Wolf, D. C.; Thoma, G. J. & Reynolds, C. M. (2006). Phytoremediation of alkylated polycyclic aromatic hydrocarbons in a crude oil-contaminated soil. *Water Air Soil Pollut*, *169*, 207-220.

INDEX

A

C

D

F

G

H

I

J

K

L

M

N

Q

R

S

T

U

V

W

X

Y

Z